国家双语教学示范课程配套教材

生物技术制药（双语教材）

Biotechnological Pharmaceutics

主编　姜　和　蔡家利　朱建伟

科学出版社
北　京

内 容 简 介

本教材根据近年国内外生物技术制药的技术发展概况，结合国内生物医药产业发展概况，以中英文双语编写而成。本教材系统地介绍了生物技术制药的基本原理、方法和实际案例以及质量控制与管理。全书共八章：生物技术制药概论、生物技术药物制备工艺、生物技术药物制剂学、生物技术药物生产的质量控制、基因工程药物、抗体药物、生物技术疫苗、核酸药物。

本教材可作为高等院校药学、制药工程、生物工程、生物技术及生物科学等专业本科生教材，有助于学生了解生物技术药物专业知识，同时熟悉专业英语，有助于跟踪迅速发展的生物医药产业前沿。本教材也可作为生物医药专业教师、科研人员、研究生及产业界人士的参考书。

图书在版编目（CIP）数据

生物技术制药＝Biotechnological Pharmaceutics：汉英对照 / 姜和，蔡家利，朱建伟主编. —北京：科学出版社，2016

国家双语教学示范课程配套教材

ISBN 978-7-03-048754-4

Ⅰ. ①生… Ⅱ. ①姜…②蔡…③朱… Ⅲ. ①生物工程－双语教学－高等学校－教材－汉、英 Ⅳ. ①Q81

中国版本图书馆CIP数据核字（2016）第131787号

责任编辑：王玉时 席 慧 / 责任校对：李 影

责任印制：吴兆东 / 封面设计：迷底书装

科学出版社 出版

北京东黄城根北街16号

邮政编码：100717

http://www.sciencep.com

北京华宇信诺印刷有限公司印刷

科学出版社发行 各地新华书店经销

*

2016年11月第 一 版 开本：787×1092 1/16

2025年12月第七次印刷 印张：21 1/2

字数：510 000

定价：69.00 元

（如有印装质量问题，我社负责调换）

《生物技术制药（双语教材）》
编委会名单

主　　编　姜　和（重庆理工大学）
　　　　　　蔡家利（重庆理工大学）
　　　　　　朱建伟（上海交通大学）
副 主 编　吴胜昔（重庆理工大学）
　　　　　　邓林红（常州大学）
　　　　　　蒋雪梅（重庆大学）
编　　者（以姓氏笔画为序）
　　　　　　卜友泉（重庆医科大学）
　　　　　　叶翠莲（重庆理工大学）
　　　　　　付爱玲（西南大学）
　　　　　　宋方洲（重庆医科大学）
　　　　　　张　芳（南京中医药大学）
　　　　　　赵　颖（重庆理工大学）
　　　　　　康　宁（天津中医药大学）
　　　　　　彭方毅（重庆理工大学）
　　　　　　潘　扬（南京中医药大学）
审　　稿　江　华
　　　　　　朱聿文（Charles Y. Zhu）
　　　　　　David Nellis

前　　言

生物技术制药作为国家双语教学示范课程已经多年，我们经过多年的教学经验积累，策划编写一本适应本科教学的双语教材。经过近三年的反复讨论，教材几经易稿，终于要与读者见面了，编者心中不免有些忐忑，因为毕竟这是一本教材，不是科技著作。著作可以任意表达作者的思想和创新，而教材是要教给学生基础的知识、共识的定律、普遍接受的理论……这不是一件容易的事情，尤其是对我们这样的从国外回来的科研工作者，要尽量避免把自己的偏爱和局部知识扔给学生。在这样的前提下，写一本生物制药的教材，具有一定的挑战性。除此之外，生物医药产业本身就在高速发展，新的生物药物品种每年飞速增长，特别是抗体药物和基因工程药物发展迅速。未来威胁人类健康的重大疾病，特别是感染性疾病、癌症以及新出现的疾病主要依靠生物药物进行防治，如艾滋病、乙型肝炎、肝癌、肺癌等。从完稿到现在，我们发现又出现了很多重要的科研成果、科学进展可以写进去，但时间有限，只能作为下一版的修订内容。

本教材具有一定的前瞻性和代表性，希望读者在学习过程中一定要带着批判的眼光，结合其他科技论文和教材来学习。但是，至少本教材是一本新颖的教材，可以帮助学生打开生物医药产业的窗户，发现 21 世纪生物医药产业的发展前景，树立对生物医药产业的信心和希望。也希望授课教师把本教材作为一个蓝本，授课时带入自己的知识、观点和心得，这样才能真正发挥教材的作用，最大限度地让学生受益。

本教材的各章节，邀请了各个领域的专家（opinion leaders）参与编写，力争做到具有专业代表性和国际水准，但也难免因作者的书写习惯和教学科研背景的差异，不能完全满足每个读者的需求。我们也希望通过这本教材，引出更多更好的著作和教材，百花齐放；激起业内专家们更多的讨论和批评，真正达到抛砖引玉的效果。我们非常欢迎业内专家和广大读者提出宝贵的建议和意见，以便在使用本书时和以后的修改时参考；也欢迎更多的专家加入到我们的作者队伍中来，把教材编得更好。

本教材已获得重庆理工大学 2013 年校级规划教材资助。美籍华裔生物医药专家江华先生、朱聿文先生 (Mr. Charles Y. Zhu) 和 David Nellis 先生对本教材进行了仔细阅读与修改，在此特别致谢！

姜　和

2016 年 4 月于重庆

Introduction

The biotechnological pharmaceutics has been our bilingual model course for years. We wanted to compile a bilingual textbook in this subject based on our previous teaching experience. After repeated discussions and revisions in the past three years, the final version of this book is about to be published, but authors of the book still have concerns. Textbooks are not scientific works, because common theory and basic concept should be taught to students by textbooks, while the authors are not supposed to express whatever creativity in their minds. It's not easy for textbook authors, as scientific researchers like us; we should try hard not to express our preference or bias to students during teaching. Besides, with the fast development of biotechnological pharmaceutics, especially in antibody-based drug and gene tech-based drug, which makes it even harder to compile a textbook. Contagion, cancer and new diseases, such as AIDS, hepatitis B, liver cancer and lung cancer, need to be treated with biotechnological pharmaceutics. Upon the finish time of this book, new research breakthroughs had been made; we'll add these new research data in the second edition due to the limited time. For this book, readers should have their own thought in the learning process, read with critique and think with creativity. This book is novel, so we hope this book can broaden readers' mind in biotechnological pharmaceutics, and we also hope teachers to use this textbook only as a directory note. Teachers' own perspectives and opinions should be expressed while teaching, and only in this way this textbook can benefit students.

Opinion leaders in many fields participated in compiling this textbook in every chapter, so we tried to set a high professional standard, and nevertheless it's impossible to meet every need from readers due to the different educational background and writing style of authors. This textbook is supposed to stimulate discussions by experts in the related fields, and critiques are very much appreciated so that we can make a better improvement and adjustment in next edition. New writers are also welcome to join us in compiling our next textbook.

This textbook was supported by the Planned Textbook Fund of Chongqing University of Technology in 2013. Many thanks to Messrs. Hua Jiang, Charles Y. Zhu and David Nellis for helping us revise and proofread this tcxtbook.

He Jiang

April 2016 in Chongqing

目　　录

Contents

第一章　生物技术制药概论

Chapter 1　Introduction of Biopharmaceuticals

1　生物技术制药基本概念

1　Basic Concept of Biotech Drugs

随着科技的进步，人类正日益感受到生物技术给生活带来的巨大变化，特别是生物技术药物对人类健康的影响。到2011年3月，有超过300个生物技术药物在美国或欧洲赢得了市场许可，全球生物技术药物市场有920亿美元的份额。至此，生物技术药物已不容人类忽视。

生物技术系指利用“生物体（含动物、植物及微生物）”来生产有用的物质或改良生物的特性，以降低成本及创新物种的科学技术。联合国条文对于生物多样性的生物技术的定义是：任何一种用生物系统、活体及其衍生体上制造、修饰有特殊用途的产品或工序的应用技术。换而言之，生命科学的技术进步用于商业化生产就是生物技术。

早期的生物技术，可追溯至远古时代。古埃及人利用酵母菌酿酒。之后，无论是传统式利用微生物发酵技术来发酵食品，或是发酵生产抗生素等，都是生物技术成功利用的例子。对于现代生物技术，在20世纪50年代DNA结构的发现以来，分子生物学急速发展，对传统的生物技术进行了一次大革命。例如利用基因克隆技术，将胰岛素（insulin）克隆到大肠

With the development of human society, more and more people feel the great influence of biotechnology on our daily lives, especially of the impact on human health by biotechnology medicine. There are more than 300 biotechnology drugs received market permission in USA or Europe until March 2011. The global market of biotechnology drugs is estimated as approximately $92 billion US dollars. Nowadays the significance of biotechnology drugs are recognized by more and more people.

Biotechnology: Biotechnology is a field of applied biology that involves the use of living organisms and bioprocesses in engineering, technology, medicine and other fields related to bio-products.The United Nations Convention on Biological Diversity defines biotechnology as “Any technological application that uses biological systems, living organisms, or derivatives thereof, to make or modify products or processes for specific use.” In other term “Application of scientific and technical advances in life science to develop commercial products” is biotechnology.

Early biotechnology can be traced back to ancient times. Ancient Egypt people used the *Yeast Saccharomyces* to make wine. Then whether traditional microbial fermentation technology was used in food fermentation, or the production of antibiotics by fermentation, are examples of the use of biotechnology. For modern biotechnology, since the discovery of DNA structure in the 1950’s, molecular biology development developed rapidly, and the traditional biotechnology get a great revolution. Such as the use of cloning technology, the insulin (insulin) cloned into *Escherichia coli*. It has opened up a path of modern biotechnology.

杆菌中生产。这开启了现代生物技术学之工业价值。

生物技术所含的主要技术范畴有：基因工程、细胞工程、酶工程、发酵工程及生化工程。基因工程是生物技术的核心和关键，是主导技术。

It contains the main technical areas: genetic engineering, cell engineering, enzyme engineering, fermentation engineering, and bio-engineering. Genetic engineering is the key of biotechnology and the dominant technology.

生物技术药物 系指应用基因工程、细胞工程、蛋白质工程等生物技术制备微生物、细胞及各种动物和人源的组织和液体等生物材料，用于人类疾病的预防、治疗和诊断的药物。现代的生物技术药物是指运用生物技术制备的医学药物。它们包括蛋白质药物（含抗体），治疗或在活体内诊断用核酸药物（含 DNA、RNA、反义寡核苷酸），以及运用其他方式（非直接）从天然非基因生物源中提取制备的产品。

Biotechnology drug or biotechnology medicine is defined here as “any pharmaceutical product used for a therapeutic or in vivo diagnostic purpose, which is produced in full or in part by either traditional or modern biotechnological means”. Biopharmaceutical refers to medical drugs produced using biotechnology. They are proteins (including antibodies), nucleic acids (DNA, RNA or antisense oligonucleotides) used for therapeutic or in vivo diagnostic purposes, and are produced by means other than direct extraction from a native (non-engineered) biological source.

生物药物 生物技术药物与天然生化药物、微生物药物、海洋药物和生物制品一起归类为生物药物。

Biological Drugs: the biological technique medicine, biochemical medicine, microbial medicine, marine drugs and biological products are classified as biological drugs.

生物技术制药 系指利用基因工程技术、细胞工程技术、微生物工程技术、酶工程技术、蛋白质工程技术、分子生物学技术等研究和开发药物，用来诊断、治疗和预防疾病的发生。现在，全世界生物制药技术的产业化已进入投资收获期，生物技术已应用和渗透到医药、保健食品和日化产品等各个领域，尤其在新药研究、开发、生产和改造传统制药工业中得到日益广泛的应用，生物制药产业已成为最活跃、进展最快的产业之一。

Biotech Drugs: it was used of genetic engineering and cell engineering, microbiological engineering, enzyme engineering, protein engineering, molecular biological technology to research and develop drugs, which used to diagnose, treat and prevent the occurrence of diseases. Now, the world bio-pharmaceuticals technology industry has entered a period of investment gains, biotechnology drugs and penetrated into medicine, health food and cosmetic products in various fields, especially in drug research, development, production and transformation has been widely used in the traditional pharmaceutical industry, bio-pharmaceutical industry has become one of the most active and fastest-development industries.

生物药物工业在全球的规模以超过 15% 的年增长率迅猛发展。它

Biopharmaceutical industry has been rapidly expanding with over 15% increased annual revenue worldwide including China. The rate of increase is higher

的增长率远高于整个制药工业的4%～5%增长率（Global 2008）。自从第一个重组蛋白胰岛素于20世纪80年代初期上市以来，已有100多个重组蛋白和单克隆抗体，以及300多个非重组生物产品包括疫苗和血液制品被美国食物和药物管理局（FDA）批准（www.fda.gov；Rader, 2010）。生物药物产品是运用现代生物技术生产的供体内诊断的临床诊断试剂、预防疾病用的疫苗以及治疗用的药品。生物药物是相对比较新的也比较复杂的大分子实体，一般由生物系统例如微生物、哺乳动物细胞、植物和动物体内生产。DNA重组技术和杂交瘤技术被用来构建生物系统生产：①重组的天然蛋白质（人生长激素、细胞因子、胰岛素等）；②天然蛋白质和生物系统的衍生物（蛋白突变体，类病毒颗粒疫苗，癌细胞疫苗，免疫毒素，抗体片段融合蛋白等）；③用作疫苗或基因治疗的带有基因或遗传信息的病毒和质粒载体，或小分子干扰核糖核酸；④用于体内诊断和治疗用的单克隆抗体。2008年1月到2011年6月，美国FDA批准了18个由微生物、动物细胞及动物体内生产的重组蛋白生物药物，以及9个从天然来源的产品，如人体血液、组织、癌细胞、减毒病毒和细菌生产的疫苗或治疗产品。18个重组蛋白中有12个是来自哺乳类细胞系统，3个来自大肠杆菌，另外3个来自昆虫病毒、酵母菌和转基因山羊。

than the overall pharmaceutical industry growth rate, 4%-5% (Global 2008). Since the approval of insulin, the first recombinant protein on market in early 1980s, there have been over 100 new recombinant protein and monoclonal antibody (MAb) entities, and more than 300 non-recombinant biopharmaceuticals including vaccines and blood products approved by U. S. Food and Drug Administration (FDA) (www.fda.gov; Rader, 2010). Biopharmaceutical products or biopharmaceuticals are clinical reagents, vaccines, and drugs produced using modern biotechnology for *in vivo* diagnostics, prophylaxis, and therapy. This type of molecules is relatively new and large complex in structure, manufactured in living systems, such as micro-organisms, mammalian cells, plants, or animals. Recombinant DNA and hybridoma technologies have been used to engineer biological system to produce ① recombinant forms of natural proteins (human growth humor, cytokines, insulin, etc.); ② derivatives of natural proteins and living systems (protein muteins, virus-like particle vaccines, cancer cell vaccines, immunotoxins, IgG fusion proteins, etc.); ③ viral and plasmid vectors and small interfering RNAs that carry genetic information for vaccination or gene-therapy; ④ *in vivo* diagnostic and therapeutic monoclonal antibodies aiming on a target in a living system. Of the 27 latest (Jan. 2008-June 2011) FDA approved biopharmaceutical entities, 18 are recombinant proteins manufactured using cells, organisms, or animals. The other 9 are vaccines and therapeutics manufactured from natural product sources such as human plasma, tissue, cancer cells, attenuated virus, and bacteria. Among the 18 recombinant products, 12 are produced using mammalian expression systems, 3 are produced by *Escherichia coli* (*E. coli*), and the remaining 3 are produced by baculovirus, yeast, and transgenic goats.

2 生物技术制药发展简史

2 History of Biopharmaceutical Industries

人类利用生物技术或者说生物体或细胞生产自己所需要产物的历史悠久，从一万年前开始，人类就利用对特定植物的耕种和指定动物的畜牧以获取稳定的粮食来源，保证人类生存和社会发展的需要。约六千年前人类又学会利用发酵技术酿酒和制作面包，以此提供美味的食材。约两千年前人类用豆腐和浆糊上的霉菌治疗伤口出血和痈疮。公元4世纪，葛洪著《肘后良方》也有用海藻酒治疗瘿病（地方性甲状腺肿）的记载。神农是中国最早的将生物材料作为治疗药物的应用者，如用羊靥治疗甲状腺肿，用紫河车（胎盘）作强壮剂。1578年明代李时珍在《本草纲目》中所载药物1892种，除植物外，还有动物药444种。1673年荷兰人列文虎克用自己制造的显微镜观察到了被他称为“小动物”的微生物世界，至此人类开始认识和确证了微生物的存在。1797年英国人琴纳首创用牛痘预防天花。随后，法国人巴斯德研制鸡霍乱、炭疽和狂犬疫苗成功，并且他还实验证明了有机物质发酵和腐败是由微生物引起的，而酒类变质是因污染了杂菌所致，还发明了沿用至今的巴氏消毒法。在19世纪最后20年间，德国学者郭霍分离培养出一大批传染病的致病菌，包括炭疽芽孢杆菌、伤寒沙门菌、结核分枝杆菌、霍乱弧菌、白喉棒状杆菌、葡萄球菌、破伤风梭菌、脑膜炎奈瑟菌、鼠疫耶氏菌、肉毒梭菌、痢疾志

The history of human using technology or living body to product what they need goes back long ago. About 10,000 years ago, human cultivated special plants and special animals to get stable foodstuff origin, which guaranteed human animation and requirement of society progress. About 6,000 years ago, human study to using fermentation technology to get wine and bread as good taste food. About 2,000 years ago, human used bean curd and mold living on paste to cure blooding and ulcerative. A Chinese medical book named *Zhou Hou Liang Fang* recorded that a Chinese man, Ge Hong, used a marine swine to cure endemic goiter in AD 4 Century, Chinese man Sheng Nong was the first original biomaterial user as therapeutic medicine. For example, he used glandula thyreoidea of sheep to cure deroncus; goiter, and he used Zhi Chehe (placenta) as invigorator. A famous Chinese ancient doctor, Li, Shizheng, recorded a large number of medicine in his book *Ben Cao Gang Mu*. There were 1892 drugs, including 444 animal medicine except plant medicine, in the book. Dutchman Leeuwenhoek used the microscope produced by himself to observe the microbiological world in 1673. Human began to realize the existence of microbiology. Englishman Jenna creatively used bovine smallpox to prevent smallpox in 1797. Then Frenchman Pasteur successfully carried out research on and obtained chicken cholera vaccine, anthrax vaccine, rabies vaccine. Pasteur proved the causes of organic matter fermentation and that their degeneration was reduced by microbial, and whoopee water deteriorating was caused by mixed bacterium. Pasteur invented the pasteurization, an effective way to kill bacteria, which has been used till today. In the last 20 years of the 19th centunry, German scholar Robert Koch isolated the culture of infective pathogenic bacteria, including *Spore-Forming*, *Eberth Ella typhi*, *mycobacterium tuberculosis*, *Bacillus comma*, *Corynebacterium diphtheria*, *staphylococci*, *Clostridium tetani*, *Diplococcus intracellularis*, *Yersinia pestis*, *Clostridium botulinum*, and *Shigella dysenteriae*. The technology of brewing wine

贺菌等。第一次世界大战时期，酿酒术扩展为更大一些的工业化生产，这种潜在的工业发酵逐步发展为生物技术。1929年，弗莱明发现了青霉菌的抑菌作用，1940年Florey获得提纯的可供临床使用的青霉素纯品。受此鼓舞，链霉素、氯霉素、金霉素、土霉素、红霉素等相继被发现，使许多由细菌引起的感染和传染病得到控制和治愈，为人类健康做出巨大贡献。抗生素产业的快速发展促进了其他发酵产品的发展，如20世纪60年代的氨基酸发酵工业，70年代的酶制剂工业等。近代生物技术产业的主要产品：医药业的抗生素、维生素、甾体激素、氨基酸，轻工业、食品业的工业酶制剂、食品氨基酸、酵母、啤酒，化工业的乙醇、丙酮、丁醇、沼气，农林业的农用抗生素等农药，环境保护业的生物治理污染等。

developed to some larger industrialization production in the period of the First World War, and the potential industrial fermentation gradually developed to biotechnology. Fleming discovered bacteriostatic action of *Penicillium* in 1929. Florey got the purified penicillinum which was used in clinical treatment in 1940. Inspirit by it , a large number of antibiotics were discovered, one after another, including of phytomycin, Chloramphenicol, Acronize, hydroxytetracycline, and erycin. These antibiotics controlled and cured a lot of bacterial infected diseases, making a great contribution to the health of mankind. The scaled up production of antibiotic progressed quickly, and promoted other fermentation production progresses, such as amino acid fermentation industry in 1960's, enzyme preparation industry in 1970's, and so on.

The principal production of biotechnological industry in modern times including antibiotics, vitamin, steroid hormone, amino acids in medical industry, enzyme preparation in food industry, food amino acids, yeast, beer, alcohol chemical engineering industry, acetone, butanol, marsh gas, pesticide, biology managing contaminate, and so on.

1953年Watson和Crick发现了遗传物质DNA的双螺旋结构，1966年Nirenberg和Crick等人破译了全部遗传密码，1969年哈佛大学成功分离出第一个基因，这些近代的重大发现使人类对构成生物体最小单位的细胞及控制细胞遗传特征的基因有更深入的了解。1974年Boyer和Cohen建立了DNA重组技术，1975年Kohler和Milstein建立了单克隆抗体技术，1977年DNA测序技术诞生，这些技术的发现可以更有效地让细胞或生物体生产我们所需要的物质，且适合工业或农业量产。因此生物技术制药从20世纪80年代开始成为一个新兴的生物科技产业，被誉为“21世纪的朝阳产业”。第一个

Watson and Crick discovered double coiled spiral structure of genetic material (DNA) in1953, while Nirenberg and Crick *et al* translated all genetic codes in 1966. The first gene was isolated in1969. These proximal important discoveries let people learn more about cell which is the most minimum unit in biosystem and gene controlling the cell hereditary feature. Boyer and Cohen established DNA recombinant DNA technique in 1974. Kohler and Milstein established monoclonal antibody technique in 1975, and DNA sequencing technique was bored in1977. These two techniques produced production we need though by cell or living body more efficiently, and fit for mass production in industry or agriculture. So a new biotechnology industry was born in1980's, which has been called a dawning industry in 21th Century. The first approved biotech drug was biosynthetic human insulin, usually abbreviated as rHI and the commercial name is humulin. This product was developed by Genentech, and put on market by Eli Lily, commercialized by humulin in

经批准可使用的治疗用途的生物技术药物是由 DNA 合成技术制备的生物合成人胰岛素。通常指 rHI，其商业品为 Humulin（人体单组分胰岛素），由 Genentech 公司研发，礼来公司得到销售许可，于 1982 年开始制造和销售该产品。1988 年 PCR 方法问世，1997 年英国培养出第一只克隆羊多莉，2001 年人类基因组草图完成。自从 1982 年美国 FDA 批准上市第一个新生物技术药物——基因重组胰岛素至今，生物制药已走过 30 余年历史，目前有近 300 种生物技术药物获批上市。1996—2005 年 10 年间，每年 FDA 批准约 16.6 个生物技术新药上市，在 1997 年有 23 个新药批准上市，其中包括 12 个重组蛋白或单克隆抗体，达到历年来获批新药的顶峰。2006—2008 年批准生物技术新药较之前有所下降，每年约 12 个。但 2009 年获批新药数字又上升至 23 个，2010 年为 15 个。

1982. PCR approach was appeared in1988, and the first clone sheep was cultured in England in1997. Human genome adumbration was finished in 2001. From the first biotech drug receiving commission going into the market to the completion of human genome adumbration, biopharmaceutical industry had undergone some 30 years of history, with about 300 biotech drugs got commission to go into market. FDA gave approval 16.6 new biotech drugs to going on the market every year from 1996 to 2005. There were 23 new biotech drugs obtained approval in 1997, including 12 recombination proteins or monoclonal antibodies, setting a record of new drug approval. The numbers of new approval decreased from 2006 to 2008, about 12 new drugs every year. But it went up to 23 in 2009, and 15 in 2010.

美国生物技术药物产业化最早，投资最大，产品数量最多。目前已批准近 300 个生物技术药物上市，适应证达 220 种，使 3.25 亿患者受益，蛋白质药物的产值和销售额已超过 200 亿美元。日本在生物技术的开发上仅次于美国，生物技术药物产业的发展居亚洲首位，目前共有生物制药公司约 600 家。欧洲各国也十分重视发展生物技术药物的产业化，英国建立了世界上效率最高、效果最好的国家临床试验体系，促进了生物技术药物的发展。英国的生物技术药物产业规模在欧洲居于首位，拥有公司的数量约为全欧洲的四分之三，其中有世界

Biopharmaceuticals was among the earliest to be industrialized, and most invested sector. American biopharmaceutical industry is the most advanced and biggest in its size. There are about 300 biotech drugs on market, with the number of indications over 220, benefiting 325 million patients. The production scale exceeded 2 billion in pharmaceutical grade protein. Japan is the second largest in progressing biotech next to American. There are about 600 biopharma companies in Japan. European countries put biopharmaceuticals industry on high priority. United Kingdom has already set up the most efficiency national clinical test system, to promote the progress of biopharmaceuticals industry. UK is the first one in European countries in biopharmaceuticals industry, taking up three quarters of biotech companies in Europe. Most of the companies are world class, including GlaxoSmithKline, Astrazeneca etc.

知名的生物制药企业，如葛兰素史克（Glaxo Smith Kline，GSK）、阿斯利康（Astrazeneca）等17家大公司。我国生物技术药物产业化以仿制为主，21世纪开始转入创新阶段，已经批准了20多个生物技术药物。1997年是我国第一个产业化的高潮。世界上生物技术药物销售额前十位的，我国已能生产8种，完全能满足国内市场需要。目前我国有20个国家生物技术药物重点实验室、3个蛋白质药物开发中心、289家生物制药企业，生物技术产业的群落和集约化已初见端倪。我国已成功开发出20多种基因工程药物和疫苗，批准上市的蛋白质药物约20种。

Biopharmaceuticals industry focuses on innovation and reform to a new creative stage in 21 century. Chinese CFDA already approved over 20 new biotech drugs. It was a climax to progress of Chinese biopharmaceuticals industry in 1997. Eight of top ten sales biotech drugs were produced in China, meeting the domestic market needs. There were approximately 20 National Key Laboratories and 3 biopharmaceutical research centers, and 290 biopharmaceuticals companies in China. China has already developed more than 20 genetically engineered drug and vaccines, and about 20 pharmaceutical grade proteins approved for market.

3 生物技术药物的分类及特征

3 Categorization and Characteristics of Biotech Drug

3.1 按药物原料来源分类

生物技术药物按药物原料来源主要分为四类：抗体药物、融合蛋白、天然生物制品和新型疫苗。

3.1.1 抗体药物

单克隆抗体药物是一个日益增长的生物技术药物产品，迄今为止，包括21个FDA认可的抗体药物，如贝伐单抗（免疫调节药物）、Panitumumab（美国安进公司）、利妥昔单抗（抗肿瘤药物，生物基因公司）、阿达木单抗（免疫调节药，艾博特公司）。

3.1.2 融合蛋白

融合蛋白是一类基因工程制造的药物，肿瘤坏死因子（TNFa）受体的

3.1 Classified by Sources

Most biopharmaceuticals today can be classified into four major categories: monoclonal antibodies, fusion proteins and native biologics as well as modern vaccine.

3.1.1 Monoclonal Antibodies

Monoclonal antibodies are a growing class of Biopharmaceutical products,including more than 21 kinds which had been approved up to date by US FDA. The list includes products such as Bevacizumab (Genentech), Panitumumab (Amgen), Rituximab (Biogen Idec/Genentech) and Adalimumab (Abbott) .

3.1.2 Fusion Proteins

Fusion proteins are a class of drugs produced by genetic engineering. For example, the extra cellular domain of the TNFa receptor was fused with the Fc

细胞表位与免疫球蛋白 IgG 的 Fc 片融合，从 1998 年发布以来，这类药就用于治疗慢性炎症，如类风湿关节炎、斑块性牛皮癣、alkylosing 脊椎炎。其他还包括阿法西普，生物基因公司生产，用于治疗多发性硬皮病，阿法西普也用于治疗类风湿关节炎。融合蛋白是创制依那西普类药物，2010 年在全球销售额达 50 亿美元。

domain of immunoglobulin IgG. Since its launch in 1998, Etanercept has been used for the treatment of chronic inflammatory diseases such as Rheumatoid arthritis, plaque psoriasis and alkylosing spondylitis. Other examples of this class of drugs include Alefacept (BiogenIdec) for treatment of multiple sclerosis and Abatecepr (Bristol-Myers Squibb) also for the treatment of rheumatoid arthritis. Fusion protein is an established class of drugs such as Etanercept projected to reach $5 billion in global sales by 2010.

3.1.3 天然生物制品

天然生物制品是一类分泌蛋白类药物（细胞因子）或替代品治疗药物。该类是美国 FDA 认可最早上市的生物治疗类药物。包括红细胞生成素、干扰素、胰岛素、凝血因子、生长激素和白细胞介素等。这类药物在生物技术药物中市场占有率居首，全球基于蛋白质药物的种类，红细胞生成素单独一种在 2004 年全球销售就接近 1000 亿美元，并仍然呈增长趋势。

3.1.3 Native Biologics

Native biologics is a class of drugs also referred to as the secreted proteins for replacement therapies. It represents the earliest category of bio therapeutic drugs approved by the US FDA. It includes erythropoietin, interferon, insulin, clotting factors, growth hormones and interleukins. By market size, this class of Biopharmaceutical products dominates the global portfolio of protein-based drugs. Erythropoietin alone drew close to $10 billion in sales worldwide in 2004 and the number is still growing.

其次为核酸药物，即用生物技术对核酸进行改造的药物。核酸衍生物以其独特的药理作用形成了一个新兴产业，正日益显示重要的治疗作用。在美国，核酸类药物近年来在抗病毒、抗肿瘤方面显示了不可代替的作用，美国 FDA 批准的治疗艾滋病药物 AZT、ddC、ddI、D4T、3TC 等都是核酸类药物。抗病毒首选药物三氮唑核苷、阿普洛韦，抗恶性肿瘤新药氟铁龙，中国开发的一类新药 8- 氯腺苷等都属于核酸药物。

The other is nucleic acid drug, which is modified by biotechnology. With its unique pharmacological effects, nucleic acid derivatives formed a new industry is increasingly important therapeutic role. In the United States, nucleic acid drugs show the irreplaceable role of antiviral and anti-tumor effection in recent years. AIDS drugs approved by FDA of United States , including AZT, ddC, ddI, D4T, 3TC, etc are all nucleic acid drugs. The first choice of Antiviral drug (triazole nucleosides, Apuluowei), antivirus drug (triazole nucleosides, apuluowei), Anti-cancer drug (furtulon), and the new drug developed by China (8-chloro-adenosine) are all belong to nucleic acid drugs.

3.1.4 新型疫苗

新型疫苗不同于传统疫苗，主要采用现代生物技术制造，包括基因工程亚单位疫苗、基因工程活载体疫苗、病毒样颗粒疫苗、DNA 疫苗和

3.1.4 Modern Vaccines

Modern vaccines, different from conven-tional vaccines, are used to prevent or treat diseases, mainly including genetically engineered subunit vaccine, virus-like vaccine, treatment vaccine and genetically engineered live carrier vaccine, DNA vaccine and

转基因疫苗等，主要用于预防或治疗感染性疾病及癌症。

transgenic vaccine *et al*. The vaccines are mainly used to prevent transmission of infectious diseases.

3.2 按药物的功能用途分类

3.2 Classified by Function and Purpose

3.2.1 治疗药物

治疗疾病是生物药物的主要功能，主要针对各种人类重大疾病，如肿瘤、艾滋病、心脑血管疾病等。

3.2.1 Treatment Drugs

Curing disease is the mayor ability of biotech drugs. Most of them focus on all kinds of significant human disease, such as cancer, AIDS, cerebrovascular disease and heart diseases.

3.2.2 预防药物

主要针对各类传染病，是控制感染性疾病传播的有效手段，常见的预防药物有各种疫苗、菌苗、类毒素等。

3.2.2 Prophylactic Agent

It focuses on all kinds of contagious disease, which is effective way to control infectious diseases. Commonly used prophylactic agents include in various kinds of vaccines, toxoid etc.

3.2.3 诊断药物

是疾病临床诊断重要手段，具有速度快、灵敏度高、特异性强的特点。常用有免疫诊断、酶诊断、放射性诊断、基因诊断试剂等。

3.2.3 Diagnostic Reagents

Biotech diagnostic reagents are important elements in clinical diagnosis. They are characterized as fast, highly sensitive, specific. Frequently used reagents are immunodiagnostic, enzyme diagnose reagent, radioactivity diagnose reagent, gene diagnosis reagent, etc.

3.3 生物技术药物的特征

3.3 Characteristics of Biotechnology Medicine

3.3.1 高特异性

许多生物药物的药理学活性与动物种属及组织特异性有关，来自人源基因编码的蛋白质或多肽类药物，其中有的与动物的相应蛋白质或多肽的同源性有很大的差别，因此对一些动物源基因编码的蛋白质或多肽类药物无药理学活性或不敏感。

3.3.1 High Specificity

The pharmacological activities of Biopharmaceuticals are related to animal species and tissue specificity. Proteins and peptides from Human genetic codes differ significantly from their animal counterparts. Therefore, human body reacts poorly to or shows no pharmacological effects by proteins and peptides of animal origins, because their homogeneity and characteristics are not similar.

3.3.2 高敏感性

治疗针对性强、疗效高，生物药物是天然存在的蛋白质或多肽，量微而活性强，用量极少就会产生显著的效应，相对来说它的不良反应较少，毒性较低，安全性较高。

3.3.2 High Sensitivity

High sensitivity usually indicates high therapeutic efficacy. Biotech drugs are often naturally occurring proteins and polypeptides therefore they are needed by small quantity, with profound effects, whereas the adverse effects and toxicities are less, with high safety.

3.3.3 稳定性差

蛋白质或多肽药物较不稳定，易变性，易失活，也易为微生物污染或

3.3.3 Low Stability

Proteins and polypeptides drugs can be instable at room temperature, deactivated easily, contaminated and enzyme is digested easily.

酶解破坏。

3.3.4 半衰期短

生物药物降解迅速，并且在体内降解的部位广泛。

3.3.5 生产要求基因稳定性

生产菌及细胞系的稳定性和生产条件的稳定性非常重要，它们的变异将导致药物生物活性的变化，产生意外的或不希望的一些生物学活性。

3.3.6 具免疫原性

许多来源于人的生物药物，在动物中有免疫原性，所以在动物中重复给予这类药品将产生抗体，有些人源性蛋白在人体中也能产生血清抗体，可能是重组药物蛋白质在结构及构型上与人体天然的蛋白质有所不同所致。

3.3.7 受体效应

许多生物技术药物是通过与特异性受体结合，信号传导机制而发挥药理作用，且受体分布具有动物种属特异性和组织特异性，因此药物在体内分布有组织特异性，有药效反应快的特点。

3.3.8 多效性和网络性效应

许多生物药物可以作用于多种组织和细胞，且在人体内相互诱导生长，相互调节，彼此协同或拮抗，形成网络性效应，因而可具有多种功能，发挥多种药理作用。

3.3.9 生产和检验的复杂性、特殊性

生产系统的复杂性，致使它们的同源性、批次间一致性及安全性的变化要大于化学产品，所以对生产过程的检测，GMP 步骤要求和质量控制的要求就更为重要和严格。

3.3.4 Short Half-life

Biotech drugs exhibit short half-life due to their quick degradation throughout the body.

3.3.5 Production Requires Gene Stability

It is essential to have constant and stable bacterial type and production cell line. The mutations may result in alterations which may lead to undesirable biologic activities.

3.3.6 Immunogenicity

Some of the biotech drugs are of human origin and they may be immunogenic in animals. Repeated dosing of such drugs may cause antibody production in animals. Human body in some cases may also produce antiserum against such drugs. Perhaps it is caused by the differences in structure between naturally occurring proteins and recombinant proteins.

3.3.7 Receptor Effect

Many biotech drugs produce pharmacologic actions though specific binding with receptors. The distribution of receptors is characterized with animal species and tissue specificities. Thus the biotech drugs may distribute inside body with tissue specificity. Their drug efficacy occurs quickly as well.

3.3.8 Multiple and Network Associated Efficacies

Many biotech drugs act upon several tissues and types of cells, resulting in induced growth and regulations among these systems. These efficacies may cause agonistic or antagonistic reactions, overall resulting in associated efficacy based on the balance of many. Thus such drugs may have multiple functions/efficacies.

3.3.9 Complexity and Specialty of Production and Controls

Complex production system may lead to bigger differences among production batches and bigger changes in safety specifications. Therefore the GMP requirement and quality control standards are usually higher and more stringent against the production processes.

4　生物技术药物研究进展

4　Research Progress on Biotech Drugs

截至2011年5月，有近300个生物技术药物在美国或欧洲赢得了市场许可。这代表全球生物技术药物市场近一千亿美元的份额。图1-1列出1984—2011年FDA批准上市的新的生物技术药物数目。

By May 2011, more than 300 biopharmaceutical products had gained market approval in the USA and/or in EU. Collectively, these represent a global biopharmaceutical market in the region of near $100 billion in 2010. A detailed list of the approved products is provided in Figure 1-1.

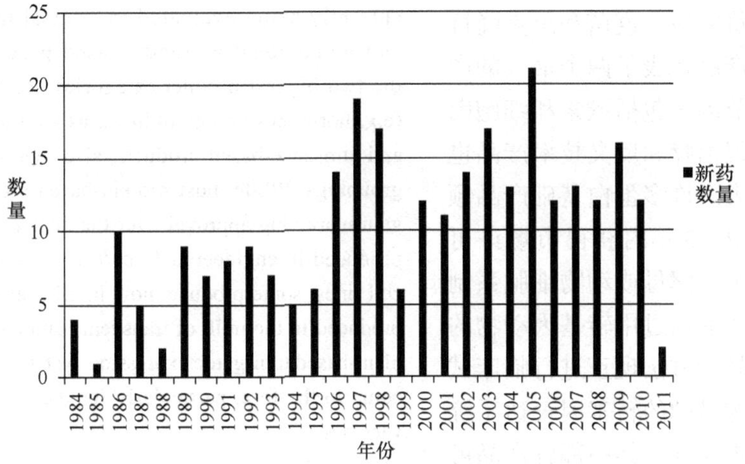

图1-1　1984—2011年FDA批准上市的新生物技术药物数目

表1-1列出明细生物药清单。这些产品包括一系列的激素产品、血液因子和疫苗等，几乎都是蛋白为基础的治疗药物。除了3种反义寡核苷酸，第一例该类药在1998年在美国被首次批准。

Table 1-1 These products include a serial of Hormone products, blood factors and vaccines, etc. Almost all are proteins as the therapeutic agents. The exceptions are 3 antisense oligonucleotide drugs, the first one being approved in the USA in 1998.

表1-1　截至2011年5月在FDA批准用于临床使用的生物药物分类计数

药物名称	获批数量	参与章节
白介素	10	第五章第1节
干扰素	11	第五章第2节
肿瘤坏死因子	2	第五章第3节
激素制剂	33	第五章第6节
疫苗	83	第七章

许多早期批准的生物药物是一些

Many of the initial biopharmaceuticals approved

简单的蛋白替代药物，包括血液因子和人胰岛素。从理论上改变了蛋白质的氨基酸顺序的能力，配以对蛋白质结构和功能的关系逐渐增加的理解，有利于更新地介绍一些设计治疗蛋白。迄今大多数被批准合成的蛋白质都由大肠杆菌、酵母或动物细胞株（如CHO中国仓鼠卵巢细胞或者幼仓鼠肾细胞）制备。

were simple replacement proteins (e.g. blood factors and human insulin). The ability to logically alter the amino acid sequence of a protein, associated with an increased understanding of the relationship between protein structure and function has facilitated the more recent introduction of several engineered therapeutic proteins. Thus far, the vast majority of approved recombinant proteins have been produced in *E. coli*, *S. cerevisiae* or in animal cell lines, most notably Chinese hamster ovary (CHO) cells or baby hamster kidney (BHK) cells.

至少有500种蛋白类生物药物现处于临床评价阶段。疫苗和单克隆抗体为基础的产品代表了两个最大的产品系列。调节因子包括激素和细胞因子，还有基因治疗和反义技术产品也是重要的代表。许多蛋白基础产品倾向于在未来2～3年内获得市场许可并由大肠杆菌、酵母或动物细胞系制备，而一些产品通过用转基因动物的牛奶获得产品的方法现正进行临床试验。另外植物基础的转基因表达系统也可能在逐步实现，其中部分产品可能成为口服疫苗。

At least 500 potential biopharmaceuticals are currently being evaluated in clinical trials. Vaccines and monoclonal antibody-based products represent the two biggest product categories. Regulatory factors (e.g. hormones and cytokines), as well as gene therapy and antisense-based products, also represent significant groupings. While most protein-based products likely to gain marketing approval over the next 2–3 years will be produced in engineered *E. coli*, *S. cerevisiae* or animal cell lines, some products now in clinical trials are being produced in the milk of transgenic animals. Additionally, plant-based transgenic expression systems may potentially come to the fore, particularly for the production of oral vaccines.

2006年6月，美国参议院健康委员会批准了《2007年生物制药产品价格竞争及创新法案》，这项法案曾经在美国国会经过多次反复讨论。根据参议院的这项法案，在仿制药公司能够提出申请，要求生产生物仿制药之前，生物制药产品的原开发者将获得12年的独家销售权；为了获得监管部门批准，仿制药公司必须至少开展一次临床试验，证明其开发的药物与原研药并无两样。许多产品专利期满是导致全球生物仿制药市场快速增长的主要动力，包括合成人生长激素、人胰岛素、促红细胞生成素、α-干扰素、粒细胞集落刺激因子等已快

In order to encourage innovation, the US Senate Health Committee approved “the Law of Biopharmaceuticals Production Price Competition and Creation in 2007” in June, 2006. The law was discussed repeatedly by the U.S. Congress before being passed. According to the Law, the original development company of the biopharmaceuticals can get 12 years solitary market right protection before any generic company can propose application for producing the generic drug. To obtain market approval, the generic company must carry out one clinical trial to testify the generic drug is the same as the original one. Large number of patent expiration of biopharmaceuticals is the mayor driving force for global growth of generic biopharmaceuticals. Examples of the generics include synthetic human growth hormone, actrapid monotard, hemopoietin, α-IF, granulocyte colony-stimulating factor, etc. Most of these biopharmaceuticals sales over $1 billion, thus was very attractive generic target for many

接近尾声。大部分这类药的每年市场额度超过10亿美元，这对许多生物技术或制药公司表现出强烈吸引力，从而掀起一波生物药仿制浪潮。公司已经生产或提出非专利生物药包括英国的Genemedix公司、美国的Sicor and Ivax公司、德国的BioGenerix公司。Genemedix公司2001年开始在中国销售集落刺激因子，后来在中国制造和销售重组红细胞生成素。Sicor公司现在东欧和许多发展中国家销售人生长激素和α-干扰素。

biotech or pharmaceuticals companies. Therefore a new wave to produce generic biopharmaceuticals appeared. The companies that produced non-patent generic biopharmaceuticals included England Genemedix, Sicor Ivax, BioGenerix, etc. Genemedix already got permissions to market colony stimulating factor (CSF) in China in 2001, and started to produce and to sell recombinant EPO. Sicor got the permission to sell human growth hormone (HGH) and α-IF in Eastern Europe and many developing countries.

中国是另一个有可能占生物仿制药较大市场份额的国家，但是大多数仍局限于胰岛素、G-CSF、干扰素和红细胞生成素。一般来说，这些产品均由国内的少数几个厂家生产，包括立志要成为国际企业的成都生物制品研究所和武汉生物制品研究所。深圳万乐药业与韩国Celltrion公司签署了一项协议，将共同开发单克隆抗体类生物仿制药。

China is potentially another bigger market of generic biopharmaceuticals. Right now most of productions seen are only insulin, G-CSF, IF, and EPO. The generic drugs in China are produced by some smaller companies, including Chengdu Institute of Biological Products and Wuhan Institute of Biological Products, Shengzhen Wanle Drug company signed a contract with Korea Celltrion about that they would exploitation a new monoclonal antibody of biotech imitation drugs together.

数据显示，截至2010年，欧洲已经上市了近10种新型生物类仿制产品。据统计，国内到2012年有超过200家企业在生产生物药品，在我国已经批准上市的13类基本工程药物和基因工程疫苗中，也有7类产品属于仿制药，其中更是有95%属于生物仿制药。但大部分都属于技术含量较低的一般类药物，且重复仿制现象突出。而在高端生物仿制药领域，不管是在研发还是生产方面，我国企业都极其缺乏竞争力。在国际市场上的国产高端生物仿制药更是基本上处于一片空白的状态。

There were about 10 generic biotech drugs on European market by 2010. A survey showed that there were over 200 companies producing biotech drugs in China by 2012 and about 13 classes of bioengineered drugs and vaccines had been approved by CFDA, 7 classes were of generic, 95% of which were biosimilars. These generic drugs were produced without much technological challenges, and were much redundant. Chinese companies are usually lack of technological competitiveness at the high end, often seen in areas of R/D, as well as in bioprocessing. The international market is yet to see high end biosimilars from China.

个人的基因密码正变得廉价和

Individual genetic codes are obtained with less

更易获得。斯坦福大学的工程师Stephen R. Quake在2009年宣布他的解码DNA新技术使他能用大约为一个家用冰箱大小的单一序列，用少于5万美元分析他自己的基因组。针对单个患者解码人类最重要的基因组的费用大幅下降。至今还没有基因治疗药被批准用在常规的医疗中。尽管基因治疗可回溯到20世纪90年代，但其结果总体还是令人失望的。美国在这一领域以健赞和辉瑞公司为首，欧洲以诺华公司为主力，基因治疗在相当长一段时间里给人们带来很大希望，尤其是在对一些罕见疾病的治疗上，但许多技术困难仍然存在，比如基因送递和调节表达等。在大多数经历的临床试验中产品的有效性不明显，在一些试验中其安全性仍然令人担忧。尽管有这些针对核酸药物令人失望的结果，未来肯定仍将发展基因治疗和核酸产品。

expensive and much ease technologies. Stanford University Engineer Stephen R. Quake at Stanford University announced in mid 2009 that his new technology for decoding DNA enabled him to make an analysis of his own genome for less than $50,000 using a unique sequencer about the size of a household refrigerator. The cost of decoding the most important sections of the human genome for an individual patient has dropped dramatically now. To date no gene therapy based product has thus been approved for general medical use. Although gene therapy trials were initiated as far back as in 1990's, the results had always been disappointing. Many technical challenges remain, especially in areas of gene delivery and regulation of expression. Product effectiveness was not apparent in the majority of clinical trials undertaken and safety concerns had been raised in several trials. Despite the disappointing results thus far generated by nucleic-acid-based products, future technical advances will almost certainly ensure the approval of gene therapy and antisense-based products in the intermediate future.

在这一领域技术发展比如基因组学、蛋白质组学和高效筛选也开始对药物发展的早期出现重大影响。比如，通过连接改变基因或蛋白的表达以针对不同的疾病状态。这些技术将从这些疾病中鉴别新药。这些目标许多是蛋白质，这些药物将被设计为特异针对个体进行干扰。它们可能是蛋白质或者更可能是低分子量的配基。

Technological developments in areas such as genomics, proteomics and high-throughput screening are also beginning to impact significantly upon the early stages of drug development. For example, by linking changes in gene/protein expression to various disease states, these technologies will identify new drug targets for such diseases. Many/most such targets will themselves be proteins, and drugs will be designed or developed specifically to interact with them. They may be protein-based or (more often) low molecular mass ligands.

干细胞研究在全球快速发展。有大量令人振奋的潜在证据表明干细胞可以治疗很多问题，从心血管疾病到神经性疾病。2009年3月9日，美国总统奥巴马签署行政命令，宣布解除对用联邦政府资金支持胚胎干细胞研究的限制，美国的胚胎干细胞研究

Stem cell breakthroughs are occurring rapidly worldwide. There is truly exciting evidence of the potential for stem cells to treat many problems, from cardiovascular disease to neurological disorders. US President Obama signed Executive Order in March 9, 2009, unleashed the banning of use of Federal funding for embryonic stem cell research. Embryonic stem cell research in the US had since gone into a new era. US NIH published a final version of the guideline for

有望进入新阶段。2009年7月6日，美国国家卫生研究院公布了最终版本的干细胞研究规范，对可以获得联邦政府资金资助的干细胞研究范围做出了界定。但是对干细胞研究仍有很大的争议，比如发育和克隆将不能获得联邦基金的赞助。2009年12月2日，美国国家卫生研究院院长弗朗西斯·柯林斯宣布，利用私人资金制成的13个人类胚胎干细胞系已获准用于该院资助的研究。这是美国总统奥巴马2009年3月解除用联邦政府资金支持胚胎干细胞研究的限制以来，美国国家科研机构首次批准将人类胚胎干细胞系用于研究。2010年10月11日，美国杰龙生物医药公司宣布，该公司已招募到一名急性脊髓损伤患者参与胚胎干细胞疗法临床试验。这是美国批准实施的首例人类胚胎干细胞疗法临床试验。该公司发表的实验结果显示当一些细胞（OPCs）从干细胞中分离出来注射到脊髓损伤的小鼠体内，小鼠很快恢复。通过该公司资料得知这些受伤后7天移植干细胞的小鼠其行走能力相较于对照组有明显改善。这种经OPC手术的动物相较于对照组动物表现出改善的前后肢协调性和体重承载能力，增加了跨步长度和更好地放置手爪。

embryonic stem cell research in July 6, 2009, defined the ranges of stem cell research that can be supported by Federal funding. There are however controversies about stem cell research/application. Development and cloning related research cannot be funded. Francis Colin, US NIH Director announced in December 2, 2009 that the 13 human embryonic stem cell lines developed by private funding received Federal funding. This was the first funding for human embryonic stem cell research by Federal government funding in the US since President Obama lifted the ban of Federal funding to support stem cell research. Geron Corporation at Menlo Park, California, had published the results of its experiments showing that when certain cells (called OPCs) derived from stem cells were injected in rats with spinal cord injuries, the rats recovered quickly. According to the data from the company, "Rats transplanted seven days after injury showed improved walking ability compared to animals receiving control transplants. The OPC-treated animals showed improved hind limb and forelimb coordination and weight bearing capacity, increased stride length, and better paw placement compared to control-treated animals."

全球趋势正积极影响生物科技的发展，911事件的发生以及结合全球流感的播散，对疫苗研究和产品研发有重大的推进作用。凭着发展生物技术产业的经验和投资优势，美国仍主导全球生物技术产业的发展。而欧洲和亚太地区各国，也在逐渐向具有区域特色产业发展。同时，美国、日本

Global trends are affecting the biotech industry in a big way. After 9/11, an emphasis was placed by government agencies on the prevention of bioterror risks, such as attacks by the spread of anthrax. This factor, combined with global concerns about the possible spread of flu, has been a significant boost to vaccine research and production. At the same time, the rapid rise of offshoring and globalization is contributing to the movement of research, development and clinical trials away from the U.S., Japan and Europe into lower cost technology centers

和欧洲在快速增长研究经费背景下将实验研究和临床试验发展进入低科技消费的印度、中国和其他地方。实际上，生物科技公司在印度、中国、新加坡和韩国快速发展，这将给老的西方国家公司在未来形成严重的竞争。

（编者 姜 和，叶翠莲）

in India, China and elsewhere. In fact, biotech firms are rising rapidly in India, China, Singapore and South Korea that will provide serious future competition to older companies in the West.

(Compiled by He Jiang & Cuilian Ye)

参 考 文 献
References

夏焕章，熊宗贵. 2006. 生物技术制药. 北京：高等教育出版社

Ernst & Young. 2007. Beyond Borders: Global Biotechnology Report 2007

Gary Walsh. 1998. Biopharmaceuticals: Biochemistry and Biotechnology. John Wiley & Sons，Ltd

IMARC. 2010. Global biopharmaceutical market report (2010—2015)

Rajesh Krishna, Lawrence Yu. 2008. Biopharmaceutics Applications in Drug Development

第二章 生物技术药物制备工艺

Chapter 2 Biopharmaceuticals Manufacturing

从2007年的最高销售额生物药品表来看，重组红细胞生成素和胰岛素的销售额超过了100亿美元/年，而抗肿瘤和治疗关节炎的单克隆抗体在2006～2008年间每年增长率超过20%，达到330亿美元（2008），成为占最大销售额的一类生物药物（Global 2008）。

In the list of recent top selling biopharmaceutical products, recombinant erythropoetins and insulins were on the top of the list with over 10 billion dollars sale annually (2007), while the global market of therapeutic monoclonal antibodies MAb for cancer and rheumatoid arthritis was increased over 20% during 2006-2008 and became the top group of biopharmaceuticals with total revenue of 33 billion dollars in 2008 (Global 2008).

生物药物是由不同生产系统来生产的，主要有微生物系统（大肠杆菌和酵母）、昆虫和哺乳动物细胞生产体系、植物或动物。大肠杆菌最初在20世纪80年代初开始作为宿主菌用来生产治疗用的人生长激素和人胰岛素等几个重组蛋白。大肠杆菌、酵母、昆虫细胞、哺乳动物细胞、植物和动物生产体系都已成功地生产出市场上销售的生物药物。百分之七十以上的产品是由哺乳动物细胞系统生产的。

Biopharmaceutical products are mainly produced using micro-organisms (bacteria and yeast), insect and mammalian cell lines, plants, or animals . *E. coli* was the first used as the host system to recombinantly produce proteins such as human growth hormone and human insulin for therapeutic purpose in the early 1980's. *E. coli*, yeast, insect cell, mammalian cells, plant, and animal, have all been successfully used to produce approved biopharmaceutical products on market, while mammalian cells have been used for producing more than 70% of current market recombinant therapeutic proteins and MAbs.

以重组蛋白质作为产品的原理是，用分子生物学技术将蛋白质的密码序列重组并转送入一个生命系统，这些遗传信息就会在这个生命系统中复制扩增进而在体内合成产物。这样的可表达产物的生命系统就会为以后的生产而被保存起来，通过发酵和若干步骤的纯化，最后制备成高质量的药物产品。

In principle, for recombinant protein as a product, recombinant DNA including the protein coding sequence is manipulated by molecular technology and transformed into a living system where the genetic signal is replicated, amplified, and the product is *in vivo* synthesized. The living system that can express the product is well preserved for future production. Product is expressed or amplified in the host system through fermentation or cell culture processes, and recovered from culture which subject to a number of purification steps to manufacture final high quality drug products.

生物药物的生产过程是相当繁复的。一则由于生产过程使用的是生物系统，再则由于最后产品的质量要求和产物本身的不稳定性以及大分子的

Manufacturing processes can be quite complicated due to the complex nature of the product and its associated regulatory requirements. Ensuring high product quality through production consistency and reproducibility which are impacted by manufacturing

复杂性造成了生产过程的复杂和困难。因此质量分析、监测和控制以保证生产产品性质的稳定性和重复性以及产品的高质量具有相当的挑战性。其结果，生物制药工业的生产过程受不同生产方法、生产控制和测定方法的影响具有不确定性。进而，从质量规程的角度来看，运用工艺分析技术（PAT）来控制生产系统和生产过程被认为对最终产品的质量具有同样的重要性。美国的FDA和中国的SFDA已经通过详细的规程和指导原则将生物药物的生产严格地规程化。从研发新药和以后的各个开发阶段，这些候选药物以及它们的生产厂家都将面临质量规程方面的挑战。从预临床实验、制备临床样品、工程放大、临床开发、进入市场以及上市后的改进各个阶段，都会有质量规程的问题要解决。本章将介绍从重组细胞株，发酵和细胞培养，到下游的纯化和最后分装的各个工艺环节，及对质量方面的要求。同时也对生物药物生产的一些不同的质量规程和药品生产质量管理规范（GMP）作一些介绍。FDA、欧洲药监局（EMA）和国际协调会（ICH）的最新规程、生物药物生产的最新指导原则等都将被包括在此章中。许多地方还引用了综合评论《大规模生物药物的生产和放大》（Langer，2007）一书来阐述科学、技术和规程方面的观点。

processes, in-process quality monitoring and control by process analytical technology (PAT) are quite crucial and challenge. Biopharmaceutical manufacturing has been closely regulated by law-enforcement agencies such as the Food and Drug Administration (FDA) in the United States and the State Food and Drug administration (SFDA) in China through detailed regulations and guidelines. Drug candidates and their sponsors face regulatory issues at each phase (pre-clinical studies, clinical manufacturing, and engineering scaling-up, clinical development, commercialization of product, and post marketing changes) after moving out of the research laboratory. This chapter is to illustrate the biopharmaceutical manufacturing processes from engineered cell line development, fermentation and cell culture, to downstream purification and fill finishing while including regulatory considerations in all aspects in the manufacturing process. Furthermore, guidelines from several regulatory authorities and principles of Good Manufacturing Practices (GMP) will be introduced to facilitate understanding the quality aspects of biopharmaceutical manufacturing. Updated information from the FDA, European Medicines Agency (EMA), and International Conference on Harmonization of Technical Requirements for Registration of Pharmaceuticals for Human Use (ICH) has been included and referenced in this chapter to provide the latest regulatory requirements in biopharmaceutical manufacturing and newest guidelines and regulations in GMP and drug approval. Comprehensive review books including *Advances in Large-Scale Biopharmaceutical Manufacturing and Scale-Up Production* (Langer, 2007) were referenced in sciences, technology and regulatory aspects in several sections of this chapter.

生物技术和生物制药技术也已在中国迅速发展，过去若干“五年计划”和“国家863高科技研发计划”中，生物技术是几个优先的发展计划之一。通过政府拨款，一些生产平

Biotechnology and biopharmaceutical technology has been rapidly developed in China through last several “Five Years Plans” and “863 National High-tech R&D Program” where biotechnology is one of several development programs with high priority. Production platforms, innovative technology, and products with huge

台、技术创新以及具有巨大的经济影响的产品已被重点发展。

中国生物制药行业正在通过“千人计划”招募来自世界各地的许多训练有素、经验丰富的技术带头人。一些新的单克隆抗体以及许多生物仿制药的分子或正在研发中，或现在已经上市几年。一系列的参考书籍，包括由朱宝泉博士主编的内有“生物制药技术”章节的《现代生物技术》系列书籍总结了在中国此领域的技术进步。鉴于中国的动态发展，此章中引用的许多参考章节还可以不断地被更新的参考文献代替。然而，这些已引用的文献已足够覆盖中国市场的生物制药战略、产品投放市场、产业技术的发展、监管和法律范围。

economic impact have been invested through government funding with favorable focused.

Chinese biopharmaceutical industry is recruiting many well-trained and experienced technical leaders from around the world through the “Thousand People Plan”. A few novel monoclonal antibodies as well as many biosimilar molecules were in pipeline and now has been on the market for a couple of years. A set of reference books referred as *Modern Biotechnology Series Book* including volume of “Biopharmaceutical Technology” edited by Dr. Zhu summarizes technical progress in the field in China. Given the dynamic nature of China, many chapters of those references could be updated today. However, the coverage of these books is sufficient from biopharmaceutical strategy, market product, and industrial technology development, regulatory and legal aspects to Chinese market.

1 生物技术药物生产平台

1 Manufacturing Platforms

在1982—2010年间FDA共批准了120多个重组蛋白和抗体，以及300多个非重组生物药品（疫苗和血液制品）（Rader，2010）。多种产品的生产需要各种生产的平台。所有重组治疗性生物产品的生产都必须遵循质量标准、安全考量和市场经济的原则，这些都由最终产品的特性和预期应用所支配。到目前为止，大多数被批准的重组生物制品由微生物（Shiloach and Rinas，2010；Zhang and An，2010；Zhu，2012）、动物细胞（Zhang，2010；Wurm，2004；Matasci，2009）、转基因植物（Franconi *et al.*，2010；Pogue *et al.*，2010；Ruffoni *et al.*，2011）

There were more than 120 recombinant proteins and monoclonal antibodies, as well as over 300 non-recombinant biopharmaceuticals (vaccines and blood products) approved by the FDA during 1982-2010 (Rader, 2010). Manufacturing of the huge variety of products involves various production platforms. All production of recombinant therapeutic biologics has to follow quality standards, safety concerns, and economic principles, which are dictated by the characteristics and the anticipated application of the final product. So far, the production of most approved recombinant biopharmaceuticals is restricted to micro-organisms (Shiloach and Rinas, 2010; Zhang and An, 2010; Zhu, 2012), animal cell lines (Zhang, 2010; Wurm, 2004; Matasci et al., 2008), and transgenic plants (Franconi et al., 2010; Pogue et al., 2010; Ruffoni et al., 2011) or transgenic animals (Redwan, 2009). Each platform has its special favorable characteristics, but also limitations

或动物体内（Redwan *et al*., 2009）生产获得。不同的生产平台有其特有的优点、局限及缺陷。没有一个系统能最佳地适用于所有的重组蛋白，因此确定由某个表达系统开发一个产品成功的准确度只限于一定的程度。判断错误将导致既费时又费人力物力。因此建议并常用的方法是同时在不同系统（平台）里评估表达目标蛋白的质与量（Gelissen，2005），在仔细并有一定深度的评估后选择一个系统用作后期临床开发和商业化生产的系统。

and drawbacks advantages and disadvantages . As there is no single system that is optimal for all recombinant proteins, prediction for successful development can only be made to a certain extent. Misjudgments may lead to costly time and resource consuming failures. It is therefore advisable and has been a common practice to assess several selected organisms or cells (platforms) in parallel for their capability to produce a particular protein in desired amount and quality (Gelissen, 2005). Following careful and intensive evaluation, one of the platforms is selected for late phase clinical development and commercial manufacturing.

选择一个专门的生产平台大多数依赖于几种因素的组合，包括合理的质量、安全性和成本。单体蛋白可能很容易地在微生物宿主里表达，而那些必须有复杂的哺乳动物类的糖基化或多个二硫键的蛋白需要高等真核生物作为宿主。有很多生物制品相对分子质量很大，结构复杂，或源于多个人类基因。例如，脑下垂体激素是从多个人类基因片段通过体重组而来的（Dingermann，2008）。许多生物治疗蛋白的生物功能，比如说促红细胞生成素（EPO）和组织型纤溶酶原激活剂（tPA），需要只有哺乳动物细胞和转基因动物才能提供的转译后修饰机制以保证接近人体内相同分子的质量和体内功效。药用蛋白的一个重要修饰是糖基化。而动物细胞具有比酵母更接近人类的糖结构。从安全的角度来说，最好选择在系统发育上远离人类的生产平台，以防止人类病原感染该系统。这就是为什么使用HEK293（人类肾脏）细胞作为表达系统的安全顾虑要比使用CHO（中国仓鼠卵巢）细胞大的多的原因。

Selection of a specific production platform is mostly dependent on a combination of several considerations including qualitative rationale, safety aspects, and economic prospective. Single-subunit proteins may easily be produced in bacterial hosts, whereas proteins that require an authentic complex glycosylation or several disulfide bonds require a higher eukaryote as a host. Many biopharmaceutical molecules are large in size, structurally complex, and derived from multiple human genes. For example, pituitary gland hormones are synthesized from a multitude of human gene fragments (Dingermann, 2008). Biological function of many recombinant therapeutic proteins such as erythropoietin (EPO) and Tissue Plasminogen Activator (tPA) requires posttranslational metabolic machinery to convey humanlike quality and efficacy *in vivo* that mammalian cells and transgenic animals may provide. One important modification on a therapeutic protein is glycosylation. and animal cells show closer glycan-structure to humans when compared to yeast, insect cells, and plants . From a safety point of view, it would be best to choose a production platform that is most phylogenetically distant from humans since human pathogens cannot contaminate such systems. As a result, safety concerns when using human embryonic kidney 293 (HEK293) cells as an expression host are greater than when using Chinese hamster ovary (CHO) cells.

微生物系统，尤其大肠杆菌，是最受生物制药研究和开发实验室青睐的用于表达细胞因子、治疗用酶、抗体片段、免疫毒素、蛋白和质粒疫苗的平台。若要快速鉴定产品是否有效，使用微生物可从相对简单的厂房和仪器里快速地生产出毫克或克级的研究用生物药物样品，从而加速生物药物的开发。然而，原核生物系统也有不小局限性。鉴于原核生物不具备转译后修饰的机制（Butler，2005；Zhang，2010），表达复杂的，比如说含有许多亚体、辅因子、二硫键和转译后修饰的大蛋白将是很大的挑战（Mahmoud，2007）。再则，大肠杆菌中表达的重组产品往往会以包含体的形式出现，它是不具活性的聚合体，其复活过程繁复且成本高。不过有一些治疗性蛋白，包括α- 干扰素、β- 干扰素、白细胞介素 -2、α- 肿瘤坏死因子等，尽管在自然人体内是受到糖基化修饰的，它们从大肠杆菌中表达的无糖基化修饰的蛋白却是具活性的药物。植物已成为越来越多的用于治疗使用的蛋白表达和单克隆抗体的平台。它提供了真核细胞的蛋白质操作环境，大大降低了生产成本和哺乳动物的病原体污染的风险。然而，植物生产的产品质量必须与源于哺乳动物系统的产品在预定的产品质量标准（特性、纯度、效力和安全性）要求上相符，这是药品监管机构的要求（Pogue *et al.*，2010）。已经证明在转基因哺乳动物中生产治疗用蛋白质的可行性，包括绵羊、山羊、牛、兔、鸡、鱼等，其中第一个从遗传工程动物（羊）而来的生物药品是重组人抗凝血酶，于

Microbial systems, especially *E. coli*, are the most popular platforms in biopharmaceutical research and development laboratories when expressing proteins, such as cytokines, therapeutic enzymes, antibody fragments, immunotoxins, and protein and plasmid vaccines. For proof of principle studies, milligram or even gram quantities of research grade product can be quickly produced with relatively simple facilities and equipment. However, prokaryotic systems have many limitations. Expression of a large complex protein containing multiple subunits, cofactors, disulfide bonds, and posttranslational modifications is quite a challenge (Mahmoud, 2007) because the posttranslational metabolic machinery is not available in prokaryotic cells (Butler, 2005; Zhang, 2010). Furthermore, recombinant products in *E. coli* are often retained as inclusion bodies, biologically inactive aggregates that require a very costly and sophisticated renaturation process. On the other hand, some proteins such as IFN-alpha, IFN-beta, interleukin-2, tumor necrosis factor-alpha and others, which are glycosylated in their natural human forms, are effective therapeutics without glycosylation after expression in *E. coli*. Plants have been reported as a platform for an increasing number of therapeutic proteins and monoclonal antibodies. It is in a eukaryotic protein processing environment with much reduced production costs and low risk of mammalian pathogen contamination. However, product quality has to be comparable to that produced from mammalian systems along with predetermined release criteria (identity, purity, potency and safety) required by pharmaceutical regulatory agencies (Pogue *et al.*, 2010). The production of therapeutic proteins has now been demonstrated in transgenic mammals including sheep, goats, cattle, rabbits, chickens and fish. The first biological product, recombinant human antithrombin made from a genetically engineered goat was approved by the FDA in 2009 (www.fda.gov). Among all the expression systems, mammalian cells and *E. coli* platforms are responsible for over 80% of market biopharmaceutical products. In the following sections we will discuss several production platforms in detail to understand their principals and applications.

2009年被FDA批准（www.fda.gov）。在所有表达系统中，哺乳动物细胞和大肠杆菌平台占市场生物制药产品的80%以上。在下面的章节中，我们将讨论几个生产平台，详细了解它们的原理和应用。

1.1 哺乳动物细胞表达平台

哺乳动物细胞由于其体内正确的蛋白折叠和装配，以及类似人类的翻译后修饰的能力，已成为生产治疗和体内诊断应用的重组蛋白的主导系统。超过50%的市场上被批准的治疗性蛋白产品是由哺乳动物细胞产生的。最近，在生物反应器里培养的哺乳动物细胞的生产力已达到10～15g/L单克隆抗体和Fc-融合蛋白（Huang *et al.*, 2010），这是通过改善细胞株、改变表达策略、优化培养基和控制过程而获得的。为了增加在哺乳动物细胞内的生产水平，很多不同的表达系统都被试验过。20世纪80年代以来最大的进步在于载体的设计和构建，包括使用强启动子、适当的信号肽、选择内含子、产品基因密码子优化、转录控制区、基因扩增的方法、合适的宿主细胞、提高转染方法、筛检工具（Aldrich *et al.*，2003；Birth and Racher，2006；Deer *et al.*，2004；Wurn，2004；Zhang，2010；Jalah *et al.*，2007）。然而，在哺乳动物细胞株中生产高水平的外源蛋白还存在几个主要的挑战。一般来说，这些细胞生长缓慢，有更挑剔的生长需求，并有被动物病毒污染的可能。构件生产细胞株通常是非常耗时的。其生产能力和表达的稳定性是决定上市产品生产

1.1 Mammalian Cell Expression Platform

Mammalian cells have become the dominant system for producing recombinant proteins for therapeutic and *in vivo* diagnostic applications because of proper protein folding and assembly, human-like posttranslational modification, as well as the ability of synthesizing proteins that are similar to those naturally occurring in humans with respect to the molecular structures and biochemical properties. More than 50% of approved therapeutic proteins are produced by mammalian cell systems. Recently, the productivity of mammalian cells cultivated in bioreactors has reached 10-15g/L in MAb and Fc-fusion protein production (Huang *et al.*, 2010) due to improvements in cell line development through expression strategies, media optimization, and process control. Various expression systems were investigated to enhance the production in mammalian cells. Vast improvements have been made since 1980's in vector design and construction, including using strong promoters, proper signal peptides, selected introns, product gene codon optimization, transcription control regions, gene amplification approaches, suitable host cells, improved transfection methods, and screening tools (Aldrich *et al.*, 2003; Birch and Racher, 2006; Deer *et al.*, 2004; Wurm, 2004; Zhang, 2010; Jalah *et al.*, 2007). However, there are several major challenges associated with producing high levels of heterologous proteins in mammalian cell lines. Generally, these cells grow slowly, have fastidious growth requirements, and can become contaminated with animal viruses. Production cell line development is usually very time-consuming. Productivity and stability of expression are the prerequisites for developing commercially viable processes. Nevertheless, in many cases expression level in a mammalian system has reached 1-5g/L or higher for product like monoclonal antibody, and with high

可行性的先决条件。有许多例子（如单克隆抗体产品）显示了在哺乳动物系统中的表达水平已达到1～5g/L或更高，并在补料细胞培养生产中达到80pg/（细胞·d）的高生产率。

productivity of 80pg/cell/day in a fed-batch cell culture production.

1.1.1 细胞株

生物制药生产过程需要细胞株能够强劲和稳定地增长以及合理的细胞倍增时间；需要稳定的外源DNA整合，从而产生在数量和质量上具有重现性的能被很好地研究其特征的重组蛋白。常用于生物制药行业的细胞株，有中国仓鼠卵巢（CHO）细胞、小鼠骨髓瘤淋巴样（NS0和Sp2/0-Ag14）细胞、人胚肾293（HEK 293）细胞、幼仓鼠肾细胞（BHK-21）等。用于生产最终将被监管机构批准的治疗用产品的细胞株必须经过仔细的鉴定，并具有一定的遗传基因稳定性，以保证长时间的培养期间其产品生产水平不变。因此，研发细胞株的最终目标是获得能够历经超长细胞代数而始终能大量分泌目标蛋白的克隆细胞株，从而保证生产的放大和成本效益。

1.1.1 Cell Lines

Biopharmaceutical manufacturing processes require cell lines to be capable of achieving robust and stable growth with reasonable cell doubling time, stably integrated heterologous DNA to yield consistent quantity, and quality of the recombinant proteins that can be well-characterized. Commonly used cell lines in biopharmaceutical industry such as CHO cells, murine myeloma lymphoblstoid-like (NS0 and Sp2/0-Ag14) cells, HEK 293 cells, and baby hamster kidney cells (BHK-21). For production of therapeutics that are subject to regulatory approval, the cell lines must be carefully tested and remain genetically stable to retain a constant level of product production for long periods of culturing. The ultimate goal of cell line development is to obtain clonal cell lines that secrete the protein of interest with high specific productivity at consistently high levels over an extended number of cell generations, allowing for scale-up and cost-efficient manufacturing.

CHO、NS0、BHK-21、HEK-293和Sp2/0细胞是表达重组蛋白的主要哺乳动物细胞株（Walsh，2006；Meyer *et al.*，2008；Birch and Racher，2006；Zhang，2010）。这些细胞的重要特征是能接受转染，具有在无血清悬浮培养基生长的能力，以及它们的高密度、大规模生产的适用性。在无血清培养基中生长的能力，不仅降低了成本，而且有利于下游加工和减少动物源性材料污染的风险。

CHO, NS0, BHK-21, HEK-293, and Sp2/0 cells have been established as the predominant systems for mammalian expression of recombinant proteins (Walsh, 2006; Meyer *et al.*, 2008; Birch and Racher, 2006; Zhang, 2010). Important characteristics of these cells are their receptiveness to transfection, their ability to grow in serum-free suspension cultures, and their suitability for high-density and large-scale production. The ability to grow in serum-free medium not only reduces cost, but also facilitates downstream processing to reduce the risk of contamination with animal-derived material.

在许多可用的系统中，有两个是常用的：用于重组蛋白和单克隆抗

Of the many available systems, two have been commonly used: CHO cell lines for recombinant proteins and MAbs and murine myeloma (NS0 and

体表达的CHO细胞株，和用于单克隆抗体表达的小鼠骨髓瘤（NS0和Sp2/0）细胞株。

1. CHO细胞株 CHO细胞是组织型纤溶酶原激活剂（TPA），首个获得监管部门批准的生产治疗性蛋白的哺乳动物细胞宿主。CHO细胞是目前在生物制药行业里最常用的细胞株，超过半数的上市重组药用蛋白由其生产。CHO-K1细胞株经诱变产生缺乏二氢叶酸还原酶（DHFR）活性的CHO-DXB11细胞株（又称CHO-DUKX或CHO-DUK-XB11）。这些细胞缺失一个*dhfr*基因并在另一个基因上有突变。随后，另一个同样衍生于CHO细胞株的脯氨酸依赖型的CHO-pro3细胞株，被诱变产生出两个*dhfr*基因都缺失的CHO-DG44细胞株。这两个DHFR缺失型细胞株需要甘氨酸、次黄嘌呤和胸腺嘧啶（GHT）才能生长，这种特性使得其能够通过外源性*dhfr*基因的表达从而在无GHT的培养基中生长，从而挑选出DHFR缺失的细胞。这种方法已成功地通过稳定转染和随后的一系列扩增步骤生产出了不少治疗性生物制品。这种基因选择方案是建立能够生产重组治疗性蛋白的稳定转染CHO细胞株的一个标准方法。另一方面，CHOK1细胞及其衍生物之一CHOK1SV，已被广泛地使用在利用谷氨酰胺合成酶（GS）作为选择标记的表达系统里。在这个系统里，CHO细胞表达的谷氨酰胺合成酶催化谷氨酸和氨的缩合反应生成谷氨酰胺，这在氮代谢中起着至关重要的作用。CHO细胞中的内源性GS被添加在细胞的生长培养基中的GS抑制剂、甲硫氨酸砜亚胺（MSX）所抑制。通过

Sp2/0) cell lines for MAbs.

1. CHO cell lines CHO was chosen as the host for the production of tissue plasminogen activator (tPA), the first recombinant therapeutic protein from mammalian cells that received regulatory approval. CHO is now the most commonly used cell line in biopharmaceutical industry from which over two-thirds of the recombinant therapeutic proteins currently on the market were made. CHO-K1 was mutagenized to generate CHO-DXB11 (also referred to as CHO-DUKX or CHO-DUK-XB11), a cell line lacking dihydrofolate reductase (DHFR) activity. These cells have a deletion of one *dhfr* allele and a missense mutation in the other one. Subsequently, the proline-dependent CHO-pro3- strain, another derivative of the original CHO cell line, was mutagenized to yield CHO-DG44, a cell line with deletions of both *dhfr* alleles. These two DHFR-minus cell lines require glycine, hypoxanthine, and thymidine (GHT) for growth, allowing the selection of DHFR-minus CHO cells via expression of an exogenous *dhfr* gene in GHT-minus medium. This approach has been successfully employed to generate a number of therapeutic biologics through stable transfection followed by a series amplification steps. This genetic selection scheme remains one of the standard methods to establish stably transfected CHO cell lines for the production of recombinant therapeutic proteins. On the other hand, the CHOK1 cell line or one of its derivatives, CHOK1SV, has been extensively used in a system utilizing glutamine synthetase (GS) as a selection marker. In this case, CHO cells express functional GS enzyme-that plays an essential role in the metabolism of nitrogen by catalyzing the condensation of glutamate and ammonia to form glutamine. Endogenous GS in CHO cells is inhibited by the addition of GS inhibitor, methionine sulphoximine (MSX), in cell growth medium. By selecting GS-CHO transfectants in the presence of MSX, only cell lines that have stably incorporated the expression vector into a transcriptionally active locus will form transfectant colonies. These cells produce enough GS enzyme to titrate out the MSX whilst leaving sufficient functional enzyme to meet

在加入MSX生长培养基中生长的选择方式，只有那些稳定纳入表达载体于活跃位点的GS-CHO转染细胞才能成为转染株。这些细胞产生足够的GS酶去结合MSX，同时留出了足够的活性酶，以满足细胞谷氨酰胺代谢的需求。重组CHO细胞株能够在翻译后高效地修饰复杂的蛋白质，而且它衍生的重组蛋白的糖基化结构和天然蛋白是相似的。

the cellular demand for glutamine metabolism. Recombinant CHO cell lines show efficient post-translational processing of complex proteins, while the glycosylation patterns of native and CHO-derived recombinant proteins are similar.

2. *鼠骨髓瘤细胞株* NS0和Sp2/0-Ag14都是从源生于BALB/c小鼠的浆细胞衍变而来的（Shulman *et al.*, 1972）。初始细胞经历了无数次的克隆，然后Sp2/0细胞与BALB小鼠的脾细胞融合，最终成为这两个亲本细胞株（Burch and Racher，2006）。两个细胞株被选中的原因是因为作为分化的B细胞他们却不能合成和分泌免疫球蛋白，而通常B细胞天生就能够高水平地合成免疫球蛋白。这两个特点有利于利用它们来生产抗体。NS0细胞的基因型特别适合用于GS表达系统。不同于其他类型的细胞，NS0细胞一定是谷氨酰胺营养缺陷型：只有谷氨酰胺合成酶（GS）基因的转染才能赋予NS0细胞的生长不依赖于谷氨酰胺。而Sp2/0-Ag14细胞突回复突变到不依赖于谷氨酰胺生长的自发频率比较高。这在NS0细胞里从未发现过。

2. Murine myeloma cell lines The murine myeloma cell lines NS0 and Sp2/0-Ag14 were both derived from a plasmacytoma cell line originating from a BALB/c mouse (Shulman *et al.*, 1972). The initial cells underwent numerous rounds of cloning and, in the case of Sp2/0, fusion with spleen cells from another BALB mouse, to become these two parental cell lines (Birch and Racher, 2006). Both the cell lines were selected for their inability to synthesize and secrete immunoglobulin proteins as differentiated B cells, which are inherently capable of high levels of immunoglobulin production. These two characteristics favor their use for manufacturing antibodies. The genotype of NS0 cells makes them particularly suited for use with the GS expression system. Unlike other cell types, NS0 cells are obligated glutamine auxotrophs: glutamine independence can be conferred upon NS0 cells following transfection with a functional GS gene. In the case of Sp2/0-Ag14, mutants that no longer require glutamine occur spontaneously with relatively high frequency. This is not observed with NS0 cells.

无论是作为一个杂交瘤技术的融合细胞，或用作生产重组蛋白的转染细胞，NS0和Sp2/0-Ag14细胞已被广泛地使用在治疗性单克隆抗体的生产上。截至2010年已经有11个持有生产许可的抗体产品是由NS0和Sp2/0生产而来，如Centocor公司的Remicade和MedImmune公司开发的

NS0 and Sp2/0-Ag14 cells have been used extensively to produce therapeutic monoclonal antibodies, either as a fusion partner in hybridoma technology or as a result of transfection for recombinant protein production. As of 2010 there have been 11 licensed antibody products manufactured using NS0 and Sp2/0 cells such as Remicade developed by Centocor and Synagis developed by MedImmune, respectively. Both of which had over one billion sales in 2007.

Synagis。二者在2007年的销售额都超过十亿美元。

1.1.2 表达系统

综上所述，构建用于生产药用蛋白质的细胞株的常见方法是依靠选择性诱导而导致的基因扩增，比如DHFR或GS（Bebbington et el. 1992；Birch and Racher 2006）。为了实现高水平的基因表达，载体通常有强启动子，如巨细胞病毒（CMV）启动子，以推动高水平的信使RNA转录（Deer and Allison 2004）。表达载体优化也可以包括最有利的RNA处理信号，如多聚腺苷酸（poly A）尾，5'和3'非翻译区，加入刺激从细胞核释放蛋白的内含子，以及去除此内含子的剪切位点，根据表达细胞种类而优化密码子，GC/AT比率的平衡，和已被证明能够加快mRNA加工和改善分泌的信号序列的优化（Jalah *et al.*, 2007）等。编码序列的优化，是避免未知的剪接位点或未知的poly A尾的产生，也可避免导致不利的mRNA折叠序列的产生。此外，靶向基因技术、染色质开放元素和附着区域已被用于表达载体的优化上，以提高最终产品的生产（Cacciatore *et al.*, 2010；Zhu, 2012）。

最常使用的用于生产治疗性单克隆抗体的表达载体系统是GS（Lonza Biologics）和*dhfr*基因表达系统。谷氨酰胺合成酶用谷氨酸和氨合成谷氨酰胺。由于谷氨酰胺是一种必需氨基酸，转染的GS表达载体能够赋予缺乏内源性GS的细胞在没有谷氨酰胺的培养基中生长的能力。DHFR的表达系统使用叶酸类似物甲氨蝶呤

1.1.2 Expression Systems

As described above, common approaches in generating cell lines for the production of therapeutic proteins rely on gene amplification induced by a selective marker such as DHFR or GS (Bebbington *et al.* 1992; Birch and Racher 2006). To achieve high levels of gene expression, vectors usually have strong promoters such as cytomegalovirus (CMV) promoter to drive high level messenger RNA transcription (Deer and Allison 2004). An expression vector may also include favorable RNA processing signals such as polyA tail, 5’ and 3’ un-translated region, an intron to encourage export from the nucleus, and a splice site to remove this intron. Codon optimization for the target cell type, GC/AT ratio balancing, and signal sequence optimization have been shown to accelerate mRNA processing and improve secretion (Jalah *et al.*, 2007). Coding sequences may also be optimized to avoid cryptic splice sites or cryptic poly A tails, or sequences that lead to unfavorable folding of the mRNA. Besides, gene-targeting technology, chromatin opening elements and attachment regions have been incorporated into vector optimization to improve final product production (Cacciatore *et al.*, 2010; Zhu, 2012).

The expression vector systems that most frequently used for the production of therapeutic monoclonal antibodies are the GS gene expression system (Lonza Biologics) and those based on *dhfr* gene. GS synthesizes glutamine from glutamate and ammonium. Since glutamine is an essential amino acid, transfection of cells that lack endogenous GS with the GS vector confers the ability to grow in glutamine-free media. DHFR expression systems use the folate analogue methotrexate (MTX) to inhibit the function of DHFR, an essential metabolic enzyme. Transfection with an expression vector containing a *dhfr* gene prevents MTX poisoning of transfected cells. The antibiotic resistance

（MTX）来抑制必须代谢酶DHFR。转染了含有*dhfr*基因的表达载体能够中和MTX对细胞的毒性。DHFR的表达载体中使用的抗生素抗性基因经常用作选择标记，*dhfr*基因的另外重要功能是促进载体的扩增。

gene frequently used in DHFR expression vectors acts as a selectable marker: the primary function of the *dhfr* gene is then to facilitate vector amplification.

基因扩增通常是通过构建扩增基因（例如胸苷激酶、腺苷脱氨酶、谷氨酰胺合成酶或DHFR）与靶向基因相连的方式来实现。转染细胞后逐步增加培养基中的酶抑制剂浓度到大大超过用于转染株选择的剂量，由于药物抑制了一种其活性对细胞的生存至关重要的酶（例如GS或DHFR），只有在细胞过量产生这种酶时它才能生存。过量生产这些酶通常是由放大基因拷贝数或是增加高效转录而导致的特定mRNA水平的提高来达到的。通常情况下，不仅仅是靶基因的DNA（大到1000 kb）能够被放大，而且当靶基因扩增时，载体上紧密相连的其他序列，如免疫球蛋白基因，也与之共同扩增。放大后的表达载体，特别是与DHFR表达系统有关联的，其高拷贝数可能会增加其生产率（Qp），但它也可能对其他的细胞特性有不利的影响。

Gene amplification is usually achieved by constructing the expression vector in a way that the genes of interest are linked to an amplifiable gene (e.g., thymidine kinase, adenosine deaminase, GS or DHFR). Transfected cells are then exposed to increasing level of a specific enzyme inhibitor at concentrations substantially higher than those used for selection of transfectants. If the drug inhibits an enzyme (e.g., GS or DHFR) that is essential for the survival of the cells, only cells that overproduce this enzyme will survive. The overproduction of the enzyme commonly results from increased levels of its particular mRNA, resulting from either an increase in gene copy number (i.e., amplification) or from more efficient transcription. Often, more DNA (up to 1000 kb) than just the target gene is amplified. Therefore, when the transfected genes are amplified, other tightly linked sequences, including the immunoglobulin genes, on the vector are co-amplified. The high copy numbers of the expression vector after amplification, especially with the DHFR expression system, may increase specific productivity (Qp) but it can also have a detrimental effect on other cellular properties.

有几篇重要的评论文章总结了常用的表达系统，包括Birch和Racher（2006）的关于GS和DHFR选择性标记来扩增转染细胞的概述；Cacciatore等（2010）的如何用基因扩增和载体工程，重点以DHFR系统为例，来实现快速和高产量的治疗性蛋白生产的论述；以及Rodrigues等（2010）的关于单克隆抗体生产系统的技术进展报告。

Several important review articles summarize commonly used expression systems including GS and DHFR selective markers for amplifying transfected cells in an overview by Birch and Racher (2006), gene amplification and vector engineering to achieve rapid and high-level therapeutic protein production with emphasis on the DHFR selection system by Cacciatore *et al.* (2010), and technological progresses in MAb production systems by Rodrigues *et al.* (2010).

1.2 微生物表达平台

生物制药行业的历史一直与表达宿主系统的研发，以及高效、准确、快速、稳定和合理成本的产品生产密切相关。相对哺乳动物表达系统而言，微生物系统具有建立生产菌株成本低的优势，同时生产周期快、工程放大可行、生产过程易于控制，而且产率高。重要的用于重组蛋白表达的微生物系统主要有大肠杆菌（原核生物）和酿酒酵母（真核生物）。容易培养的大肠杆菌因为其遗传序列大部分已知而成为最常见的生产重组蛋白的微生物宿主。事实上，大肠杆菌的完整基因组序列已在1997年9月被威斯康星-麦迪逊大学的遗传学实验室的研究人员完成分析，4 639 221碱基对包含了4403个基因（www.accessexcellence.org/WN/SUA11/ecoli997.php）。酿酒酵母（面包或啤酒酵母）具有与大肠杆菌生产系统类似特点，因此也被经常用于传统的发酵生产。其基因组在1996年被全部测序，此前酿酒酵母已被人类使用了数千年。20世纪第一个被批准的重组蛋白胰岛素（humulin）是Eli Lilly公司于20世纪70年代利用重组DNA（rDNA）转化大肠杆菌而制造的。最早的从酵母系统表达和生产的产品也是胰岛素（Novolin），它由Novo Nordisk公司在同一时期内使用重组DNA转化的酿酒酵母开发并得到批准。其他的微生物表达系统也都在20世纪70年代得到开发：*Pichia* 酵母和非大肠杆菌的细菌，如枯草芽孢杆菌，但它们还未被充分应用到人的治

1.2 Microbial Expression Platforms

The history of biopharmaceutical industry is closely associated with development of expression host systems and their ability to express products efficiently, accurately, quickly, consistently, and cost-effectively. Microbial systems have the advantage of low cost in establishing a production strain, quick production cycle, engineering scale-up, easy in-process control, and high productivity as compared to mammalian expression systems. Important microbial expression systems for recombinant proteins have been *E. coli* (a prokaryote) and *Saccharomyces cerevisiae* (*S. Cerevisiae*, a eukaryotic microbe). The easily cultivated *E. coli* has been the most common microbial source for recombinant protein production because the most of its genetics is known. In fact, the complete genome sequence of *E. coli* was determined in September 1997 by a team of researchers in the Laboratory of Genetics at the University of Wisconsin-Madison. The 4,639,221 base pairs of the genome contain 4,403 genes (www.accessexcellence.org/WN/SUA11/ecoli997.php). For many of the same reasons that *E. coli* is popular among bacterial production systems, *S. cerevisiae* (baker's or brewer's yeast) is commonly used in traditional fermentation processes. In addition to having its full genome sequenced in 1996, *S. cerevisiae* has been used by people for thousands of years. The first approved recombinant protein insulin (humulin) was manufactured by Eli Lilly & Co in the 1970's using recombinant DNA (rDNA) transformed *E. coli* and approved in the early 1980's. The earliest product expressed by yeast was another insulin product (Novolin) developed and approved in that same time frame from Novo Nordisk using rDNA transformed *S. cerevisiae*. Other microbial expression systems were also developed in the 1970's: *Pichia* yeast and non–*E. coli* bacteria such as *Bacillus subtilis*, but their adoption for human therapeutics has yet to be fully realized.

疗用药生产中。

超过20%的上市生物治疗药物来自于微生物表达平台。许多源于大肠杆菌和酵母生产系统的蛋白质和生物产品（市场及临床试验）包括：B型肝炎抗原，流感抗原，脊髓灰质炎抗原；人类生长激素和胰岛素；抗体，抗体片段和免疫毒素；干扰素和白细胞介素；血液成分如人血清白蛋白；组织纤溶酶原激活剂；和质粒疫苗。

Over 20% of market biotherapeutic drugs are produced in microbial expression platforms. A wide range of proteins and other biopharmaceuticals on the market and in clinical trials produced in *E. coli* and yeast includes: antigens to hepatitis B, influenza, and polio; human growth hormone and insulin; antibodies, antibody fragments and immunotoxins; interferon and interleukin; blood components such as human serum albumin; tissue plasminogen activator; and plasmid vaccines.

不过，原核生物如大肠杆菌系统也有种种限制。一个主要的因素是此系统缺乏翻译后修饰（Mahmond, 2007）。许多重组蛋白，比如组织型纤溶酶原激活剂（tPA）和促红细胞生成素（EPO）需要有翻译后修饰如糖基化，才能具备其生物功能（Sasaki *et al*., 1987）。微生物表达系统要么没有糖基化修饰，要么寡糖结构与人类大不相同。同样的，单克隆抗体的糖基化对其最佳生物学功能和药代动力学是至关重要的（Beck *et al*., 2008； Jefferis, 2005）。

However, there are various limitations for prokaryote systems like *E. coli*. One major factor is that the system lacks of the posttranslational modifications (Mahmoud, 2007). Many recombinant proteins, for instant, tPA and EPO, need to have posttranslational modification such as glycosylation, for its biological function (Sasaki *et al*., 1987). In microbial expression systems either there is no glycosylation or glycol-structure is quite different from the human glycan-structure . Likewise, glycosylation of MAb is important for optimal biological function and pharmacokinetics (Beck *et al*., 2008; Jefferis, 2005).

1.2.1　大肠杆菌表达系统

革兰氏阴性菌大肠杆菌之所以成为生产重组蛋白胰岛素的第一个系统，是因为其遗传背景清楚，易于遗传操作以及行之有效的发酵过程。20世纪70年代以来，大肠杆菌一直是许多用于治疗、诊断和工业应用的重组蛋白的首选宿主。由于受产物难于分泌和缺乏应有的糖基化的限制，大肠杆菌表达平台只是广泛地使用在其活性不需要复杂糖基化的药用蛋白质的大规模生产上。从2008至2011年间，有三个获得批准的产品是利用大肠杆菌生产的，包括由Savient公司开发的

1.2.1 *E. coli* Expression Systems

The Gram-negative bacterium *E. coli* was the first organism to be employed for recombinant protein insulin production because of its clear genetic background, the ease of genetic manipulations as well as the availability of well-established fermentation procedures. Over the past four decades, *E. coli* has been the primary host of choice for the expression of a wide variety of recombinant protein for therapeutic, diagnostic, and industrial applications. Due to the limitations in secretion and lack of glycosylation which impose restrictions on general use, the *E. coli* expression platform has been most extensively used for the large-scale production of therapeutic proteins that do not require complex glycosylation for bioactivity. In recent years from 2008 to 2011, *E. coli* host was used for three license products: Pegloticase, a Pegylated urate oxidase developed by Savient, Crtmzia

聚乙二醇尿酸氧化酶 Pegloticase，由 Union Chimmupue Belge 公司开发的 TNF alfa- 肿瘤坏死因子抑制剂 Crtmzia（Certolizumab Pegol），和由 Amgen 公司开发的血小板生成素短肽与 Fc 的融合抗体 Nplate（Romiplostin）。

(Certolizumab Pegol), an antibody fragment Fab as TNF alfa inhibitor developed by Union Chimique Belge (UCB), and Nplate (Romiplostin), a thrombopoietin peptibody Fc fusion developed by Amgen .

微生物宿主细胞表达系统的设计主旨在于产生大量所需蛋白的拷贝数。为了做到这一点，表达载体被转入到宿主细胞。这个载体包含了所有必要的表达蛋白质的基因编码，包括适合于宿主细胞的启动子、转录终止序列、核糖体结合蛋白编码序列。W. F. Studier 和 B. A. Moffatt 在 1986 年发明了高度受噬菌体 T7 RNA 聚合酶选择的 RNA 聚合酶表达系统。最初的系统涉及两种不同的方法来维持细胞中的 T7 RNA 聚合酶的一定数量，一个方法是用λ噬菌体来插入带有 T7 RNA 聚合酶编码的基因，另一个方法是把 T7 RNA 聚合酶基因直接插入宿主染色体中（Studier *et al.*, 1986）。由于它能够大量地表达蛋白质，此后被称为 pET 的该表达系统被广泛使用（Unger, 1997）。这个系统具有只有 T7 RNA 聚合酶才能结合的 T7 启动子的特异性，其设计便于操作与控制蛋白质表达量和表达时间。

Expression systems are designed to produce many copies of a desired protein within microbial host cells. In order to accomplish this, an expression vector is inserted into a host cell. This vector contains all of the genetic coding necessary to produce the protein, including a promoter appropriate to the host cell, a sequence which terminates transcription, and a sequence which codes for ribosome binding factor. One expression system was developed in 1986 by W. F. Studier and B. A. Moffatt, who created an RNA polymerase expression system which was highly selective for bacteriophage T7 RNA polymerase. The initial system involved two different methods of maintaining T7 RNA polymerase in the cell - in one method, a lambda bacteriophage was used to insert the gene which codes for T7 RNA polymerase, and in the other, the gene for T7 RNA polymerase was inserted into the host chromosome (Studier *et al.*, 1986). This expression system has become known as the “pET Expression System”. It is now widely used because of its ability to mass-produce proteins, its specificity of the T7 promoter to which only T7 RNA polymerase binds. It is also the design of the system which allows for the easy manipulation of how much of the desired protein is expressed and when the expression occurs (Unger, 1997).

通常通过加入化学物质或温度变化的机制来诱导蛋白质的表达。在大肠杆菌表达系统中最广泛使用的强调控启动子是噬菌体 T7 启动子、噬菌体的 PL 启动子和色氨酸与乳糖启动子相结合的 *tac* 启动子。虽然氨苄青霉素抗性基因（β- 内酰胺酶基因）作为表达选择标记在实验室里很普遍，但是根据法规指导原则，它在 GMP 生产中却受严格限制（www.fda.gov）。

Expression is usually regulated by expression induction mechanism through either chemicals addition or temperature shift. The most widely used strong regulatable promoters in *E. coli* expression system are those bacteriophage T7 promoter, PL promoter from bacteriophage l, and the *tac* promoter that combines *trp* and *lac* promoters. While ampicillin resistance gene (β-lactamase gene) is quite popular in research laboratories as expression selection marker, it is strictly restricted in GMP manufacturing according to regulatory guidelines (www.fda.gov). Instead, kanamycin resistance marker is the choice of large

与之相反的是，卡那霉素是生物制药中大多数药用蛋白质大规模生产的选择抗性标记。

scale manufacturing for most therapeutic proteins in biopharmaceutical industry.

在普遍使用的T7启动子表达系统中，构建的质粒转化进入含有T7 RNA聚合酶的宿主从而表达所需的蛋白质。pLysS质粒中含有的T7溶菌酶基因（*LysS*）是T7 RNA聚合酶的天然抑制剂，它可减少在细胞未诱导的境况下靶基因的转录能力。异丙基-β-D-硫代吡喃半乳糖苷（IPTG）加入到培养基中后，T7 RNA聚合酶便大量增加从而克服了*LysS*对其的抑制。靶基因的表达因此可在pLysS系统中严格受到控制。

For a popularly used T7 promoter expression system, the established plasmids were transformed into a host bearing the T7 RNA polymerase for expression of the protein of interest. pLysS is a plasmid that has the T7 lysozyme gene (*LysS*), a natural inhibitor of T7 RNA polymerase, upon which to reduce the ability of transcribing target genes in un-induced cells. When an expression inducer Isopropyl β-D-1-thiogalactopyranosides (IPTG) is added, the amount of T7 RNA polymerase increases and overcome the inhibition by *LysS*. Expression can be tightly controlled in expression hosts with pLysS.

pET载体是细菌质粒，旨在利用强大的T7启动子，一旦激活便快速地大量生产蛋白质。该质粒包含几个重要的元素：一个编码lac阻遏蛋白的*LacI*基因，一个只对T7 RNA聚合酶（不是细菌的RNA聚合酶，此酶并不存在于原核生物中）有亲和力的T7启动子，一个可阻止转录的乳糖操纵子，一个用于基因插入的聚链接位点，一个通常是抗生素的抗性基因的遗传选择标记，和质粒复制的源点。

A pET vector is a bacterial plasmid designed to utilize strong T7 promoter expression to enable a quick production of a large quantity of any desired protein when activated. This plasmid contains several important elements: a *lac I* gene which codes for the *lac*-repressor protein, a T7 promoter which is specific to only T7 RNA polymerase (not bacterial RNA polymerase) which does not present anywhere in the prokaryotic genome, a *lac* operator which can block transcription, a polylinker for transgene insertion, a genetic selection marker normally an antibiotic resistance gene, and an origin of replication.

在大肠杆菌系统的治疗性蛋白的生产中，编码基因的cDNA被插入质粒，然后转化进入B或K12株系列的宿主细胞。B株细胞系列有B12，BL21（DE3），BL21（DE3）pLysS，BLR（DE3）pLysS和B21-AI，K株细胞系列包括大肠杆菌HMS174。BLR细胞株是从BL21（DE3）衍生的*recA*基因缺失细胞株，用于提高质粒的稳定性。BL21-AI大肠杆菌是通过把受阿拉伯糖诱导araBAD启动

To produce a therapeutic protein in *E. coli* system, the coding cDNA is constructed in a plasmid that transformed into host cells either B strain such as commonly used BL 21, BL21(DE3), BL21(DE3) pLysS, BLR(DE3) pLysS, and BL 21-AI or K strains including *E. coli* HMS174. The BLR strain is a *recA*-derivative of BL21(DE3) that may improve plasmid stability. The *E. coli* BL21-AI strain was constructed by inserting a chromosomal copy of the T7 RNA polymerase gene (T7 RNAP) under control of the arabinose-inducible *araBAD* promoter. The *araBAD* promoter tightly regulates the expression of the T7 RNA polymerase and hence expresses the gene of interest in T7-mediated expression systems without

子控制下的T7 RNA聚合酶基因插入染色体构建而成。araBAD启动子严密控制T7 RNA聚合酶的表达，在没有DE3溶原菌的情况下控制所需蛋白的表达。当发现目标蛋白的表达受到抑制时，可考虑比较几个这样的菌株来衡量产品的表达以及表达的调控（Vyas *et al*., 2011）。从法规事务的角度考虑，无论是大肠杆菌BLR还是BL21-AI细胞株，它们都比BL-21（DE3）细胞株减少了在大肠杆菌中发生的低量的噬菌体污染的危险。

DE3 lysogen. When a therapeutic protein expression is initiated, several such strains may be evaluated on expression of the product as well as expression regulation (Vyas *et al*., 2011). In regulatory affair consideration, both *E. coli* BLR and BL21-AI reduce the risk of low-level phage contamination that was occasionally observed in *E. coli* strain tested with the BL-21(DE3) strains.

1.2.2 酵母菌表达系统

酵母系统已多年应用于工业和生物药物的蛋白质的大量生产中。它可以产生胞内的和胞外的两种外源蛋白。在酵母中制成的药物蛋白的例子包括：细胞因子、酶、乙型肝炎疫苗、人血白蛋白（人造血浆的一个组成部分）。通常情况下，酵母表达载体包含能使它们在细菌和酵母里都具功能的部分。通常一个酵母表达载体包括大肠杆菌复制点和抗细菌抗生素的基因，使其可以先在大肠杆菌中扩增而后转化入酵母。类似于大肠杆菌，酵母非常容易在基因水平上操纵，有较快的增长速度，也能在化学成分确定的培养基中高细胞密度生长，还能提供高达每升菌液几百毫克到克蛋白的表达量。在酵母中表达的重组蛋白既可以胞内表达也可以分泌到发酵液里回收。然而，由于存在于酵母细胞内和分泌到细胞外的蛋白酶对蛋白的快速降解，使得从酵母培养液中回收高表达蛋白质变得很困难。酵母细胞属于真核生物，因此，它们有着复杂的多细胞生物体的生物学结构并拥有蛋白质转译

1.2.2 Yeast Expression Systems

Yeast systems have been used for many years to produce large amount of proteins for industry and biopharmaceutical use. It can produce both intra- and extracellular heterologous proteins. Examples of pharmaceutical proteins made in yeast include: cytokines, enzymes, vaccine against hepatitis B, and human serum albumin, and a component for an artificial blood plasma. Typically, yeast expression vectors contain features that allow them to function in both bacteria and yeast. An *E. coli* origin of replication and bacterial antibiotic resistant genes are usually included in the vector, enabling all manipulations to first be performed in *E. coli* before the vector is transformed to yeast for expression. A range of yeast promoters is available for protein expression. Some like ADH2, SUC2 are inducible and others like GAPDH are constitutive in expression. Similar to the selection markers, a wide variety of combinations of promoters, markers, and expression systems are commercially available (Rudolph *et al*. 2006). Similar to bacteria, yeast is easy to manipulate genetically, has a rapid growth rate, and can achieve high cell densities in culture, including well-defined medium, delivering up to several hundred milligrams to gram of protein per liter of culture. Recombinant proteins in yeast can be intra-cellular or secreted to fermentation broth for recovery. However, recovering high amounts of desired protein from yeast cultures can be difficult due to rapid degradation of expressed proteins by endo- and exoproteases that present in yeast. Yeast cells are eukaryotic; therefore, they share the complex biology

后修饰（PTMs）的能力，如N-糖基化修饰、蛋白加工和脂化。由酵母菌分泌的蛋白质含有很多的在特殊糖基化位点的修饰、尤其是大量的甘露糖糖基化，而其糖链结构可能跟其他表达系统，包括人类不尽相同。这可能会导致酵母系统未必适合那些在体内半衰期短的蛋白质，制成的药可能活性欠佳，甚至免疫原性较低。

广泛用于有效蛋白生产中的酵母包括毕节酵母（*Pichia pastoris* 或 *P. pastoris*）、酿酒酵母（*Saccharomice scerevisiae*）、粟酒裂殖酵母（*Schizosacchoromyces pombe*）和多型汉逊酵母（*Hansanuela polymorpha*）。然而，到目前为止，只有酿酒酵母被用于上市的包括细胞因子和疫苗等生物制品的生产。市场上使用的来自酿酒酵母的产品要么不需要糖基化要么不受酵母糖基化的影响。毕节酵母（*P. pastoris*）常被用来表达药用蛋白并已进入了临床阶段（Bollok *et al*., 2009; Macauley *et al*., 2005）。蛋白酶缺陷的 *P. pastoris* 酵母已成为最流行的用于研究和临床研发的酵母表达系统之一，因为它能提供高细胞密度，拥有高表达水平和一个可控制的生产过程，并且具有稳定性和耐久性。*Pichia* 也能在化学成分确定的培养基中有效地分泌外源蛋白。利用 *Pichia* 酵母表达系统生产的几个生物制药已完成了早期临床开发阶段，他们很有可能被批准上市。运用基因工程将人类糖基化方式引进酵母的最新进展为在 *Pichia* 酵母中生产治疗蛋白提供了进一步的发展前途。基于其拥有复制人的糖基化模式的能力，酵母表达平台以其便宜

of multicellular organisms and are able to produce proteins that carry out post translational modifications (PTMs) such as N-glycosylation, proteolytic processing, and lipidation. Proteins secreted by yeasts are heavily glycosylated at consensus glycosylation sites with high content of mannose, while glycan structure may not be the same as the other expression systems such as humans. This may result in that the yeast system may not be suitable for pharmaceutical applications due to a short *in vivo* half-life of a protein produced and possible less efficacious or even less immunogenic.

Widely used yeasts for the production of functional proteins are *Pichia pastoris* (*P. pastoris*), *Saccharomyces cerevisiae* (*S. cerevisiae*), *Schizosaccharomyces pombe*, and *Hansanuela polymorpha*. However, so far, only *S. cerevisiae* has been used for market biological products including cytokines and vaccines. Products on the market using *S. cerevisiae* are either not glycosylated or are not impaired by yeast glycosylation. *P. pastoris* has been used to express therapeutic proteins that have entered clinical trials (Bollok *et al*., 2009; Macauley *et al*., 2005). Protease-deficient *P. pastoris* is becoming one of the most popular yeast expression systems in research and clinical development because it offers high cell density, superior expression, a controllable process, stability and durability. *Pichia* also efficiently secretes heterologous proteins in defined media. Several biopharmaceuticals produced using *Pichia* expression system have completed early stage clinical development and they are will be potentially approved to the market. Further development in the field of therapeutic glycoprotein production with *P. pastoris* strains is expected due to recent advances in genetic engineering of human glycosylation pathways into yeasts. With the ability to replicate certain human glycosylation patterns, yeast-based expression platforms offer an attractive alternative to current mammalian or insect cell culture processes due to a variety of additional advantages with to cheaper operating costs, simple chemically defined media and no viral contamination.

的运营成本、简单的化学成分确定的培养基和无病毒污染的优点，将提供一个有吸引力的能替代目前哺乳动物或昆虫细胞培养方式的平台。

1.3 其他的表达平台

1.3.1 昆虫表达系统

哺乳动物系统以其能提供最相关的生物修饰的能力而固然成为最可取的生物产品生产系统，然而它表达量低、细胞株开发耗时、资本投入相对高，而且整体质量要求高，这就使得其他如昆虫细胞表达系统成为更有经济效益的生物制药系统。此外，昆虫细胞系统大大降低了导致人体临床应用安全问题的人类病毒污染。

昆虫细胞培养提供了类似哺乳动物细胞糖基结构的糖基化修饰，和人源性蛋白的三维折叠以满足被表达蛋白的生物活性的需要。用于蛋白表达的昆虫细胞有鳞翅目昆虫属（*Lepidoptera*），包括蝴蝶和飞蛾昆虫细胞。其他常用的昆虫细胞类型，包括草地夜蛾（秋黏虫 *Spodoptera frugiperda*）、家蚕（桑蚕 *Bombyx mori*）、夜蛾（白菜尺蠖 *Trichoplusia ni*）和森林帐篷毛虫（*Malacosoma disstria*）。所有属于该组的昆虫细胞都容易受到杆状病毒（*baculovirus*）感染，因而这种昆虫病毒的遗传物质被用作载体把外源基因引进昆虫细胞内。

大多数 *baculovirus* 的表达载体基于夜蛾多个核型多角体病毒（AcMNPV）。AcMNPV 的病毒粒子带有 120～150 双链基因组 DNA 的碱基对（kb），可容纳高达 100kb 的外来

1.3 Other Expression Platforms

1.3.1 Insect Cells Expression

While the mammalian system is the most desirable system for biotherapeutic manufacturing in order to deliver fully functional proteins with the most relevant biological modifications, relative low expression titer with time-consuming cell line development, relative large capital investment, and overall high quality requirement makes other alternative systems such as insect cell expression system became economically valuable in biopharmaceutical manufacturing. In addition, insect cell culture substantially reduces risk of human virus contamination that cause related safety concern in human clinical applications.

Insect cell culture addresses the need of glycosylation with similar glycan base structure as mammalian cells, and three-dimensional folding among human-derived proteins that are required for biological activities. Proteins are expressed in insect cells such as *Lepidoptera*, the insect genus that includes butterflies and moths. Other commonly used insect cell types include *Spodoptera frugiperda* (fall army worm), *Bombyx mori* (silkworm), *Trichoplusia ni* (cabbage looper), and *Malacosoma disstria* (forest tent caterpillar). All the insects belonging to this group are susceptible to infection by *baculovirus*, and the genetic material of this insect virus serves as the vector for inserting foreign genes into the insect cells.

Most of the baculoviral expression vectors are based on *Autographa californica* multiple nuclear polyhedrosis viruses (AcMNPV). The AcMNPV baculoviral particle carries 120-150 kilobases (kb) of double-stranded genomic DNA and can accommodate an additional 100 kb of foreign genetic material,

遗传物质，为单个或多个基因的表达提供了无限的容量。最常用的杆状病毒启动子之一是多面体的启动子，能在杆状病毒的生命周期的后期阶段调节多面体杆状病毒的蛋白表达。多面体蛋白质通常可以达到菌体总蛋白的50%，它通常从杆状病毒的表达载体被删除，然后插入一个受多面体启动子控制的人们感兴趣的基因（GOI）。昆虫与杆状病毒表达系统可以生产1g/L 表达的蛋白。表达于昆虫细胞的蛋白拥有与哺乳动物细胞类似的蛋白质转移后的修饰（信号肽切割、磷酸化、脂质修饰、糖基化）以及蛋白折叠（二硫键的形成）。昆虫细胞培养（SF9、SF21 和其他细胞）很容易放大其规模，且可在 27℃生长附着和悬浮细胞。杆状病毒只感染昆虫而对人类无致病性。

杆状病毒 - 昆虫细胞技术提供了一个频繁使用的可生产具活性 / 免疫原性的重组蛋白的真核表达系统，为越来越多的人类药物的应用造福。其中尤其是重组 AcMNPV 被广泛使用于单个蛋白的大规模生产上。蛋白质高级结构也可以在此表达系统形成，如酶复合物、病毒源的基因载体和病毒蛋白自组装的病毒类颗粒（VLPs）等（Pillayet *et al.*, 2009）。第一个从杆状病毒表达系统生产的产品是针对子宫颈癌的病毒类颗粒疫苗（CervarixTM，英国 GlaxoSmithKline 公司）。

1.3.2 植物表达系统

在过去的几年中，植物已成为一个越来越具有吸引力的重组蛋白的生产平台。植物表达系统的优点包括大规模放大、上游成本低、不具人类或

which presents practically limitless capacity for insertion of single or multiple gene expression constructs. One of the most commonly used baculoviral promoters is the polyhedron promoter that regulates expression of baculoviral polyhedron protein during the late phase of the baculovirus life cycle. Polyhedron protein usually can comprise up to 50% of the total cellular proteins and it is usually deleted from a baculoviral expression vector, which allows a gene of interest (GOI) be inserted under the control of the polyhedron promoter. Baculoviral-insect expression system can reach production of up to 1 g/L of the protein of interest (POI). Protein expression in insect cells allows protein post-translational modification (PTM) including signal peptide cleavage, phosphorylation, lipid modification, glycosylation, and folding (disulfide bond formation, oligomerization) similar to that in mammalian cells. Insect cell cultures (SF9, SF21, and others) are easy to scale-up and can be maintained both as adherent and suspension cells at 27℃. *Baculoviruses* infect only insects and are nonpathogenic to humans.

Baculovirus-insect cell technology provides a very frequently used eukaryotic expression system for the production of functional and/or immunogenic recombinant proteins, which are used more and more for applications in human medicine. In particular, recombinant AcMNPV is widely used for the large scale production of individual proteins. Higher order protein structures can also be produced in this expression system, such as enzyme complexes, virus-like particles (VLPs) that self-assemble from structural viral proteins, and viral-derived gene delivery vehicles (Pillay *et al.*, 2009). The first licensed human vaccine produced with the baculovirus expression system is VLP-based and directed against cervical cancer (CervarixTM; GlaxoSmithKline UK).

1.3.2 Plant Production System

In the last few years, plants have become an increasingly attractive platform for recombinant protein production. The advantages of plant-based expression systems include high scalability, low upstream costs, lacking of human or animal

动物的病原体，并能生产具有所需结构和生物功能的目标蛋白质。在植物细胞内也能进行大多数的翻译后修饰。使用转基因或瞬时表达的整个植物或植物细胞培养，已经生产了各种重组亚单位候选疫苗和治疗性蛋白质，包括单克隆抗体、食用蛋白质等。这些产品有的已在早期临床试验，并显示其安全性和有效性。

转基因植物具有巨大的生产治疗性抗癌抗体和重组治疗性蛋白的潜力。植物能产生功能齐全的重组人或动物的抗体。农业规模培育的转基因植物将提供几乎无限量的药用重组蛋白。组合库技术是治疗性抗体在植物表达前用于优化表达的一个关键工具。优化抗体的表达可以在建立转基因悬浮细胞或植物株之前用瞬时表达分析来迅速验证。高表达株和稳定性优化的蛋白质表达株可以用瞬时表达方法识别亚细胞目标信号并建立植物表达株。当高表达者被选中后，下一步是研发高效提纯方法，以回收尽可能多的活性抗体。工业规模生产的抗体可以用植物细胞悬浮培养，或在土壤上生长稳定转化的植物。重组蛋白可产生于整个植物、种子或块茎，其蛋白质和生产系统可长期储存（Yusibov *et al.*, 2011）。

重组蛋白可以产生于以植物为基础的多样化的系统，包括生长在土壤的整个植物，或生长在生物反应器中用化学成分清楚的合成培养基培养的植物悬浮细胞。当植物生产蛋白被用于医疗用途（植物获得药用蛋白）时，它们受制于与所有其他来源的药物相同的监管指引法规。当进入临床开发生产阶段，它需要按照良好生产

pathogens, and capability of producing target proteins with desired structures and biological functions. Plants perform the majority of post-translational modification. Using transgenic and transient expression in whole plants or plant cell culture, a variety of recombinant subunit vaccine candidates, therapeutic proteins, including monoclonal antibodies, and dietary proteins have been produced. Some of these products have been tested in early phase clinical trials, and show safety and efficacy.

Transgenic plants have great potential for the production of therapeutic anti-cancer antibodies and recombinant therapeutic proteins. Plants make fully functional recombinant human or animal antibodies. Cultivating transgenic plants on an agricultural scale will produce almost unlimited supplies of recombinant proteins for uses in medicine. Combinatorial library technology is a key tool for the generation and optimization of therapeutic antibodies ahead of their expression in plants. Optimized antibody expression can be rapidly verified using transient expression assays in plants before the creation of transgenic suspension cells or plant lines. Subcellular targeting signals that increase expression levels and optimize protein stability can be identified and exploited using transient expression to create high expresser plant lines. When high expresser lines have been selected, the final step is the development of efficient purification methods to retrieve functional antibody. Antibody production on an industrial scale is then possibly using plant suspension cell culture in fermenters, or by the propagation of stably transformed plant lines in the field. Recombinant proteins can be produced either in whole plants or in seeds and tubers, which can be used for the long-term storage of both the protein and its production system (Yusibov *et al.*, 2011).

Recombinant proteins can be produced in a diverse array of plant-based systems, ranging from whole plants growing in the soil to plant suspension cells growing in a fully-defined synthetic medium in a bioreactor. When the recombinant proteins are intended for medical use (plant-derive pharmaceutical proteins), they fall under the same regulatory guidelines as the drugs from all other sources. When such proteins enter clinical development the requirement for production needs follow good manufacturing practice (GMP). In principle, GMP regulations that apply to pharmaceutical proteins

规范（GMP）的要求来生产。原则上，所有应用于细菌和哺乳动物细胞产生的药用蛋白的GMP法规，都直接适用于植物生产系统。在实践中，原来的GMP标准对于细胞特定的无菌环境的要求只适用于生物反应器中培养的植物细胞。利用整个植物生产蛋白质的系统必须有重大调整才能适应这些法规的要求。

produced in bacteria and mammalian cells are directly transferrable to the production using plants. In practice, the cell-specific terminology and the requirement for a contained, sterile environment mean that only plant cells in a bioreactor can fully meet the original GMP criteria. Significant changes are required to adapt these regulations for proteins produced in whole-plant systems.

1.3.3 转基因动物系统

经基因工程改造过的或称为“转基因”的动物是带有重组DNA的细胞并且其DNA序列会传递到其后代的一种动物。动物基因改造技术正运用在两个不同的方向：人用医药和农业生产。

运用转基因动物技术生产生物药品显示了巨大潜力，而且可能比传统的蛋白质生产系统更符合成本效益。动物中（鱼、鸡、牛、兔、绵羊和山羊）生产的重组蛋白一部分是使动物更好地抵御疾病，也为消费者产生质量更好的动物产品（牛乳、鸡蛋和肉类），但大多数转基因动物的研究和发展是为了生产用于临床试验的药物。现已证明在这些转基因动物中能够生产药用蛋白质。目前，大多数产生于动物细胞的生物药物需要一个复杂的设施和庞大的资金成本，导致许多患者负担不起昂贵的药物。遗传工程动物可能在未来成为提供这些蛋白质药物的重要来源，因为从转基因动物的乳、血液或卵中生产重组蛋白的工艺成本比动物细胞生产药用蛋白要低得多。

到目前为止只有两个转基因动物系统——乳和蛋已被证明能够生产重

1.3.3 Transgenic Animal Systems

A genetically engineered (GE) or “transgenic” animal is an animal that carries a known sequence of recombinant DNA in its cells which passes that DNA onto its offspring. Genetically engineered animals are being produced for two distinct applications: human medicine and agriculture.

The technology of transgenic drug-producing animals shows great potential in the production of biotherapeutics, and can potentially be more cost-effective than conventional protein-production engines. While recombinant proteins may be expressed in animals (fish, chicken, cattle, rabbits, sheep and goat) to enable animals to better resist diseases and produce better quality of animal products (milk, eggs and meat) for consumers, most transgenic animal research and development is to produce medicines for clinical applications. The production of therapeutic proteins has now been demonstrated in those transgenic animals. Most biotherapeutic molecules that produced in animal cells require a sophisticated facility with huge capital cost which result in expensive drug impossibly for many patients to afford. Genetically engineered animals may provide an important source of these protein drugs in the future, because the production of recombinant proteins in the milk, blood, or eggs of transgenic animals presents a potentially much less expensive approach than producing therapeutic proteins in animal cells.

A few transgenic animal systems have been shown capable of producing recombinant proteins but presently

组蛋白。目前正在研究用几种哺乳动物（兔、猪、绵羊、山羊和奶牛）的乳生产重组蛋白。不同的产生转基因动物的方法包括直接微量注射慢病毒（Lentiviral）载体于胚胎，以及精子与外源基因一起培养而后再注射到卵母细胞受精。若要蛋白可靠地表达，理想的转基因必须包含启动子、增强子、绝缘子、内含子和转录终止子。乳中蛋白的成功表达是使用了来自乳蛋白基因的启动子。使用包含被表达基因天然启动子的较长基因组 DNA 片段大大增强了外源 cDNA 的表达（Houdebine，2009）。从山羊收集和过滤的表达产物，例如，在羊乳中生产 ATryn（抗凝血酶蛋白）通过亲和层析、纳米过滤等步骤来清除病毒。最终通过进一步纯化柱层析步骤，纯化的产品被分装和冻干用于临床应用（http://www.abopharmaceuticals.com/ATryn.shtml）。

人类第一个来自奶山羊遗传工程的治疗性蛋白——抗凝血酶Ⅲ（ATryn 具有 GTC 的生物疗法），被 EMEA 和 FDA（www.ema.europa.eu；www.fda.gov）批准用于患者治疗遗传性抗凝血酶缺乏症。要制造这一产品，将编码人类抗凝血酶蛋白质的 DNA 植入到一个单细胞山羊胚胎。胚胎放入代孕山羊，其后此山羊孕育能够在其乳中生产抗凝血酶蛋白的转基因后代。乳蛋白被收集和纯化从而生产 ATryn（Lavine，2009）。

Rhucin（重组 C1 酯酶抑制剂）——另一种治疗遗传性血管水肿的治疗性蛋白——已被 Pharming 公司在转基因兔中生产，通过使用了在功能上与编码人 C1 抑制剂基因相连的牛的特

only two systems, milk and egg, have started being implemented. Several mammalian species (Rabbits, pigs, sheep, goats and cows) are currently being studied to produce recombinant proteins in their milk. Different methods to generate transgenic animals include direct microinjection into embryo using lentiviral vector, and incubation of sperm with foreign genes before is injected into oocyte cytoplasm for fertilization. To be expressed in a reliable manner, a transgene must ideally contain a promoter, enhancers, insulators, introns, and a transcription termination. Expression in milk is achieved successfully with promoters from milk protein genes. Using long genomic DNA fragments containing the promoter of interest generally greatly enhances the expression of foreign cDNA (Houdebine，2009). The expressed products, for example, ATryn (antithrombin protein) produced in milk is then collected from goats and filtered, purified through an affinity purification step followed by Nano-filtration to remove viruses. Further purification through column chromatography steps and finally purified bulk is vialed and lyophilized before clinical use (http://www.abopharmaceuticals.com/ATryn.shtml).

The first human therapeutic protein, Antithrombin Ⅲ (ATryn, GTC Biotherapeutics), derived from the milk of genetically engineered goats was approved by the EMEA and FDA (www.ema.europa.eu; www.fda.gov) for treatment of patients with hereditary antithrombin deficiency. To create the product, DNA codes for the human antithrombin protein was implemented into a single-celled goat embryo. The embryo was placed into a surrogate goat, which produced a transgenic offspring capable of producing the antithrombin protein in its milk. The protein was collected and purified from the milk to make ATryn (Lavine，2009).

Rhucin (recombinant C1 esterase inhibitor) – another therapeutic protein for the treatment of hereditary angioedema (HAE) – has been produced in transgenic rabbits by Pharming. High protein level was achieved by including an extra bovine milk-specific promoter sequence (alpha-S1 casein) that functionally linked to the

定的启动子序列（α-S1 酪蛋白）得到高产量的表达。家兔每一年可产生高达 10L 的乳。转基因兔每升乳能够生产 10～12g 的蛋白质，而高度优化的细胞培养通常每升培养液只可以产生 0.2～1g 的蛋白质。

gene encoding human C1 inhibitor. Rabbits can produce up to 10 liters of milk a year.The transgenic rabbits are able to produce between 10 and 12 grams of protein per liter of milk, as compared to highly optimized cell cultures that can typically generate 0.2 to 1 grams of protein per liter of culture medium.

美国 FDA 根据联邦食品、药品和化妆品法案（FFDCA）动物药品方面的新规定对工业界转基因遗传工程动物发布了行业指导，名为"包含遗传重组 DNA 载体基因工程动物规例"，澄清了 FDA 的法定及监管权力，并为基因工程动物的生产商提出了建议，以帮助他们履行自己的义务和法律规定的职责（www.dgra.de/studiengang/PDF/master_schmit t_eh.pdf）。

The U.S. FDA issued guidance for industry on the regulation of genetically engineered animals under the new animal drug provisions of the Federal Food, Drug and Cosmetic Act (FFDCA). The guidance, titled "The Regulation of Genetically Engineered Animals Containing Heritable rDNA Constructs," clarifies the FDA's statutory and regulatory authority, and provides recommendations to producers of GE animals to help them meet their obligations and responsibilities under the law (www.dgra.de/studiengang/PDF/master_schmit t_eh.pdf).

2 生物技术药物生产工艺

2 Manufacturing Process

前面的章节描述了各种生命系统，如大肠杆菌、酵母、哺乳动物细胞，作为生产生物药物的平台。生产生物药物过程就是以这样的生命体系为基点的。这些能够表达产品的生命系统应该妥善保存，以用于日后的生产。重组蛋白、单抗和疫苗的基本生产工艺包括建立表达宿主（细胞库）、上游生产（发酵和细胞培养）、提取回收、下游纯化、制剂及最终产品的分装，这些将在本节中加以讨论。

As mentioned previously in the Production Platform section, biopharmaceutical production processes are based on living systems such as bacteria, yeast, and mammalian cells. The living system used for manufacturing that can express the desired product should be well preserved for future production. Generally, manufacturing of recombinant therapeutic proteins, therapeutic MAb and vaccines involves establishment of expression hosts (cell banks), upstream production (fermentation and cell culture), recovery, downstream purification, formulation, and final product fill and finish , all of which will be discussed in this section.

产品通过发酵或细胞培养过程在宿主系统中扩增或表达，并从这样的培养液经一系列纯化步骤，制备高质量的最终产物。上游工艺过程是为了在最佳条件下培养这样的生命系统以

Product is expressed and amplified in the host system through fermentation or cell culture. Recovery from culture is subject to a number of purification steps to manufacture final high quality drug products. Upstream manufacturing is designed to grow the living system to generate a product of interest at a sufficient scale while

产生所要的产物，而下游工艺过程则是为了将所表达的产物提取分离出来，得到高质量的治疗用药品。宿主细胞在生物反应器中进行纯种培养以确保最终产物的质量和生产过程的重复性。同样，下游加工一般在洁净区进行，以防止尘埃粒子和其他杂质可能对最终产品质量的影响。工业化生产重组生物制剂需要各种技术和工程能力，以维持宿主生命系统的稳定，均一和重复地在无菌的环境中生产大量产品。为了满足市场需求以及提高成本效益和生产效率，生物药物的生产或可放大到2万立升的生物反应器中进行（通常为单克隆抗体）以及下游每批生产若干千克的成品。由于这样大规模生产用的是生命系统和对最终产品的质量要求，由于大分子产物的复杂性质及不稳定性，也由于大规模生产的工程问题等，生物药物的生产工艺通常是非常复杂的。带有过程控制、监测和定性分析的制造过程对最终产品的质量将有显著影响。质量分析、监测和控制能确保生产的稳定性和可重复性，以及产品的高品质。这些也是相当大的挑战。

downstream manufacturing recovers a product of interest from a living system or its culture broth to final pure product with high quality for therapeutic usage. The living system is cultured in a bioreactor under aseptic conditions to ensure a pure source and quality of a final product as well as reproducibility of the production. Likewise, downstream processing is routinely conducted in a classified clean area to prevent contamination with dust particles or other impurities that may have an impact on quality of the final product. Manufacturing recombinant biologics at industrial levels requires technologies and engineering capacities that are able to stably maintain the producing organism and consistently and aseptically generating product in large quantities. To meet deliverables and market demand while maintaining cost and production efficiencies, manufacturing a therapeutic protein or MAb can be scaled up to 20 000L bioreactor scale (usually for MAb) to obtain kilogram yields of product per batch for downstream processing. Such a manufacturing process can be quite complicated due to involvement with living systems, final product quality requirements, and unstable and complex nature of the large molecular and large scale process engineering issues. Manufacturing processes with in-process control, monitoring, and characterization will have notable impact on final product quality. Quality analysis, monitoring, and control are to ensure production consistency and reproducibility as well as product high quality, all of which are challenging.

1. 过程分析技术（PAT） 过程分析技术（PAT）已被美国食品和药物管理局（USFDA）定义为在药品生产过程中，通过测量影响关键质量因素（CQA）的关键工艺参数（CPP）来设计、分析和控制药品生产的系统。美国FDA概述了过程分析技术的实施指南（www.fda.gov/downloads/AboutFDA/CentersOffices/CDER/UCM174306.pdf），这是一个对创新药物的开发和制造质量保证的框架的一个重要部分。

1. Process Analytical Technology (PAT) Process analytical technology (PAT) has been defined by the USFDA as a system for designing, analyzing, and controlling pharmaceutical manufacturing through the measurement of Critical Process Parameters (CPP) which affects Critical Quality Attributes (CQA). The USFDA outlined a regulatory guideline for PAT implementation, which is a quality assurance framework for innovative pharmaceutical development and manufacturing (www.fda.gov/downloads/AboutFDA/CentersOffices/CDER/UCM174306.pdf).

过程分析技术的目的是增进了解和控制生产过程；鼓励开发更好的分析方法作为工具来监控生产过程；以及尽量减少在创新药物或创新工艺的开发过程中的风险。“产品的质量不能仅由产品的分析测试而得到保证，更应该设计和内置于生产产品的过程之中（www.fda.gov）。”过程分析技术正是由这样的理念转化为运行机制从而在药物生产中保证质量，减少在制造过程中的失误和偏差。过程分析技术作为在生产过程中的控制手段，可以更好地控制和预测关键工艺和产品质量相关的因素，这是符合 cGMP 在生产中实施过程分析技术要求的，因为这样的控制程序进一步验证了制造工艺。FDA 和业内专家预计使用 PAT 比传统制造方法的优越性包括：达到更高的最终产品的质量，更高的生产效率，较低的运营成本，更好的加工能力和较少的次品。

The goal of PAT is to enhance understanding and control of the manufacturing process, to develop better analytical methods as PAT tools for monitoring the production process, and to minimize risks during development of an innovative product and process. The statement of “quality cannot be tested into products but should be built-in or should be by design” (www.fda.gov) is now being applied with PAT as the operating mechanism of assuring quality, reducing failures, and minimizing deviations during manufacturing. With PAT as in-process control tools, one can have better control and prediction of relevant critical process and product attributes. This is consistent with CGMP requirements, as such control procedures validate the performance of the manufacturing process. Both the FDA and industrial experts expect benefits over conventional manufacturing practices: higher final product quality, increased production efficiency, decreased operating costs, better process capacity, and fewer rejections.

一个例子是用高分辨率超声波谐振器和拉曼光谱作为一种新型的自乳化药物传递系统（SEDDS）（Stillhart and Kuentz，2012）。两位作者在测定了散装的含有不同浓度的药物配方的超声速度和衰减测量后，在药物定量过程分析上对这两种技术作了比较。浸入式探头用于产品原液，而多光纤传感器用于自乳化药物传递系统的硬胶囊的测量。每种方法都在预测以及量化的限制和平均回收率相对标准误差计算上进行了评估，最后，作者得出结论：超声波谐振技术和拉曼光谱都可作为自乳化药物传递系统的药物定量分析方法，这样研究使得过程分析技术作为定量化分析的方法得到合适应用。

One example was the comparison of a high-resolution ultrasonic resonator technology and Raman spectroscopy as a novel PAT for drug quantification in self-emulsifying drug delivery systems (SEDDS) (Stillhart and Kuentz, 2012). After measuring ultrasound velocity and attenuation in the bulk formulation containing drug at different concentrations, the authors used both technologies. An in-line immersion probe was used for the bulk formulation and a multi-fiber sensor was used for measuring through hard-gelatin capsules that were filled with SEDDS. Each method was assessed by calculating the relative standard error of prediction as well as the limit of quantification and the mean recovery. In the end, the authors concluded that ultrasonic resonator technology and Raman spectroscopy constitute suitable methods for drug quantification in SEDDS, which is promising for their use as process analytical technologies.

在单克隆抗体的生产中使用了在线近红外（NIR）透射光谱测量大分子量（MW）产品，即单克隆抗体以及 CHO 细胞的代谢曲线。这一方法适用于从 2.5～1000L 的不同规模以及不同细胞株的生产。用在线获得的近红外光谱能够计算出多元过程中的轨迹。这些已被用来检测不同生产规模工艺过程；评估各生产批次的变化；以及评估对不同来源的生产过程偏差的相对重要性。这一方法对 CHO 细胞生产过程工艺开发以及现有的大规模验证过的工艺具有潜在的应用价值（Hakemeyer *et al*., 2012）。

In MAb production, at-line near-infrared (NIR) transmittance spectroscopy can be used to measure large molecular weight (MW) product MAb as well as metabolic profiles for MAb manufacturing CHO cells. This method is suitable for scales spanning from 2.5 to 1000 L with different cell-lines. Multivariate process trajectories were computed from NIR spectra acquired at-line. These were used to enhance process understanding across different scales up to industrial scale, to assess batch-to-batch variability, and to examine the relative importance of different sources of process variability. The approach is potentially useful throughout process development as well as to existing large-scale validated CHO bioprocesses (Hakemeyer *et al*., 2012).

2. 平台技术　平台技术指的是可以用于一个以上的生物制药的生产操作过程。这可以减少为每个产品的生产过程而开发以及为生产本身进行的重复投资，有益于未来新产品的研发工作。单克隆抗体制造平台是一个典型的例子。抗体生产的下游单元操作序列或多或少已经标准化了，例如澄清收获液体、抗体吸附、精制去除宿主细胞蛋白和残余 DNA，最后配成原液制剂和封装操作。在这个平台过程中，单克隆抗体产品通常是通过切向流过滤进行浓缩提取的培养液，接着经过蛋白质 A 柱，或者澄清上清液直接上蛋白质 A 柱，再用低 pH 的缓冲液洗脱用作病毒灭活步骤。然后经过两步柱层析，进一步纯化和清除内毒素、残留的 DNA 和病毒，以及聚合或降解产物。为了产品安全性，添加了纳米过滤作为在最终产品原液之前清除病毒的最后一步。有关详细纯化过程可参考以后章节的介绍。如果一家公司生产多种单克隆抗体，他们可能会对所有

2. Platform Technologies Platform technologies are manufacturing operations that should be applicable to more than one biopharmaceutical and can ease future research work with the desired effect of eliminating process re-invention for each product and reducing manufacturing investments. MAb manufacturing is the primary example of platform development with more or less standard unit operation sequences being used in downstream purification: feed stock clarification, antibody capture, removal of host cell proteins and residual DNA via polishing, and formulation of the bulk active product prior to fill and finish operations. In this platform process, MAb product is normally recovered from conditional medium through either TFF concentration followed by a protein A column step or direct application to a protein A column with low pH elution serving as viral inactivation step. Post-protein A solution are passed through two column chromatography steps for removal of endotoxin, residual DNA, virus, and aggregated or degraded product. Adding Nano-filtration serves as a virus removal step before final formulation for safety purpose. The details of the purification process are described in later sections. If a company produces several MAbs, they may use the same procedure with minor tweaks across different products, which will speed up development time, minimize capital equipment investment, eliminate unnecessary validation effort, and simplify operator training. We may give another example of platform technology in

的产品使用相同的工艺或做细微调整，这样会加快开发速度，减少设备资本投资，消除不必要的验证工作，及简化操作人员的培训。类似的另一个例子是药物制剂。如果一个眼滴液使用某些聚合物作为辅料，这样一个平台技术可以建立在优化的聚合物上，使其他研究人员可以用于不同的活性药物，将较少的资金和时间投入在辅料的选择上，生产出一系列全新的配方及用量的药物。

drug formulation: if an eye drop solution uses certain polymers as excipients, then a platform technology can be established whereby other researchers can use the same optimized polymers for different drugs, leading to the development of a new formulation with less money and time investment.

2.1 生产上游工艺

发酵和细胞培养是生物制药生产过程中的关键因素。不同的生产平台，生产工艺显然不同。为确保每一批生产产品的一致性，在生产之前，需要通过DNA与宿主细胞的基因操纵，建立并保存一个被称为“细胞库”的生产种子。通过DNA序列，在细胞库遗传特性上进行严格测试，必须是不含噬菌体（在大肠杆菌中的情况下）或无支原体和无病毒感染（在哺乳动物细胞的情况下）。其生产通常会以一小瓶冷冻细胞库开始，细菌宿主或哺乳动物细胞接种到一个摇瓶，在环境控温的摇床或二氧化碳培养箱中培养生长，细胞扩增成为种子细胞。在大规模生产（超过1000L生物反应器体积）情况下，需要1～3个中间细胞扩张阶段以累积足够的种子用于生产罐。经常会看到在哺乳动物细胞培养放大时10倍的培养体积扩增，和细菌发酵50～100倍规模的扩增。三个阶段的种子培养后，种子接种在超过1000L的大规模生产生物反应器。

2.1 Upstream Manufacturing Process

Upstream including fermentation and cell culture are crucial processes in the manufacturing of biopharmaceuticals. Production procedures may significantly change depending on the production host system. To ensure batch consistency during production, a preserved stock of the living system called “cell bank” is usually required to be established through genetic manipulation of the host cell before production. Cell banks are tested intensively in genetic identity through DNA sequencing and must be free of phage (in the case of *E. coli*) or free of mycoplasma and adventitious agents (in mammalian cells). Production normally starts with a vial of a frozen cell bank. Bacterial or mammalian cells are inoculated into a flask and incubated in either an environmental shaker or CO_2 incubator for cell expansion purposes. For large scale production (larger than 1000L bioreactor volume), one to three intermediate cell expansion stages are required for accumulating enough seed for the production tank. Ten-fold or 50-100-fold scale-ups are common in mammalian culture or bacterial fermentation, respectively. After a three stage of seed culture process, inoculation into a large production bioreactor at over 1000L scale.

生产的工艺过程可以是一个批

A production process can be operated as batch

生产工艺，这是一个简单的在培养过程中无需添加更多的营养物质到培养系统中的工艺。在许多情况下，一个批生产过程具有开发时间短、设置简单的优点，可以“封闭”在一个封闭的系统，减少交叉污染，及减少清洁处理时间。对于相对较小的规模和早期临床开发，这样的工艺可以在较短的时间内提供产品。为了优化生产系统中的产量以及生产能力，许多上游工艺中需要添加营养成分以维持培养系统表达产品，这是补料生产工艺过程。补料生产工艺代表了大多用CHO/NS0和大肠杆菌或其他宿主系统生产单克隆抗体和其他生物药物的生产工艺。由于其操作简便可靠，以及实施操作的灵活性，这种工艺被广泛采用。更复杂的工艺过程是一个连续的细胞培养，称为“灌注培养”的过程。不断提供新鲜培养基于生物反应器中供细胞培养，同时从培养系统不断地收获。细胞不断地从收获管道经过滤，返回到反应器或保持在生物反应器内，以保持恒定的培养体积。这样连续的过程，可以达到高细胞密度15×10^6～50×10^6个细胞/ml，并至少保持2周以上。与批次或补料分批培养的结果不同，这个工艺会达到相当高的生产能力，并保持一个较长的生产时期。在产品蛋白质不稳定性的情况下，或如果细胞株的表达量较小时，灌注工艺被视为最佳工艺方案，这样一个恒定的输出也显示出更好的经济效益。生产系统中使用的中空纤维的生物反应器是一个典型灌注工艺，并有许多成功案例（Whitford and Cadwell，2011，Usuludin *et al.*，2012）。然而，在这样

culture operation, which is a simple process with no additional nutrient addition into the system during culturing. In many cases, a batch process has advantages of short development time, easy setup, the possibility of a closed system to reduce cross-contamination, and quick turn-over processing time. For small scale and early clinical development, batch-processes are able to deliver product within short period of time. To optimize a production system's expression yield and productivity, many upstream processes use fed-batch processes, which are characterized with periodic nutrient addition to maintain culture longer production. Most large scale processes that produce MAbs and other biotherapeutic products with CHO/NS0 cell lines and *E. coli* or other production systems use fed-batch culture for its operational simplicity, reliability, and implementation flexibility in multipurpose facilities. A more complicated process is the perfusion culture process, in which cells are continuously supplied with fresh medium to the bioreactor while spent media is continuously harvested. The cells are continuously filtered from the harvest stream and returned to the bioreactor or kept inside the bioreactor to maintain a constant culture volume. This continuous process allows the cells to reach high densities of 15-30 million cells per ml which can be routinely reached and maintained for at least 2 weeks. This results in a highly productive cell culture process that can produce for a longer period of time as opposed to batch or fed-batch cultures. Perfusion is regarded as the best option when the stability of the protein is poor or if the expression of the cell line is minimal such that a constant output becomes more economical. Production systems using a hollow-fiber bioreactor is common and has been successful many times (Whitford and Cadwell, 2011; Usuludin *et al.*, 2012). However, maintaining sterility during long periods of time and the technical requirements for operating such a system are major challenges. Additionally, more medium is needed to run the perfusion culture, which in many cases increases expenses.

的培养系统中长期操作并保持无菌的技术要求是很大的挑战。此外，灌注培养需要相对较多的培养基，在许多情况下，这是相当昂贵的。

在上游生产系统中往往是通过SDS-PAGE电泳、免疫印迹、琼脂糖凝胶电泳（质粒DNA）、高效液相色谱分析、酶联免疫吸附分析的过程来监测细胞的生长和代谢概况，以及检测产品蛋白质或抗体及其他生物药物产品的表达情况。测试结果可能反馈到生产用于调整过程控制，以使产品数量和质量达到最佳的输出。在生产过程中，无论是在线或现场测试的结果都可以决定上游过程的收获时间。

While upstream process is monitored with cell growth and metabolic profiles, expression of therapeutic proteins, MAb, and other biopharmaceutical products is analyzed through SDS-PAGE, Western-blot, agarose gel (plasmid DNA), analytical HPLC, and ELISA. Testing results may serve as feedback to adjust in-process control to reach optimum outputs in product quantity and quality. End point of upstream process is often decided by such either on-line or on-site testing results during production.

2.2 生产下游工艺

生物制药下游制造工艺过程是从有生命的宿主，如植物、细胞、组织或培养液系统回收和纯化得到药用生物药物。这是在生产生物药物如重组细胞因子、激素、抗体、疫苗诊断和治疗用酶中的重要步骤。

2.2.1 生产下游工艺的步骤

为了把产品从它的自然状态，即从组织、细胞或发酵液中，分离及逐步增加纯度和浓度，下游工艺过程分为几个单元操作。下游工艺过程的第一步是将其表达的产物和表达系统分离开来，它通常被称为“提取”。它涉及将细胞、细胞碎片或其他颗粒物从发酵液中分离掉而得到含有产物的液体，实施这一步骤的典型单元操作是过滤、离心、沉淀等。在表达产品在细胞质内积累的情况下，如大多数大肠杆菌，细胞包括植物细胞和动物组织要被匀浆破碎以获得产品。

2.2 Downstream Manufacturing Process

Biopharmaceutical downstream manufacturing process is the recovery and purification of a therapeutic bio-product from living systems such as plants, cells, tissues, or culture broth. It is an essential step in the manufacture of biopharmaceuticals such as recombinant cytokines, hormones, antibodies, vaccines, and enzymes used in diagnostics and therapeutic applications.

2.2.1 The Procedure of Downstream Manufacturing

A downstream process is categorized into several functional operation groups in order to bring a product from its natural state, a component of a tissue, cell or fermentation broth, through progressive improvements in purity and concentration. Separation from its expression system is the first step and it is usually called “recovery” in the downstream process. It involves capture of the product in a liquid where cells, cell debris or other particulate matter from fermentation broth is separated or removed. Typical operations to achieve separation are filtration, centrifugation, and precipitation. In the case that a product accumulates within the system (for example, within cellular cytoplasm), as is observed for most of *E. coli* expression, cells have to be homogenized to recover the product.

产品纯化是去除主要杂质和提高产品纯度。因此分离纯化步骤的目的是设法去除这些杂质，减少要处理材料的量和浓缩最终产品。通过超滤、沉淀和不同的机理，包括离子交换、疏水、亲和，或分子筛柱层析来达到产品纯化的目的。产品纯化要着眼于分离那些与产品在物理和化学性质非常接近的杂质。由于需要敏感和精良的设备，在这个阶段的纯化步骤是相当昂贵的，包括亲和柱层析、分子筛柱层析、制备规模的反相高压液相色谱（HPLC）系统，这些占了整个下游工艺开支的相当大部分。

产品精制步骤包括了最后一步单元操作以得到稳定的产品，利于最后的包装运输。虽然许多产品，例如单克隆抗体，可以稳定在2～8℃液体形式长期储存，首选用于储存的单元操作是冷冻干燥和喷雾干燥。根据产品和其使用目的，精制可以包括除去可能危及安全的微量污染物（产品的多聚合体）和产品的无菌操作。这些操作可以包括在分装前的清除病毒和无菌过滤。

在下游工艺的挑战之一是设计可靠的工艺步骤从复杂的发酵液中有效地分离纯化得到不稳定的产品，如重组蛋白，并达到最终药品所需的纯度。下游工艺通常占50%～80%的总制造成本，因此下游生产工艺的优化被认为是最佳工艺设计中的核心要素之一。一个纯化过程的优化是两个以上不同机制的纯化单元操作的有效结合。每个单元操作应考虑在操作容量和效率、最有效除去杂质和成本效益等方面进行优化。在整体工艺设计上，工程师

The purpose of product purification is removing major impurities while increasing product purity. As such, the isolation steps are designed to remove these impurities, reduce the volume of the material to be handled, and concentrate the final product. Capturing product by means of ultrafiltration, precipitation, and various methods of column chromatography such as ion-exchange, hydrophobic interaction, affinity, or size exclusion is used for this purpose. Product purification aims to remove contaminants that resemble the product very closely in physical and chemical properties. Downstream steps in this stage are expensive to carry out due to the need for sensitive and sophisticated equipment. This stage contributes a significant fraction of the entire downstream processing expenditure. Examples of such operations include affinity, size exclusion, and preparative scale reversed phase high pressure liquid chromatography (HPLC) system.

Product polishing steps include the final processing operations that end with packaging of the product in a form that is stable and conveniently transportable. While many products, for instance MAbs, can be stably stored at liquid form in 2-8℃ for long periods of time, lyophilization and spray-drying are preferred unit operations for products that do not possess such stability. Depending on the product and its intended use, polishing may also include operations to sterilize the product and remove trace contaminants (aggregates) which might compromise product safety. Such operations might include viral removal and sterile filtration prior to vialing.

One of the challenges in downstream processing is designing robust process steps to efficiently isolate unstable products such as recombinant proteins from complex fermentation broths to the final product with required pharmaceutical degree of purity. Downstream operations usually represent 50%-80% of the total processing costs. Therefore, optimization of downstream processing technologies is considered one of the central elements in appropriate process design. Purification process optimization typically involves combining various unit operations that are suitable for the desired product. Each unit operation is optimized in operational capacity and efficiency, maximal impurity removal, and cost effectiveness. In the overall process

必须考虑不同方面，如除去杂质、层析柱容量、缓冲液的要求、收率、生产时限、工艺的可放大性、自动化、生产线的生产能力和药品法规等，需要在每一步单元操作和整体产量，以及产品在单元操作之间的中间体质量和产品原液的质量方面进行优化，每一步骤的规模应根据设计的整体产量进行调整，每个单元操作应在产物的产量、纯度和稳定性方面进行优化并验证。

design, the engineer has to consider elimination of contaminants, the load capacity of the columns, buffer requirements, yields, process timelines, process scalability, automation, capacity of the production line, and regulatory compliance. Each step of the purification process needs to be optimized in terms of individual and overall yield, and the product quality of the intermediates in the hold steps and of the bulk. The scale of the individual steps of the purification process has to be adjusted according to the intended product quantity. Every single step of the process has to be optimized and validated in terms of yield, purity, and stability.

2.2.2　常用的下游工艺单元操作

以下列出了几种常用的下游工艺单元操作，其中低压柱层析法是目前最流行的工艺步骤，它可以用不同的机制，包括离子交换层析（IEX）、亲和层析、疏水作用层析（HIC）和切口流过滤来纯化生物活性大分子。

1. *离子交换层析法*（IEX）　离子交换层析依赖于蛋白质和树脂（珠）之间电荷 - 电荷的相互作用。离子交换可分为：阳离子交换层析，带正电荷的蛋白质与带负电荷的树脂结合；阴离子交换层析，带负电荷的蛋白质与带正电荷的树脂结合。知道蛋白质的等电点，就可以使用不同 pH 的缓冲液对蛋白质的电荷进行调节，以改变蛋白质的电荷。被分离的蛋白质结合在树脂上后，再用平衡缓冲液冲洗掉未结合的物质，然后增加洗脱缓冲液的离子强度或用不同 pH 的缓冲液洗脱结合在层析柱上的蛋白。

离子交换树脂是相对便宜的层析介质，因此被广泛用于生物药物蛋白纯化步骤。除了纯化，它还经常被用在浓缩体积，以及去除病毒和内毒素等步骤。

2. *亲和层析*　亲和层析分离蛋

2.2.2 The Commonly Operation of Downstream Manufacturing

In the following sections, several commonly used unit operations in downstream processing are listed. Among them, low pressure production chromatography is the most popular process used to purify a bioactive molecule with different mechanisms and include affinity, ion exchange (IEX), hydrophobic interaction chromatography (HIC), and gel filtration chromatography.

1. Ion Exchange Chromatography (IEX)　Ion Exchange Chromatography (IEX) relies on charge-charge interactions between the protein of interest and the column resin. IEX can be subdivided into cation exchange chromatography, in which a positively charged protein of interest binds to a negatively charged resin, and anion exchange chromatography, in which a negatively charged protein of interest binds to a positively charged resin. The charges on the protein can be manipulated by using buffers of different pH to alter the charge on the protein based on its isoelectric point (pI). Once the protein of interest is bound, the column is washed with equilibration buffer to remove unattached compounds. The bound protein is then eluted off using an elution buffer with increased ionic strength or different pH.

Ion exchange resins are relatively inexpensive in the chromatography media available and are therefore almost always used as a step in biopharmaceutical protein production purification. In addition to purification, IEX is frequently used to concentrate a process solution and remove virus and endotoxins.

2. Affinity Chromatography　Affinity chromatography

白是基于蛋白质和耦合在树脂上的抗体（或配体）之间的可逆的亲和作用。由于具有高选择性、高分辨率、高容量，通常可以达到几千倍的蛋白纯化倍数。生物分子蛋白和蛋白之间的相互作用，也可能来源于静电相互作用、范德华力和（或）氢键的作用。通过改变pH或离子强度可以互逆地将在亲和珠上的所要的蛋白洗脱。

虽然亲和纯化是非常有效的和有选择性的，但制备所要的蛋白质的抗体并不便宜。因此，亲和层析法用于生产步骤并不是最经济的选择。然而，在单克隆抗体生产工艺中，产品是抗体，抗体通过和耦合在层析珠上的蛋白A的结合而得到纯化。亲和层析是为单克隆抗体生产设计的通用的下游工艺中的第一步纯化步骤。

3. *疏水作用层析*（HIC） 疏水层析（HIC）在生物药物生产上的应用有增加的趋势（Jiang *et al.*，2012）。由于疏水层析的分子机制依赖于其独特的结构特点，它是与亲和层析、离子交换、凝胶过滤层析不同的层析选择。疏水层析的介质通常有高容量和经济且稳定的特点。蛋白在高盐条件下吸附，而在低盐浓度下洗脱。

4. *切向流过滤*（TFF） 切向流过滤（TFF）使用横向流，以减少在膜表面滞留物的堆积。使用正确的分子量截流的TFF过滤膜，所要的蛋白质可以从带有其他蛋白质和其他分子中的培养液中分离出来。这个单元操作也经常用于在制备最终成品制剂时交换缓冲液和调整最终产品的浓度。

separates the protein of interest on the basis of a reversible interaction between the protein and an antibody or ligand that is coupled onto a chromatography bead. With high selectivity, resolution, and loading capacity for the protein of interest, purification factors on the order of several thousand are achievable. Biological interactions between the antigen and the protein of interest can result from electrostatic interactions, van der Waals' forces and/or hydrogen bonding. To elute the protein of interest from the affinity beads, the interaction can be reversed by changing the pH or ionic strength.

While affinity purification is highly effective and selective, generating antibodies specific to the protein of interest is expensive. Therefore, affinity chromatography is the least economical choice for production chromatography except when manufacturing MAb, in which antibody is purified via binding to protein A resin. In such a case, affinity chromatography is commonly used as the first purification step in a generic downstream process.

3. Hydrophobic Interaction Chromatography (HIC) HIC is becoming more widely used in bioprocessing (Jiang *et al.*, 2012). Since the molecular mechanism of HIC relies on unique structural features, it serves as a non-redundant option to ion exchange, affinity, and gel filtration chromatography. HIC media usually have high capacity and are economical and stable. Adsorption takes place in high salt condition and elution in low salt concentrations.

4. Tangential Flow Filtration (TFF) TFF uses crossflow to reduce buildup of retained components on the membrane surface. Using TFF with the appropriate molecular weight cut off filters, one can separate proteins of interest from impurities in the clarified medium. This unit operation is regularly used in final formulation steps along with buffer exchange and adjustment of final product concentration.

2.2.3 下游生产工艺的多样性

下游生产工艺的单元操作可以随不同的表达平台而有相当的区别。以大肠杆菌为宿主的生产，在大多数情况下，由于表达的蛋白质在细胞内，由细胞匀浆破壁为起点，包含体制备和蛋白复性纳入纯化过程，之后再有几步柱层析。此外，尤其是用大肠杆菌生产的工艺过程，使用阴离子的交换树脂或膜去除内毒素是常规的步骤。

在哺乳动物细胞表达系统中的产品通常会以可溶的形式表达且分泌在培养基中。所以纯化过程以离心或过滤去除细胞和细胞碎片为始。TFF 可减少初始收获的培养液的体积。下游工艺开发的结果往往是几种不同机理的层析的最佳组合以达到纯化目的。而对于大多数单克隆抗体的生产来说，常见亲和层析柱后再通过离子交换或疏水层析，有时再加分子筛凝胶过滤作为精制步骤。

整个下游过程用多种分析测试来密切监测。除了产品特定的分析，下游的分析测试主要集中在产品质量、产品特性，以及产品相关的杂质分布和宿主细胞的 DNA、宿主细胞蛋白质的残留上。有时在生产过程中，也密切监察特定产品的生物活性和翻译后修饰，包括糖基化，以确保最终产品的质量。

2.2.3 Downstream Processing Variation

Downstream processing can vary based on changes in upstream unit operations. For an *E. coli* based production, cell disruption by homogenization is a common starting point since the protein of interest is in the cells. An inclusion body preparation and product refolding are incorporated into the purification, which is normally followed by several column chromatography steps. Endotoxin removal steps using either anion IEX resin or membrane is routinely included in the process in *E. coli* based manufacturing.

Product expressed in mammalian systems is usually soluble and secreted in a culture medium. Purification processes start with cell and cell debris removal through either centrifugation or filtration. Large volumes of initial culture may be reduced by TFF. For purification, an optimal combination of several chromatography steps with different mechanisms is the goal of downstream process development. For most MAbs production, an affinity column followed by either IEX or HIC is common and may be followed-up with size-exclusion gel filtration step as a polishing step.

Entire purification processes are closely monitored through a number of assays. Besides product specific assays, downstream analytical testing is mainly focused on product quality, product characterization, product-related impurity profile and host cell DNA and protein residue. Specific biologic activity of the product and post-translational modification including glycosylation are also closely monitored to ensure the final product quality.

2.3 大肠杆菌表达系统的生产工艺

大肠杆菌已被广泛用于生产近 30% 的市场上的生物药物。多年来，人们在提高这一系统的生产效率方面已经取得了许多进展，这包括使用

2.3 Manufacturing Process by *E. coli* Expression System

Manufacturing therapeutic proteins in *E. coli* is widely used and accounts for nearly 30% of biopharmaceutical products on the market. Over the years, many improvements such as the usage of a strong T7 promoter, optimization of inclusion body refolding, and

T7 强表达启动子，优化包含体复性或可溶性表达，以及高密度发酵。高密度发酵和高效地从大肠杆菌包含体纯化得到重组蛋白是生产效率的两个主要瓶颈。

soluble expression, have increased productivity. High cell density fermentation and high throughput purification of the recombinant proteins from *E. coli* inclusion bodies are the two major bottlenecks for cost effective production of therapeutic proteins.

发酵过程必须控制那些影响细胞的生长以及产物表达的关键化学、物理和生物学参数。这些参数包括最佳工艺控制温度、溶氧（DO）、pH 和营养成分的供应。一个完善的工业发酵过程还需要考虑培养基的组成和成本、补料策略和工艺放大。大规模生产过程应满足法规监管的指导，如培养基成分的要求，对工艺过程特征和验证实施过程分析技术（PAT）和质量源于设计（QbD）的理念。许多监管机构对所有大肠杆菌生产系统要求无动物成分和无 β- 内酰胺类抗生素的培养基。化学成分确定的培养基受到推崇，但是开发这些培养基可能消耗相当的时间和资源。一些成功生产的例子，包括生长因子、抗体片段和 Fc- 融合蛋白，都在生产上使用了成分或半确定或复杂的培养基（Huang *et al.*，2012）。然而，由于成分不清的培养基会在批号与批号之间有很多变化，在发酵中使用复杂成分的培养基，可能会影响到产品产量和质量重复性，这是在生产药用蛋白质时非常不希望见到的，因为生产过程被认为是产品质量不可分割的一部分。不断提供所需的营养来维持和控制细胞生长经常被用来达到高密度生长和高生产效率。在大多数情况下，大肠杆菌细胞产生多种代谢产品，如醋酸等，对细胞的生长和蛋白质生产有负面影响。一个最佳的补料过程必须控制这

Fermentation process has to be controlled over key chemical, physical and biological parameters that affect cell growth, as well as product production. These include optimum process control of temperature, dissolved oxygen (DO) level, pH and nutrient supply. A robust industrial fermentation process would also need to consider the composition and cost of the media, feeding strategies and scale-up process. A large scale production process should be designed to meet regulatory guidelines, such as requirement for medium component, implementation of PAT and Quality by Design (QbD) for process characterization and validation. Animal component-free and beta-lactamase-free media are required by many regulatory agents for all *E. coli* production system. Chemical defined media are recommended, however development of such media may be time and resources consuming. A number of successful productions of therapeutics, including growth factors, antibody fragments and Fc-fusion proteins have been produced using either semi-defined or complex media (Huang *et al.*, 2012). However, fermentation using complex ingredients can show inconsistent performance that affects product yield and quality because of the lot-to-lot variation associated with poorly defined components. This issue is highly undesirable for protein therapeutics because processes are considered an integral part of the product. A fed-batch process by continuously providing required nutrients to sustain and control cell growth is often used to achieve high cell density and high productivity. In most cases, *E. coli* cells generate several metabolic by-products such as acetate that have negative effects on cell growth and protein production. An optimum feeding procedure must be applied to control the concentration of these by-products and maintain proper cell growth and protein production.

些副产品的浓度，并维持正常的细胞生长和蛋白质的生产。

对于商业化生产的药用蛋白，发酵工艺通常在10～30L的小规模开发，以确定合适的增长和产品表达条件。然后将小试工艺放大到中试水平（100～1000L）并建立操作参数，最终放大到生产规模（超过1000L）。任何药用蛋白的生产过程应着眼于大肠杆菌的生长、蛋白质质量的重现性以及生产效率。在放大到超过1000L规模，一些工程问题可能会出现，如在物料混合动力学上，与规模较小相比较，在大规模发酵罐中显示较慢的传质速率和不同的补料营养梯度。因此，在大肠杆菌大规模发酵工艺放大时采用某些策略，如在不同规模的发酵罐上尽量保持一个或多个参数不变。这些参数包括单位体积功耗或氧气的传输速率、氧传质系数（kLa）、搅拌叶尖速度和叶轮雷诺数。

For commercial production of a therapeutic protein, the fermentation process is usually developed at small scale of 10-30L to identify suitable growth and product expression conditions. The process then is transferred to pilot level (100-1000L) to establish operating parameters and finally to manufacturing scale (over 1000L) for manufacturing. The scale-up process for any therapeutic protein should aim for *E. coli* to grow and for high productivity with consistency in the protein quality and specific yield. Some engineering issues may associated with scaling-up to over 1000L, such as different mixing kinetics as compared to smaller scale, result in slower mass transfer rate and different gradient in feeding nutrients. Therefore, several strategies have been used as principles to scale-up *E. coli* fermentation to minimize the differences between scales by keeping one or more parameters constant from small to manufacturing bioreactors. These parameters include power consumption per liquid volume or oxygen transfer rate, oxygen mass transfer coefficient (kLa), impeller tip speed, and impeller Reynold's number.

用大肠杆菌生产重组蛋白往往会导致其中包含聚合蛋白质的包含体的产生，而这些包含体大部分处于无活性构象状态。这些聚集的蛋白质（非折叠的蛋白质）需要在变性的条件下溶解在缓冲液中，然后复性成有活性的且稳定的构象状态。人们一直致力于复性机制的研究，以理解复性机理和提高复性效率。然而，大部分工业生产使用类似的复性条件，包括高pH，在含有不同比例的氧化和还原的谷胱甘肽中溶解包含体以促进折叠复性。可在复性前纯化包含体，有利于提高复性收率，高pH可以加速溶解包含体，然而，这也具有形成重组蛋白

The recombinant production of proteins by *E. coli* often leads to inclusion bodies which contain aggregated proteins. Majority of inclusion bodies are in inactive conformational states. These aggregated proteins need to be dissolved in a buffer under denatured condition (unfold the protein), then refolded to active and stable conformational state. Great effort has been devoted to refolding mechanism research that leads to understand and to improve refolding efficiency; however, most of industrial processes employed more or less similar refolding condition including high pH to dissolve an inclusion body followed by different ratio of oxidized and reduced glutathione to facilitate refolding. Inclusion body may be purified prior to refolding, which improves refolding yield. High pH may speed up dissolving aggregates; however, it is also risk of forming deamidation of a recombinant protein (Nellis *et al*., 2011).

的脱酰胺反应的风险（Nellis *et al.*, 2011）。

2.4 治疗抗体的生产工艺

单克隆抗体的临床和商业上的成功，导致需要大规模哺乳动物细胞的培养生产。这使得全球制造能力迅速扩张，包括增加生物反应器的规模（达到 20 000L）、大幅度提高工艺过程的效率和降低制造成本。在 1990 年至 2005 年间，上游部分特别成功。由于工程细胞表达技术的改进和工艺的优化，尤其是高产细胞株的开发及上游工艺的补料培养的开发，细胞培养的生产率提高了 100 倍。除了改善工艺 / 成本效益，第二个关键领域是减少了开发工艺过程和生产最初临床试验以作“原理论证”所需的材料的时间。细胞株的建立往往是在这个阶段发展过程中最慢的一步（Birch and Racher，2006）。

目前，市场上批准用于治疗用途的单克隆抗体超过 20 个。这些抗体多数是通过重组 DNA 和杂交瘤技术来生产的。常用哺乳动物细胞表达系统包括 CHO 细胞或小鼠淋巴瘤细胞系（例如 NS0 和 Sp2/0-Ag14）来生产重组抗体。在临床试验中，大多数产品都是在哺乳动物细胞表达系统生产的完整抗体，但也有一些抗体片段，它们可在微生物，如大肠杆菌中表达。例如，CIMZIA 是一种聚乙二醇修饰的肿瘤坏死因子抑制剂单抗 Fab 片段，在微生物系统表达生产并得到美国 FDA 的批准。

工业界继续寻找新技术和新工艺开发战略来减少新产品和新工艺开发

2.4 Therapeutic Antibody Production Process

The clinical and commercial success of monoclonal antibodies has led to the need for very large-scale production in mammalian cell culture. This has resulted in rapid expansion of global manufacturing capacity, an increase in size of reactors (up to 20 000 L) and a greatly increased effort to improve process efficiency with manufacturing cost reduction. This has been particularly successful in the upstream part of the process where productivity of cell cultures has improved 100 fold. This success has resulted from improvements in expression technology and from process optimization, especially the development of high productivity cell line, and of upstream fed-batch cultures. In addition to improving process/cost efficiencies, a second key area has been reducing the time taken to develop processes and produce the first material required for clinical testing and proof-of-principle. Cell line creation is often the slowest step in this stage of process development (Birch and Racher, 2006).

Currently, there are more than 20 monoclonal antibodies approved for therapeutic use. The majority of these antibodies are produced by recombinant DNA and hybridomas technologies. The recombinant antibodies are produced in mammalian cell expression systems using CHO or murine lymphoid cell lines (e.g., NS0, Sp2/0-Ag14). Most products in clinical trial are whole antibodies made in mammalian cell systems but some are antibody fragments, which can be made in microorganisms such as *E. coli*. For example, CIMZIA is a pegylated Fab fragment of a humanized TNF inhibitor MAb made in a microbial system and was approved by the FDA.

The industry continues to look at new technologies and process development strategies that will reduce

时间。由此产生的新工艺必须是容易放大的、稳定的、符合质量和安全标准的。缩短开发时间的方法之一是细胞培养过程中使用更好的技术，例如使用标准的培养基、补料和生长条件。细胞株的构建和筛选往往是一个关键环节，需要及时完成，又不能降低质量要求。理想的情况下，快速构建可用于长期生产的高产率的细胞株，避免以后在开发后期需要制备改进的第二代细胞株。

timelines for new product and new process development. The resulting processes must be easily scalable, robust and meet quality and safety criteria. One approach to shortening the timelines is the use of platform technologies for cell culture processes, for example using standard media, feeds and growth conditions. Cell line construction and selection is often a critical path activity and needs to be completed rapidly without compromising quality criteria. Ideally, one would like to rapidly create a highly productive cell line that could be used for long-term manufacture obviating the need to create an improved second generation cell line at a later stage of development.

近年来，哺乳动物细胞工艺过程的生产力显著提高，现代细胞培养过程可以实现超过5g/L的抗体表达水平。目前，大多数的上游工艺流程是补料分批培养，特别是补料工艺的进展，显著地增加了抗体产量。高产率的细胞株来自于使用有所需特征的宿主细胞、适当的表达系统和良好的转染与筛选技术。有些表达系统具备产生高产率（Qp）细胞株的潜在能力。真正面临的挑战不仅是建立具有很高产率（Qp）的细胞株，而是还要具备细胞稳定生长的特点，这才导致在生产过程中有高产率的结果。稳定的生产细胞株，在悬浮培养中使用确定化学组分的无动物成分培养基，达到所需的翻译后修饰能力，这些都是筛选高产细胞株的关键因素。

Productivity of mammalian cell processes has improved dramatically in recent years and modern cell culture processes can achieve antibody concentrations exceeding 5 g/L. Most current upstream processes are based on fed-batch culture and the development of feeds in particular has made a significant contribution to increased antibody yields. Highly productive cell lines result from using a host cell line that has the desired characteristics, an appropriate expression system, and a good transfection and selection protocol. A number of expression systems with the potential to produce cell lines with high specific production (Qp) rates are available. The challenge is to create cell lines that not only have high Qp but also have the growth characteristics that will lead to high productivity in the manufacturing process. The resulting cell lines are also selected for stability, growth in suspension culture using a chemically defined, animal component-free medium and the ability to make the desired post-translational modifications.

一个典型的单克隆抗体制造过程如下所示。将工作细胞库接种到摇瓶中生长初始种子。然后转移到带搅拌的种子罐或波浪生物反应器中，从最初的10～50L到150L、750L、5000L逐步增大种子培养的体积。种子培养从冷冻小瓶到大罐种子需大约6周时间。在搅拌罐内的生产要求悬浮细

A typical MAb manufacturing process is illustrated as followed. A couple of vials of the Working Cell Bank are inoculated into flask to grow initial seed. Then the seed transferred to a stirred-tank or a Wave bioreactor for seed build-up with increasing volume from initial 10-50L to 150L, 750L and 5000L seed bioreactors. The seed train takes about 6 weeks from frozen vials. Production in a stirred tank requires the cell line suspension adapted and normally serum-free medium is used. Largest mammalian cell culture production is 20 000L for MAb production

胞和无血清培养基。最大的哺乳动物细胞培养生产单克隆抗体的规模达20 000L，表达水平达1～5g/L，所以每一批生产可能得到含有15～75kg产物的培养液供下游纯化。从细胞培养液中回收产品可用于离心或过滤，除去细胞和细胞碎片。

with expression titer 1-5g/L, so each batch of production may generate 15-75kg harvested material for downstream purification. Centrifugation or depth filtration may be used as product recovery from the cell culture broth by removing cell and cell debrides.

纯化策略是基于通常包括蛋白A亲和层析、离子交换、疏水层析、分子筛等的柱层析步骤。它也包括病毒灭活和去除；切向流过滤和离心等工艺步骤。随着上游表达量的增加，人们越来越多关注改善下游工艺效率。

Purification strategies are based on chromatographic procedures usually including a protein A affinity step IEX, HIC, and size exclusion chromatographs. Besides, other unit operations may be also involved, such as viral removal and inactivation; TFF, and centrifugation. With increasing upstream concentrations, significant attention is now being paid to improving downstream efficiency.

纯化工艺通常先将收获的培养上清液浓缩以减少体积，或直接加载在第一个蛋白A亲和层析柱上，然后在低pH下洗脱并作为病毒灭活步骤。经过几个小时的低pH处理后（取决于产品在低pH下的稳定性），调整pH到接近中性，然后用疏水或离子交换作为第二步柱层析。在纯化过程的后期，通过纳米过滤去除病毒是常用的程序。然后最终精制步骤可以用离子交换QFF柱或膜去除宿主细胞的DNA、内毒素，以及进一步减少病毒。最后的超滤或凝胶过滤柱纯化取决于产品所需蛋白浓度和配方。如果产品的稳定性数据允许的话，产品可以液态分装，而在大多数情况下，一般是以冻干形式分装最后成品。

Harvested culture supernatant may be reduced in volume or directly load on the first column protein A affinity chromatography. Product is eluted out from the protein A column with low pH that can be served as viral inactivation step. After low pH treatment for hours (depend on how stable of a product in that low pH), pH is adjusted back to near neutral for second and columns that are either HIC or IEX. In this late stage of a purification process, a viral removal through nano-filter is commonly used procedure. Then final polishing step can be QFF column or membrane flow-through to remove host cell DNA, endotoxin if there is any, and virus further reduction. The purified bulk is formulated by an ultrafiltration or gel filtration column depend upon required final protein concentration and formulation. Formulated bulk is vialed as liquid form if product stability data supports or lyophilized form in most cases.

2.5 生产设备和工艺验证

2.5 Manufacturing Facility and Process Validation

根据FDA的指导文件，验证是一个有记录的文件化程序，它提供了一个高度保证：即一个特定的生产过程会持之以恒地生产出满足其预定规格和质量的产品。有效的过程验证会充

According to the FDA guideline, validation is a documented program which provides a high degree of assurance that a specific process will constantly produce a product meeting its predetermined specifications and quality attributes. Effective process validation contributes significantly to assuring drug quality. The basic principal

分保证药品质量。质量保证的基本原则是药物应符合其使用目的。这个原则的具体理解是：①质量、安全性和有效性源于设计或包含在产品的生产中；②仅靠过程中和成品的检验或测试不能充分保证质量（《FDA指导行业工艺验证：一般原则与实践2011》，www.fda.gov）。为了满足验证要求，生产设施的设计应符合良好生产规范的总体布局和符合人员和物资的流动规定，而验证的公用工程应包括注射用水（WFI）、洁净的蒸汽、采暖、通风、空调（暖通空调）、废弃物处理，以及清洁区域与监测验证相关的表面处理。

of quality assurance is that a drug should be produced for its intended use. This principle incorporates the understanding that ① Quality, safety, and efficacy are designed or built into the product; ② Quality cannot be adequately assured merely by in-process and finished-product inspection or testing (*FDA Guidance for Industry Process Validation: General Principles and Practices 2011*, www.fda.gov). To meet validation requirements, a manufacturing facility should be designed in consistent with GMP for overall layout, flow of personnel and materials, and validated utilities including water for injection (WFI), clean steam, heating, ventilation, and air-conditioning (HVAC), waste treatment, as well as surface finishes that will support cleaning validation by area monitoring.

工艺验证的目的是保证具体生产过程依据预先设定的程序进行控制。事先充分考虑工艺过程的变量和可接受的范围，然后确定并测试。工艺验证的形式和程度与产品的使用目的是一致的。所有在临床研究或上市的产品必须在符合GMP的条件下生产，因此必须经过验证的过程。虽然验证概念是对任何GMP生产的要求，然而FDA也认可，在验证资料的准备方面，对一个上市产品和临床样品的生产过程，可要求不同程度的详细信息。例如，直到三期临床试验样品的生产前，可能不会需要生产过程和分析方法验证，虽然生产设施需要进行验证。因此，在三期临床试验进展之前，并不要求准备一个完整的工艺过程验证材料。

Objective of process validation is the assurance that the specific production process being employed is under predefined control. Process variables and acceptable limits have been duly considered, identified, and tested. The form and degree of process validation is consistent with the intended use of the product. All products that will be introduced into clinical studies or marketed must be made under GMP conditions and therefore must be made with a validated process. While conceptually validation is required for any GMP production, it is recognized by the FDA that process validation of manufacturing clinical material is different from that of a production process for a market product in terms of the detailed information package. For instance, production process and analytical method validation may not be required until the production of the material for Phase Ⅲ clinical trials, although the manufacturing facility is required to be validated. Thus, developing a complete process validation package before the product advances to phase III trials would be not warranted.

在生产工厂，那些对产品的人体安全使用有直接影响的系统，必须在药品临床使用之前验证。这些系统包括洁净蒸汽、纯净水、高压灭菌器、生物反应器灭菌和清洁。应该验证及

Those systems in a manufacturing facility that have a direct impact on the product's safe use in human must be validated prior to clinical usage. These include clean stream, purified water, autoclaves, bioreactor sterilization and cleaning. Process steps critical to product safety should be validated including final product (terminal)

将验证重点放在对产品的安全性、有效性、稳定性有影响的关键工艺步骤，包括最终产品（终端）的无菌过滤、清除病毒、去除宿主细胞的DNA和多余的蛋白质。那些不能在最终产品中测试的性能应给予特别强调。此外，那些保护人员和环境的系统是至关重要的，必须在去污的高压灭菌器，高效率尘埃颗粒过滤器（HEPA），以及空调系统等的使用之前进行验证。

sterilization, viral clearance, and removal of host cell DNA and extraneous protein. The focus should be on validation of step that affects the safety, potency, and stability of the product. Special emphasize should be given to those properties that cannot be tested in the final product. In addition, those systems that are critical to the protection of personnel and the environment must be validated prior to their use in the decontamination of autoclaves, High Efficiency Particulate Arresting (HEPA) filters, HVAC systems etc.

对于制造过程中使用哺乳动物细胞如CHO或HEK293，美国FDA和中国国家食品药品监督管理局出于产品使用安全的考虑要求病毒清除验证。

For a manufacturing process using mammalian cells such as CHO or HEK 293, virus clearance validation is required by the FDA and SFDA for product safe use consideration.

从真核细胞生产产品的工艺验证的一个重要部分是评估纯化过程中对潜在的病毒污染物去除/灭活的能力。使用哺乳动物细胞的生产过程有一定的病毒污染的风险，如用染有内源性病毒细胞株用来表达蛋白产品（例如小鼠逆转录病毒），主种子库从外部源获得的外源性病毒（例如EB病毒），在制造过程中通过外来污染（例如细小病毒）或通过受污染的原材料将病毒外源因子引入产品。许多病毒污染事件发生在生物制药的生产设施上，包括GSK公司由于猪病毒污染造成轮状病毒疫苗在2010年停产。波士顿的Genzyme公司，在其Allston工厂的生物反应器中发现了病毒 *Vesivirus* 2117，因此两种药物停止生产（Bethencourt，2009）。虽然该病毒并没有造成人类感染，但损害表达生物药物的CHO细胞的生长。因此，为了保证产品安全，有效的病毒清除步骤必须是制造过程中的一部分。

An essential component of process validation studies for products derived from eukaryotic cells is to assess the capacity of the purification process of removing/inactivating potential viral contaminants. Production process using mammalian cells may have certain risk of virus contamination. Endogenous viruses exist as a part of the cell line used to express the protein product (e.g., *murine retrovirus*). Non-endogenous viruses are viruses from external sources present in the MCB (e.g., *Epstein Barr virus*). Adventitious viruses are introduced into the product during the manufacturing process either through the addition of contaminated raw materials or through extraneous contamination (e.g., *parvovirus*). Many virus contamination incidents happened in biopharmaceutical manufacturing facilities including GlaxoSmithKline's Retavirus vaccine that was suspended for pig virus contamination in 2010 and Genzyme Corp in Boston had to halt the production of two drugs after *Vesivirus* 2117 was discovered in bioreactor at its Allston plant. Although the virus does not cause human infections, it impairs growth of the biologics-producing CHO cells (Bethencourt, 2009). Therefore, to assure product safety, validated virus clearance steps have to be a part of the manufacturing process.

必须在临床I期产品用于人体开始之前验证病毒清除步骤。通常生物技术公司使用老鼠的细胞株验证对鼠白血病病毒进行去除/灭活病毒清除步骤，这已作为对小鼠细胞表达系统验证反转录病毒样颗粒特定的模型。可以使用热、pH、洗涤剂、射线照射、溶剂来灭活病毒，通过过滤和层析来清除病毒。

Viral clearance steps must be validated before Phase I products may be administered to humans. It is common for biotech manufacturers to use rodent cell lines to validate virus clearance steps for the removal/inactivation of *murine leukemia virus*. The virus has been served as a specific model virus for the retrovirus-like particles produced by murine cell expression systems. Virus inactivation can be achieved by heat, pH, detergents, irradiation, and solvents. Virus removal can be accomplished by filtration and chromatography.

2.6 生物技术药物生产的一次性使用技术

2.6 Single-Use Technology in Biopharmaceutical Manufacturing

2.6.1 一次性使用的技术和优势

一次性使用技术最早于20世纪90年代引进到生物药物的生产上。它之所以被称为“一次性使用”，意指使用后弃用的特点。

第一，一次性使用技术的最大优点在于减少交叉感染，这是生物制品安全性方面的主要风险。利用这一技术减少了清洁和消毒验证、相关的文件以及操作上的训练，而这正是最容易受到审查机构警告和开出罚单的地方。第二，这一技术减少了生产设备的安装和保养，也就加速了工艺的开发和生产。减少的清洁和消毒验证也相应减低了人力需求。第三，运行成本降低。对一个使用一次性技术的生产厂家，尽管需要一定的场地来存放大量的袋子，但大宗设备和公用工程的需要却会远低于一般工厂，而且操作成本也会有所降低（Kostyukovsky，2007）。

2.6.1 Single-Use Technology and Advantages

Single-Use or disposable technology was first introduced into biopharmaceutical manufacturing in the 1990's. It is called “single-use” or “disposable”, indicating the characteristic of being discarded after application.

Firstly, the biggest advantage of disposable technology is reducing cross-contamination, one of the major regulatory risks for biological product safety. It reduces cleaning and associated validation, documentation, and operator training that commonly get warned and citations from regulatory agencies. Secondly, this technology provides an opportunity to speed up process development and manufacturing. Installation and maintenance of equipment are minimized. Cleaning process and cleaning validation are decreased. Labor for these activities is reduced. Thirdly, the overall operation cost can be lowed. For a manufacturing facility using disposable technology, the capital equipment and utility can be much less even though the footprint may increase due to large quantity of bags. Operation time and labor are reduced as well (Kostyukovsky, 2007).

越来越多的公司开始在生产中使用这一技术以缩短生产周期从而获得更大的利润。合同生产公司以及刚起步的小公司也从运用这一技术中获得极大的好处。当一工艺过程在一密封

More and more companies started to apply this technology to their product production which enabling them to be more competitive by shortening the production cycle and profitability. Contract manufacturing organization (CMO) and startup/small companies are beneficial tremendously from disposable technology.

系统（在袋中加入原材料、液体以及内装搅拌装置）中进行时，车间空气洁净要求可以降低，几种不同产品也能在同一空气系统中同时进行生产操作。Xcellerex公司的生产设施是个很好的例子。

When a process is in a closed system (bags with the closed addition of raw materials and liquids with integral mixing devices), the air system classification can be less stringent, and multi-product processing may occur under one air handling system, which is the case in Xellerex (Xellerex Inc.) facility.

2.6.2 一次性使用技术在研发和生产中的应用

一次性使用技术几乎能应用在每一步的生物制品生产中。最常见的生产步骤有发酵（生物反应器）、收获、分离、切线流过滤、柱层析和无菌灌装。一次性抗低温小瓶被用来在−80℃或液氮中储存细胞株或细胞库。细胞放大培养能在波浪袋或搅拌式反应器中进行，波浪袋反应器的温和混合方式为动物和昆虫细胞生长提供了低度的切剪力和适量的氧气供给。截至2012年，大型的一次性使用细胞反应器已达1000～2000L（Xcellerex, Thermo Scientific和Sartorius），一次性使用感应器（测定溶氧或酸碱度）也可供微生物发酵和细胞培养使用。各式各样的一次性使用软管可用来传送液体，例如缓冲液、培养液和中间产物，用于储存大体积的缓冲液、培养液或中间产物的一次性袋子体积可达500L，若有固体支架支撑的话，其体积可达2500L。

2.6.2 Single-Use Technology in Process Development and Manufacturing

Single-use technology can be employed in almost every unit operation of biological manufacturing. Commonly used bioprocess includes fermentation using a bioreactor, harvest, clarification, depth or tangential flow filtration (TFF), chromatography and aseptic filling. Starting from seed or cell banks storage, disposable cryovials are used for storing cell banks at −80℃ or in liquid nitrogen. Cell expansion can be achieved in disposable bioreactors including wave or stirred-tank reactors (STR). Wave technology provides low shear stress and adequate oxygen for animal and insect cell culture by gentle rocking mixing. By 2012, large size disposable bioreactor is available for cell culture up to 1000-2000L (Xcellerex, Thermo Scientific and Sartorius). Disposable probes (DO, pH) are available for fermentation, cell culture and other bio-processing. Multi-type of tubing is available for transfer liquid material, such as buffer, medium and in-process product. Large size plastic bags up to 500L (without mechanical support) and up to 2500L (with rigid support) are available to support buffer or medium storage and material transfer.

培养基交换和流加技术能用连续式离心法进行（例如Pneumatic Scale Angelus的Unifuge离心机）。细胞分离可用微滤（孔径在0.1μm以上）或深层膜过滤法进行，一次性的膜层析（Mustang或Sartobind）也可用于产品的回收步骤上，这些膜结合了过滤和层析的特点而被用来快速分离生物产

Medium exchange and perfusion can be achieved by using single-use centrifugation (Unifuge from Carr Centritech Separation System). Cells can be removed by microfiltration (more than 0.1 μm pore size TFF) or depth filtration. There are single-use membrane chromatography (Mustang or Sartobind) available for product capture step. These membranes combine filtration and chromatography for quick bioseparations. Although it is not common to use single-use column due to obvious economic reason, there are products so

品。尽管一次性层析柱由于价格的原因而不被广泛采用，市场上还是有类似的已装好的柱产品，称为即用层析柱（RTP，GE），这些分离层析柱可在同一个产品上使用许多次后再扔掉。

1990年前分离碳水化合物和手性对映体的技术“模拟移动床层析法”（SMB）现被应用于生物大分子的纯化，被称为“生物模拟移动床层析技术”（BioSMB，生产厂家Tarpon）。多个分离柱事先装成比一次分离柱要小得多的柱子，受程序控制，整个柱子群是呈连续式的运作，而每一根柱子却处在不同的分离阶段，例如平衡、上样、洗脱和清洁。柱子中的树脂被高效地利用，因此树脂的需求量较正常的层析柱大为降低，缓冲液和清洁溶液的体积也相应减少，这些小柱子在多次使用后即可丢弃。

在无菌软管连接中，有不同的连接体和连接口供选择。不同的厂家生产大小不同的可放大的无菌和去病毒过滤膜。制成原液后的产品若是少量但很昂贵的话，可用一次性装瓶系统来灌装。

2.6.3 一次性使用技术的缺点

一次性使用技术给生产的操作和法规遵从带来一定的隐患。不那么结实的袋子（或装置组成部分）将是生产操作中的最不可靠的部分，这些袋子只能承受不超过68.9kPa的压强，大于500L的袋子必须有钢架支撑。袋子发现有孔、有洞，或密封部位泄漏是很有可能发生的，一旦这种事发生，整批生产就可能报废。这也解释了为什么最大的一次性细胞反应器体积不超过2000L。一次性使用技术同

called Ready-to-Process (RTP, GE) column that provide customers the pre-packed column. These columns can be used many times and dedicated to one product before disposal.

Simulated Moving Bed (SMB) chromatography, a technology used in hydrocarbons and chiral compounds production before 1990's, is now started to apply to bioproducts downstream processing and is hence called BioSMB (Tarpon). Multi-columns are pre-packed in much smaller size than one batch size column. The columns are programmed to run in a continuous mode with each at different phase of chromatography, such as equilibration, load, elution or cleaning. Resin in columns is effectively utilized and the overall resin demand is much less than that in a batch column. The buffer and cleaning solutions volumes are proportionally significantly reduced.

There are different disposable tubing connectors or fittings available for aseptic tubing connection. Sterile and viral removal filters at scalable sizes are manufactured by different companies. Formulated product at low volume but high value can be aseptic filled by single-use filling machine tubing assemblies.

2.6.3 Disadvantage of Single-Use Technology

Disposable technology poses risks on operation and regulatory compliance. Low structure strength of bags (or equipment component) is one main risk for operation reliability. These bags can only handle up to 10 psi pressure. Bags at more than 500L need a solid steel holding support. There is always risk associated with punch, hole and leak on sealing. Once this happens, a whole batch could lose. This explains why the largest disposable bioreactor can only support 2000L cell culture scalc currently. Single-use technology increases material handling cost as well. Significant solid waste is generated.

时也增加了物料处理的费用，可观的固体垃圾也相应产生。

尽管说是清洁验证工作量被降低了，生产厂家还得在引进一次性使用技术后再检验自行的清洁程序。另一个与塑料容器有关的药品法规方面的风险是浸出性和提取性。FDA 和 ICH 指南都强调指出凡碰触任何物料，中间产物和最后产品的装置表面组成都不得改变药物产品的安全性、成分、有效性、质量和纯度。可浸出物质是指那些从盛装药物产品的有弹性的或塑料的容器和（或）封闭操作系统中沥滤出来的化合物，可提取物质是指当有机溶解液接触到这些盛装药物产品的容器或封闭操作系统时从中萃取的化学物质，从塑料里释放的化学物质主要是有机物质。最常见袋子制作法是用多层薄膜结构，聚乙烯（PE）和乙烯 - 醋酸乙烯（EVA）通常用作最靠近溶液的一层膜，而乙烯 - 乙烯醇（EVOH）则用作阻隔层。这些塑料薄膜同时含有添加剂，以利于袋子的生产，增加牢固性、优化可操作性和延长其寿命。这些都将增加滤取性和提取性物质进入溶液和产品中间体的可能。容器的消毒，包括加热和伽马射线辐射，以及化学处理会进一步增加滤取性和提取性物质的量。生产企业应和供应商一起解决接触溶液和中间体的容器或物质的滤取和提取问题。这些物质的测试方法在美国药典（USP）、国际标准化组织（ISO）、FDA 及其他机构均有描述。

Although cleaning validation is reduced, re-evaluation of in-house cleaning procedure after incorporation of a disposable system is required for the user. Another big regulatory risk associated with plastic container is leachable and extractable. FDA and ICH guidelines emphasize that the equipment surfaces that contact raw material, intermediates and product should be constructed from materials that do not alter the safety, identity, strength, quality and purity of the drug product. Leachable are compounds that leach into the formulation from elastomeric or plastic components of the drug product container and/or closure system. Extractable are compounds that can be extracted from elastomeric or plastic components of the container closure system when in the presence of solvent. Compounds released by plastics are mainly organic. Most commonly used material for bags is multiweb film consists of polyethylene (PE) or ethylene-vinyl acetate (EVA) for solution contact, ethylene-vinyl alcohol (EVOH) as a barrier resin. The plastic films also contain additives that aid in processing, better strength, optimized performance and prolonged shelf life. All of these potentially increase the possibility of leachable or extractable to solution and product intermediates. Sterilization on containers (bag or bottle) by heat, gamma irradiation or chemical agents can enhance profile of leachable or extractable components. Manufacturer needs work together with vendors on evaluation on leachable and extractable of container/material that has intermediate or product contact. Testing methods are described by various agencies, including U.S. Pharmacopoeia (USP), International Organization for Standardization (ISO), FDA and others.

2.6.4 一次性使用技术在生物制药的前景

一次性使用技术在走过了较长的引进过程后才被迅速地用于生物制品的

2.6.4 Future of Single-Use Technology in Biopharmaceutical Manufacturing

The disposable technology for biopharmaceutical applications has been rapidly applied since a long way of introduction. It brings tremendous benefit from regulatory

生产。从规程、生产速度和生产成本的角度来看，它给生产带来了极大的好处，当然它也有缺点。其中一个争议是关于环境保护方面的，当一次性使用技术产生了多余的固体垃圾时，它却又减少了液体垃圾，诸如用来清洁容器和系统的注射用水以及化学品，同时也降低了电力资源的损耗。许多公众评论认为一次性使用技术在环境保护上功大于过。至于一个企业是否使用一次性使用技术纯粹依赖于公司的重点和需求。几乎生物制品生产每一个环节都可利用一次性使用技术，一次性使用技术的时代将不可避免地到来。

（编者　朱建伟）

point of view, as well as the process speed and operation cost. There is certainly disadvantage associated. One debate is on the environmental concern. While disposable technology generates additional solid waste, it greatly reduces liquid waste including water for injection (WFI) and chemicals that used for cleaning vessels/systems. Energy consumption is reduced when use single-use technology for a production. Many public views consider the environmental advantages of single-use technology overweight the generation of solid waste. Whether an entity uses disposable technology or not is solely depending on the priorities and needs in a manufacturing facility. Clearly disposable technology is available for almost all steps of bioprocess for manufacturer to choose. The era of single-use technology has evitably arrived.

(Compiled by Jianwei Zhu)

参 考 文 献
References

Aldrrich TL, Viaje A, Morris AE. 2003. EASE vectors for rapid stable expression of recombinant antibodies. Biotechnol Prog 19: 1433-1438

Athmaram TN, Saraswat S, Santhosh SR, *et al.* 2011. Yeast expressed recombinant Hemagglutinin protein of Novel H1N1 elicits neutralizing antibodies in rabbits and mice. Virology J 8: 524-536

Bebbington CR, Renner G, Thomson S, *et al.* 1992. High level expression of a recombinant antibody from myeloma cells using a glutamine synthetase gene as an amplifiable selectable maker. Biotechnol 10: 169-175

Beck A, Wagner-Rousset E, Bussat MC, *et al.* 2008. Trends in glycosylation, glycoanalysis and glycoengineering of therapeutic antibodies and Fc-fusion proteins. Curr Pharm Biotechnol 9: 421-422

Birch JR, Racher AJ. 2006. Antibody production. Adv Drug Deliv Rev 58: 671-685

Bollok M, Resina D, Valero F, *et al.* 2009. Recent Patents on the *Pichia Pastoris* Expression System: Expanding the Toolbox for Recombinant Protein Production. Recent Patents on Biotechnol 3: 192-201

Butler M. 2005. Animal cell cultures: recent achievements and perspectives in the production of biopharmaceuticals. Appl Microbiol Biotechnol 68:283-291

Cacciatore JJ, Chasin LA, Leonard EF. 2010. Gene amplification and vector engineering to achieve rapid and high-level therapeutic protein production using the Dhfr-based CHO cell selection system. Biotechnol Adv. 28:673-681

Deer JR, Allison DS. 2004. High-level expression of proteins in mammalian cells using transcription regulatory sequences from the Chinese hamster EF1- gene. Biotechnol Prog. 20: 880-889

第三章 生物技术药物制剂学

Chapter 3 Formulations of Biopharmaceuticals

1 生物技术药物的结构特点和理化性质

1 Structural Characteristics and Physicochemical Properties of Biopharmaceuticals

生物技术药物主要包括蛋白及多肽类药物，此类药物化学结构复杂、理化性质特殊，由此产生的稳定性问题对于生物技术药物的制剂研究、生产、质量控制、贮存等环节有非常重要的影响。因此，在设计与评价这类药物制剂时，首先应充分了解其结构特点和理化性质。

Biotechnological drugs mainly include protein and polypeptide drugs. These drugs have complex chemical structures and the special physical & chemical properties. The stability of proteins plays an important part in the research of preparation of biotechnological drugs, production, quality control and stockpile, etc. Therefore, in the design and evaluation of these drug delivery systems, we should first fully understand the structural characteristics and physicochemical properties.

蛋白质药物的结构复杂，其分子结构可分为一、二、三、四级结构。蛋白质一级结构是空间结构的基础，一级结构相似的蛋白质，其基本构象及功能也相似。此外，蛋白质分子只有在其立体结构呈特定构象时才具有活性，蛋白质的空间构象是其功能活性的基础，构象发生变化，其功能和生物活性也随之改变。

The protein drugs have complex chemical structures; their molecular structures can be divided as primary structure, secondary structure, tertiary structure and quaternary structure. The primary structure is the base of spatial structure. The basic conformation and functions of the protein which is like the primary structure are similar too. In additional, the protein's spatial structure is the base of its functional activity. The conformation changes, the functional activity changes too.

蛋白质药物相对分子质量大，为两性电解质，具有胶体溶液性质、紫外吸收、旋光性等性质。由于结构上的特殊性，此类药物性质常不稳定。引起蛋白质药物不稳定的因素比较多，如共价键与非共价键的破坏与生成，均可导致蛋白类药物的不稳定而失活。由共价键改变引起蛋白质不稳定的化学反应有氧化、水解和消旋化等，有时可能几种反应同时进行；由非共价键引起的蛋白质失活作用主要有聚集、

The relative molecular mass of the protein is huge, and it has the properties of ampholyte, colloidal dispersion, ultraviolet absorption and optical activity. As the special structure, the nature of these drugs is often unstable. There are many reasons for the instability of proteins, such as the destruction and formation of covalent bond and non-covalent bond. The chemistry reactions, which cause the instability of proteins because of the change of covalent bone, are oxidation, hydrolysis, and racemization. Sometimes, there are several reactions happening together. There are several types of irreversible inactivation of protein. They are aggregation, precipitation, surface adsorption, and protein denaturalization. This phenomenon often brings a lot of

沉淀、表面吸附与蛋白质变性几种类型。这些现象在开发蛋白与多肽类药物制剂中常会带来许多困难。

difficulties when developing the protein and polypeptide drugs.

蛋白质中具有芳香侧链的氨基酸，可在一些氧化剂（氧自由基、过氧化氢、过氧酸、碘等）的作用下发生氧化反应，巯基的氧化在碱性条件下，尤其是在金属离子（如 Cu^{2+}）的存在下容易发生，蛋白与多肽类药物发生氧化后通常会导致生物活性的降低或消失。蛋白类药物可被酸、碱和蛋白酶催化水解，使蛋白质分子断裂成分子量大小不等的氨基酸和肽段。在强酸或强碱条件下通常发生完全水解，水解产物是各种氨基酸的混合物，当蛋白质碱水解时会使某些氨基酸产生消旋作用，其中 Asp 残基最易发生消旋反应，蛋白质中氨基酸的消旋作用一般能使氨基酸成为非代谢的形式。不完全水解是在酶或稀酸等较温和的条件下进行，不破坏氨基酸，也不引起消旋作用，水解产物有肽段与氨基酸。此外，蛋白质的特有反应二硫键的断裂与交换能改变蛋白质的三级结构，从而影响其生物活性。二硫键对稳定蛋白质构象具有重要作用。通常蛋白质中二硫键的数目越多，其结构稳定性越强，如生物体中的角蛋白。二硫键交换可由硫醇类物质催化。在某些蛋白质中，二硫键一旦被破坏，蛋白质构象随即发生改变，蛋白质的生物活性即丧失。

Oxidation reaction may take place by oxidant (eg., O^-, H_2O_2, peroxyacid, I_2) in an amino acid which has aromatics side chain in a protein, such as methionine, cystein etc. The oxidation of mercapto group may easily take place under the condition of alkalescence. Especially, there exists metalion, for example Cu^{2+}. The oxidation reactions of protein and polypeptide drugs often lead to the reduction or loss of biological activity. The protein drugs can be catalyzed and hydrolyzed by acid, base and protease to rapture the protein molecular, which becomes amino acids and peptide segments of different molecular weights and sizes. The protein drugs could be completely hydrolyzed in the condition of strong acid or base. When the base hydrolysis of protein method is used, certain amino acids produce racemic role, especially Asp residues. Racemization of amino acids in proteins can make themselves into nonmetabolizable forms. Incomplete hydrolysis happens under the mild condition of enzyme or acid, etc., hydrolysales are peptide segments and amino acid. Besides, there are still special reaction of proteins, that is, the crack and exchange of disulfide bond. It can change the proteins' tertiary structure, which will influence the biological activity. Disulfide bonds play an important part in the conformation of stable proteins, the more number of the disulfide bonds in the protein, the stronger structural stability is. One example is the keratin in the organisms. The bisulfide bond exchange can be catalyzed by mercaptan. In some proteins, if the disulfide bond is destroyed, the biological activity will be lost.

蛋白质分子凝聚从溶液中析出的现象称为蛋白质沉淀。当破坏蛋白质的水膜或中和蛋白质的电荷，如在溶液中加入高浓度盐类、有机溶剂或某些生物碱试剂等，蛋白质胶体溶液就

The phenomenon of the protein molecular condensation being separated out from the solution is called albumen precipitation. When we destroy the hydrated shell of protein or cause protein charge to undergo neutralization, the protein colloid solution will become instable and precipitated. Some reagents such as high concentration salts, organic solvents and some

不稳定而出现沉淀现象。蛋白质沉淀可导致产品外观改变、效价降低等，影响这一类制剂的使用。蛋白质因受物理或化学因素影响，其分子的空间结构改变，导致其理化性质、生物活性等发生改变，这种现象称为蛋白质的变性。变性蛋白质一般易于沉淀。蛋白质变性时，由于其空间构象被破坏，引起功能活性丧失，变性蛋白质在复性后，构象复原，生物活性即能恢复。防止蛋白质变性能有效地保存蛋白质类药物制剂。

alkaloids can cause precipitation. Proteins precipitation may lead to change in the product appearance and potency decreased, affecting the use of these preparations. The molecular spatial structure of protein will change if it is affected by the physical and chemical factors, then the physical and chemical properties, biological activity will change too. This phenomenon is called protein denaturation. Denatured proteins precipitated easily. When the protein is denatured, the functional activity will be lost as the spatial conformation is destroyed. However, when the denatured protein is renatured, the conformation may be recovered; so can the activity. The protein formulation may be preserved by means preventing protein from denaturing.

2 生物技术药物制剂的处方与工艺

2 Formulation Recipes and Processing of Biopharmaceuticals

生物技术药物制剂的研制首先应保证生物利用度及产品的市场化能力，关键是解决这类药物的稳定性和有效性。蛋白和多肽类药物在体内外具有不稳定性，非注射途径给药存在生物利用度低等问题，而注射给药能避免药物被物理破坏和酶降解，保证药物通过血液和淋巴系统到达机体特定靶区。因此，目前临床主要剂型是注射剂，包括水针剂和冻干粉针。本节主要讨论蛋白和多肽类药物注射剂的处方和工艺。

Biotech drug preparation should first ensure the bioavailability and the market capacity of the product. The key is to solve the stability and effectiveness of such drugs. Protein and polypeptide drugs in vivo could be instable; there exists low bioavailability in non-injection and low bioavailability problems. Injection drugs should be able to avoid physical damage and enzymatic degradation, so to deliver the drug through blood and lymphatic system to the specific target tissues. Therefore, the current clinical formulation is primarily for injection, including the injection solution and lyophilized powder. This section focuses on the recipes and processing of injection formulation of protein and polypeptide drugs.

2.1 处方前研究

2.1 Preformulation Study

2.1.1 处方设计时需要考虑的理化因素

2.1.1 The Physical and Chemical Factors When Prescribing the Design

在确定生物技术药物的给药途径和剂型后，针对药物的基本性质及制剂的基本要求，选择适宜辅料，设计合理的处方与工艺，将其制成质量可

After determining the route of administration of biotech drugs and dosage forms, according to the basic properties of drugs and for preparation of the basic requirements, we may select appropriate materials, design rational prescription and use right technology to make the pharmaceutical product with reliable quality

靠、使用方便、成本低廉的药物制剂。蛋白类药物的性质独特、结构复杂、分子量大、热不稳定、易发生聚集。蛋白一旦发生聚集，随后发生化学和物理性质改变，不稳定性急剧增加。处方前研究开始前，需对活性蛋白、多肽的理化性质进行广泛研究。研究过程中，应尽早发现潜在问题，评价各种可能，并稳定处方，最好能形成产品。处方前研究应考虑生物药物的不稳定性，所涉及的物理化学因素列于表 3-1。

and low cost, and easy to use. The protein drugs have unique nature, complex structure, high molecular weight, and are heat unstable and easily precipitated. Protein aggregation may lead to changes of chemical and physical properties with significantly instable drug. Before preformulation, one needs to research and understand the physical and chemical properties of active protein or polypeptide thoroughly. During study, early detection of potential problems should be noted and all possibilities should be considered. Besides, one needs to make the formulation stable and best for formulating the products. The instability of protein drugs in the previous chapters is summerized. Preformulation studies involving physical and chemical factors are listed in Table 3-1.

表 3-1 蛋白类药物处方设计时需要考虑的理化因素

蛋白药物的结构	影响稳定性的因素	影响溶解度的试剂	其他
等电点	pH，温度		多晶型
分子量	光照度，氧	去垢剂	立体异构体
氨基酸组成	金属离子，冻融	盐	过滤介质相容性
二硫键	机械力	金属离子	剪切
光谱性质		pH	表面变性

2.1.2 需要考查的初期变量

在处方前研究初期，可能需针对不同蛋白浓度考察 5～8 个变量，包括缓冲盐种类、pH 范围、离子强度、温度、最初剪切力、表面变性和搅拌等。

2.1.3 处方前研究的一些试验条件

1. 蛋白浓度 考查不同浓度范围的蛋白药物活性，各种蛋白的浓度范围从纳克、微克到毫克水平不等。

2. pH 范围和缓冲液 药剂学研究中制剂的 pH 控制很重要。选择正确的 pH 是制备得到稳定产品的重要因素之一。pH 对蛋白类药物溶解度和稳定性影响较大，具体 pH 应在最终处方研究中确定。缓冲液应选择蛋白和多肽类药物最

2.1.2 The Need to Examine the Initial Variables

5-8 variables in the formulation may need to be considered before the beginning of the study, including the buffer salt type, pH range, ionic strength, temperature, initial shear stress, surface degeneration and stirring, etc. Different protein concerntrations should also be studied.

2.1.3 The Test Conditions before the Initial Prescription

1. Protein concentration Different protein concentration levels ranging from nanograms, micrograms to milligrams should be used to test the activities of protein drug.

2. pH value range and buffer In general, the control of pH value of the preparation is very important. Choosing the right pH is an important factor to ensure a stable product. pH value imposes a great influence on the solubility and stability of protein drugs. The specific pH value should be determined in the final formulation study. One should select the buffer system which has the best pH range of physiological buffer for protein and polypeptide drugs. In general, the pH range of buffer is: phosphate (6.2-8.2), citrate (2.0-6.2), Tris (7.1-9.1), etc.

佳 pH 范围的生理缓冲液。常用缓冲液及其 pH 范围：磷酸盐（6.2～8.2），枸橼酸盐（2.0～6.2），Tris（7.1～9.1）等。

3. 螯合剂　蛋白发酵及大批量纯化过程中与不锈钢、铜或铁表面的接触，在其表面可能有残留金属离子，因此，有必要使用螯合剂去除残留的金属离子，抑制药物氧化。常用螯合剂为 EDTA 盐，剂量范围为 0.01%～0.05%。

3. Chelator The formulation may contain residual metal ions from protein fermentation and large purification process inside the stainless steel, copper or iron surface. Therefore, the use of chelator is necessary. Commonly used chelator is EDTA salt. The common dosage of chelator ranges from 0.01% to 0.05%.

4. 抗氧化剂　氧化是蛋白质降解的主要原因之一，所以在处方中需要使用一些特殊的抗氧化剂，如抗坏血酸、α- 生育酚、硫代甘油等。常用抗氧化剂的剂量范围为 0.05%～0.1%。

4. Antioxidants Oxidation is one of the main causes of protein drug degradation. One needs to use special antioxidants such as ascorbic acid, α -tocopherol, and thio-glycerol. The dosage of antioxidants often range from 0.05% to 0.1%.

5. 防腐剂　在多剂量制剂中需加入防腐剂。防腐剂必须能杀死或抑制污染和掺杂在产品中的细菌，防腐剂的筛选必须与蛋白及处方中其他辅料相容。蛋白类药物制剂中常用的防腐剂有苯酚（0.3%～0.5%）、苯甲醇（1.0%～3.0%）、三氯叔丁醇（0.3%～0.5%）、间甲酚（0.3%～0.5%）、尼泊金类等。

5. Preservatives Preservaties are often used for formulations of multiple dosages. Preservatives should be able to kill or inhibit pollution and dope the bacteria in the product. The screening of the preservative and the prescription of protein must be compatible in materials. The common preservatives in protein drugs are phenol (0.3% -0.5%), benzyl alcohol (1.0%-3.0%), trichloro-tert-butyl alcohol (0.3% -0.5%), M-cresol (0.3%- 0.5%), and Nipagin.

6. 降解产物与降解机制研究　在不同温度条件下进行的初期稳定性试验过程中，观察和记录下蛋白的物理变化和化学降解。在有关物质分析及降解机制研究中，应选择可定量和定性区别物理变化和化学降解的方法。从处方前研究总结的数据中，可获得以稳定性为基础的潜在先导物，研发的下一阶段（处方研究）应考虑所有参数，后者可能获得一个或多个工业化的稳定处方并市场化。

6. Study on degradation products and degradation mechanism During the initial stability test at different temperature levels, one should observe and record the physical changes of protein and chemical degradation. In the analysis of related substances and study on the degradation mechanism, the researchers must select those quantitative and qualitative differences between physical changes and chemical degradation methods. According to the data obtained from the preformulation study, one may get potential lead compound based on its stability. The research and development of the next stage (prescription) should involve tests for all the parameters, leading to one or more formulation of commercialization and market stability.

2.2 处方设计

处方设计就是设计出药物和不同辅料混合制备的剂型以便于患者使用。处方设计应根据处方前研究获得的支持性数据来进行。在设计时需考虑的主要因素有：活性药物与辅料的初步相容性试验；处方前研究中的初步降解产物和有关物质；影响稳定性的因素；稳定性的分析方法等。掌握足够多的性质后，再选择大批量活性药物进行处方设计，进一步了解制剂的均匀性和批间重复性。

2.2.1 处方组成

蛋白多肽类药物注射剂，可用于肌内注射或静脉输注等途径，其处方组成和要求与一般注射剂基本相同。

2.2.2 注射剂中蛋白类药物的稳定化

由于蛋白类药物结构的特殊性，可能出现氧化、脱酰胺、水解、二硫键交换等一系列化学反应和聚集、沉淀、变性等物理不稳定性。因此对于蛋白类药物的稳定化研究很重要，通过在处方中加入各类稳定剂是生物药物制剂中常用的稳定化方法。蛋白类药物的常用稳定剂有以下几类。

1. *缓冲液* pH对蛋白类药物的稳定性和溶解度均有重要影响。大多数蛋白类药物在pH 4～10范围内是比较稳定的，在等电点时最稳定。因此，应采用适当的缓冲系统，提高蛋白质在溶液中的稳定性。常用的缓冲剂包括枸橼酸盐缓冲液、磷酸缓冲液等。缓冲液类除了影响蛋白质的稳定性外，其浓度对蛋白质的溶解度与聚集也有很大

2.2 Recipes Design

The recipes design should be based on the supporting data obtained from the research of the formulation. The main factors to be considered include: the active drug and a preliminary compatibility test of the articles; Preliminary preformulation studies of degradation products; the factors to affect the stability; and the operation of the stabilization. After understanding enough nature, one can choose large quantities of active drug to design the recipes, learn more about preparations uniformity and reproducibility between batches.

2.2.1 Composed of Prescriptions

The injection formulation of protein and polypeptide drug can be used in many ways, such as intramuscular injection or intravenous infusion. The prescription requirements and the general composition are the same as for general injection.

2.2.2 The Stability of Protein Drugs in Injection

Because of the special structure of protein drugs, protein may occur such as oxidation, de-amide, hydrolysis, disulfide exchange and aggregation of a series of chemical reactions, precipitation, denaturation and other physical instability. So the stabilization of protein drugs study is an important step. By adding all kinds of materials, changing the nature of protein drugs is common stabilization assays used in pharmaceutical preparations. The categories of the stabilizer of protein drugs as follows:

1. Buffer pH value has an important effect on protein stability and solubility of drugs. Most protein drugs are relatively stable in the range of pH 4-10, the most stable value is at its isoelectric point. Therefore, an appropriate buffer system should be used to enhance the stability of the protein in solution. Commonly used buffers include citrate buffer salt, phosphate buffer salt, etc. Buffer salts in addition to affecting the stability of the protein, its concentration on protein solubility and aggregation have a significant impact. The most common buffer salt concentration should be in the range of 10-20mmol · L^{-1}, the increasing buffer salt

影响。最常用的缓冲液浓度应在 10～20mmol·L^{-1} 范围内，缓冲液浓度的增加会增强注射时的疼痛感。

concentration will increase the pain when injected.

2. *表面活性剂* 蛋白类药物对表面活性剂非常敏感。含长链脂肪酸的离子型表面活性剂或其中性化合物均可引起蛋白质的变性。所以，在蛋白质药物如 G-CSF、干扰素 α-2b、组织纤溶酶原激活素等制剂中加入少量的非离子表面活性剂可防止蛋白质表面吸附、聚集和沉淀，究其机理可能是表面活性剂倾向于排列在气 - 液界面上，从而使蛋白质离开界面来抑制蛋白质的变性。常用的表面活性剂有吐温 80、SDS、Pluronic、泊洛沙姆 407 等。

2. Surfactants The protein drug is very sensitive to surfactants. Ionic surfactant containing long-chain fatty acids and neutral compounds can cause protein denaturation. Therefore, protein drugs, such as G-CSF, α -2b interferon, tissue plasminogen activator factor fibrinolytic agents such as adding a small amount of non-ionic surfactant, can prevent protein aggregation and precipitation. The mechanism may be due to surfactants to be arranged in the gas-liquid interface, allowing the protein to leave the interface to inhibit protein denaturation. The non-ionic surfactants commonly used are Tween 80, SDS, Pluronic, Poloxamer 407 and so on.

3. *糖和多元醇* 糖和多元醇属于非特异性蛋白质稳定剂，在低浓度下迫使更多的水分子围绕在蛋白质周围，从而提高蛋白质药物的稳定性。常用的糖类有蔗糖、葡萄糖、海藻糖、麦芽糖等，多元醇有甘油、甘露醇、肌醇、山梨醇（浓度 1%～10%）等。在蛋白质药物冻干过程中常作为冻干保护剂使用。糖和多元醇对冻结的影响取决于蛋白质的浓度与种类。

3. Sugar and polyol Sugar and polyol are non-specific protein stabilizing agents, which can force more water molecules around the protein in low concentration and increase the protein stability of drugs in the water. Common sugars are sucrose, glucose, trehalose and maltose, etc., polyols are glycerol, mannitol, inositol, sorbitol (concentration 1% -10%), and so on. Sugar and the stabilizing role of polyols is closely related to its concentration, the stability of different sugars and polyols depends on the types of protein.

4. *无机盐类* 无机盐类对蛋白质的稳定性和溶解度有比较复杂的影响。无机盐是起到稳定蛋白质的作用还是破坏其稳定性，主要取决于盐的浓度、种类、离子相互作用的性质及蛋白质的电荷。低浓度的盐通过非特异性静电作用来提高蛋白质的稳定性。如 Na^+、K^+、$(CH_3)N^+$、NH_4^+、SO_4^{2-}、HPO_4^{2-}、$CHCOO^-$ 等能增加溶液的离子强度，提高疏水作用，降低疏水基团溶解度，使蛋白质发生盐析；它们还能使水分子聚集在蛋白质

4. Inorganic salts Inorganic salts have more complex effects on the stability and solubility of proteins. Salt can play a role in stabilizing protein; can sometimes damage its stability, which depends on concentration, salt type, nature of the interaction and protein ion charge. The salt of low concentration can increase protein stability by non-specific electrostatic interaction. Such as Na^+、K^+、$(CH_3)N^+$、NH_4^+、SO_4^{2-}、HPO_4^{2-}、$CHCOO^-$ etc., can increase the ionic strength, improve the hydrophobic effect, and reduce the solubility of hydrophobic groups to make proteins salting. They also can make water molecules surrounding the protein preferentially hydrated, which are closer to the protein stability. Inorganic salts commonlly used is NaCl.

周围被优先水化，使蛋白质更加紧密稳定。常使用的无机盐为NaCl。

5. 氨基酸及其盐类 某些氨基酸如组氨酸、甘氨酸、亮氨酸、谷氨酸、天冬氨酸钠等可以增加蛋白质药物在一定pH下溶剂中的溶解度，抑制聚集，提高其稳定性。如：在组织纤溶酶原激活剂中加入带正电荷的精氨酸能增加其在所需pH值下的溶解度。

5. Amino acid and salts Certain amino acids such as histidine, glycine, leucine, glutamic acid, aspartic acid and sodium can increase the solubility of the protein drug in a certain pH, inhibit aggregation and enhance its stability. For example: the positively charged arginine added to tissue plasminogen activator can increase its solubility in the desired pH value.

6. 大分子化合物 某些大分子化合物具有稳定蛋白质的作用。其机制可能是通过大分子的表面活性、蛋白质之间相互作用的空间隐蔽以及提高黏度来限制蛋白质运动或通过优先吸附于大分子起到稳定作用，如重组人血清白蛋白（rHSA）在许多蛋白质类药物制剂中常作为稳定剂使用，主要是通过处方中的HSA优先被吸附从而保护蛋白药物免遭吸附。近年来也有采用环糊精制成包合物来增加蛋白质药物的溶解度，如用2-羟丙基-β-环糊精（2-HP-β-CD）包合药物来抑制人体生长激素（hGH）的界面变性、增加羊生长激素（OGH）的稳定性、抑制重组人角化细胞生长因子（rhKGF）的聚集等，且其本身又是增溶剂，可静脉注射，因此2-HP-β-CD是较有前途的稳定剂。

6. Macromolecular compounds Many macromolecular compounds play significant role in protein stability. The mechanism may be that many macromolecular compounds conceal space of macromolecular surface activity and protein interactions, increase the viscosity to restrict protein movement and preferential adsorption to the molecule. For example: recombinant human serum albumin (rHSA) has acted as a stabilizer in many biotechnologies derived proteins in pharmaceutical preparations, which are adsorbed preferentially through HSA in the formulation to protect the protein drugs. In recent years, cyclodextrins (CD) has also been made into clathrates to increase the solubility of protein drugs. For example, One can use 2- hydroxypropyl-β-CD (2-HP-β-CD) inclusions to inhibit interface variability of Human Serum Albumin (hGH)、increase the stability of the ovine growth hormone (OGH) and inhibit the aggregation of recombinant human keratinocyte growth factor (rhKGF). 2-HP-β-CD is a solubilizer itself and can be used as intravenous injection, so 2-HP-β-CD is a more promising stabilizer.

7. 聚合物类 一定浓度的聚乙二醇类（PEG）、聚乙烯比咯烷酮类（PVP）常作为蛋白质的低温保护剂和沉淀结晶剂，抑制蛋白聚集。不同分子量的PEG和PVP产生的作用不同，如PEG300可抑制rhKGF的聚集；PEG200、400、600和1000可用

7. Polymer The polyethylene glycol and polyvinylpyrrolidone of certain concentration are often used as cryoprotectant and precipitation crystallization agent to inhibit protein aggregation. The results show that PEG and PVP of different molecular weight have different effection, such as, PEG300 inhibiting aggregation of rhKGF, PEG200, 400, 600 and 1 000 to stabilizing BSA and lysozyme.

于稳定 BSA 和溶菌酶。

8. 抗氧化剂 氧化反应是蛋白降解的主要原因之一，所以在蛋白类药物制剂处方中需要使用抗氧化剂，常用的抗氧化剂有抗坏血酸、硫代甘油、α- 生育酚等。浓度范围为 0.05%～0.1%。

8. Antioxidants Oxidation reaction is one of the main reasons of protein degradation, so the antioxidants are needed in Protein preparations prescription drugs. The common antioxidants are ascorbic acid, thioglycerol, α-tocopherol, etc. The concentrations range from 0.05% -0.1%.

9. 还原剂 几种巯基化合物可抑制二硫键形成，防止聚集。

9. Reductant Several thiol compounds can inhibit the formation of disulfide bond to prevent aggregation.

10. 金属离子 某些金属离子，如钙、镁、锌与蛋白质结合，可起到稳定蛋白构象的作用，使蛋白质结构更加紧实、稳定。不同金属离子的稳定作用因离子的种类、浓度不同而不同，实际应用中应根据稳定性实验结果来选择金属离子的种类和浓度。

10. Metalions Certain metal ions, such as calcium, magnesium, zinc banding with protein, make the whole protein structure more tight and stable. The stabilizing effect of different metal ions varies depending on ion type, concentration. The type and concentration of metal ions should be selected by stability test in practical application.

表 3-2 总结了蛋白、多肽类药物处方设计中常涉及的稳定剂及其作用。

Table 3-2 summarizes the common stabilizers and their roles involved in formulation design of protein and polypeptide drugs.

表 3-2 蛋白处方中使用的稳定剂

稳定剂	作用
缓冲液	
枸橼酸盐缓冲液，磷酸缓冲液，Tris，琥珀酸盐	稳定剂；阻聚剂
表面活性剂	
吐温 20，吐温 80，SDS	阻聚剂
泊洛沙姆 407	防止变性，使混浊稳定
糖和多元醇	
山梨醇	阻聚剂，防止变性
甘露醇	冻干保护剂
甘油	抗氧化剂
蔗糖	稳定剂，冻干保护剂
葡萄糖	强化构象
丙烯乙二醇	阻聚剂
乙烯乙二醇	
无机盐类	
NaCl	稳定剂

续表

稳定剂	作用
氨基酸	
甘氨酸	稳定剂
丙氨酸	增溶剂
精氨酸	增溶剂
亮氨酸	阻聚剂
谷氨酸	热稳定剂
门冬氨酸	同分异构抑制剂
大分子化合物	
人血清白蛋白（HSA）、重组人血清白蛋白（rHSA）	阻止表面吸附，构象稳定剂，络合剂，冷冻保护剂
2- 羟丙基 -β- 环糊精	增溶剂，阻聚剂，防止变性
聚合物	
聚乙二醇	稳定剂
聚乙烯比咯烷酮 10，24，40	阻聚剂
抗氧化剂	
抗坏血酸，盐酸半胱氨酸，1- 硫代山梨醇，谷胱甘肽	延缓氧化
还原剂	
几种巯基化合物	抑制二硫键形成，防止聚集
金属离子	
Ca^{2+}，Mg^{2+}，Mn^{2+}	稳定蛋白构象

2.3 蛋白和多肽类药物注射剂的工艺研究

2.3.1 溶液型注射剂的制备工艺

蛋白和多肽类药物注射剂的制备工艺与一般注射剂基本相同。制备工艺流程包括注射用水的制备、安瓿的前处理、注射液的配制、无菌过滤、灌封、检查、印字包装、成品等步骤。

在制备工艺过程中，影响成品质量的因素较多，提高该类制剂稳定性的方法有：采用温和的生产条件如对温度、机械搅拌强度和有机溶剂的选择，对无菌条件的控制，容器的吸附

2.3 Processing of Protein and Polypeptide as Injection Drugs

2.3.1 The Preparation of Injection Formulation

The preparation process of protein and polypeptide injection drugs is based on the same process as for general injection. The preparation process includes the preparation of WFI, pre-treatment of ampoules, the injection preparation, sterile filtration, filling and sealing, inspection, printed packaging and finished products and other steps.

In the preparation process, there have a lot of factors can affect the quality of finished products.Methods to improve stability preparations as follows: use of mild conditions of production such as temperature, mechanical agitation intensity and the choice of organic solvent ,the control of the sterile condition , container adsorption effects, moisture control, low temperature refrigeration,

效应，水分控制，低温冷藏等。在制备过程中需要注意的单元操作是：无菌蛋白、多肽药物制剂仅能通过无菌过滤和γ射线照射制备，而不能使用高压蒸汽灭菌，所以膜过滤对于生物药物产品的开发至关重要。可选择具有最小吸附且与产品具有最大相容性的膜滤器。蛋白处方最常使用的过滤介质有纤维素醋酸硝酸混合酯、聚砜、尼龙 66 等。

ect. We should note the unit operations in the preparation process: sterile proteins and polypeptide drugs products only go through sterile filtration and preparation of γ-ray irradiation, and can not use high-pressure steam sterilization. Therefore, the membrane filter is essential for the development of bio-pharmaceutical products. We can choose the membrane filters with low absorption and the high compatibility.The common filter mediums are cellulose acetate nitrate mixed ester, polysulfone and nylon 66, etc.

2.3.2 蛋白质药物冻干制剂的制备工艺

蛋白和多肽类药物在有水参与的条件下可发生水解、脱酰胺等一系列化学反应，因此，在一些蛋白药物不能制备成溶液型制剂时，往往可采用冷冻干燥的工艺使其形成固体状态以延缓溶液中常见的分解作用，解决这类制剂的稳定性问题。

制备冻干无菌粉末前药液的配制与溶液型注射剂基本相同，药液在冻干前需经药液配制、无菌过滤、分装等处理过程，再进行冷冻干燥。由冷冻干燥原理可知，冻干无菌粉末的制备工艺可分为预冻、减压、升华、干燥等过程。用冷冻干燥法制备蛋白质类药物制剂时，需考虑配方组成对冻结、干燥过程的影响和对保证蛋白质药物在干燥状态下的长期稳定性的影响；预冻时间和预冻速率、最高与最低干燥温度、干燥时间等因素对产品的外观和质量的影响。因此，在工艺研究中应对这些工艺参数进行重点考察。

冻干状态与溶液状态的稳定机制不同，冻干状态的稳定机制可能是除去水化层后，辅料作为水的替代物结合在蛋白质上。虽然冻干可以使大多

2.3.2 The Preparation of Lyophilized Formulations

Protein and polypeptide drugs in the solution may have a series of chemical reactions such as hydrolysis, amide, etc. Therefore, drying can improve stability of these drugs significantly. When some protein drugs can not be made into soluble preparations, we often use the freeze-drying (lyophilizing) process to form a solid state to improve the stability of such preparations.

The preparation of solution before being lyophilized sterile powder is the same as that of injection solution. The solution needs to be under sterile filtration and filling before lyophilization. According to the principle of lyophilization, the process can be separated into pre-freezing, vacuum, distillation, and drying processes. One should consider two questions by using lyophilization techniques to produce protein drugs. First, one needs to choose appropriate recipe so that it meets the long-term stability requirements in the dry state. Second, one has to consider the process of lyophilization potentially exerting untoward effects on protein parameters, such as the time of pre-frozen, the rate of pre-frozen, the high and low drying temperature, drying time, lyophilized product appearance.

The stability mechanisms are different for freeze-dried state and solution state. In the process for lyophilization substances are required as substitute for water binding to proteins after the removal of hydration layers. Although most protein drugs can be stable after

数蛋白类药物稳定，但在冻干过程中有些蛋白质药物反而失去活性，主要原因是：在预冻阶段，从液态到固态的相变过程中，蛋白质易因包在周围的水分子被除去而失活；高浓度的盐、缓冲组分的结晶或缓冲液 pKa 对温度敏感而导致 pH 变化以及浓缩时蛋白质溶解度有限等均能导致蛋白质药物的二、三级结构发生变化而失活。因此，在蛋白和多肽类药物冻干过程中常加入冻干保护剂来改善产品外观和稳定性，如甘露醇、山梨醇、蔗糖、葡萄糖、右旋糖酐等糖和多元醇类物质，此类物质可以取代水与多肽形成氢键来稳定多肽构象，并能提高冻干产品的玻璃化温度。其他保护剂还有聚维酮、聚乙二醇、吐温 80、氨基酸等。此外，在选择冻干产品的缓冲剂时，需考虑到温度对 pH 和缓冲系统溶解度的影响。单剂量的蛋白质药物的量一般很少，冻干过程中和除去水分后，瓶内只剩很少的物质，也不会形成冻干饼块，因此需用填充剂来改善产品外观。

freeze-drying process, some protein drugs may lose certain activity. The main reasons include: the removal of water around the protein may change the activity in the transition from liquid to solid phase; high concentrations of salt and the crystallization of buffer components; buffer pKa sensitive to the temperature to cause pH changes, and the limited solubility of the protein also leading to protein inactivation. Therefore, in the process of lyophilization protective agents may be added to improve the appearance and stability of products. The protective agenst include mannitol, sorbitol, sucrose, glucose, and dextran. Sugars and polyols can replace the water in the polypeptide to form hydrogen bonds thus stabilizing the polypeptide conformation, and can improve the glass transition temperature of lyophilized products. There are other protective agents such as PVP, PEG, Tween-80 and amino acids. In addition, the buffer system for lyophilization should be selected by taking into account the temperature effects on the pH and the solubility. The amount of a single dose of protein drugs is usually small; there are only a few substances in the bottle after lyophilization. Fillers may be used to improve the appearance of lyophilized products.

溶液中的成分也可以影响冷冻干燥过程中与热有关的工艺参数。冻干过程中需要控制与热有关的参数，例如制剂的冷冻温度、最高和最低干燥温度、使产品发生降解的温度、使饼状物熔化或坍塌的温度等，从而使产品冷冻适度、不降解，饼状物不熔化、不坍塌。此外，还应考虑药物的含水量与饼状物的物理状态（无定形或晶形）。其物理状态与冷冻过程的温度及添加辅料种类有关。在冷冻过程和贮存过程中辅料结晶现象会改变冻干固体中的水分分布，导致药物所

The composition of the solution can also influence the process of freeze-dried and heat-related parameters. One need to control the process of lyophilization and heat-related parameters, such as the temperature of freezing preparation; the temperature of degradation; the temperature of the pie melting or collapsing, so that the product freezes moderately without degradation, and the pie will not melt and collapse. In addition, the water content of the drug and physical state of pie (not qualitative or crystal) should be considered. Its physical state is related to the freezing temperature and the added accessories. During the process of freezing or storage, accessories crystallization will change the distribution of water in the freeze-dried soilds and lead to the moisture content increase. The increase of water will reduce the physical stability of the pie material during storage and

处微环境的水分含量增加，水分的增加会降低饼状物的物理稳定性并可能导致在贮存过程中饼状物坍塌，影响产品外观，同时药物的稳定性下降。为提高冻干产品中药物的稳定性，可以选择能保持无定形状态的辅料如麦芽糊精，严格控制产品的含水量，保证产品质量。

may lead to the collapse of pie thing, affect the appearance of the product, while lead to the stability of the drugs decrease. To improve the stability of the lyophilized drug, one may also choose certain fittings such as maltodextrin to maintain the amorphous state. Water content should be strictly controlled to ensure product quality.

2.3.3 鲑降钙素注射剂制备实例

【处方】鲑降钙素（sCT）50U，0.1mol/L 乙酸水溶液适量，0.1mol/L NaOH 溶液适量。

【制备工艺】配制 0.1mol/L 乙酸溶液及 0.1mol/L NaOH 溶液，将两种溶液按处方中比例混合，并调节 pH 至酸性。在无菌操作条件下用 G_6 垂熔漏斗抽滤，得滤液。取少量滤液溶解 sCT，然后将药液经一次性除菌器过滤至滤液中，摇匀、灌封，即得成品。

【用途】用于治疗老年骨质疏松症、绝经后骨质疏松症、骨转移癌致高钙血症。

2.3.3 Case Analysis: Preparation of Salmon Calcitonin (sCT) Injection

【Prescription】

sCT	50U
0.1mol/L HAc	moderate
0.1mol/L NaOH	moderate

【Preparation】

Prepare 0.1mol/L HAc and 0.1mol/L NaOH, mixing the two solutions according to the ratio of formulation, adjust the pH. Under aseptic conditions, collect the filtrate from the filter. Take a small amount of filtrate and dissolve sCT, then pass it through microporous membrane in to filtrate, shake and filling.

【Use】

sCT injection used for the treatment of senile osteoporosis, postmenopausal osteoporosis, metastatic bone cancer induced hypercalcemia.

3 生物技术药物新型给药系统

3 New Biotech Drug Delivery System

随着生物技术的发展，生物技术药物在人类疾病治疗中的地位日趋重要。与传统的化学合成药物和中草药提取物相比，其与体内正常生理物质接近，更容易被机体吸收，药理活性高、毒性低。但其同时具有体内半衰期短、清除率高、分子量大、不易通过生物屏障、易受体内酶破坏、非注射给药生物利用度低等特点。对于此类药物研究工作的重心首先应该“保

With the development of biotechnology, biotech drugs have become more important in the treatment of human disease. Compared with the traditional synthetic chemical drugs, biotech drugs use similar materials to molecules inside the human body, therefore are easier to be absorbed by the body and have high pharmacological efficacy and low toxicity. Biotech drug is also characterized as having a short half-life in vivo, high clearance rate, and large molecular weight making it hard to go through the biological barrier, easy to be degraded by enzymes, and low bioavailability after non-injection administration. Therefore, the basic formulations of

证可开发性"，因此，此类药物的基本剂型主要为溶液型注射剂和冻干粉针剂。

these drugs are mainly injection solution and lyophilized powder.

随着制药技术、药用高分子材料、药物新剂型的发展，目前药物制剂研究已经进入了一个全新的药物传递系统（drug delivery system，DDS）时代。由于新型给药系统用于药物的临床给药形式具有令人非常满意的特性，例如靶向传递至特定组织，降低毒性，降低用药量，降低给药频率，更好的安全性、有效性和稳定性，注射时疼痛感减少或消失等，一直是世界上许多大型科研机构和制药企业的研究重点。生物技术药物新型给药系统的研究也一直是药剂学领域的热点之一，并在多种新剂型、新技术上取得了长足的进步。在注射给药途径方面，如缓控释微球、植入剂、原位微球、脂质体等剂型均取得了一系列进展；而在非注射给药途径方面，如胃肠道、黏膜、肺部、皮肤等途径给药的新技术、新剂型的研究开发亦非常活跃。经过近年来的不断发展，有的生物技术药物新型给药系统已经进入实用化阶段，大大拓展了其临床应用领域。相信在不久的将来会开发出更多安全、高效、低毒、稳定、使用方便的生物技术药物制剂，提高其生物利用度和患者的生活质量。

With the development of medicinal polymer materials and new formulation technologies, the development of drug formulations has entered a new era in drug delivery systems (DDS). Because the new DDS has showed very satisfactory characteristics in clinical administration, such as target delivery to specific tissues, less toxicity, lower dosage, less frequency of dosing, better safety, efficacy and stability, and less painful at injection. These advantages have made DDS the research priorities for many large research institutes and pharmaceutical companies in the world. The new DDS of biotech drugs has also been one of the hot fields of pharmacy, and has made a considerable progress in new formulations and technology. Aspects of the route of injection administration, such as sustained (controlled)-release microspheres, implants, in situ forming microspheres, liposomes and other formulations have made a series of progress; In terms of non-injection routes of administration, such as the gastrointestinal tract, mucous membrane, lung, skin and other means of new drug delivery technologies, new forms of research and development are very active. After years of continuous development of biotech drugs, some new DDS has entered the practical stage, which greatly expands its clinical applications. We believe that in the future we will have more new pharmaceutical preparations of biotechnology developed, which are more safe, efficient, low toxicity, stable, easy to use to improve its bioavailability and the quality of patients' life.

3.1 新型注射给药系统

3.1 New Injection Drug Delivery System

生物技术药物普通注射剂需频繁注射，患者顺应性差，限制了其在临床上的广泛应用。近年来，随着制剂新技术及新型药物载体材料尤其是生物可降解型材料的发展，

Regular biotech drugs need inject frequently and the patient compliance is poor, which limit the clinical application. In recent years, with the development of new technologies and new materials, especially biodegradable material, new forms of drugs like microspheres are widely used in the injection of new biotech drugs, which improve

使得微球等药物新剂型被更多地用于生物技术药物新型注射剂的制备，提高了生物技术药物的稳定性和生物利用度，使药物在治疗上更加有效和安全，病人也更易使用。高分子材料的发展带动了生物技术药物新型注射给药系统的发展。

the stability and bioavailability of biotech drugs. The drug is more effective and safe and easier to be used by the patients. So the development of polymer materials leads to the development of new injection DDS of biotech drugs.

3.1.1 缓控释微球注射给药系统

微球（microspheres）是药物溶解或分散于高分子材料中形成的微小球形或类球形实体，粒径范围一般为1～500um。1986年法国Ipsen生物技术公司上市了第一种缓释多肽微球制剂——促黄体生成素释放激素（LHRH）类似物曲普瑞林（triptorelin），可缓释1个月，此后又陆续生产了醋酸亮丙瑞林（leuprolide acetate）、布舍瑞林（buserelin）和meterelin肌内注射微球，临床上用于治疗一些激素依赖性疾病，如前列腺癌、乳腺癌、子宫肌瘤、子宫内膜异位等，使注射用缓释微球得到了迅速发展。

3.1.1 Controlled Release Microspheres Patenteral Drug Delivery System

Microspheres is a small spherical or solid spherical with drug dissolved or dispersed in the polymer material, the particle size range is 1-500μm. In 1986, Ipsen Biotechnology in France first listed a sustained-release polypeptide microsphere formulation, luteinizing hormone releasing hormone (LHRH) –triptorelin, which can release for a month continuously. Leuprolide acetate, buserelin and meterelin intramuscular microspheres were produced after triptorelin for the treatment of a number of clinically hormone-dependent diseases, such as prostate cancer, breast cancer, uterine fibroids, endometriosis, etc., which made sustained release microspheres for injection developing rapidly.

1. *常用载体材料* 微球的处方主要由药物、载体材料和稳定剂三部分组成，其中载体材料是处方设计的核心，对微球的成型和药物的缓释起到关键作用。微球载体材料的类型按其来源可分为天然高分子材料、半合成高分子材料和合成高分子材料。其中合成高分子材料中的聚酯类是目前研究最多、应用最广泛的生物可降解型材料，常用的有聚乳酸（PLA）或乙酸/羟基乙酸共聚物（PLGA）。PLGA无毒，由乳酸及羟基乙酸聚合而成，在人体内无积聚问题，具有良好的生物相容性和生物

1. Common carrier materials Microspheres prescription is mainly composed of drug, carrier material and stabilizer. Carrier material is the core of formulation design, which plays a key role in the shape of the microspheres and sustained release of drugs. According to their source, the type of microspheres carrier materials can be divided into natural polymer materials, semi-synthetic polymer materials and synthetic polymer materials. Polyester of synthetic polymer materials is most widely studied and used. Commonly biodegradable materials are polylactic acid (PLA) or poly (lactic-*co*-glycolic acid) (PLGA). PLGA is non-toxic, which is polymerized by lactic acid and glycolic acid without the accumulation of problems in the human body, and it has good bioavailability and biodegradation. Changes in the ratio or molecular weight of lactic acid and glycolic acid may obtain the biodegradable nature of the material at different

降解性。改变乳酸与羟基乙酸的比例或分子量，可得到不同时间生物可降解性质的材料。目前采用PLGA作为骨架材料研究缓释微球注射剂的蛋白和多肽类药物有胰岛素、奥曲肽（octreotide）、γ-干扰素（rhIFN-γ）、白介素（IL-α）、人粒细胞巨噬细胞集落刺激因子（GM-CSF）、过氧化物歧化酶（SOD）等。

times. Currently there are some protein and polypeptide drugs such as insulin, octreotide, γ-interferon (rhIFN-γ), interleukin (IL-α), human granulocyte macrophage colony Stimulating factor (GM-CSF), superoxide dismutase (SOD) have already become a model drug in the study of sustained-release microspheres injection by using PLGA as a reinforcing material.

2. *制备方法*　用于制备生物大分子药物缓释微球的方法主要有相分离法、乳化液中干燥法、喷雾干燥法、低温喷雾提取法、超临界流体技术等，根据载体材料的不同性质可选择不同的制备方法。

2. Preparation　The methods of preparing sustained release microspheres for biological macromolecules include phase separation method, in-liquid drying process, spray drying method, low temperature spray extraction, supercritical extraction technology, etc. One can choose different preparation methods according to different nature of the carrier materials.

（1）相分离法　这是在药物与材料的混合溶液中，加入另外一种物质或不良溶剂，使药物与材料的溶解度降低，在一定条件下自溶液中产生新相析出的制备方法。目前相分离法制备微球的工艺普遍存在一些问题，如微粒聚集和粘连，因此采用此法制备微球时应注意工艺过程和参数条件的控制。

(1) Phase separation method　Phase separation is adding another substance or poor solvent into a mixture of drugs and materials to make the solubility of drug and material reduced, which generate a new phase from solution under certain conditions. At present, microspheres prepared by the phase separation process still have some problems, such as particle aggregation and adhesion, and processing and controlling parameter should be emphasized.

（2）乳化-液中干燥法　又称为乳化-溶剂蒸发法，是指从乳状液中除去分散相中的挥发性溶剂以制备微球的方法。按操作可分为连续干燥法、间歇干燥法和复乳法。液中干燥法是制备生物大分子药物微球的常用方法，其特点为药物包封率较高、微球粒径均匀、药物活性损失小、设备和工艺简单，但缺点为首日突释效应明显、难放大生产。目前用此法研究制备的药物有亮丙瑞林、白细胞介素、人生长激素、环孢素、促红细胞生长素等。

(2) Emulsion in-liquid drying process　It is also known as emulsion solvent evaporation method, which refers to the method of preparing microspheres by removing the volatile solvents in the dispersed phase from the emulsion. According to the operation, it can be divided into continuous drying, intermittent drying and complex emulsion. In-liquid drying is the common method of preparation of biological macromolecules microspheres, which has the features of high drug encapsulation efficiency, less activity loss, simple equipment and technology, but the drawback is that the first day of the burst effect is obvious, and more difficult to enlarge the production. At present, leuprolide, interleukin, human growth hormone, cyclosporine, and erythropoietin (EPO) and so on have been prepared by this method.

（3）喷雾干燥法　将生物大分子药物及其稳定剂的混合粉末（或水溶液）与溶有高分子聚合物的有机溶液混合形成混悬液（或乳浊液），将此混悬液（或乳浊液）经喷嘴雾化干燥制得微球，粒径范围通常为5～600μm。如用于治疗帕金森病的药物溴隐亭（Bromocriptine）PLGA微球注射剂（商品名：Parlodel LAR，Sandoz），即是用喷雾干燥法制备微球制剂的早期范例。喷雾干燥法的工艺影响因素包括混合液的黏度、均匀性、药物及聚合物的浓度、喷雾速率、喷雾方法及干燥速率等。微球的干燥过程中常在囊材中加入聚乙二醇作抗黏剂，可降低微球带电而减少粘连。文献曾报道采用双喷嘴喷雾干燥装置，一个喷嘴喷出药物和生物可降解材料PLGA的乳浊液，另一喷嘴同时喷出5%的甘露醇水溶液，甘露醇作为抗黏剂，将微球包层，这样可避免普通喷雾干燥装置制备微球过程中易黏附器壁的缺陷，增加了微球的收率，有效减少生物大分子药物的活性损失。

（3）Spray drying method Biological macromolecular drugs and its stabilizer powder mixture (or solution) and dissolved organic polymer solution are mixed to form suspension (or emulsion). This suspension (or emulsion) is transformed into microspheres by the atomization drying system, whose particle size is commonly in the range of 5-600μm. Bromocriptine PLGA microspheres injection that is used for the treatment of Parkinson's disease (brand name: Parlodel LAR, Sandoz) prepared by spray drying method is an early example of microspheres preparation. The factors of spray drying process include the viscosity of the mixture, uniformity, drug and polymer concentration, spray rate, spray method and the drying rate and so on. Drying process for microspheres often adding PEG as anti-adherent in the capsular material can reduce the adhesion of charged microspheres .In order to reduce the loss of biological activity of macromolecular drugs, reported in the literature, by using double-nozzle spray-drying device, a nozzle spray drug and PLGA polymer emulsion, while the other spray nozzle 5% mannitol solution, cladding the microspheres, which can avoid microspheres easily adhering to the wall in the common spray drying device, increasing the yield of microspheres.

（4）低温喷雾提取法　将生物大分子药物与保护剂的均匀粉末加至溶解PLGA的有机溶剂中超声分散形成混悬液，将此混悬液经喷头雾化后喷入氮气保护下的冰冻的乙醇溶液中（−80℃），在低温状态下，聚合物载体中的有机溶剂在乙醇中不断扩散完全，包裹生物大分子药物的PLGA微球析出，真空过滤分离微球，低温干燥除去乙醇得到粉末状微球。该方法的优点为药物的包封率高、微球粒径均匀、工艺稳定、可实现产业化；缺

（4）Low temperature spray extraction The biological macromolecular drugs and the homogeneous powder of protective agent are added to the organic solvent of biodegradable polymer PLGA to form suspension. This suspension is atomized by the nozzle, then spraying the freezing ethanol, at low temperature (−80℃) state, the organic solvent in the polymer spread completely in the ethanol, separate microspheres, low-temperature drying to remove ethanol to get powder microspheres. The method has the advantages of the high encapsulation efficiency, concentration of microsphere size, process stability, which can achieve industrialization. The disadvantages are that the loss of biological activity of macromolecular drugs is larger than that of in-liquid drying. Recombinant human growth hormone (rhGH)

点为生物大分子药物活性损失较液中干燥法大。用此方法成功开发了可缓释1个月的基因重组人生长激素（rhGH）微球注射剂（商品名：Nutropin Depot）。

microsphere injection (brand name: Nutropin Depot) that can release a month continually has been successfully developed by this method.

（5）超临界流体技术　超临界流体技术（SFT）是从20世纪80年代逐渐发展起来的一门新技术。常用超临界流体为CO_2，它具有无毒、价廉、液态时溶解能力强、临界温度（T_C=31.3℃）和临界压力（P_C=7.38MPa）低，特别适用于生物大分子这一类对热敏感的药物，可降低设备成本，增加安全性。利用超临界流体（supercritic fluids，SCF）的溶解-增强分散能力，可以用于制备微球。在制备微球时根据聚合物及药物的溶解特性又分为超临界溶液快速膨胀（rapid expansion of supercritical solution，RESS）技术和气体反溶剂（gas antisolution，GAS）技术。

（5）Supercritical fluid technology (SFT)　SFT has been gradually developed into a new technology from 1980s. Commonly used supercritical fluid is CO_2, it has some characteristics of non-toxic, low cost, high solubility in liquid state, low critical temperature (T_C=31.3℃) and pressure (P_C=7.38MPa), which is especially fit for this type of biological macromolecules heat-sensitive drugs, can reduce equipment costs and increase security. By using the capability of solution-enhanced dispersion of supercritical fluids (SCF), microspheres can be prepared. According to the polymers and dissolution characteristics of drugs in the preparation of microspheres, they are divided into rapid expansion of supercritical solution (RESS) and gas antisolution (GAS) technology.

3.1.2　新型植入给药系统（IDDS）

植入给药系统是一类由药物与赋形剂或不加赋形剂经熔融压制或模制而成的一种供皮下植入用的无菌控释制剂。植入剂的发展已有几十年的历史，作为药物载体，植入剂可实现药物的零级释放，载药量较大，可避开首过效应，注射前无需用溶剂溶解，可达到局部或全身治疗的目的。随着生物降解型材料如聚酯和壳聚糖的开发与应用，植入剂逐渐成为一种具有广阔发展前景的剂型。Zenica公司采用生物降解材料，将多肽类药物戈舍瑞林研制成生物降解型埋植剂，将其直接注射于皮下或肌内，材料在体内可逐渐降解，并缓慢释放药物，彻底

3.1.2　New Implantable Drug Delivery System (IDDS)

IDDS is a class of sterile controlled release formulations for subcutaneous implant, which is made by melt pressing or molding the drugs with excipients or without excipients. Implants has been developing for decades, as a drug carrier, an implant can achieve zero order release of drugs. The drug loading is large, which can avoid the first pass effect. Without using solvent prior to injection, it can achieve the purpose of local or systemic administration. With the application of biodegradable materials such as polyester and chitosan, implants have broad prospects for development. The Zenica adopts the biodegradable materials to make the polypeptide drug goserelin into biodegradable implants, which can be directly injected into subcutaneous or intramuscular. The gradual degradation of polymer in vivo will completely change the situation which means that implants must be removed by surgery or implantation in the past. The Ala has developed DUROS system, which uses the titanium alloy as the material to develop implant

改变了以往植入剂必须经手术途径植入或取出的局面。Ala公司以植入剂技术为基础，开发了DUROS系统，以钛合金为材料，包括半透膜、渗透泵、活塞、药物储库和出口5个部分，该系统能保证药物在体内不降解，长时间保持稳定，为生物技术大分子药物的有效传递提供了新的途径。由美国三角研究院研制的Capronor是一种释放左旋18-甲基炔诺酮（LNG）的生物降解型皮下植入剂，可持续释放2年，达到长效避孕的效果。醋酸亮丙瑞林埋植剂（Viadur）已被批准用于治疗前列腺癌，药物可脉冲释放达1年之久。此外，新型植入给药系统的研究还有植入式药物芯片等。

technology, including the five parts of semi-permeable membrane, osmotic pump, piston, drug reservoir and exports. The system can ensure the stability of the drug in vivo for a long time; it provides a means of drug delivery for the macromolecular drugs of biological activity. Capronor that is developed by the American Academy of Capronor triangle is a subcutaneous implant of biodegradable that can release L-18-levonorgestrel (LNG), and LNG can sustainablly release for 2 years to reach long-term contraceptive effect. Leuprolide acetate implant (Viadur) has been approved for the treatment of prostate cancer, drug release can be pulsatiled release to 1 years. In addition, the study on new implantable DDS also has implanted drug chips, etc.

3.1.3 原位微球

原位微球属于可注射给药的原位成型给药系统中的一种类型，是将药物与可生物降解型材料共同溶解在特定的溶剂中形成注射剂，当注入体内后，聚合物随着溶剂的扩散析出并逐渐固化，包裹药物形成微球，从而达到控制释药的目的。原位微球与常规的注射微球和皮下植入剂相比，具有制备工艺简单、控释可靠、使用方便等优点。FDA批准了Atrix Lab公司生产的Leuprolde缓释注射液（商品名Elgard），它可通过小口径针头皮下注射后在体内形成微球，缓慢释放出Leuprolde，30mg的Elgard（I）已用于晚期前列腺癌的治疗。

3.1.3 In Situ Forming Microspheres

An in situ forming microsphere is one type of injectable in situ forming drug delivery system, which dissolves the drug and biodegradable polymer in a solvent to form a particular injection. When injected into the body, the polymer with the solvent diffusion and curing, to form microspheres to control drug release purposes. Compared with microspheres and subcutaneous implants, in situ forming microspheres have a simple process, and the advantages of reliable controlled release and easy to use. Recently, FDA has approved the Atrix Lab company's Leuprolde release injection (trade name: Elgard), the microspheres can be formed through small diameter needles injected in the body, which can release Leuprolde, 30mg Elgard (I) for palliative care of advanced prostate cancer.

3.1.4 脂质体

脂质体（liposomes）系指将药物包封于类脂质双分子层内而形成的微型泡囊体。脂质体具有在生物体内可降解，无毒性、无免疫原性，作为药

3.1.4 Liposomes

Liposome is a kind of micro-vesicular formed by drug encapsulated in the lipid bilayer. Liposomes can biodegradable in vivo, non-toxic, and non-immunogenic. Eespecially, as a drug carrier of targeting, it can reduce the drug dose and side effects. Phospholipids, cholesterol and so on are often used to be such materials of

物载体具有靶向性，能减少药物剂量，降低毒副作用等特点。制备脂质体常用的材料有磷脂、胆固醇等。脂质体的常用制备方法有薄膜法、机械分散法、逆相蒸发法、复乳法、注入法、冷冻干燥法、表面活性处理法。目前，对热敏脂质体、pH 敏感脂质体、柔性脂质体、磁性脂质体、长循环脂质体、靶向脂质体等的研究较活跃。将脂质体用一些具有保护性的聚合物如聚乙二醇（PEG）、聚乙烯醇（PVA）等进行表面修饰，它们通过在脂质体表面构成空间位阻，使脂质体半衰期显著延长，从而达到长效目的。免疫脂质体的制备方法也同样应用于生物技术药物。为了增加长效脂质体的靶向性，通常将抗体或其他配体连接于长效脂质体表面的聚合物（如 PEG）链的末端上，从而避免 PEG 链对靶位识别的干扰，得到新型长效免疫脂质体。以脂质体为载体的生物技术药物现已涵盖核酸、激素、酶、疫苗、质粒、基因物质等多种类型。

3.1.5 无针粉末注射给药系统

无针粉末注射（powder needleless injection，PNI）是一种主要利用高速气体喷射，使药物微粉瞬时加速至足够的速率，并将药物递送至皮内、皮下或黏膜部位，从而发挥药效作用的给药技术，具有无针、无痛、无交叉感染、便捷、微量、高效、安全等特点，因而特别适合于需长期注射用药、有恐针感的妇女、儿童和老人等特殊人群的临床注射给药、大规模的预防接种和野外作业条件下的疾病防治。目前，正在研发的药物有流感疫苗、艾滋病疫苗、胰岛素等。无针粉

preparation. The common liposomes preparation methods are film method, mechanical dispersion method, reverse-phase evaporation vesicles (REV), complex emulsion method, solvent injection, freeze-drying method, detergent removal. At present, thermosensitive liposomes, pH-sensitive liposomes, flexible liposomes, magnetic liposomes, long circulating liposomes, targeted liposomes and other researches are active. The liposomes with some protective polymers such as PEG、PVA process surface modification, constitute steric hindrance in the liposome surface, making the half-life of liposomes significantly longer to achieve long-term objectives. Immune liposome method is also applied to biotech drugs. In order to increase targeting of long-term liposomes, the antibodies or other ligands connected to the long-term lipid surface of the polymer (e.g. PEG) end of the chain, can avoid disturbance on the targets identified by the PEG chain to form new long-term immunity liposomes. Liposomes as the carrier of biotech drugs now covers nucleic acids, hormones, enzymes, vaccines, plasmids, genetic material and other types.

3.1.5 Powder Needleless Injection (PNI)

PNI is a drug delivery technology that is mainly using high-speed gas jet to make the instantaneous acceleration of drug powder to the adequate rate, and make the drug deliver to intradermal, subcutaneous or mucous membranes, which plays a pharmacodynamic drug delivery technology. With no needles, no pain, no cross-infection, convenient, traceable, efficient, and safe, it is particularly suitable for women, children and elderly people fear of needles and needing long-term injection drug in special clinical injection, large scale vaccination and disease prevention under field operating conditions. Currently, drugs are being developed including flu vaccine, AIDS vaccine and insulin. The core of PNI technology is the acceleration of solid powder technology, and the accelerated method of solid powder is the flow acceleration. At present, the evaluations of needle-free powder injection system mainly are powder speed,

末注射给药技术其核心是固体粉末的加速技术，实现固体粉末加速的方法主要为气流加速技术。对无针粉末注射给药系统的评价主要为粉末速度、粉末透射深度、粉末导入效果三个方面。无针粉末注射技术未来的发展主要是进一步提高安全性，广泛筛选、研制适用于无针粉末注射的生物药物、疫苗粉末，推进无针粉末注射技术的普及。

powder penetration depth and powder into effect. In the future, PNI need enhance safe and extensive screening, to be developed for PNI of biotech drugs, vaccines powder, to popularize PNI technology.

3.2 非注射给药系统

为解决某些疾病需长期用药的问题，研究生物技术药物非注射途径的给药系统，可克服注射剂的不便和缺点，将有益于增加病人的顺应性。生物技术药物的非注射给药途径包括胃肠道、黏膜、皮肤和肺部等。其中，口服给药是最受欢迎和最方便的给药途径，口服给药系统的研究是生物技术药物最热门的研究开发方向之一，但难度很大。

3.2 Non-injection Drug Delivery System

To solve some diseases requiring long-term medication, the research of non-injected biotech drugs DDS can overcome the inconvenience and weaknesses of injections and increase the patient's compliance. Non-injecting route of administration biotech drugs include the gastrointestinal tract, mucous membranes, skin and lungs and so on. Among them, oral administration is the most popular and convenient route of administration. The research of oral administration DDS is the most popular one in biotech drugs, but it is very difficult.

3.2.1 口服给药系统

蛋白多肽类药物口服给药生物利用度低主要问题在于：①分子量大，脂溶性差，对胃肠道黏膜的透过性差；②存在化学和构象不稳定，在胃内酸催化降解；③易被胃肠道中的酶降解，生物活性消失；④易被肝脏消除（首过效应）。因此，改进其生物利用度的焦点集中在如何提高其生物膜透过性和抵抗酶降解两个方面。提高其生物膜透过性的常用方法是使用吸收促进剂，如表面活性剂、胆酸盐、水杨酸、脂肪酸、酰基肉碱、蛋白毒素等；抵抗酶降解的方法有 PEG 修饰、使用酶抑制剂如甘

3.2.1 Oral Administration

The main problems for protein and polypeptide drugs with low bioavailability by oral administration are: ① large molecular weight, lipophilic poor, poor permeability through the gastrointestinal mucosa; ② chemical and conformational instability and degradation in the stomach acid; ③easy enzymatic degradation in the gastrointestinal tract, disappearance of biological activity; ④easily removed by the liver (first pass effect). Therefore, to improve its bioavailability focuses on how to improve their membrane permeability and resistance to enzymatic degradation. The method of improving membrane permeability is to use absorption enhancers such as surfactants, bile salts, salicylic acid, fatty acids, acyl carnitine, zonula occludens toxin (ZOT), etc.; Methods of resistance to enzymatic degradation have PEG modification, enzyme inhibitors such as sodium cholate Gan, camostat mesilate, bacitracin, aprotinin, soybean trypsin inhibitors. In addition, preparation

胆酸钠、camostat mesilate、bacitracin、aprotinin、大豆胰酶抑制剂等。此外，采用制剂技术和新型口服给药系统也是克服蛋白多肽类药物口服吸收障碍的有效途径。目前研究较多的口服给药系统主要有以下几种。

technology and new formulations of oral drug delivery system is an effective way to overcome the barriers to oral absorption of protein and polypeptide drugs. At present, oral drug delivery systems are studied mainly in the following.

1. 口服微粒给药系统 微粒有可能替代吸收促进剂成为可行的生物技术药物口服给药载体。口服后微粒在胃肠道存在3种吸收机制：①跨膜通道转运；②上皮细胞的胞饮作用；③派尔集合淋巴结M细胞的吞噬作用。多数学者认为微粒的吸收主要发生在肠道淋巴组织。微粒在胃肠道的吸收主要与微粒的性质、粒径和表面特性有关。小于10μm的微粒几乎全由派尔集合淋巴结M细胞吸收，疏水微粒比亲水微粒更易吸收。口服微粒的给药载体主要有聚合物微粒和稳定性更高的脂质体，如PLG微球、聚丙烯淀粉微球、海藻酸微粒、葡聚糖纳米球、DODPC聚合膜脂质体等。目前认为蛋白多肽类药物微粒给药系统最大的问题是口服后其生物利用度仍然较低。提高微粒吸收效率的方法有对微粒表面修饰靶向分子如凝集素［ulex europaeus（UEA-Ⅰ）］或抗体，或通过选用黏附性聚合物材料制备微粒降低胃肠道的微粒转运，延长口服微粒在肠道的滞留时间。

1. Oral microparticles drug delivery system Microparticles may be alternative absorption enhancers in the feature and become the oral delivery carrier of biotechnology drugs. Microparticles in the gastrointestinal tract after oral administration have three kinds of absorption mechanism: ① transmembrane transport channel; ②endocytosis of epithelial cells; ③phagocytosis of the Peyer's lymph nodes M cells. Most scholars believe that the absorption of microparticles mainly occurs in the intestinal lymphoid tissue. Particulate absorption in the gastrointestinal tract is mainly associated with the nature of particles, particle size and surface characteristics. Microparticles smaller than 10μm are almost entirely absorbed by the Peyer's lymph nodes M cells, the hydrophobic particles more easily absorbed than the hydrophilic particles. The drug carrier of oral particles are polymer microparticles and a higher stability of liposomes, such as PLG microspheres, polypropylene starch microspheres, alginate microparticles, dextran nanoparticles, DODPC polymerized liposomes and so on. At present, the biggest problem of protein and polypeptide drugs microparticles delivery system is that the bioavailability after oral administration is still low. The ways to improve the efficiency of microparticle absorption are modifying targeting molecules on the particle surface such as lectins (ulex europaeus, UEA-Ⅰ) or antibody, or choosing the adhesion of polymer particles preparation to reduce particles transit in the gastrointestinal tract and to extend intestinal retention time of oral particles.

2. 自微乳化释药系统（SMEDDS） 自微乳化释药系统处方主要由药物、油相、乳化剂和助乳化剂等组成。在室温及温和搅拌的条件下遇水能自发形成微乳，增加亲脂性药物和难溶性药物的溶解度和溶出速率，从而改善体内吸收，提高口服生物利

2. Self-microemulsifying drug delivery systems (SMEDDS) The prescription of self-microemulsifying drug delivery system（SMEDDS）including the drug, oil phase, surfactant, cosurfactant, and spontaneously. It can forming microemulsions with water under the conditions of room temperature and mild Stirring conditions, increasing the solubility and dissolution rate of lipophilic drugs and insoluble drugs, thereby improving

用度。将生物技术药物制备成口服自微乳剂可以改进其吸收。1994 年 5 月德国上市的环孢菌素 A（Neoral™）微乳化浓缩液软胶囊，口服后能吸收胃肠道的水分进行自微乳化，其生物利用度较口服溶液剂高，使肾移植排斥反应发生率降低。这一产品的成功问世使得 SMEDDS 作为药物载体的研究越来越受到重视，发展非常迅速，已进入临床研究的有胰岛素口服微乳胶囊等。但是此类系统最大的问题在于处方中高浓度乳化剂和助乳化剂的毒性以及对胃肠黏膜的刺激性，因此，今后 SMEDDS 的研究重点主要有两方面，一是采用工艺优化降低乳化剂和助乳化剂的用量；二是寻找和开发高效低毒的乳化剂和助乳化剂。

3. 口服结肠定位给药系统（OCDDS） 口服结肠定位给药系统是通过适宜的药物制剂手段和靶向定位技术，使药物口服后避免在胃、十二指肠、空肠和回肠前端释放药物，转运至回盲肠或结肠部位，以速释（脉冲）、缓释或控释方式发挥局部或全身治疗作用的一种给药系统。该类系统非常适合生物技术药物的制备，主要基于以下吸收机制：①结肠不能主动吸收糖、氨基酸和小分子肽等物质，但其内容物在结肠内滞留的时间较长（20～30h），可发挥其吸收功能；②结肠内大量消化酶已失活，结肠黏膜的渗漏点——派尔集合淋巴结（Peyer's patch）可将药物聚积，并以完整的结构被丰富的淋巴组织吞噬；为口服大分子药物特别是蛋白多肽类药物的吸收提供一条有效途径。用于 OCDDS 制备的方法很多，包括前体药物法、凝胶骨架法、

the absorption and oral bioavailability. Biotech drugs being prepared to microemulsion can improve its oral absorption. In May, 1994, cyclosporine A (Neoral™) micro-emulsion concentrate capsules are listed in Germany, which absorbed water in the gastrointestinal tract after oral administration to make the micro-emulsion. The bioavailability is higher than oral solution, but the kidney transplant rejection rate is lower. Because the research of SMEDDS is paid attention and develops very rapidly, oral insulin microemulsion capsules has entered the clinical study .But the biggest problem of SMEDDS is that high concentrations surfactant and cosurfactant in the prescription can bring toxicity and irritation of the gastrointestinal mucosa. So in the future the main research focus of SMEDDS has two aspects, first, reduce the amount of surfactant and cosurfactant by process optimization; second, it is to find and develop efficient and low toxic surfactant and cosurfactant.

3. Oral colon-specific drug delivery system (OCDDS) OCDDS is a drug delivery system that can avoid release of the drug in the stomach, duodenum, jejunum and ileum front-end after oral administration of drug by means of appropriate pharmaceutical preparations and targeted positioning technology, and that can be transported to the back part of the cecum or colon and play the role of local or systemic treatment by immediate-release (pulse), sustained release or controlled release. OCDDS is ideal for the preparation of biotech drugs absorption mechanism based on the following: ① the colon can not be active in absorption of sugar, amino acids and small peptides and other substances, but its contents remain in the colon for a longer time (20-30h), can be absorbed; ② Although a large number of digestive enzymes in the colon have been deactivated, the leakage of colonic mucosa point—Peyer's patches can accumulate drugs, and the complete structure will be swallowed by rich lymphoid tissue; rich lymphoid tissue in colon provide an effective way of the absorption of oral macromolecular drugs especially to protein and polypeptide drugs. Preparations for OCDDS have many ways, including prodrugs, gel matrix method, silica gel covalently technology and coating methods. Among them,

硅胶共价技术及包衣等方法。其中包衣方法与其他方法相比，所需技术与设备条件不高、操作简单、生产成本比较低、适于工业大生产。因此，选择适宜的载体材料，通过包衣技术设计的具有 pH 依赖与时滞效应释药机制的 OCDDS，是目前最为实用的释药系统。对胰岛素胶囊或微丸采用肠溶材料丙烯酸树脂包衣可减少胃酸和部分酶的破坏，达到定位释放。在 pH 7.5～8.0 释放药物，大鼠口服 16IU 胰岛素的降糖效果与腹腔内注射 4IU 胰岛素的效果类似，相对生物利用度为 9.3%～12.7%。

compared with other methods, coating method requires low technology and equipment conditions is simple, low cost, and is suitable for industrial production. Thus, the pH dependent of OCDDS by coating technology designed is the most practical drug delivery systems. Insulin capsules or pellets with enteric coating materials can reduce the damage of acrylic acid and some enzymes, to locate the release. Release of the drug in pH7.5-8.0, the rat oral hypoglycemic effect of insulin 16IU is similar to 4IU intraperitoneal injection of insulin, the relative bioavailability of 9.3% -12.7%.

3.2.2 鼻腔黏膜给药系统

鼻腔给药系统指经由鼻腔给药，发挥局部或全身治疗或预防作用的一类制剂。此类系统尤其适用于给药剂量小、口服难以吸收、在胃肠道中不稳定、肝脏首过作用强的药物等。鼻腔黏膜面积约 200cm^2，呼吸区细胞表面具有大量微小绒毛，上皮细胞间隙较大，血管和淋巴管丰富，鼻腔黏膜的穿透性较高而酶相对较少，血流速度快等均是药物经鼻腔吸收的有利条件。1978 年，Hirai 等将含表面活性剂的胰岛素制剂经犬和大鼠鼻腔给药，吸收显著提高，从此鼻腔给药新剂型的研究非常活跃，尤其是蛋白和多肽类药物，鼻腔给药将成为取代长期注射的有效方法。目前，已有十几种鼻腔给药制剂上市和应用于临床。现已上市的产品有鲑降钙素（calcitonin）、那法瑞林（nafarelin）、高浓度去氨加压素（desmopressin，即 DDAVP）、催产素（oxytocin）鼻腔喷雾剂和胰岛素（商品名 Nazlin）鼻用制剂。

3.2.2 Nasal Mucosal Drug Delivery System

Nasal drug delivery system refers to a class of agents that play local or systemic treatment or prevention by intranasal administration. The systems are especially suitable for small-dose, oral difficult to absorb, instability in the gastrointestinal tract and strong effect drugs of liver first pass effect. Nasal mucosa area is about 200 cm^2, cells in breathing zone with a large number of tiny hair, epithelial cells in the larger space, rich in blood vessels and lymphatic vessels. Nasal mucosa has relatively high penetration and low enzyme, faster blood flow, which become an advantage of nasal drug delivery system. In 1978, Hirai administered surfactant-containing insulin formulations in the nasal of dog and rat and the absorption increased significantly. The research of nasal administration is very active, particularly protein and polypeptide drugs. Nasal delivery will be an effective way to replace the long-term injection. At present, there are dozens of nasal administration formulations for public and clinical application. Products have been listed with salmon calcitonin nasal spray, nafarelin nasal spray, high concentration of desmopressin nasal spray, oxytocin and insulin nasal spray (trade name: Nazlin).

为了提高蛋白质类药物鼻腔给药的生物利用度，常采用加入吸收促进剂、酶抑制剂、前药设计、使用生物黏附材料（延长滞留时间）、改变给药剂型等方法。但一些吸收促进剂具刺激性，长期应用可损害鼻黏膜和纤毛。其中，梭链孢酸衍生物如牛磺酸二氢梭链孢酸盐和环糊精类可能是较安全有效的促进剂，值得深入研究。

To improve the bioavailability of nasal delivery of protein drugs, we often add absorption enhancers, enzyme inhibitors, the former drug design, bioadhesive materials (extended detention time), changed drug formulations. However, some absorption enhancers have irritating and long-term using can damage the nasal mucosa and cilia. Among them, the derivatives of fusidic acid such as taurine dihydrogen fusidic acid salt and cyclodextrin may be safer and more effective promoters, which deserve further research.

鼻腔疫苗的研制也是近年来鼻腔给药研究的一个新方向。近年来，已有少数鼻腔给药的疫苗制剂上市，如美国FDA已经批准的MedImmune公司的鼻腔喷雾给药流感病毒活疫苗（influenza virus vaccine live，商品名FluMist）等，就是通过鼻黏膜给药从而直接发生黏膜免疫反应产生抗体来治疗流感。但是疫苗鼻腔给药具有易受酶降解、黏膜刺激性强、无法保持缓释长效作用等问题，而以微粒（微球、脂质体）为载体的疫苗微粒鼻腔给药系统（vaccine microspheres nasal drug delivery system，VMNDDS）可部分解决这些问题。作为一种新型非侵入性疫苗给药系统，VMNDDS具有方便、安全、病人顺应性好、可以实现疫苗的缓释长效作用、提高免疫效果并诱导黏膜免疫等特点，特别适合用于那些经黏膜感染的疾病和流行性疾病，易进行大规模人群的免疫，如流感、白喉等的预防。

Nasal vaccine is a new research direction of intranasal administration. In recent years, small number of nasal administration of the vaccine formulations are listed, for example, the U.S. FDA has approved MedImmune's nasal spray influenza virus vaccine live(brand name: FluMist), which delivers the drug through the nasal mucosa to directly lead to mucosal immune responses to produce antibodies to treat influenza. But the nasal vaccine has some problems of vulnerable to enzymatic degradation, mucosal irritation intensity, unable to release long-lasting effect and so on, while the vaccine microspheres nasal drug delivery system(VMNDDS) can solve these problems partly. As a new non-invasive vaccine delivery system, VMNDDS has the advantages of convenience, safe, patient compliance, which can achieve long-term slow-release vaccines, and improve the immune effect and induction of mucosal immunity and other characteristics, especially suitable for those who have mucosal infectious diseases and epidemic diseases. It is easy to carry out large-scale population immunity, such as influenza, diphtheria prevention.

3.2.3 肺部给药系统

肺部吸收面积巨大，总表面积可达100～150m^2，远大于鼻腔等除胃肠道以外的其他途径的吸收面积，肺泡囊壁由单层上皮细胞所构成，这些细

3.2.3 Pulmonary drug delivery system

Pulmonary absorption area is huge, with a total surface area 100-150m^2, is much larger than the nasal cavity, etc. Alveolar wall is composed of epithelial cells from the monolayer, and these cells are close to the dense capillary network. The thickness of cell wall

胞紧靠着致密的毛细血管网，细胞壁或毛细血管壁的厚度只有0.5～1μm，药物到达肺泡囊即可迅速吸收显效，因此肺部给药是药物吸收很有前景的途径之一。自1925年德国首次将胰岛素气雾剂运用于糖尿病患者并观察到血糖的下降后，蛋白多肽类药物肺部吸入给药技术已成功应用于制药领域中。肺部吸入系统的制剂类型主要为喷雾剂、定量吸入气雾剂（metered dose inhaler，MDI）和干粉雾剂（dry powder inhaler，DPI）。一些多肽或蛋白质在形成气溶胶微粒时容易发生变性，因此，一般首选小剂量的粉雾剂系统，干粉吸入避免了溶液剂中药物的降解，微粉化技术保证了药物不在气管或支气管滞留，顺利进入肺部组织，选择合适的给药装置是解决肺部给药的关键。近年来胰岛素、白介素、鲑降钙素吸入粉雾剂肺部给药已取得重大进展，现已进入临床实验阶段。

or the capillary wall is only 0.5-1μm, and drugs can quickly reach the alveolar sacs and absorb markedly, so pulmonary delivery is one of promising ways of drug absorption. Since 1925 Germany firstly treated diabetes patients with insulin aerosol and observed a decline in blood glucose, the pulmonary drug delivery system of protein and polypeptide drugs has been successfully used in the pharmaceutical field. Main types of lungs preparation system have nebulizer, metered dose inhaler(MDI) and dry powder inhaler(DPI). Some polypeptides or proteins in the formation of aerosol particles are susceptible to degeneration; therefore, we generally prefer system of small doses of DPI. Dry powder inhalation avoid degradation of the drug in solution, while micronization technology ensure that the drug will not retent in the trachea or bronchial, smoothly into the lung tissue. So selecting the appropriate delivery device is the key to solve the pulmonary drug delivery. In recent years, insulin, interleukin, pulmonary delivery of salmon calcitonin inhalation DPI has made significant progress and now entered the clinical trials.

目前蛋白多肽类药物肺部给药系统存在的主要问题是：长期给药后安全性评估；肺吸收药物粒子大小的限制；促进肺部吸收的措施；吸收和有效的重现性问题；使大分子药物稳定化的处方设计等。一旦这些问题获得解决，肺部给药系统也是这类药物的适宜途径。

At present, the main problems of protein and polypeptide pulmonary drug delivery system are: long-term safety assessment after administration; molecular size limit of pulmonary absorption; promoting the absorption of measures; the reproduction of the efficiency and absorption; the prescription design methods of the stable macromolecular drugs. Once these questions are resolved, pulmonary drug delivery system is the appropriate way to these drugs.

3.2.4 经皮给药系统

经皮给药系统（transdermal drug delivery system，TDDS）是指经皮肤贴敷方式给药，药物以一定的速率通过皮肤经毛细血管吸收进入血液循环并达到有效血药浓度，实现疾病治疗或预防的新制剂。对于生物技术大分子药物，皮肤的穿透性

3.2.4 Transdermal Drug Delivery System (TDDS)

TDDS is administered through the skin applicator means. Drugs are absorbed through the skin into the blood to reach effective blood concentration at a certain rate to achieve the treatment or prevention of disease. The low percutaneous is the main obstacle for absorption of macromolecules of biotech drugs, but the skin's hydrolytic activity is very low, which creates favorable conditions for protein and polypeptide transdermal drug. To improve the transdermal absorption of biotech drugs,

低是该类药物透皮吸收的主要障碍，但皮肤的水解酶活性相当低，给蛋白多肽类药物经皮给药创造了有利条件。为改善生物技术药物的透皮吸收，需要采用促进药物经皮吸收的方法，常用方法有加入透皮吸收促进剂；采用物理促渗技术；将脂质体和微乳作为生物大分子药物透皮给药的载体等。

we need to promote the use of drug penetration absorption method, common methods are as followed: adding the penetration enhances; physical penetration techniques; the liposomes and microemulsions as drug carriers, etc.

离子导入指电荷或中性分子在电场作用下通过汗腺和毛孔等途径迁移进入皮肤的过程。采用离子导入技术结合脂质体的方法也可使脂质体进入角质层。该项技术的应用使大分子量、荷电和亲水性的多肽和蛋白质类药物能透过皮肤角质层。物理促渗技术的发展使得一些生物技术药物如胰岛素、TRH、LHRH、精氨酸加压素等的透皮吸收都取得了一定研究进展。

Iontophoresis means the process that the charge or neutral molecules in the electric field enter the skin through the sweat glands, pores of the skin and other means. Iontophoresis technology combined with the method of liposomes can also be driven into the stratum corneum. High molecular weight, charged and hydrophilic polypeptide and protein drugs can go through the skin stratum corneum by the application of the technology. Physical penetration technology makes some biotech drugs such as insulin, TRH, LHRH, arginine vasopressin and other transdermal absorption studies have made some progress.

将脂质体和微乳作为生物大分子药物透皮给药的载体也能有效地改善此类药物在皮肤的透过性。脂质体的促透机制主要包括水合机制、穿透机制、融合机制。脂质体作为生物药物透皮给药的载体已有大量报道，如胰岛素、干扰素、胶原蛋白、肝素、超氧化物歧化酶、组织生长因子等。脂质体的主要材料是磷脂，磷脂分子在一定条件下形成双分子层，与角质层细胞间隙的类脂双分子层有相同结构，推测这种结构有利于两者的亲和，从而改进药物的渗透速率。微乳的促透机制主要包括增溶作用、改变皮肤的脂质途径、增加皮肤水合作用、经毛囊透皮吸收等。

The liposomes and microemulsions as carriers for transdermal delivery of biological macromolecular drugs can be effective to improve the permeability of drugs in the skin. The mechanism of liposomes to promote penetration includes hydration mechanism; penetration mechanism; fusion mechanism,etc. Liposomes as carriers for transdermal delivery of biotech drugs have been numerous reported, such as insulin, interferon, collagen, heparin, SOD, tissue growth factor, etc. The main material of liposomes is phospholipids, phospholipids molecules under certain conditions form a double molecular layer with the same structure of stratum corneum intercellular lipid bilayers, suggesting that this structure is conducive for affinity, which can improve drug penetration rate. The microemulsions promote penetration including solubilization; changing skin lipid approach; increasing in skin hydration; transdermal absorption by the hair follicles.

3.3 生物技术药物新型给药系统应用实例

实例 1：破伤风类毒素微球

【处方】破伤风类毒素 90μl；PLA 0.2g；聚乙烯醇水溶液 1ml（$10g \cdot L^{-1}$）

【制备工艺】将 90μl 破伤风类毒素加至 1ml 含 0.2g PLA 的有机相中，分散，制成初乳。然后将该乳液在搅拌速度为 $900r \cdot min^{-1}$ 的条件下加至聚乙烯醇水溶液中，乳化 3min（室温），再继续搅拌 4h 使有机溶剂挥发完全。所得微球通过滤器收集，用蒸馏水洗涤 5 次，真空干燥 72h，即得成品。

【用途】用于预防破伤风。

【处方及工艺分析】

（1）破伤风类毒素（tetanus toxoid，TT）为主药，PLA 为载体材料。

（2）采用该法制备的微球呈圆形，不粘连，表面光滑，流动性好，分布均匀。

（3）微球制备工艺的影响因素有 PVA 浓度、乳化转速、有机相 / 水相相比、TT/PVA 投料比、有机相浓度等。PVA 浓度增大、乳化转速增大、有机相 / 水相相比减少均使微球粒径减少；PVA 浓度增大，包封率略增加；TT/PVA 投料比、有机相浓度增大，均使包封率增加。

实例 2：肺吸入胰岛素粉雾剂

【处方】胰岛素 100U；癸酸钠适量；乳糖 25mg。

【制备工艺】胰岛素于压缩氮气流中在 Airfilco 射流粉碎机中粉碎成平均粒径为 2.4μm。癸酸钠用压缩氮气粉碎至粒径为 1.6μm。微粉化后的

3.3 Preparation Examples of Biotech Drug Delivery System

Example 1: Tetanus toxoid microspheres

[Prescription]

tetanus toxoid	90μl
PLA	0.2g
PVA solution	1ml（$10g \cdot L^{-1}$）

[Preparation]

90μl TT added to 1ml organic phase containing 0.2g PLA, distributed and made colostrum, and then mixing the emulsion under the conditions of speed $900r \cdot min^{-1}$ added to the PVA solution, emulsion 3min (room temperature), then continue stirring 4h until the organic solvent completely evaporation, collect the microspheres from the filter, washed 5 times with cold distilled water, vacuum drying 72h.

[Use]

Tetanus toxoid microspheres injection used for the prevention of tetanus.

[Prescription and process analysis]

（1）Tetanus toxoid（TT）is drug, PLA as a carrier material.

（2）The microspheres prepared by this method is round, no adhesion, smooth surface, mobility is good, uniform distribution.

（3）Preparation of microspheres factors are PVA concentration, speed of emulsification, the organic phase/water phase phase ratio, TT/PVA weight ratio, organic phase concentration etc. PVA concentration increases, the emulsion speed increases, the organic phase/water phase is reduced compared to the microsphere size decreased; PVA concentration increases, the encapsulation efficiency increases slightly; TT/PVA weight ratio and the concentration of organic phase increases, the encapsulation efficiency increases.

Example 2: Insulin powder inhalation (lung inhalation)

[Prescription]

Insulin	100U
Sodium decanoate	moderate
Lactose	25 mg

[Preparation]

Insulin in the compressed nitrogen stream is smashed in the Airfilco jet mill until the mass average

胰岛素与相当于其质量10%～25%的癸酸钠混合均匀后，再与处方量的乳糖混合均匀，装入胶囊中，置粉末吸入装置吸入给药。

【用途】用于糖尿病的治疗。

【处方及工艺分析】

（1）胰岛素为主药，癸酸钠为吸收促进剂，乳糖为载体。

（2）微粉化技术为该粉雾剂制备的关键。胰岛素具有生物活性，应避免粉碎过程对主药的破坏。稳定的配方可保证药物的疗效。在粉雾剂处方中需要加入一定的载体，以防止药物因微粉化后表面能较大而聚集。载体还具有增强药物在储存时的稳定性的作用。常用的载体为乳糖。

（3）胰岛素经肺吸入给药，治疗的重现性与皮下注射相仿，而吸收要比皮下注射迅速，可减少胰岛素依赖性糖尿病所需的多次注射，提高了患者的顺应性。

实例3：环孢菌素A微乳浓缩液

【处方】环孢菌素A 200mg；三甘油辛/癸酸酯470mg；聚氧乙烯氢化蓖麻油930mg；无水乙醇150mg；丙二醇250mg。

【制备工艺】室温下，将环孢菌素A溶于无水乙醇和丙二醇的混合溶剂中，并在搅拌的同时将三甘油辛/癸酸酯和聚氧乙烯氢化蓖麻油加入到混合溶剂中，充分搅拌直到形成均一澄明的微乳浓缩液。

【用途】广泛用于器官移植术后的抗排斥反应和自身免疫性疾病的治疗。

particle size is 2.4μm. Sodium decanoate is crushed by compressed nitrogen to the diameter of 1.6μm. Insulin after micronization mixes with sodium decanoate of 10%-25% mass, mixed evenly, then mixes with the amount of lactose in prescription, encapsulated, last inhalation powder inhalation device is set.

【Use】

Insulin powder inhalation used for the treatment of diabetes.

【Prescription and process analysis】

（1）Insulin is drug, while sodium decanoate is penetration enhance and lactose is the carrier.

（2）Micronization technology is the key for the preparation. Biological activity of insulin should avoid major damage during grinding, Stable formula can ensure drug efficacy. In the prescription of powder inhalation needs to add certain carrier to prevent the larger surface energy from concentrating after micronization. Carrier also has the function of enhancing the stability of drugs in the role of storage time. Common carrier is lactose.

（3）Pulmonary inhalation of insulin, the reproducibility of treatment is similar to subcutaneous injection, while rapid absorption than subcutaneous injection, which can reduce the multiple injections required by insulin-dependent diabetes and increase patients compliance.

Examples 3: Cyclosporin A (CyA) microemulsion preconcentrate

【Prescription】

CyA	200mg
Triglycerol caprylate/caprate	470mg
Cermophor RH40	930mg
Ethanol	150mg
1,2 - Propylene glycol	250mg

【Preparation】

At room temperature, the CyA is dissolved in ethanol and 1,2 - Propylene glycol in the mixed solvent, and stirring at the same time, adding triglycerol caprylate/caprate and cermophor RH40 in the solution, stir completely until form the microemulsion preconcentrate of the uniform and clear.

【Use】

CyA microemulsion preconcentrate is widely used anti-rejection after organ transplantation reactions and autoimmune diseases.

【处方及工艺分析】

（1）环孢菌素 A（cyclosporin A，CyA）为主药，它是由 11 个氨基酸组成的环状多肽物质，是一种强效选择性抑制剂。三甘油辛/癸酸酯为油相。聚氧乙烯氢化蓖麻油为表面活性剂。无水乙醇为助乳化剂，丙二醇为助表面活性剂。

（2）本品是由油相、亲水性表面活性剂、无水乙醇和丙二醇等辅料与 CyA 经适宜工艺配制而成的微乳浓缩液，制备工艺简单。微乳浓缩液属于自乳化释药系统，处方中包括药物、油相、表面活性剂、助表面活性剂，在室温及温和搅拌的条件下遇水能自发形成微乳，增加亲脂性药物和难溶性药物的溶解度和溶出速率，从而改善体内吸收，提高口服生物利用度。CyA 改为 O/W 微乳浓缩液软胶囊后，生物利用度为 W/O 型乳剂软胶囊的 170%～233%，平均剂量减少 16%，排斥反应发生率由 54% 下降到 40%。

（编者 赵 颖）

【Prescription and process analysis】

（1）CyA is drug，it is a cyclic peptide composed of 11 amino acids ,which is a potent selective inhibitors. Triglycerol caprylate/caprate as oil phase, cermophor RH40 as surfactant, ethanol as the emulsifier, 1,2 - Propylene glycol as co-surfactant.

（2）This product is a microemulsion preconcentrate that is made of vegetable oil, hydrophilic surfactant, ethanol and 1,2 - Propylene glycol and other accessories with CyA from the appropriate preparation process, which is a simple process. Microemulsion preconcentrate belongs to self-emulsifying drug delivery systems, prescription including the drug, oil phase, surfactant, cosurfactant, and spontaneously forming microemulsions with water under the conditions of room temperature and mild Stirring conditions, increasing the solubility and dissolution rate of lipophilic drugs and insoluble drugs, thereby improving the absorption and oral bioavailability. After CyA is changed to O/W microemulsion preconcentrate soft capsules, the bioavailability is 170%-233% of W/O emulsion soft capsules, the average dose is reduced to 16%, and rejection rate is reduced from 54% to 40%.

（Compiled by Ying Zhao）

参 考 文 献
References

陈庆华，刘晓华．2002．多肽、蛋白质类药物微囊化的两种新型实用设备．中国医药工业杂志，33（3）：142-144

陈庆华，瞿文．2000．多肽、蛋白质类药物缓释剂型的研究进展．中国药学杂志，35（3）：147

崔福德．2008．药剂学（第六版）．北京：人民卫生出版社

董素英，马红勇，王笃年．2009．对蛋白质变性中一些问题的探究．化学教育（8）：68

梅兴国．2004．生物技术药物制剂基础与应用．北京：化学工业出版社

宋思扬，楼士林．2010．生物技术概论（第三版）．北京：科学出版社

王炳和，T. 萨哈恩，R. A. 索特洛编；金义光，杜丽娜译．2008．药物传递：原理与应用．北京：化学工业出版社

徐国恒. 2010. 二硫键与蛋白质的结构. 生物学通报，45（5）: 5-7

Chen L,Apte R N, Cohen S.1997.Characterization of PLGA microspheres for the controlled delivery of IL-1 for tumor immunotherapy. J Control Release, 43:261–272

Iwata M, McGinity J W.1992. Preparation of multi-phase microspheres of poly (lactic acid) and poly (lactic-co-glycolic acid) containing a w/o emulsion by a multiple solvent evaporation technique. J Microencapsul,9:201–214

Jacobi G H, Wenderoth U K, Ehrenthal W,et al.1988.Endocrine and clinical evaluation of 107 patients with advanced prostatic carcinoma under long-term pernasal buserelin or intramuscular decapeptyl depot treatment.American Journal of Clinical Oncology.1988;11 Suppl 1:S36-43

Pettit DK,Lawter JR,Huang WJ,et al.1997.Characterization of poly (glylcolide-co-DL-lactide)/poly (DL-Lactide) microspheres for controlled release of GM-CSF.Pharm Res,14(10):1422

Rubin J, Ajani J, Schirmer W, et al.1999. Octreotide acetate long-acting formulation versus open-label subcutaneous octreotide acetate in malignant carcinoid syndrome. J Clin Oncol,17:600–606

YANG J,Cleland J L.1997.Factors affecting in vitro release of recombinant human interferon- γ from PLGA microspheres. J Pharm Sci, 86 (8): 908

第四章　生物技术药物生产的质量控制

Chapter 4　Quality Aspect of Biopharmaceutical Manufacturing

生物制药工业是严格质量受控的领域。自20世纪70年代后期以来，重组DNA和融合技术的科学基础发展很快。自20世纪80年代早期第一个重组DNA药物胰岛素以来，全球市场上已有200多个生物药物产品。这些产品的广泛性和创新性给监管机构提出了相当有力的科学性挑战。在这期间，必须建立管理这些产品的框架以保证它们的质量和安全。对许多国家来说，不管它们是否很早就已经实施了药品注册管理，其在20世纪60年代和70年代期间出现了很大的改进，包括在逐步健全新药安全性、质量和有效性的评估方面的改进，在质量报告和法规方面的改进，以及在控制和指导原则方面改进。那个时期的制药工业正在日益国际化并寻求全球市场。在这样的背景下，1990年，形成了人用药品注册技术要求国际协调会（ICH）。

Biopharmaceutical industry is a heavily regulated product industry. Since the late 1970's, the scientific bases for recombinant DNA (rDNA) and hybridoma technologies were developed very rapidly. There are more than 200 biopharmaceutical products on the global market since the first rDNA medicine human insulin in early 1980's. The diversity and novelty of these products present the regulatory authorities with an enormous scientific challenge. During this period, a framework for the regulation of these products needed to be developed by regulatory authorities to assure quality and safety. For most countries, regardless of whether or not they had initiated product registration controls earlier, the 1960's and 1970's presented a rapid increase in laws, regulations, and guidelines for reporting and evaluating the data on safety, quality, and efficacy of new medicinal products. At the time, the industry was becoming more international and seeking new global markets. In 1990, the International Conference of Harmonization (ICH) was formed.

在过去的25年中，在重组DNA技术和细胞杂交技术方面的进步，导致许多现代生物技术药物的开展。在这段时间内，监管当局认识到了生物制药分子作为产品的复杂性，和制造过程对产品质量的影响。为了保证生物药物质量和安全，“生产过程决定产品”和“质量不能由测试产品而得知，应该是设计在和内置于产品的生产之中”已经在生物制药行业中成为一个被广泛接受的概念。换言之，制造过程是一个产品的生产过程。

Over the last 25 years, advances in recombinant DNA and hybridoma technologies have led to the development of many modern biotechnology medicines. During this time, regulatory authorities realized complexity of biopharmaceutical molecules as products, and quality impact by manufacturing processes. To assure quality and safety, “the process defines the product” and “Quality cannot be tested into products but should be built-in or should be by design” become a widely accepted concept in biopharmaceutical manufacturing industry. In other words, manufacturing process is a product.

1 药品生产质量管理规范（GMP）

1 Good Manufacturing Practice (GMP)

药品生产质量管理规范（GMP）是一部法律法规。GMP是一套质量体系，涵盖了所有药品和生物制品、原料和成品、药物有效成分、疫苗和疫苗佐剂、药品制剂添加剂以及其他的诊断试剂、食品和医用仪器的生产和测定。GMP也是一套操作标准系统，是由各国有关部门通过技术要求而制定的指导原则。而药物和生物药品正是在这些原则的指导下进行生产与测试。

Good manufacturing practice (GMP) is regulations and laws. GMP is a quality system covering the manufacture and testing of pharmaceutical and biopharmaceutical dosage forms, drug substances and products, active pharmaceutical ingredients, vaccines and vaccine adjuvant, excipients, diagnostics, foods, and medical devices. GMP is also a set of operational standards defined by individual countries through guidelines that outline the aspects of production and testing that can impact the quality of a product.

GMP也被称作“cGMP”或“CGMP”。“c”代表“目前”，提醒生产厂家要运用合乎法规的最新的技术和系统。cGMP提供了一套系统以确保对生产过程和设施的恰当设计、监测和控制。遵守cGMP法规要求药品制造商严格控制生产过程，以保证产品的成分、效力、质量和纯度。cGMP包括建立质量管理体系以确保原材料的质量，建立可靠的操作程序，检测和研究产品质量的偏差，维持可靠的实验室。这种正规的控制系统如若实施，将有效地防止污染、样品混淆、偏差、失控及错误，这也就确保药品符合其质量标准。进而，这也就防止了消费者买无效药甚至引发危险的情况。药品若不能遵守GMP规程，在许多国家会导致非常严重的后果，诸如药品撤回、罚款，甚至于制造商获刑入狱。

GMP is sometimes referred as “cGMP” or “CGMP”. The “c” stands for “current”, reminding manufacturers that they must employ technologies and systems which are up-to-date in order to comply with the regulation. cGMPs provide systems that assure proper design, monitoring, and control of manufacturing processes and facilities. Adherence to the cGMP regulations assures the identity, strength, quality, and purity of drug products by requiring that manufacturers of medications adequately control manufacturing operations. CGMP includes establishing strong quality management systems, obtaining appropriate quality raw materials, establishing robust operating procedures, detecting and investigating product quality deviations, and maintaining reliable testing laboratories. This formal system of controls, if adequately put into practice, helps biopharmaceutical manufacturers to prevent instances of contamination, mix-up, deviation, failure, and error. This assures that the drug products meet their quality standards and protects the consumers from purchasing a product which is not effective or even dangerous. Failure of firms to comply with GMP regulations will result in very serious consequences including recall, seizure, fines, and jail custody in many countries.

GMP是一部法规，由相应的

GMP regulation is a set of law enforced by the

执法部门如中国国家食品和药物管理局（SFDA）、美国的食品和药物管理局（FDA）以及欧洲药品局（EMA），根据各国的法律依法执行。监管机构（包括FDA、SFDA和许多欧洲国家的监管机构）被授权进行突击巡查，虽然有一些是事先安排的。FDA的国内常规检查通常是突击的，但必须按照监管要求在“合理时间”执行。欧洲联盟内是由国家监管机构进行GMP检查的。每个检查员进行例行GMP检查，以确保药品生产的安全和正确性。此外，许多国家进行新药物上市审批前的GMP检查（PAI）。

administrative law enforcement department such as “State Food and Drug Administration (SFDA)” in China, “Food and Drug Administration (FDA)” in the United States, and “European Medicines Agency (EMA)” under the authority of drug administration laws by individual countries. Regulatory agencies (including the FDA, SFDA, and regulatory agencies in many European nations) are authorized to conduct unannounced inspections, though some are scheduled. FDA routine domestic inspections are usually unannounced, but must be conducted according to regulation which requires that they are performed at a “reasonable time”. Within the European Union, GMP inspections are performed by National Regulatory Agencies. Each of the inspectorates carry out routine GMP inspections to ensure that drug products are produced safely and correctly; additionally, many countries perform pre-approval inspections (PAI) for GMP compliance prior to the approval of a new drug for marketing.

所有的仪器和设备，包括生产设施、生产系统、制造工艺、分析方法，生产过程控制系统，生产过程分析技术，以防止污染、混合和错误等系统，在硬件和软件上都必须满足当前工业的专业水平，又称为“顶级水平”或“国家艺术”级。必须指出的很重要的一点是cGMPs只是最低要求。许多制药厂在美国和欧盟已经实施了综合性、现代化的质量体系和风险管理方法，均已超过这些最低要求。

Systems including manufacturing facility, production systems, manufacturing processes, analytical methods, in manufacturing process control, in process analytical technologies, instruments and equipment used to prevent contamination, mix-ups, and errors will have to meet current industrial professional standard which may be “top-of-the-line” and “state-of-art” in hardware and software. It is important to note that cGMPs are minimum requirements. Many pharmaceutical manufacturers in US and EU are already implementing comprehensive, modern quality systems and risk management approaches that exceed these minimum requirements.

GMP文件与记录保存、员工训练和考核、环境消毒和清洁、设备校验、工艺验证以及投诉处理等已有一系列原则性指导。绝大多数GMP的要求是提纲性的，这就给予了生产厂家一定的机动性，能够自己按照实际情况作出最好的操作控制以符合GMP标准。GMP提供的相当的灵活性能够让生产厂家基于自己的商业目标解释并执行这些GMP的要求。

GMP regulations address issues including documentation and recordkeeping, personnel training and qualifications, sanitation, cleanliness, equipment verification, process validation, and complaint handling. Most GMP requirements are very general and open-ended, allowing each manufacturer to decide individually how to best implement the necessary operations to meet GMP standards. This provides much flexibility, but also requires that the manufacturer interprets the requirements in a manner which makes sense for each individual business.

1.1 生物技术制药GMP关键点

1.1.1 建筑和设施

（1）用于生产和储存生物技术药物的建筑应大小适宜，设计和构造合理，适应设备无障碍放置，材料有序存放，容易操作、清洁及保养。

（2）地面、墙壁和天花板的构造要求平滑，容易清洁表面和保持干净，便于维修。

（3）固定装置、通风管道和排水的安装要适应水滴和冷凝水的排泄，不能污染生物制药的材料、产品和设备的表面或产品的灌装。

（4）灯光和通风设备要容易操作和充分满足人员的舒适性。

（5）供水、冲洗、卫生间、地面排泄和污物处理系统等都应充分供应清洁设备设施和器皿等，以便于使用者需要和进行个人的清洁。

1.1.2 设备

（1）用于生产、储存、转运和灌装的设备器皿设计合理，材料和工艺防止交叉污染，材料的包扎避免掺杂油污、脏污或消毒剂。

（2）器皿、输送管道和设备的产品接触面要保持干净和间歇性的消毒。

（3）清洁和消毒完的便携式设备和器皿应小心存储和放置，设备的产品接触面进行覆盖。避免碰撞、尘埃和其他污染。

1.1.3 人员

操作人员进行培训和规范化操作。

（1）人员的监管、操作生产或进行生物制药产品的质量控制都需要进行教育培训或具有经验才能进行相关

1.1 GMP Key Points in Biopharmaceutical Manufacturing

1.1.1 Building and Facilities

(1) Buildings used in the manufacture or storage of biopharmaceutical products are of suitable size, design and construction to permit unobstructed placement of equipment, orderly storage of materials, sanitary operation, and proper cleaning and maintenance.

(2) Floors, walls and ceilings are constructed of smooth, easily cleanable surfaces and are kept clean and in good repair.

(3) Fixtures, ducts and pipes are installed in such a manner that drip or condensate does not contaminate biopharmaceutical materials, product contact surfaces of equipment, or finished products in bulk.

(4) Lighting and ventilation are sufficient for the intended operation and comfort of personnel.

(5) Water supply, washing and toilet facilities, floor drainage and sewage system are adequate for sanitary operation and cleaning of facilities, equipment and utensils, as well as to satisfy employee needs and facilitate personal cleanliness.

1.1.2 Equipment

(1) Equipment and utensils used in processing, holding, transferring and filling are of appropriate design, material and workmanship to prevent corrosion, buildup of material, or adulteration with lubricants, dirt or sanitizing agent.

(2) Utensils, transfer piping and product contact surfaces of equipment are well-maintained and clean and are sanitized at appropriate intervals.

(3) Cleaned and sanitized portable equipment and utensils are stored and located, and product contact surfaces of equipment are covered, in a manner that protects them from splash, dust or other contamination.

1.1.3 Personnel

Operators are trained to carry out and document procedures.

(1) The personnel supervising or performing the manufacture or control of biopharmaceutical products has the education, training and/or experience to perform the

工作。

（2）直接与生物制药材料接触的人员，完成产品包装或接触生物制药材料的整个表面都必须防止生物药品的杂质污染，穿戴适宜的外套、手套和保证头发的整洁等，保持人员的个人卫生。

（3）在指定区域内严禁饮食、吸烟或使用抗生素等。

1.1.4 原材料

（1）原材料和包装材料要储存妥当、便于操作，防止混杂、微生物或其他化学品污染，放置时避免暴露在过热、寒冷、阳光照射和受潮的环境下。

（2）材料容器要密闭，包装材料离地面储存。

（3）包装材料贴标签，便于鉴定、分批和质控。

（4）在相应的生产流程中进行材料抽样和检验，确保不受污物、微生物或其他外部物质的污染，这是防止终产品不受污染所必需的。特别要注意有哪些用于生物制药的原材料要进行冷藏，防止污物或微生物的污染。

（5）对于那些不被认可的专用的材料进行适宜的鉴定和控制，防止它们应用于生物制药的生产。

1.1.5 生产

生产流程要有明确的规定和质量控制，所有工艺具有严格的明文规定，确保与专业化的持续性和一致性，严格控制生产流程，任何的改变要进行评估。对药品质量具有影响的改变必须进行认证。

检查生产和质量控制是否建立并

assigned functions.

(2) Persons coming into direct contact with biopharmaceutical materials, finished products in bulk or biopharmaceutical contact surfaces, to the extent necessary to prevent adulteration of biopharmaceutical products, wear appropriate outer garments, gloves, hair restraints etc., and maintain adequate personal cleanliness.

(3) Consumption of food or drink, use of tobacco, or use of antibiotics is restricted to appropriately designated areas.

1.1.4 Raw Materials

(1) Raw materials and primary packaging materials are stored and handled in a manner which prevents their mix-up, contamination with microorganisms or other chemicals, or decomposition from exposure to excessive heat, cold, sunlight or moisture.

(2) Containers of materials are closed, and bagged or boxed materials are stored off the floor.

(3) Containers of materials are labeled with respect to identity, lot identification and control status.

(4) Materials are sampled and tested or examined in conformance with procedures assuring the absence of contamination with filth, microorganisms or other extraneous substances to the extent necessary to prevent adulteration of finished products. Pay particular attention to materials of those used in the manufacture of biopharmaceuticals by cold processing methods with respect to contamination with filth or microorganisms.

(5) Materials not meeting acceptance specifications are properly identified and controlled to prevent their use in biopharmaceutical production.

1.1.5 Production

Manufacturing processes are clearly defined and controlled. All critical processes are validated to ensure consistency and compliance with specifications. Manufacturing processes are controlled, and any changes to the process are evaluated. Changes that have an impact on the quality of the drug are validated as necessary.

Check whether manufacturing and control have been

有明文规定，例如制剂、加工、转运和灌装指导、生产控制方法等，确定整个生产过程是否达到要求。

（1）生产设备、转运和灌装容器以及原材料容器、成批的材料是干净的，处于良好的保修和无污染条件。

（2）仅使用的是经过认证的材料。

（3）明文规定的指导原则和生产流程是清楚的，不具模糊的语言（良好的文件实施管理）。

（4）抽样是合理的，包括在生产期间和生产后、转运或灌装的检验要适当的混合或其他形式的生产加工，都不应存在有害的微生物或化学物的污染，并与其他认可的专业化一致。

（5）每一个操作步骤，称重、计量、计算等需经第二人检查。

（6）用于生产、灌装、存储生物药物的主要的设备，如转运线、容器和反应罐都需进行鉴定，明确内容、批次、质量控制情况和其他的认可的信息。

（7）在标注前检查标签，进行鉴定，避免混淆。

（8）生产、储存、运输和灌装设备的批次的标注，标明鉴定、批次和质量控制状态。

（9）成品的包装保证永久性的编码标记。

（10）复查产品是否变质或污染。

1.1.6 实验室控制

（1）原材料、半成品样本和成品都需检验，鉴定它们的特征是否合格，包括物理化学性质、微生物污染、有毒有害物质和其他未预测到的化学污染。

（2）在特定的时间段预留签发的

established and written instructions, i.e., formulations, processing, transfer and filling instructions, in-process control methods etc., are being maintained. Determine whether such procedures require that:

(1) The equipment for processing, transfer and filling the utensils and the containers for holding raw and bulk materials are clean, in good repair and in sanitary condition.

(2) Only approved materials are used.

(3) Instructions and procedures are written in clear and unambiguous language (Good Documentation Practices).

(4) Samples are taken, as appropriate, during and/or after processing, transfer or filling for testing for adequacy of mixing or other forms of processing, absence of hazardous microorganisms or chemical contaminants, and compliance with any other acceptance specification.

(5) Each step of operation, weighting, measuring, calculating etc., are checked by a second person.

(6) Major equipment, transfer lines, containers and tanks are used for processing, filling or holding biopharmaceuticals are identified to indicate contents, batch designation, control status and other pertinent information.

(7) Labels are examined for identity before labeling operations to avoid mix-up.

(8) The equipment for processing, holding, transferring and filling of batch is labeled regarding identity, batch identification and control status.

(9) Packages of finished products bear permanent code marks.

(10) Returned cosmetics are examined for deterioration or contamination.

1.1.6 Laboratory Controls

(1) Raw materials, in-process samples and finished products are tested or examined to verify their identity and determine their compliance with specifications for physical and chemical properties, microbial contamination, and hazardous or other unwanted chemical contaminants.

(2) Reserve samples of approved lots or batches of

保存样本或原材料和成品，储存在无污染或不变质的环境，继续对特征重新检验。

（3）水的供给，特别是用作生物制药成分的水按规则进行化学成分分析和微生物特征鉴定。

（4）成品的新鲜样本和预留的样本检验要保鲜存储，防止出现在各个环节的微生物污染，包括用户储存在合理的可预测的条件下和消费者使用时。

1.1.7　记录

在生产期间做好人工或仪器记录，以证明所有的步骤都是按既定的规程要求进行的，按指南进行实际操作，药物的定量和定性按预期进行。对误差进行调查和行文。生产记录（包括分发）复核能够追踪一个批次的历史，可以用便于理解和方便的表格形式记录。

（1）原材料和原包装材料、废弃材料的记录。

（2）生产批次文件记录：①所用材料的种类、批次和批量；②生产、加工、转运、存储和灌装；③抽样、质检、校正和复核；④批次和成品的编码标记。

（3）成品、抽样、单个实验室检验结果和质量控制状况。

（4）分发记录，包括最初的省份之间的运输记载、编码和收货人。

1.2　中华人民共和国生物药物质量管理

现在 GMP 已经是在全世界范围内被公认的一个有效的、可操作的、可强制执行的能确保药物安全性和高质

raw materials and finished products are retained for the specified time period, are stored under conditions that protect them from contamination or deterioration, and are retested for continued compliance with established acceptance specifications.

(3) The water supply, particularly the water used as a biopharmaceutical ingredient, is tested regularly for conformance with chemical-analytical and microbiological specifications.

(4) Fresh as well as retained samples of finished products are tested for adequacy of preservation against microbial contamination which may occur user reasonably foreseeable condition of storage and consumer use.

1.1.7 Records

Records are made, manually or by instruments, during manufacture that demonstrate that all the steps required by the defined procedures and instructions were in fact taken and that the quantity and quality of the drug was as expected. Deviations are investigated and documented. Records of manufacture (including distribution) that enable the complete history of a batch to be traced are retained in a comprehensible and accessible form.

(1) Raw materials and primary packaging materials, documenting disposition of rejected materials.

(2) Manufacturing of batches, documenting the: ① kinds, lots and quantities of material used; ② processing, handling, transferring, holding and filling; ③ sampling, controlling, adjusting and reworking; ④ code marks of batches and finished products.

(3) Finished products, documenting sampling, individual laboratory controls, test results and control status.

(4) Distribution, documenting initial interstate shipment, code marks and consignees.

1.2 Quality Control of P.R.China

Today GMP is internationally recognized as an effective, practical, and enforceable system for ensuring drug safety and drug quality. China is engaged in bring its pharmaceutical sector to a level where it can compete effectively in the world

量的体系。在中国，GMP 标准基本与世界卫生组织（WHO）的 GMP 指导原则相近。WHO 的标准从 20 世纪 80 年代引进，于 2004 年强制执行。中国国家食品和药物管理局（SFDA）依据中华人民共和国药品管理法制定 GMP 标准，并于 1999 年和 2001 年内几番修订。但 SFDA 的职责不包括监督工厂自行出口的药物中间体的生产。GMP 的实施应该能够强化中国的制药工业。根据新的 GMP 管理准则，药品生产企业必须设立特别行政区办事处，以监督生产和产品质量。管理人员必须是拥有经验的制药行业人员，而负责质量检测的技术人员必须接受专业培训。药品监督和管理总局于 1999 年 9 月颁发“药品生产质量控制框架公约”。该公约规定符合 GMP 标准的药品生产的各种准则。它规定了药品核查和验证，包括设施、设备的安装、操作，属性和产品的有关规定。然而尽管有政府的规定，GMP 的强制执行仍是困难重重，低效的生产和不合格的产品质量在制药厂家时有发生。虽然地方政府正在着手改变，但资金短缺、僵化的经营机制和思想阻力仍持续在一些厂家存在。中国制药企业中合格人才的短缺更进一步拖延了整个国家的 GMP 的实施。不合格的企业缺乏通晓 GMP 的高级管理人员，而且也难以找到训练有素的 GMP 检查员对 GMP 的实施给予公正、客观和准确的评价。一些公司对中国卫生部（MOH）颁发的 GMP 标准的解释含糊不清。Gai 等人的坦诚报道和直率评论讨论了 GMP 对中国制药工业可能的负面影响（Gai *et al.*, 2007）。然而，这些不健全

market. The World Health Organization (WHO) initiated the GMP system in the 1960s, and China adopted it in the early 1980s and made it compulsory in 2004. GMP standard was set by SFDA under the authority of The Drug Administration Law of the People's Republic of China followed with several revisions in 1999 and 2001. However, the SFDA is not responsible for regulating pharmaceutical ingredients manufactured for exporting by chemical companies in China. Under new GMP management guidelines, pharmaceutical producers must set up special administrative offices to supervise production and product quality. Administrative personnel must be pharmaceutical professionals with prior experience, and technicians responsible for quality testing must receive professional training. The State Administration for the Supervision and Management of Medicine issued the “Quality Control Convention in Drug Production” in September 1999. This convention provides guidelines for various kinds of drug manufacture in keeping with GMP standards. It states provisions concerning drug verification and authentication, including facility and equipment installation, operation, property and products. Difficulty in GMP enforcement has allowed inefficient production and substandard quality to persist in the majority of pharmaceutical factories, despite the government's regulations. Fund shortage, rigid operation mechanisms, and ideological resistance among some producers have contributed to the continuing problem, although local governments are working to initiate change. A shortage of qualified personnel in China's pharmaceutical enterprises further delays national GMP implementation. Substandard companies find a lack of senior managers who are aware of GMP, as well as difficulty in finding well-trained GMP inspectors that are able to give a fair, objective and accurate appraisal of GMP results. Augmenting the problem, companies have discovered some ambiguity in their interpretations of GMP standards issued by the Chinese Ministry of Health (MOH). A candid observation and frank discussion on GMP in Chinese pharmaceutical industry was presented by Gai *et al.* as some negative impact on the industry observed (Gai *et al.*, 2007). However, GMP implementation

的GMP管理近年来正在迅速地得到改观，许多企业通过GMP认证，完善了企业内的质量管理，不符合GMP标准的制药企业被淘汰，医药行业的结构得到进一步的调整。GMP的严格实施必将加强中国的医药行业。

分析家观察和分析比较了在用于人类药物产品活性药物成分（API）生产上中国的GMP和ICHQ7规则之间的差异（Goliath，2007）。这一分析报告侧重于准则、人员、质量管理、建筑设施、物料管理和生产管理上的差异。经过5年以上几轮修订，在2011年2月12日，卫生部公布了中国新的药品GMP的最终版本。2011年3月1日起生效的2010年版GMP引进了美国和欧盟的许多重要的概念，采纳了WHO的许多GMP指导原则，极大地提高了中国的GMP水准（中国新药品GMP2011）。

新GMP的主要变化包括以下内容：①加强药物生产质量管理体系的建立；②要求巩固整体员工素质；③细化操作规程、生产记录等文件，如管理规则，以加强指导经营能力；④进一步提高了对应措施，以确保药物的安全性。

1.3 美国及欧洲共同体生物药物质量管理

1.3.1 美国

制药和生物制药产业受美国法律的严格控制。《美国联邦食品、药品和化妆品法》（FDCA）是在1938年国会通过的一系列法律，授权美国FDA来监督食品、药品和化妆品的安全。大部分包括GMP法规在内的药品相

was expected to strengthen China's pharmaceutical industry. Through GMP certification, China has eliminated pharmaceutical enterprises which failed to meet GMP standards; improved quality management among enterprises; and promoted structural adjustment in the pharmaceutical industry.

Differences between Chinese GMP and ICH Q7 guidelines in manufacture of active pharmaceutical ingredients (APIs) for the use of human drug products were observed, analyzed, and compared (Goliath, 2007). The comparisons were focused on guidelines, personnel, quality management, building facility, material management, and production management. On February 12, 2011, after rounds of revisions that took more than 5 years, the MOH published the final version of China's new drug GMP ("GMP 2010"). GMP 2010, effective on March 1, 2011, introduced many key concepts in the GMP regulations of the U.S, and EU, adopted numerous important standards in the GMP guidelines of WHO, and significantly raised the GMP standards in China (China's New Drug GMP 2011).

Key changes in the New GMP: ① strengthens the establishment of a drug manufacturing quality management system; ② bolsters the overall employee quality requirements; ③ refines document management rules such as operating procedures, manufacturing records, etc., in order to enhance guidance and operating capabilities; and ④ further more improves the measures to be taken in order to ensure drug safety.

1.3 Quality Control of United States and Europe

1.3.1 United States

Pharmaceutical and biopharmaceutical industries are closely regulated by Federal laws in the United States. *The United States Federal Food, Drug, and Cosmetic Act* (FDCA) is a set of laws passed by Congress in 1938 giving authority to the US FDA to oversee the safety of food, drugs, and cosmetics. Most pharmaceutical related regulations such as GMP regulations are included in Title 21 of the

关法规描述见于联邦法规第21卷。此规定用“当前良好的生产质量管理规范”（cGMP）来描述这些准则。cGMP的设立有一定的灵活性，以便每个制造商能够通过合理的科学设计、加工方法和测试程序来单独决定如何最好地实施必要的控制措施。这些法规的灵活性允许企业使用现代技术和创新的办法，实现质量的不断改善。

Code of Federal Regulations. The regulations use the phrase “current good manufacturing practices” (cGMP) to describe these guidelines. The cGMP requirements were established to be flexible in order to allow each manufacturer to decide individually how to best implement the necessary controls by using scientifically sound design, processing methods, and testing procedures. The flexibility in these regulations allows companies to use modern technologies and innovative approaches to achieve higher quality through continual improvement.

FDA使用在科学技术和cGMP法规上训练有素的个人检查全球药品制造商的生产设施，其任务是评估这些公司是否遵循cGMP法规。FDA还依赖从公众和工业界的潜在缺陷药品报告来评估这些公司。FDA通常会使用这些报告，以确定是否有必要对某一公司视察或调查。被检查的大多数公司都是完全符合cGMP的规定的。

FDA inspects pharmaceutical manufacturing facilities worldwide using scientifically and cGMP-trained individuals whose job is to evaluate whether the company is following the cGMP regulations. FDA also relies upon reports of potentially defective drug products from the public and the industry. FDA will often use these reports to identify sites for which an inspection or investigation is needed. Most companies that are inspected are found to be fully compliant with the cGMP regulations.

如果一家公司不符合cGMP法规，根据法律规定它生产的任何药品将被视为“掺假”。这种掺假意味着该药物是在不遵守cGMP的条件下制造的。这并不意味着此药是不安全的或无效的。但即使没有违反特定的法规要求，只要生产过程没有按照行业标准执行，法院在理论上就可能不放行该药物产品。

If a company is not complying with cGMP regulations, any drug it makes is considered “adulterated” under the law. This kind of adulteration means that the drug was not manufactured under conditions that comply with cGMP. It does not necessarily mean that there is something wrong with the drug. Courts may withhold adulterated drug products even if no specific regulatory requirement was violated as long as the process was not performed according to industry standards.

对于正在使用没有遵从cGMPs的公司药品的消费者，FDA通常建议他们不要中断他们的药物治疗，因为这可能会对他们的健康造成严重影响。消费者应设法从他们的健康护理专业人员那里就是否停止或改变药物治疗获得建议。针对生产过程不符合FDA的监管标准的公司而采取的监管行动是一项预防措施。通过侧重

For consumers currently taking medicines from a company that was not following cGMPs, FDA usually advises these consumers not to interrupt their drug therapy, which could have serious implications for their health. Consumers should seek advice from their health care professionals before stopping or changing medications. Regulatory actions against companies with poor cGMPs are taken as a preventive measure because the manufacturing processes do not meet the FDA's regulatory standards. By focusing on the procedures

于监管生产药物的使用程序和进程，FDA能确保药品符合质量标准从而是安全和有效的。cGMP违规行为的影响取决于这些违法行为的性质和所涉及的具体药物。在违反cGMP原则下生产的药物很可能仍然满足其标示的规格，此时这种药物是不安全或无效的可能性是微乎其微的。因此，FDA建议保健专业人员应具体问题具体分析，最好能够平衡风险和收益并为他们的病人作出正确的决定。

and processes used to make these drugs, FDA is working to ensure that drugs meet their quality standards and are safe and effective. The impact of cGMP violations depends on the nature of those violations and on the specific drugs involved. A drug manufactured in violation of cGMP may still meet its labeled specifications, and the risk that the drug is unsafe or ineffective could be minimal. Thus, FDA's advice will be specific to the circumstances, and health care professionals are best able to balance risks and benefits and make the right decision for their patients.

《美国食品和药物管理局现代化法案（1997）》（FDAMA）修订了《美国联邦食品、药品和化妆品法》（1938年）。这个法案与FDA监管食品、药品、设备和生物制品有关。这些变化有利于FDA适应21世纪的经营方式的变化。变化的重点在于对技术、贸易和公共健康的复杂性的确认。2003年2月20日，美国食品和药物管理局（FDA）发布了主要涉及药品质量监管的第一份进度报告。经两年的倡议，《21世纪的制药cGMP：一个基于风险的方法》（制药cGMP的倡议）于2002年8月21日推出。它适用于人类药物、生物药物产品和兽药，其目标有：①鼓励制药行业尽早采用新的技术；②促进制造业运用现代质量管理技术，全面保证药品生产和质量；③鼓励实施风险评估，让工业和监管机构把重点放在关键领域；④确保监管的审查和检验是基于最先进的制药科学；⑤加强FDA的药品质量监管方案的一致性和协调性，其中一部分是将更高质量的系统方法纳入监管机构的业务流程和有关审查、检查活动的监管政策中。

The United States FDA Modernization Act of 1997 (FDAMA) amended *the Federal Food, Drug, and Cosmetic Act* (1938). This act is related to the regulation of food, drugs, devices, and biological products by the FDA. These changes were made in order to recognize the changes in the way that the FDA would be operating in the 21st century. The main focus of this is the acknowledgment in the advancement of technological, trade, and public health complexities. On February 20, 2003, the Food and Drug Administration (FDA) released its first progress report on a major initiative concerning the regulation of drug product quality. The 2-year initiative, *Pharmaceutical cGMPs for the 21st Century: A Risk-Based Approach* (the Pharmaceutical cGMP initiative), which was launched on August 21, 2002, applies to human drug and biological drug products and veterinary drugs and has several objectives: ① to encourage the early adoption of new technological advances by the pharmaceutical industry; ② to facilitate industry application of modern quality management techniques, including implementation of quality systems approaches, to all aspects of pharmaceutical production and quality assurance; ③ to encourage implementation of risk-based approaches that focus both industry and Agency attention on critical areas; ④ to ensure that regulatory review and inspection policies are based on state-of-the-art pharmaceutical science; ⑤ to enhance the consistency and coordination of FDA's drug quality regulatory programs, in part, by integrating enhanced quality systems approaches into the Agency's business processes and regulatory policies concerning review and inspection activities.

1.3.2 欧洲共同体

欧洲药品管理局（EMA）是欧洲医药产品的评审机构。它与美国FDA大致平行，但没有FDA集权式的风格。EMA成立于1995年，由欧洲联盟和制药业投资，并得到会员国间接资金补贴，旨在统一（但不是代替）现有国家药品监管机构的工作。此举不仅仅是为了减少制药公司为赢得各会员国单独批准而花费的为数三亿五千万美元的年度成本，也是为了消除各国的保护主义倾向，例如会员国不愿批准那些已在其国内制药公司生产的新药物。欧盟目前每年给世界市场带来大约三分之一的新药。

如此的集中手续允许公司提交一个单一的申报并从欧洲委员会获得一个集中的在所有欧盟和欧洲经济区、欧洲自由贸易联盟国家（冰岛、列支敦士登和挪威）都有效的营销权。此集中程序对所有药物都是强制性的，包括从生物技术和其他高新技术的过程中获得的所有药品，如用于治疗艾滋病毒/艾滋病、癌症、糖尿病、神经退行性疾病、自体免疫和其他免疫功能失调疾病和病毒的人类药物，以及促进增长或增强收益率的兽药。集中程序也对其他产品开放，例如带有显著治疗效果、具有科学或技术创新的产品，以及与病人或动物健康有关联的产品。因此，大多数真正的新型药物都是通过EMA授权的。

每个公司都会对它的合格或需要集中审批的产品向EMA提交营销授权申请。人用医药产品委员会（CHMP）或兽用药品委员会（CVMP）对属于它领域的产品进行

1.3.2 Europe

The European Medicines Agency (EMA) is a European agency for the evaluation of medicinal products. Roughly parallel to the U.S. FDA, but without FDA-style centralization, the EMA was set up in 1995 with funding from the European Union and the pharmaceutical industry, as well as indirect subsidy from member states, in an attempt to harmonize (but not replace) the work of existing national medicine regulatory bodies. The hope was that this plan would not only reduce the $350 million annual cost drug companies incurred by having to win separate approvals from each member state but also that it would eliminate the protectionist tendencies of states unwilling to approve new drugs that might compete with those already produced by domestic drug companies. The EU is currently the source of about one-third of the new drugs brought onto the world market each year.

The centralized procedure allow companies to submit a single application to the Agency and to obtain from the European Commission a centralized marketing authorization that is valid in all EU and EEA-EFTA states (Iceland, Liechtenstein and Norway). The centralized procedure is compulsory for all medicines derived from biotechnology and other high-tech processes, as well as for human medicines for the treatment of HIV/AIDS, cancer, diabetes, neurodegenerative diseases, auto-immune and other immune dysfunctions, and viral diseases, and for veterinary medicines for growth or yield enhancers. The centralized procedure is also open to products that bring a significant therapeutic, scientific or technical innovation, or is in any other respect in the interest of patient or animal health. As a result, the majority of genuinely novel medicines are authorized through the EMA.

For products eligible for or requiring centralized approval, a company submits an application for a marketing authorization to the EMA. A single evaluation is carried out through the Committee for Medicinal Products for Human Use (CHMP) or Committee for Medicinal Products for Veterinary Use

评估。如果相关委员会的结论是充分证明了其药用产品的质量、安全性和疗效，EMA将采纳此正面的评价并将此结论上呈给欧盟委员会，再将其转化为在整个欧盟都有效的营销授权。

CHMP和CVMP有责任在210d内做出决定，但如果要求申请人澄清或提供进一步支持数据时，这时间不计算在内。与此相比，美国FDA却要花平均500d的时间来审批一个新药。

1.4 世界卫生组织（WHO）生物药物质量管理

世界卫生组织（WHO）版本的GMP在全球一百多个国家，主要在发展中国家被药品监管机构和制药行业所使用。欧盟（EU）、世界卫生组织（WHO）和美国的FDA对GMP有类似的要求。类似的GMP也用于其他国家，如澳大利亚、加拿大、日本、新加坡及其他具有复杂GMP要求的国家。在英国，《药品法》（1968）涵盖了GMP规则的方方面面，通常被称为"橙色指南"。它的命名是因为它的封面颜色，它也被正式称为《药品制造商和推销商的规则和指导》。

自从ICH 1999年出版了适用于活性药物成分（API）的GMP以来，GMP现已广泛应用于那些与ICH有合约的国家或地区（欧盟、日本和美国），而在另一些国家（如加拿大、澳大利亚、新加坡）则采用ICH指导原则来制造和测试具有活性的原料药。

(CVMP). If the relevant Committee concludes that the quality, safety and efficacy of the medicinal product are sufficiently proven, EMA adopts a positive opinion. This will be sent to the European Commission to be transformed into a marketing authorization valid for the whole EU.

The CHMP and CVMP are obliged by the Regulation to reach decisions within 210 days, though the clock is stopped if it is necessary to ask the applicant for clarification or further supporting data. This compares well with the average of 500 days spent by the U.S. FDA.

1.4 Quality Control of World Health Organization (WHO)

The World Health Organization (WHO) version of GMP is used by pharmaceutical regulators and the pharmaceutical industry in over one hundred countries worldwide, primarily in the developing world. The European Union's GMP (EU-GMP) enforces similar requirements to WHO GMP, as does the Food and Drug Administration's version in the US. Similar GMPs are used in other countries, with Australia, Canada, Japan, Singapore and others having highly developed/sophisticated GMP requirements. In the United Kingdom, *the Medicines Act* (1968) covers most aspects of GMP which is commonly referred to as "The Orange Guide". It is named so because of the color of its cover; and it is officially known as *Rules and Guidance for Pharmaceutical Manufacturers and Distributors*.

Since the 1999 publication of *GMPs for Active Pharmaceutical Ingredients* by the ICH, GMPs in now applied in those countries and trading groups that are signatories to ICH (the EU, Japan and the U.S.), and in other countries (e.g., Australia, Canada, Singapore) which adopt ICH guidelines for the manufacture and testing of active raw materials.

2 GMP 的指导原则

2 GMP Guidelines

GMP 对确保高品质的药品生产是必不可少的。GMP 指导原则有利于制造厂家在制造过程的所有阶段中实施 GMP 从而消除可能的风险，而这些风险往往在临床试验中不易发现。对于生物制药产品来说，由于生物分子的复杂性和有限的分析手段，在生产过程中的控制和规程制度尤为重要。

"Good manufacturing practices (GMP)" is essential to ensure drugs are produced to a consistently high quality. GMP guidelines are meant to guide manufacturers to eliminate the possibility of risks that may not be detected during clinical trials. For biopharmaceutical products, due to the complexity of bio-molecules and limitation of analytical methods, regulations in in-production process control become even more crucial.

如果没有正确贯彻 GMP 指导原则，质量低劣和不合格药品有可能上市并损害消费者的健康。此外，劣质药物还会导致浪费政府的财政和消耗个人的钱囊。

If GMP guidelines are not correctly implemented, there is a risk that poor quality and substandard pharmaceuticals could make it on to the market and potentially cause damage to the health of members of the public. Poor quality medicines also cause governments and individuals to waste their money.

2.1 WHO 指导原则

2.1 WHO Guidelines

WHO 是严谨的 GMP 原则的强有力支持者，它认为若没有 GMP，生产同样质量的药品是不可能的。WHO 指出了若干项由于不正确遵从药品生产 GMP 原则而可能造成的风险，包括产品污染，损害公共健康而导致死亡，甚至不正确使用标签以至于病人使用错误的药品。如果没有遵从药物活性组分（API）的 GMP 生产指导原则，那么用低标准生产出来的药品可能会治疗无效或产生副作用。因此，生产系统必须同时正确遵从生产（API）和最终产品的 GMP 原则。以 WHO 的观点，实施 GMP 指导原则是对优质药品的一种投资，这将改善个人和整个社会的医疗状况，同时也有益于医药工业和卫生行

The WHO is a strong supporter of stringent GMP guidelines, without which it claims it would be impossible to ensure that all drugs have been manufactured to the same standards. The WHO identifies a number of possible risks which can occur when GMP guidelines for pharmaceuticals are not correctly implemented. The risks include the contamination of manufactured products or patients receiving the wrong medication, which may lead to fatalities. If GMP guidelines for APIs are not followed, there is also a risk that the incorrect active ingredient could be present in medicines, causing treatment to become ineffective or creating adverse effects. Pharmaceutical manufacturing systems must demonstrate that GMP guidelines for pharmaceuticals and for API have been properly followed. According to the WHO, the implementation of GMP guidelines is "an investment in good quality medicines. This will improve the health of the individual patient and the community, as well as benefiting the pharmaceutical industry and health professionals." The WHO recommended GMP guidelines are adopted in a

业。WHO推荐的GMP指导原则已经为一些国家所采用或协调使用，其中有一些国家或机构，包括欧洲共同体、东南亚国家联盟和药品检查协会等，协调使用了WHO和它们自己的指导原则。

number of countries, while a number have harmonized their guidelines including the European Union, the Association of South-East Asian Nations and the Pharmaceutical Inspection Convention.

2.2 FDA指导原则

美国FDA认为制药的GMP指导原则会影响每一个美国人。按FDA的定义，cGMP指导原则要求公司以最新的系统和实践来执行。FDA解释道："十几年到二十年前的可谓最先进的防止污染、防止混淆和防止错误的系统和仪器以今天的目光来看已不合适了。" 但是FDA又强调制药的GMP指导原则应有一定的弹性，使得制造厂家能最恰当地控制生产过程，这是制药工业必须遵从的最低标准。FDA派出专业人员进行厂房检查，保证规范被正确执行。

2.2 FDA Guidelines

The United States Agency FDA takes the view that GMP guidelines for pharmaceuticals "affect every American". Current GMP (cGMP) guidelines, as defined by the FDA, require companies to use the most up-to-date systems and practices in order to comply with the regulations. "Systems and equipment that may have been 'top-of-the-line' to prevent contamination, mix-ups, and errors ten or twenty years ago may be less than adequate by today's standards," the FDA explains. However, the FDA stresses that its GMP guidelines for pharmaceuticals, which are flexible to allow each manufacturer to create the most appropriate controls for its processes, are the minimum standards that the pharmaceutical industry should meet. To ensure that regulations are followed correctly, the FDA sends trained professionals to inspect facilities.

2.3 ICH指导原则

人用药品注册技术要求国际协调会指导原则（ICH）是人用药物注册的技术要求(Ref: http://www.ich.org/products/guidelines)。它把来自欧洲联盟、日本和美国三方面的规范专职人员和工业界专家集合在一起，讨论产品注册的科学和技术问题。具体有六个单位参加：①欧洲联盟，有25个国家成员，由欧洲药品局负责具体事务；②欧洲制药工业药协会，由29个制药工业协会及45家从事研发和生产药品的大公司组成；③美国食品与药物管理局；④美国药物研究的生

2.3 ICH Guidelines

The International Conference on Harmonization (ICH) was a result of technical requirements for registration of pharmaceuticals for human use (Ref: http://www.ich.org/products/guidelines). ICH brings the regulatory authorities and industry experts from Europe, Japan and the United States together to discuss scientific and technical aspects of product registration. Six organizations participated in the conference: ① European Union (EU). EU represents 25 member countries and European Medicines Agency is responsible for implementing much of the new legislation; ② European Federation of Pharmaceutical Industries Association (EFPIA) including 29 association members and 45 large pharmaceutical companies; ③ US FDA; ④ Pharmaceutical Research and Manufacturers of America, PRMA; ⑤ Ministry of Health, Labor and Welfare (MHLW); ⑥ Japan Pharmaceutical

产联合会；⑤日本厚生省；⑥日本制药工业协会。

Manufacturers Association (JPMA).

协调会的宗旨是通过推荐技术指导和产品注册的统一性来降低新药研发过程中的重复测试。这种协调将更为有效地利用人、动物和物质资源，减少不必要的国际新药研发和注册的延迟，同时确立其质量、安全和有效的公共健康规范。协调会的指导条例已经为几个国家采纳作为法律文件，但在美国仅限于美国FDA用来作为指导文件。

The purpose of ICH is to reduce the need to duplicate the testing during the research and development of new medicines by recommending harmonization in the technical guidelines and requirements for product registration. Harmonization would lead to a more economical use of human, animal and material resources, and the elimination of unnecessary delay in the global development and availability of new medicines while maintaining safeguards on quality, safety, and efficacy, and regulatory obligations to protect public health. ICH guidelines have been adopted as law in several countries, but are only used as guidance for the U.S. FDA.

协调会的专题共分四个类别，指导条例是以这些类别分别归类编号。与生物药品有关的指导条例大都在质量和安全性两大类中。对制药工业指导的GMP条例编号为Q7A。

The ICH topics are divided into four categories and ICH topic codes are assigned according to these categories. Biopharmaceutical manufacturing related guidelines are mostly in Quality Guidelines and Safety Guidelines. Specific GMP guidance for industry to manufacture API is in Q7A.

1. 质量方面的技术要求原则 在质量领域的协调成就表现在进行稳定性试验，确定对杂质测定的最低量，以及基于GMP风险管理之上的保证药物质量的灵活方法。

1. Quality Guidelines Harmonization achievements in the Quality area include pivotal milestones such as the conduct of stability studies, defining relevant thresholds for impurities testing and more flexible approaches to pharmaceutical quality based on GMP risk management.

2. 安全性方面的技术要求指导原则 协调会形成了一套综合性的安全指导要求。这些技术要求指导条例分析了潜在的风险，诸如致癌性、遗传毒性和生殖毒性。最近的突破是人用药延迟心室复极化（QT间期延长）潜在作用的非临床评价指导原则，这已成为近年来撤销药物申报的最重要的原因。

2. Safety Guidelines ICH has produced a comprehensive set of safety guidelines to uncover potential risks like carcinogenicity, genotoxicity and reprotoxicity. A recent breakthrough has been a non-clinical testing strategy for assessing the QT interval prolongation liability: the single most important cause of drug withdrawals in recent years.

3. 有效性的技术原则 这类原则包括临床试验的有效性方案设计、方案实施与研究报告。它也包括了生物技术生产的新型药物以及运用药物遗传和基因组技术来生产更好的靶药。

3. Efficacy Guidelines The work carried out by ICH under the Efficacy heading is concerned with the design, conduct, safety, and reporting of clinical trials. It also covers novel types of medicines derived from biotechnological processes and the use of pharmacogenetics/genomics techniques to produce better targeted medicines.

4．综合类的技术要求原则　这类指导原则包括了所有不适于任何一个诸如质量、安全和有效性类别的原则，如医学术语、通用技术文件（CTD），以及注册资料传递所需的电子代码。

4. Multidisciplinary Guidelines Those are the cross-cutting topics which do not fit uniquely into one of the Quality, Safety and Efficacy categories. It includes the ICH medical terminology (MedDRA), the Common Technical Document (CTD) and the development of Electronic Standards for the Transfer of Regulatory Information (ESTRI).

3 药　典

3 Pharmacopoeia

药典从字面上看指“药品-制造”，用现代技术来释义，是一本含有鉴定样品成分和制备复方药剂的指南。它是由政府授权者，或医疗和医药行业发布。在更广泛的意义上，它是药品规格的参考书。

Pharmacopoeia, pharmacopeia, or pharmacopoea (literally, “drug-making”), in its modern technical sense, is a book containing directions for the identification of samples and the preparation of compound medicines, and it is published by the authority of a government or a medical or pharmaceutical society. In a broader sense it is a reference work for pharmaceutical drug specifications.

3.1 国际药典

国际药典（IP）包括一系列推荐的药用物质（活性成分和辅料）、剂型的分析和质量规范程序的集合。它作为参考资源，为任何希望建立制药规范要求的世卫组织会员国提供服务。第一届世界卫生大会（WHA）统一药典专家委员会成立于1948年，并于1950年首次出版由世界卫生大会批准的《国际药典》，其最新版本是2015年发布的第五版。有人建议，国际药典的目的不是用作任何一个国家的法律药典，除非该国药典机构采用此药典。

国际药典的基本工作是整体促进药品的安全性和疗效的质量控制。入选国际药典的标准是基于特定疾病的需要以及这些疾病所需要的基本药物，并且同时也是根据目前世卫组织

3.1 International Pharmacopeia

The International Pharmacopoeia (IP) comprises a collection of recommended procedures for the analysis and quality specifications for the determination of pharmaceutical substances (active ingredients and excipients) and dosage forms. It intended to serve as source material for reference or adaptation by any WHO Member State wishing to establish pharmaceutical requirements. First World Health Assembly (WHA) established Expert Committee on Unification of Pharmacopoeia in 1948 and the first publication of *Pharmacopoeia Internationalis* (International Pharmacopeia) was approved by WHA in 1950, While the latest edition (the fifth edition) was published in 2015. It was recommended that the Pharmacopoeia Internationalis was not intended to be a legal pharmacopoeia in any country unless adopted by the pharmacopoeia authority of that country.

The activities related to The International Pharmacopoeia are essential elements in the overall quality control and assurance of pharmaceuticals contributing to the safety and efficacy of medicines. The selection of monographs for inclusion in The International Pharmacopoeia recognizes the needs of specific disease programs and the essential medicines nominated under these

基本药物标准清单来制定的。

自1979年以来，出现在IP的药物是从世卫组织专家委员会的第一次报告中的基本药物遴选清单中选定。专论提供列于世界卫生组织基本药物标准清单规格中出现的基本药物的鉴定、纯度和含量。大多数的抗生素和合成小分子药物被列入名单，大部分生物制药不在列表中，如天门冬氨酸和血液制品第八因子治疗酶。

programs; it is based primarily on those substances included in the current WHO Model List of Essential Medicines.

Since 1979, the drugs appearing in the IP have been selected from the list of essential drugs based on the first report of the WHO Expert Committee on the Selection of Essential Drugs. Specifications are provided in the monographs for the identification, purity, and content of the essential drugs appearing in the WHO Model List of Essential Drugs, and their updates. While most antibiotics and synthesized small molecular drugs were included into the list, most biopharmaceuticals are not in the list except for a couple of therapeutic enzymes such as asparaginase and blood products such as factor Ⅷ.

3.2 美国药典

3.2 United State Pharmacopeia

美国药典（USP）是一个非政府的，为在美国生产或销售的处方药和非处方药品以及保健产品设定的公共标准权威。美国药典规定对公共卫生至关重要产品的质量、纯度和强度的标准。USP标准被世界上130多个国家和地区认可并使用。

美国药典规定用于药品、食品添加剂、膳食补充剂产品和原料的书面（文件）和实物（参考）标准品。监管机构和制造商使用这些标准品以确保这些产品具有适当的成分、效价、质量、纯度和重现性。

只要USP制定的这些标准存在，在美国的处方和非处方药品必须根据联邦法律规定符合USP-NF的公共标准。许多其他国家使用USP-NF作为官方药典，或政府药典的补充。

美国药典-国家处方集（USP-NF）是一个公共的药典标准。它包含标准的药品、制剂、原料药、辅料、医疗器械和膳食补充剂。USP-NF是两部正式的汇编组合，即美国药典

U.S. Pharmacopeia (USP) was founded in 1820 and is a non-governmental, official public standards-setting authority for prescription and over-the-counter medicines and healthcare products manufactured or sold in the United States. USP sets standards for the quality, purity, strength, and consistency of these products-critical to the public health. USP's standards are recognized and used in more than 130 countries around world.

USP establishes written (documentary) and physical (reference) standards for medicines, food ingredients, dietary supplement products and ingredients. These standards are used by regulatory agencies and manufacturers to help to ensure that these products are of the appropriate identity, as well as strength, quality, purity, and consistency.

Prescription and over-the-counter medicines available in the United States must, by federal law, meet USP-NF public standards, where such standards exist. Many other countries use the USP-NF instead of issuing their own pharmacopeia, or to supplement their government pharmacopeia.

The United States Pharmacopeia-National Formulary (USP-NF) is a book of public pharmacopeia standards. It contains standards for medicines, dosage forms, drug substances, excipients, medical devices, and dietary supplements. The USP-NF is a combination of two official compendia, the United States Pharmacopeia (USP) and

（USP）和国家处方集（NF）。特色原料药和制剂的色谱在USP里出现。食品添加剂和组分的色谱则以一个单独的章节出现在USP里，辅料归在NF里。

药典标准包括药品成分或制剂的名称、定义、包装、储存、标签要求和规范。该规范包括一系列的测试、测试的程序和验收标准。这些测试和程序需要使用正式的USP标准品。如果测试的药品或其组分与标准品和其他有关的一般描述相符，其生产出来的药用成分和产品将会符合美国药典规定的浓度、质量和纯度。

the National Formulary (NF). Monographs for drug substances and preparations are featured in the USP. Monographs for dietary supplements and ingredients appear in a separate section of the USP. Excipient monographs are in the NF.

A monograph includes the name of the ingredient or preparation; the definition; packaging, storage, and labeling requirements; and the specification. The specification consists of a series of tests, procedures for the tests, and acceptance criteria. These tests and procedures require the use of official USP Reference Standards. Medicinal ingredients and products will have the stipulated strength, quality, and purity if they conform to the requirements of the monograph and relevant general chapters.

3.3 欧洲药典

欧洲理事会的欧洲药典列出了多种用于欧洲药物生产的活性物质和辅料。它包括2000多个具体和一般标准，包括各种化学物质、抗生素、生物物质、人用或兽用疫苗；以及免疫血清、放射性药物制剂、草药、顺势疗法制剂和顺势疗法的药用原液。它还包含有剂型、一般标准、材料和容器、手术缝合线等。此药典描述了268种带有图示的一般方法或层析图谱，以及2210种试剂。此药典标准是所有在欧洲使用的主要药物的质量标准。在36个欧洲药典成员国内销售的所有药品必须符合这些质量标准，使消费者从药店和其他合法供应商获得的产品得到保证。

欧洲药典是由欧洲药品质量决策者（EDQM）制定的，并是法国斯特拉斯堡欧洲议会的一部分。它

3.3 European Pharmacopoeia

The European Pharmacopoeia (*Ph. Eur.*) of the Council of Europe is a pharmacopoeia, listing a wide range of active substances and excipients used to prepare pharmaceutical products in Europe. It includes more than 2000 specific and general monographs, including various chemical substances, antibiotics, biological substances; vaccines for human or veterinary use; immunosera; radiopharmaceutical preparations; herbal drugs; homoeopathic preparations and homoeopathic stocks. It also contains Dosage forms, General monographs, Materials and Containers, Sutures. There are 268 general methods with figures or chromatograms and 2210 reagents described. The monographs give quality standards for all the main medicines used in Europe. All medicines sold in the 36 Member States of the European Pharmacopoeia must comply with these quality standards so that consumers have a guarantee for products obtained from pharmacies and other legal suppliers.

The European Pharmacopoeia is developed by the European Directorate for the Quality of Medicines (EDQM) and is a part of the Council

是自1964年以来由《欧洲药典公约》创建的。虽然药典常常被视为欧洲理事会中部分协定的一部分，严格地说它不是一个部分协议。它有一个和其他不同的法律依据，是一项条约的结果。其他的都是通过由部长委员会通过的"法定决议"来决定的。药典由EDQM用英语和法语出版，也有正式翻译成德语和西班牙语的版本。

of Europe, Strasbourg, France. It has been created by the *Convention on the elaboration of a European Pharmacopoeia* from 1964. Although the Pharmacopoeia is often listed as one of Council of Europe partial agreements, it is not, strictly speaking, a partial agreement. It has a different legal basis from the others, being established as the result of a treaty. The others were established by a "statutory resolution" adopted by the Committee of Ministers. It is published by EDQM in English and French Official national translations are available in German and Spanish.

4 疫苗的安全性和注册管理

4 Vaccine Safety and Regulatory Controls

开发疫苗产品比其他小分子产品更为困难。第一，灭活或减毒活疫苗的体积和复杂性要远比那些分子量不超过100kDa的单个蛋白的小分子生物药物来的大。举例来说，HPV的VLP疫苗的体积是蛋白分子的10 000倍（Buckland，2005）。这就需要在整个药物开发过程中使用复杂的分析方法检测多组分结构的疫苗。考察研发的历史告诉我们，生产过程本身决定产品的质量。细胞生长条件决定疫苗的结构，例如膜的组成和表面抗原的化学计量分布。纯化过程保证最终产品的安全纯度。质量控制部门必须对产品的开发过程进行严密的监视。第二，不像其他用于病人治病的药物产品，疫苗是用于健康人体的，包括成人和儿童。仅在美国，每年有约四百万儿童注射常规疫苗。疫苗的安全性因此要求极高。必须有包括所有使用仪器和每一步骤的研发及生产的完整文件。被批准上市后的疫苗仍需

Developing a vaccine product is as or even more challenge as compared to many therapeutic biomolecules. Firstly, killed or attenuated vaccine is far bigger and complex than biomolecule drugs which is single protein (in most cases, such as antibodies) with usually not more than 100K Dalton molecular weight. For example, HPV's VLP vaccine is about 10 000 times larger than protein molecule (Buckland, 2005). This requires sophisticated analytical method to characterize the multi-components structure of a vaccine through its developing life cycle. A historic trend is observed in development that the process of producing the vaccine itself defines the product quality. Cell growth condition may impact the structure of these vaccines, such as component of membrane and stoichiometric distribution of the surface antigens. Purification ensures a safe purity of the final product. A close monitoring on process development and product manufacturing by a quality management system is necessary. Secondly, unlike other biopharmaceutical products which may be applied to patients who are suffering from disease, vaccines are targeted on healthy and large population, including adults and children. Just in the United States, about 4 million children are routinely taken recommended childhood vaccines. The demand for vaccine safety is hence extremely high. A thorough documentation is needed for vaccine process development and manufacture including specific instrument and each

通过监视和观察的方式继续评估其安全性。第三，由于许多疫苗是从具致病性的菌或病毒生产而来的，有关操作人员在疫苗的研发和生产过程中的安全性必须严加控制。

step of process. Continuous evaluation on vaccine safety after licensure is also needed through surveillance and observational studies. Thirdly, safety to employee during vaccine development and manufacturing need be carefully controlled since many vaccine candidates are pathogenic selected agents.

疫苗的全球性规范化被发达和发展中国家所共识。世界卫生组织（WHO）对疫苗产品的关键要求提供了指导原则（http://www.who.int/topics/vaccines/en）。WHO 识别和加强了对于关键法规项目的最新认知，用提供对常规和特殊产品的指导文件方式来与各国家管制机构（National Regulatory Authorities，NRA）和生产厂家进行交流。通过这个方式，NRA 监视发达国家的已批准产品和发展中国家的新产品，以保证全球疫苗产品的安全性和有效性。100 多个国家和地区已采用了世卫组织要求作为自己国家的要求，包括非洲、亚洲、南美和北美、澳大利亚和欧洲地区。这些国家和地区的管制机构需要迅速地：①建立监管战略，以规范在本国正在进行的临床试验；②审批临床试验申请；③评估申请执照的临床数据和产品特性。

Regulation on vaccine is recognized worldwide by developed and developing countries. The WHO provides guidance regarding acceptance criteria of vaccine products (http://www.who.int/topics/vaccines/en). WHO identifies and consolidates current consensus opinions on key regulatory issues, and communicates them to National Regulatory Authorities (NRA) and manufacturers through guidance documents addressing both general issues and specific products. Through this mechanism, NRA oversights on not only the vaccines that were licensed in developed counties, but also on new vaccines that developed for developing world and sometimes exclusively in these countries, to ensure the global manufacturing of high quality and efficacious vaccines. More than 100 countries have adopted WHO requirements as their own national requirements. These countries include regions of Africa, Asia, South and North America, Australia and Europe. The NRAs of these countries need to be rapidly able to ① establish regulatory strategies to regulate clinical trials taking place in their countries; ② perform a regulatory review of clinical trial applications; ③ assess clinical data and product characteristics to respond to license application.

在美国，负责疫苗监管的机构是 FDA 的生物制品评估和研究中心（CBER）。CBER 同时与其他的政府机构相关部门，包括疾病控制与预防中心（CDC）和国立卫生研究院（NIH）密切合作。CDC 对疾病的流行病学监测和免疫项目的支持负责，它的免疫操作咨询委员会（Advisory Committee on Immunization Practices, ACIP）推荐疫苗的使用。像其他的生物产品一

In the Unites States, the organization responsible for vaccines regulation is the Center for Biological Evaluation and Research (CBER) of the FDA. CBER also works closely with its counterparts in other government agencies, including Center for Disease Control and Prevention (CDC) and the National Institute of Health (NIH). The CDC is responsible for epidemiological surveillance of disease and for support of immunization programs. Its Advisory Committee on Immunization Practices (ACIP) makes recommendations for vaccine use. Like many other biologics, vaccine development goes through Investigational New Drug Applications

样，疫苗的开发也需经历新药临床批文（IND）、三阶段的临床试验、生物产品执照申请（BLA）和上市后的监测（Vaccine Product Approval Process, CBER; Marshall and Baylor，2011）。IND需要生产厂家提交有关产品的化学、生产和控制的信息（CMC），在安全性上的和引发免疫反应能力的动物毒理试验结果，以及临床试验的规划。一期临床试验主要是在一小组健康人员范围内测定其安全性和免疫原性，二期临床在几百人身上测定剂量范围，三期临床在上千人中继续测定安全性并检查其有效性。如果所有临床试验都成功，厂家将向FDA申报美国的BLA。FDA只有在全面评估了产品的风险与利益，以及检查了厂家上市产品的生产厂房后，才会授予厂家销售执照。申请欧盟、日本、中国或其他国家的疫苗执照必须遵从他们各自的规定和审查机构，例如欧盟的EMA、日本的JP和中国的SFDA。

(IND) which include three phases clinical trials, Biological Licensure Application (BLA), and post marketing surveillance (Vaccine Product Approval Process, CBER; Marshall and Baylor, 2011). The IND requires the applicant to submit information package called “Chemistry, Manufacturing and Control (CMC)” describing the production, animal study results on safety and capability of eliciting immune response, and a protocol for clinical trials. Phase Ⅰ clinical trial targets safety and immunogenicity on a small group of healthy people. Phase Ⅱ studies look for a dose-response on hundreds of subjects. Phase Ⅲ trials test effectiveness and further safety on thousands of people. Upon successful of clinical trials, the sponsor will submit BLA to the FDA for licensure in the United States. A BLA will be granted after FDA fully evaluates on the risk/benefit of the submitted vaccine candidate and inspection of the facility where the vaccine is going to be manufactured. Application for vaccine license in Europe, Japan, China or other countries follows regulations correspondently, such as regulation for European Union (EU), JP’s regulations for Japan, and SFDA’s regulations for China.

一旦产品被FDA或其他机构批准，上市后的产品监视将成为重要的检测疫苗安全性的阶段。所有的疫苗都或多或少地有一些副作用。注射部位的疼痛和发炎都属于小的副作用。中等和严重的副作用如热型惊厥和过敏反应只有上千分之一至几百万分之一的发生概率（Maldonado，2002）。在产品执照批准前未被发现的不良反应在此阶段将被严密关注，且生产厂家必须继续其安全性报告。

Once the product is approved by FDA or other authorities, post-marketing surveillance becomes an important stage to monitor vaccine safety. All vaccines have some extent of adverse reaction. Among them, pain and inflammation at the injection site are considered as minor. Moderate to severe reactions such as febrile seizures and allergic reactions only happen at one out of thousands or millions of dose administered (Maldonado, 2002). Adverse event that was not detected during pre-licensure will be closely watched, and ongoing reports on safety are required from manufacturer.

（编者　朱建伟）

(Compiled by Jianwei Zhu)

参 考 文 献
References

Bonanni P. 1999. Demographic impact of vaccination: a review. Vaccine, 17 (suppl 3):S120-125

Buckland BC. 2005. The Process Development Challenge for a New Vaccine, Nature Med Suppl, 4:S16-S19

Center for Disease Control and Prevention, USA. 2009. Licensure of a Haemophilus influenzae Type b (Hib) Vaccine (Hiberix) and Updated Recommendations for use of Hib Vaccine. Morb Mortal Wkly Rep, 58:1008-1009

Ding XY, Becht S, Gu X. 2007. Vaccine Characterization Using Advanced Technology. BioPharm Aug 2

Gerdil C. 2003. The Annual Production Cycle for Influenza Vaccine. Vaccine, 21:1776-1779

Kistner O, Howard K, Spruth M, et al. 2007. Cell culture (Vero) derived whole virus (H5N1) vaccine based on wild-type virus strain induces cross-protective immune responses. Vaccine, 25(32): 6028-6036

Liu MA. 2011. DNA Vaccines: An Historical Perspective and View to the Future. Immunol Rev, 239:62-84

Marshall V, Baylor NW. 2011. Food and Drug Administration Regulation and Evaluation of Vaccines. Pediatrics, 127:S23-S30

Matthews J T. 2006. Egg-Based Production of Influenza Vaccine: 30 Years of Commercial Experience, The Bridge, 36:6

Research and Markets Company. 2010. Vaccines Review and Outlook 2010-Market Research Report, http://www.researchandmarkets.com/reports/1370489

Ulmer J B, Valley U, Rappuoli R. 2006a. Vaccine manufacturing: Challenges and Solutions. Nature Biotechnol, 24:1377-1383

第五章 基因工程药物

Chapter 5 Gene Engineering Drugs

1 白 介 素

1 Interleukin

1.1 白介素概述

白细胞介素（Interleukin，IL）简称白介素，最初因由白细胞产生又在白细胞间发挥作用而得名。在1979年第二届国际淋巴因子专题的讨论会上，人们将来自单核-巨噬细胞、淋巴细胞所分泌的参与免疫调节、造血以及在炎症反应中起作用的因子，正式命名为白介素，并沿用至今。然而，目前发现许多白介素却来自单核-巨噬细胞和淋巴细胞以外的其他细胞。因此，并不能完全从最初的产生来源来定义是否为白介素。准确地说，白介素是一类可由多种细胞产生并作用于不同类型细胞的细胞因子，是一组结构和功能各异的可溶性蛋白。它与血细胞生长因子相互协调、相互作用、共同完成造血和免疫调节功能，在传递信息、激活与调节免疫细胞、介导T淋巴细胞和B淋巴细胞活化、增殖与分化，以及在炎症反应中起重要作用。迄今为止，已正式报道的白介素有30多种，今后可能有更多的白介素被继续发现。对于具体的某一种白介素，其命名方式是在IL符号后加相应的数字，如白介素1，记为IL-1。

1.1 Introduction of Interleukin

Interleukin, referred to as IL, originally was produced by white blood cells, and play a role in the white blood cells, hence named interleukin. At the Second International Symposium about Lymphocyte in 1979, the cytokines had been formally named as interleukin which are secreted from monocot/macrophage, lymphocyte, and take effect in terms of unspecific immunoregulation and inflammatory response. But many known interleukins come from other cells than monocytes, macrophages and lymphocytes. Therefore, it cannot be completely generated from the original source to define IL. Precisely, interleukins are that generated by a variety of cells and act on a variety of cell type cytokines, and they are group of soluble proteins with different structures and functions. Like the blood cell growth factor, IL also belongs to the cytokines, and they coordinated interaction together with each other to complete hematopoietic and immune regulation. Interleukins have many important functions such as signal transduction, activation and accommodation of immunocytes, mediation of T and B cells' activation, proliferation, and differentiation, and inflammatory responses. Currently, there are about 30 kinds of IL reported, and there may be more interleukins to be found in the future. Every specific representation of interleukins, IL symbols added in the corresponding figures, such as interleukin-1, denoted by IL-1.

大多数白介素可由许多不同类型的细胞产生。到目前为止，至少有17种不同类型的细胞能产生IL-1，有10种不同类型的细胞可产生IL-8，但IL-2、IL-9、IL-13却只能由T淋巴细胞产生。除了IL-1之外，这些起调节作用的白介素大多为糖基化多肽，其分子量均在15～30kDa之间。少数白介素具有更大分子量，如高度糖基化的IL-9，分子量约为40kDa。除了IL-1有一种形式是细胞缔合之外，几乎所有的白介素都是可溶性分子，它们通过与靶细胞表面的特异性受体结合引发生物学反应。大多数白介素具有旁分泌活性，即它所作用的靶细胞紧邻其生成细胞；也有一些具有自分泌活性，如IL-2能够刺激其产生细胞的生长和分化；其他的白介素更多地显示出系统内分泌效应。通过白介素诱导的生物学效应广泛多样且非常复杂。这些细胞因子参与调节许多生理和病理过程，包括正常细胞与恶性肿瘤细胞的增殖、免疫反应以及炎症反应等。表5-1给出了现有的研究较清楚的主要白介素的相关信息。

Most of the interleukins can be produced by a number of different cell types. For example, IL-1 can be secreted by at least 17 different cell types, and IL-8 is produced by at least 10, however, IL-2, IL -9 and IL -13 are produced only by T lymphocytes. Except for IL-1, nearly all the interleukins are glycosylated, and display a molecular mass in the range 15-30 kDa. A few interleukins display a higher molecular mass, e.g. IL-9, the heavily glycosylated, has the molecular of 40 kDa. Nearly all ILs are soluble molecules (one form of IL-1 is cell-associated). They promote their biological response by binding to specific receptors on the surface of target cells. Most ILs exhibit paracrine activity, i.e. the target cells are in the immediate vicinity of the producer cells. While some display autocrine activity, e.g. IL-2 can stimulate the growth and di erentiation of the cells that produce it. Other ILs display more systematic endocrine e ects, e.g. IL-1. The sum total of biological responses induced by the ILs is large, varied and exceedingly complex, and these cytokines regulate a variety of physiological and pathological conditions, including: normal and malignant cell growth, all aspects of the immune response, regulation of inflammation. Table 5-1 shows some information of some major human interleukins.

表5-1 部分主要人白介素的相关信息列表

名称	来源	靶细胞	作用
IL-1	单核-巨噬细胞，B细胞，树状细胞	辅助性T细胞	联合刺激
		B细胞	成熟，增殖
		NK细胞	活化
		巨噬细胞，内皮细胞	炎症，少量可引起急性期反应，大量可引起发热
IL-2	1型辅助性T细胞	活化的T细胞，B细胞，NK细胞，巨噬细胞，少突胶质细胞	刺激T细胞生长和分化，用于免疫疗法治疗癌症或抑制接受移植病人的不良反应；临床上用来增加HIV阳性患者体内CD4细胞数量
IL-3	活化的辅助性T细胞，肥大细胞，NK细胞，嗜酸性粒细胞，内皮细胞	造血干细胞	骨髓祖细胞向红细胞、粒细胞的分化和增殖

续表

名　称	来　源	靶　细　胞	作　用
IL-4	2 型辅助性 T 细胞，活化的初始型 $CD4^+$ 细胞，记忆性 $CD4^+$ 细胞，肥大细胞，巨噬细胞	活化的 B 细胞	在过敏反应中起重要作用，IgG1 和 IgE 的合成、增殖及分化
		T 细胞	增殖
		内皮细胞	促进血管内皮细胞黏附分子的表达，加重炎症
IL-5	2 型辅助性 T 细胞，嗜酸性粒细胞	嗜酸性粒细胞	产生
		B 细胞	分化，IgA 的产生
IL-6	巨噬细胞，2 型辅助性 T 细胞，B 细胞，星形胶质细胞，内皮细胞	活化的 B 细胞	向浆细胞分化
		浆细胞	分泌抗体
		造血干细胞	分化
		T 细胞	诱导急性期反应，造血作用，分化，炎症反应
IL-7	骨髓基质细胞，胸腺基质细胞	前 B 细胞，前 T 细胞，NK 细胞	淋巴祖细胞的分化和增殖，包括 B 细胞、T 细胞和 NK 细胞的存活、发育和体内平衡；上调促炎症细胞因子
IL-8	巨噬细胞，淋巴细胞，上皮细胞，内皮细胞	中性粒细胞，嗜碱性粒细胞，淋巴细胞	中性粒细胞趋化性
IL-9	2 型辅助性 T 细胞，特别是辅助性 $CD4^+$ 细胞	T 细胞，B 细胞	增强 IgM、IgG 和 IgE 的作用，刺激肥大细胞
IL-10	单核 - 巨噬细胞，2 型辅助性 T 细胞，$CD8^+$T 细胞，肥大细胞，B 细胞	巨噬细胞	产生细胞因子
		B 细胞	活化作用
		肥大细胞	
		1 型辅助性 T 细胞	抑制 1 型辅助性 T 细胞因子的产生（IFN-γ，TNF-α，IL-2）
		Th2 细胞	刺激
IL-11	骨髓基质	骨髓基质	产生急性期蛋白，破骨细胞的形成
IL-12	树状细胞，B 细胞，T 细胞，巨噬细胞	活化的 T 细胞	向细胞毒性 T 细胞分化，上调 IFN-γ、TNF-a，下调 IL-10
		NK 细胞	上调 IFN-γ、TNF- α
IL-13	活化的 2 型辅助性 T 细胞，肥大细胞，NK 细胞	2 型辅助性 T 细胞，B 细胞，巨噬细胞	刺激 B 细胞的生长和分化，抑制 1 型辅助性 T 细胞和巨噬细胞炎症细胞因子产生（IL-1、IL-6）
IL-14	T 细胞，某些恶性 B 细胞	活化的 B 细胞	控制 B 细胞的生长和增殖，抑制 Ig 的分泌
IL-15	单核吞噬细胞，特别是受病毒感染的巨噬细胞	T 细胞，活化的 B 细胞	诱导 NK 细胞的产生
IL-16	淋巴细胞，上皮细胞，内皮细胞，$CD8^+$T 细胞	$CD4^+$T 细胞	$CD4^+$ 化学引诱物
IL-17	17 型辅助性 T 细胞	上皮细胞，内皮细胞	破骨细胞生成，血管生成，上调炎症细胞因子
IL-18	巨噬细胞	1 型辅助性 T 细胞，NK 细胞	诱导 IFN-γ 的产生，上调 NK 细胞的活性
IL-21	活化的辅助性 T 细胞，NK 细胞	所有淋巴细胞，树状细胞	刺激 $CD8^+$ T 细胞的活化和增殖，增强 NK 细胞的细胞毒性，增强 $CD4^-$ 细胞催化 B 细胞的增殖、分化和同种型转换，促进 17 型辅助性 T 细胞的分化

免疫系统的功能在很大程度上依赖于白介素才能发挥，一些少见的缺乏白介素的情况会导致自身免疫性疾病或免疫缺陷。其中一些白介素，尤其是那些能调节转化细胞生长以及具有免疫刺激特性的白介素，在临床的应用中受到格外关注。同其他细胞因子一样，DNA 重组技术的出现使得白介素能够大量地生产以满足实际与潜在的医学需求。虽然在未来几年里可能有其他的白介素将得到认可，但是只有综合临床试验才能揭示这个细胞因子家族真正的临床意义。

The function of the immune system depends in a large part on interleukins, and rare deficiencies of a number of them have been described, all featuring autoimmune diseases or immune deficiency. Several ILs, particularly those capable of modulating transformed cell growth, as well as those exhibiting immunostimulatory properties, enjoy significant clinical interest. As with other cytokines, the advent of recombinant DNA technology facilitates production of these molecules in quantities su cient to meet actual/potential medical needs. While it is likely that several additional IL preparations will gain regulatory approval over the next few years, only comprehensive clinical trials will reveal the true medical significance of this family of cytokines. Interleukin-11.

1.2 白介素 11

作为白介素家族的一种，与其他白介素相比，白介素 11（interleukin-11，IL-11）虽然发现的时间不长，但有关对 IL-11 的研究却已很深入。IL-11 是一种多效性的细胞因子，具有多种生物功能，如刺激巨核细胞的成熟和分化，升高血小板计数；保护胃肠道黏膜；下调致炎性细胞因子。这预示着重组人 IL-11 在临床上有好的应用前景。重组人 IL-11 是 1997 年 10 月美国食品与药物管理局（FDA）批准上市、用于临床第一个促进血小板生成的细胞因子，用于防止化疗引起的再生性血小板减少症。另外国内也已开发出同类产品，并已进入临床试用。因此，IL-11 在预防和治疗化疗所致的血小板减少症及非造血系统有良好的应用前景。

1.2 Interleukin-11

As a member of interleukin family, compared with others, interleukin-11 (IL-11) had been found for just several years, but the researches on IL-11 are deeply enough. IL-11 is a kind polyphenic cytekine, with many biological functions such as stimulating megakayocyte mature and differentiation to increases platelet counts; protecting gastrointinal tract mucous membrane; down-regulating inflammatory cytokine production. Therefore, recombinant IL-11 has good application prospect in clinical therapy. Recombinant human IL-11 was approved for clinical use in October 1997 by FDA, and it was used for the prevention and treatment of thrombocytopenia. Besides, many congeneric products have been used in China. IL-11 has shown promising potential for applications in the prevention and treatment of thrombocytopenia and disorders in non-hematopoietic systems.

IL-11 最初是美国学者 Paul 及其同事在灵长类动物骨髓基质细胞株 PU-34 培养上清中发现的一种细胞间调节因子，这种生长因子可刺激 IL-6

IL-11 was first found as a cell regulatory factor by American scholar Paul and his colleagues in primate marrow stroma cell line PU-34 culture medium supernate. This type of growth factor stimulates IL-6-dependent mouse plasmocytema

依赖的小鼠浆细胞瘤细胞系的生长，并且这种刺激不能被 IL-6 单克隆中和抗体所消除。此外，这种因子与来自骨髓基质通道细胞的脂肪形成抑制因子（adipogenesis inhibitory factor, AGIF）是同类物质。IL-11 主要是由间充质来源的黏附细胞产生，如骨髓基质细胞、基质成纤维细胞、人胚胎成纤维细胞、胚胎滋养层细胞、肺成纤维细胞、滑液细胞、原代成骨细胞、关节软骨细胞。另外胶质母细胞瘤细胞、黑色素瘤细胞、骨肉瘤细胞和甲状腺癌细胞等也能产生 IL-11。

cell line growth, which can not be eliminated by IL-6 monoclone neutralizing antibody. Besides this factor is same material with adipogenesis inhibitory factor which comes from bone marrow origin stroma passage cell. IL-11 is primarily produced by mesenchyme origin adherent cells, such as marrow stroma cells, substance fibroblasts, human embryo fibroblasts, embryo trophoblastic cells, lung fibroblasts, synovial fluid cells, primary osteoblasts and articular cartilage cells. PU-34 cells can secrete IL-11 at large quantity upon IL-1 stimulation. Besides glioblastoma cells, melanoma cells, osteosarcoma cells and oncogenic thyroid cells also secrete IL-11.

人 IL-11 基因是由杨育中等首先发现并克隆的。人 IL-11 基因位于 19 号染色体长臂 13 区（19q13.3—q13.4）（图 5-1），基因长约 7kb，含有 5 个外显子（分别编码 3、57、29、54 和 56 个氨基酸残基）和 4 个内含子，第 4 个内含子中含有 4 个 30bp 的重复序列。IL-11 的基因 5′端非翻译区（5′ untranslation region, 5′-UTR）含有与 RNA 聚合酶 Ⅱ 结合的 TATA 框；3′-UTR 含有多个 ATTTA 重复序列，可能与降低 mRNA 的稳定性有关。另外 IL- Ⅱ 基因的 5′-UTR 与其他细胞因子的 5′端调节序列相似，含有可被 PMA 诱导的 AP1、AP3 结合序列和 IL-1α 反应元件（ACATGGCAAAACCC）、干扰素反应元件。IL-11 的 mRNA 为 1.5kb 和 2.5kb 两种转录物，3′端非编码区含有

Yang first identified and cloned human IL-11 gene. The human IL-11 gene is localized at 19q13.3—q13.4 (Figure 5-1), and the gene length is 7kb in size. The gene has 5 extrons (codes 3, 57, 29, 54 or 56 amino acids) and 4 introns, the fourth intron has four repetitive sequences of 30bp. IL-11 gene 5′ UTR has TATA box which linked RNA polymerase II; 3′ UTR has multiple ATTTA repeats that may be important in the regulation of mRNA stability. Besides IL-11 gene 5′ terminal untranslated region has the similar 5′ terminal regulatory sequence with other cytokines, in which the AP1 and AP3 combine sequences can be induced by PMA and IL-1 α response element (ACATGGCAAAACCC), as well as by interferon response element. IL-11 mRNA contains transcripts of 1.5kb or 2.5kb, and multiple copies of AU and Alu structures in 3′ terminal noncoding region, making mRNA unstable. This region may be the significant regulatory component of cytokine gene expression.

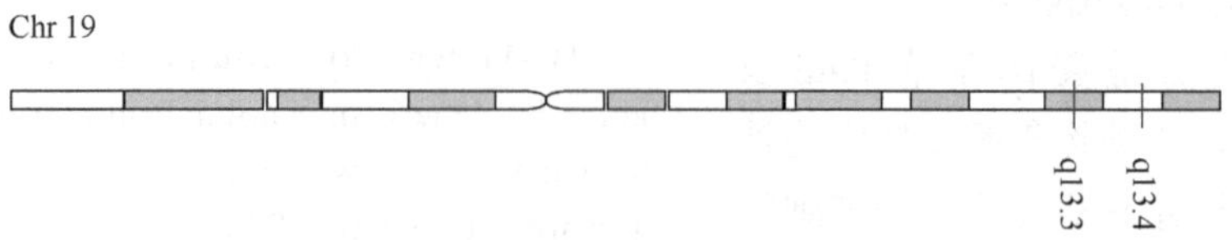

图 5-1 人 IL-11 在 19 号染色体上的定位

可使mRNA不稳定的重复AU和Alu结构，它们可能是细胞因子基因表达的重要调节成分。

1.2.1 IL-11分子结构及其受体结构

人类IL-11的前体是199肽，其中含21个氨基酸残基的信号肽。成熟IL-11分子有178个氨基酸残基，分子量约为23kDa，没有二硫键（即无半胱氨酸残基）和糖基化位点，但富含脯氨酸（12%）、亮氨酸（23%）和其他带正电荷的氨基酸（14%）并且呈高度螺旋化。中性环境中成熟IL-11呈高度碱性，带7个正电荷，等电点为11.7。23%亮氨酸使IL-11具有强的疏水特征；12%脯氨酸集中在N端1/3处，可能与IL-11的功能有关，但去除N端序列并不显著影响IL-11的活性。IL-11有4个α螺旋，构成2个受体结合位点：C端α螺旋（螺旋D）和甲硫氨酸-59邻近区构成结合IL-11Rα链的位点1，N端α螺旋（螺旋A，精氨酸-15～亮氨酸-35）构成结合gp130链的位点2（图5-2）。

IL-11受体含有两条链（图5-3）。IL-11Rα链是含有432个氨基酸残基的分子量为150kDa的低亲和力受体的跨膜蛋白，属于造血因子受体超家族，是受体与IL-11特异性结合的部分；另一条链gp130是与IL-6、白血病抑制因子（leukemia inhibitory factor, LIF）、抑瘤素M（oncostatin M, OSM）和睫状神经营养因子（ciliary neurotrophic factor, CNTF）等受体共用的信号传导链。因此IL-11与这些细胞因子有相似的生物学活性。

IL-11对热稳定性强，在中性磷酸盐溶液中80℃处理也不丧失活

1.2.1 IL-11 Molecule Structure and Receptor

Human IL-11 gene encodes a 199-amino-acid precursor polypeptide with a 21 residue signal sequence. The mature IL-11 is composed of 178 amino acids with molecular weight of 23kDa without disulfide bond or glycosylation site. IL-11 has abundant Pro(12%), Leu(23%) and other positively charged amino acids and is highly spiral. In neutral environment, the mature IL-11 is highly basic, with 7 positive charges. Its isoelectric point (pl) is 11.7. About 23% Leu results in highly water drain property; and 12% Pro locate inside the N terminal region. The activity of IL-11, however, was not influenced after removing N terminal sequence. There are 4 α -helix in IL-11, constructing 2 receptor binding sites, site 1 being consisted of C terminal α -helix (helix D) and Met 59 adjacent domain which binds to IL-11Ra, and site2 being consisted of N terminal α -helix (helix A, Arg15~Leu35) which binds to gp130 (Figure 5-2).

IL-11 receptor has two chains (Figure 5-3), IL-11R α chain is a transmembrane protein with low affinity receptor IL-11, which contains 432 amino acids. It belongs to a hemopoiesis factor receptor superfamily with molecular weight at 150 kDa. IL-11R α is the IL-11 specific binding part on the receptor. The other chain of the receptor is gp130, which uses the same signal conduction chain with IL-6, leukemia inhibitory factor (LIF), oncostatin M (OSM) and ciliary neurotrophic factor (CNTF). This is why IL-11 exerts the similar biological activities with these cytokines.

As IL-11 is highly thermostable, it cannot be inactivated in neutral phosphate solution at 80℃.

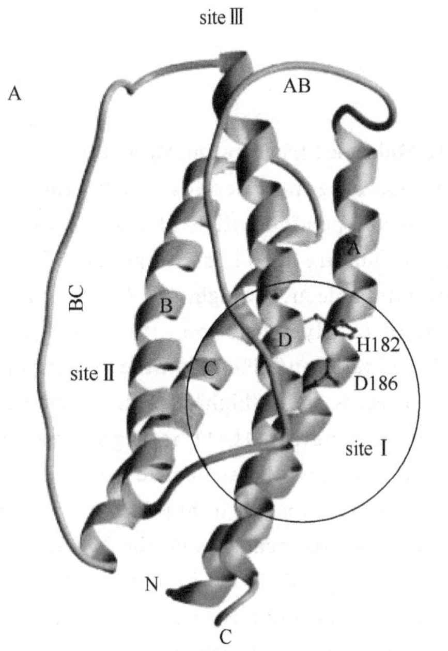

图 5-2 IL-11 的分子结构［来自 Biochem J, 2003, 375（Pt1）: 23-32］

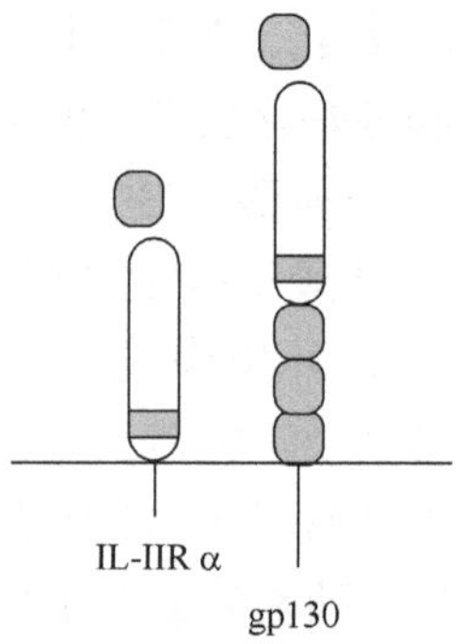

图 5-3 IL-11 的受体结构

性。在疏水表面、液气交界面和机械剪切下可引起 IL-11 的结构改变，导致自身凝聚和沉淀。IL-11 在碱性环境中稳定，但易与保存管管壁结合而降低活性，在低浓度时尤其明显。在酸性环境下，IL-11 的两个天冬氨酸-脯氨酸结构（Asp_{12}～Pro_{13} 和 Asp_{134}～Pro_{135}）极易被水解而灭活。Met_{59} 的氧化易使 IL-11 失去活性。

But Hedrophobic surface, fluid-air boundary and machinery shearing can cause structural changes of IL-11, leading to condensation and precipitation. IL-11 is stabe in basic environment, but it may easily bind to holding tubal wall to loss activity, especially at low density. In acidic environment, the 2 Asp-Pro (12-13 and 134-135) structures can be easily hydrolyzed resulting in inactivation of the molecule. The Met59 oxidative solution may also result in inactivation of IL-11.

1.2.2 生物学功能

IL-11 自发现以来，对其生物学功能已经进行了大量研究。目前已发现其对造血细胞的生长和分化、炎症因子的分泌、急性相关蛋白的产生和神经细胞的分化等均有作用，与 IL-6 的功能有部分重叠，但又有很多不同。

1. IL-11 促进造血作用

（1）祖细胞的生长和分化。IL-11 与 IL-3、IL-4、IL-7、IL-12、IL-13 和

1.2.2 IL-11 Biological Functions

Since its discovery, large number of studies has been conducted about IL-11 biological functions. Numerous research reports had confirmed that IL-11 exerts effects on proliferation and differentiation of hematopoietic cells, induction of secretion of inflammatory factors, production of acute related proteins, and differentiation of nerve cells. Some of these functions overlap with the functions of IL-6, whereas still exist some different features.

1. Hematopoietic effects of IL-11

(1) Progenitor cells: In synergy with IL-3, IL-4, IL-7, IL-12, IL-13, stem cell factor (SCF),

肝细胞因子（stem cell factor, SCF），以及粒细胞/巨噬细胞集落刺激因子（granulocyte-macrophage colony stimulating factor，GM-CSF）有协同作用，刺激各阶段和各系统祖细胞的生长。IL-11刺激早期干细胞、多潜能和定向祖细胞（来源于脐血、骨髓及外周血）在各种培养体系的扩增，这种增生是由于这些细胞的静息群体（处于G_0期的细胞）进入细胞周期，缩短了这些细胞的细胞周期。IL-11与造血微环境中的其他细胞因子协同作用可加速原始干细胞向多系祖细胞分化，并刺激定向祖细胞的增殖和分化。

flt3 ligand, and granulocyte-macrophage colony stimulating factor (GM-CSF), IL-11 stimulates the proliferation of primitive stem cells, multipotential and committed progenitor cells. These cells come from various sources including cord blood, bone marrow, and peripheral blood. This proliferation may be resulted from the entry of a quiescent (G_0) population of these cells into active cell cycle as well as shortening of the cell-cycle time in some cells. In combination with other cytokines present in hematopoietic microenvironment, IL-11 may induce more committed primitive stem cells into the multilineage progenitor differentiation, and stimulates proliferation and differentiation of committed progenitor cells.

（2）巨核细胞系的生成和血小板的生成。IL-11与IL-3、促血小板生成素（thrombopoietin，TPO）或SCF协同作用可刺激小鼠和人不同时期巨核细胞生成和血小板生成。体内试验表明，使用IL-11处理啮齿类、狒狒和人的骨髓（bone marrow，BM）细胞可刺激巨核细胞生成、分化和成熟。抗TPO抗血清可使IL-11刺激引起的巨核细胞集落形成降低90%，而抗IL-3抗血清则降低此效应28%。上述结果表明，IL-11引起的巨核细胞集落形成或血小板生成有可能由TPO介导。IL-11和TPO协同刺激静息的多系祖细胞可缩短其G_0期，若加入ACK2（中和C-kit的抗体），则可完全消除此作用，提示IL-11和TPO对多系祖细胞的协同作用可能受SCF/C-kit互相作用的影响。

(2) Megakaryocytopoiesis and thrombocytopoiesis: IL-11 acts synergistically with IL-3, thrombopoietin (TPO) and SCF to stimulate various stages of megakaryocytopoiesis and thrombopoiesis in both murine and human bone marrowl cells. In vivo treatment with IL-11 results in marked stimulation of megakaryocytopoiesis in rodents, nonhuman primates, and human, including the production, differentiation, and maturation of megakaryocytes. Anti-TPO antiserum can reduce IL-11 stimulated megakarocyte colony formation by 90%, whereas anti-IL-3 antiserum results in a 28% reduction in colony formation. These studies suggest that IL-11 effects on megakaryocytopoiesis and thrombopoiesis may be mediated in part via TPO. IL-11 and TPO can also synergistically stimulatc the proliferation of dormant multilineage progenitors by shortening G_0, and this effect can be completely abrogated by addition of ACK2, a neutralizing antibody to c-kit, the receptor of SCF, suggesting that the synergistic effects of IL-11 and TPO on multilineage cells may be mediated in part by SCF/c-kit interactions.

（3）促红细胞生成。IL-11单独或与其他细胞因子［如IL-3、SCF或红细胞生成素（erythropoietin，EPO）］联合使用，可刺激人和小鼠

(3) Erythropoiesis: L-11 alone or in combination with other cytokines [IL-3, SCF, or erythropoietin (EPO)] can stimulate multiple stages of erythropoiesis using murine and human BM cells. In vitro effect of IL-11 on burst-forming unti-erythroid (BFU-E)

骨髓细胞中不同时期的红细胞生成。体外实验显示，IL-11 对红细胞系爆裂样生成单位的生成作用不能通过使用 SCF、IL-3 或 GM-CSF 抗体而抵消，提示 IL-11 对人和鼠的红系祖细胞有直接作用。体内试验研究发现，IL-11 和 SCF 可促进多系细胞分化为红系细胞，而 IL-11 和 EPO 还可促进红系细胞的进一步扩增。另外，IL-11 和 SCF 还可引起 BM 的红系细胞进入脾脏后重新分布。

formation cannot be abrogated by antibodies against SCF, IL-3, or GM-CSF, suggesting a direct effect of IL-11 on human and murine eythroid progenitors. In vivo studies of cytokine administration indicate that IL-11 and SCF may increase the input from a mutilineage cell compartment into the erythroid lineage, whereas IL-11 and Epo may stimulate further amplification of erythroid cells. Moreover, IL-11 and SCF may lead to a redistribution of erythroid cells from BM to spleen.

（4）髓系细胞的生成。IL-11 还可调节髓系祖细胞的分化和成熟。IL-11 和 SCF 联合可刺激白鼠 $Lin^{-}/Seal^{+}$BM 细胞形成髓系集落。该集落主要由粒细胞和骨髓粒系群体组成。IL-11 和 IL-13 或 IL-4 联合使用，可使髓系集落中粒细胞和骨髓粒系群体的比例减少，巨噬细胞的比例升高。对新生大鼠联合使用 IL-11、SCF 和 G-CSF 可显著增加周围血中的中性粒细胞数。

(4) Myelopoiesis: IL-11 modulates the differentiation and maturation of myeloid progenitor cells. IL-11 in combination with SCF stimulates myeloid colony formation from murine $Lin^{-}/Seal^{+}$ BM cells. These colonies are composed mostly of granulocytes and myeloid blasts. The combination of IL-11 with IL-13 or IL-14 can reduce the proportion of granulocytes and blasts in myeloid colonies, with a concomitant increase in macrophages. Combination treatment with IL-11, SCF, and G-CSF in the newborn rat has been shown to significantly increase peripheral neutrophil counts.

（5）淋巴细胞的生成。IL-11 和 SCF 或 IL-4 联合使用，可有效地促进 B 细胞的增殖，该 B 细胞来自经五氟尿嘧啶（5-FU）处理后的小鼠骨髓细胞。IL-11 和 IL-4 还可以逆转 IL-3 对早期 B 淋巴细胞发育的抑制作用。

(5) Lymphopoiesis: IL-11 in combination with SCF or IL-4 effectively supports the generation of B cells in primary cultures of BM cells from 5-fluorouracil (5-FU)–treated mice. IL-11 and IL-4 can also reverse the inhibitory effect of IL-3 on early B-lymphocyte development.

（6）对造血微环境的影响。IL-11 最先来源于造血微环境（hematopoietic microenvironment，HM）中的基质细胞，是微环境中的旁分泌和自分泌的生长因子。将 IL-11 加入长期培养的人 BM 中，可显著增加黏附细胞数，抑制黏附细胞中脂肪堆积，增强造血作用。在再生障碍性贫血（aplastic anemia，AA）患者的骨髓细胞培养液中加入 IL-11，可显著促进黏附基质层的形成，

(6) Effects on hematopoietic microenvironment: IL-11 was originally isolated from cells derived from the hematopoietic microenvironment (HM) and may act as a paracrine or autocrine growth factor in this environment. Addition of IL-11 to human long-term BM culture (LTBMC) significantly increases the cellularity of the adherent cells, inhibits adipose accumulation in adherent cells, and leads to enhanced hematopoiesis. Addition of IL-11 to bone marrow cultures derived from aplastic anemia (AA) patients significantly enhances the formation of an adherent stromal layer, suggesting that IL-11 may have

提示IL-11对造血微环境缺陷的AA患者可能有治疗价值。巨核细胞可刺激BM成纤维细胞的生长。骨髓纤维化的形成通常与异常的巨核细胞生成有关，IL-11可调节巨核细胞依赖型骨髓纤维化。体内和体外实验均证实，IL-11及其他细胞因子能动员造血干/祖细胞进入外周血。用IL-11和SCF治疗能够增强其移动作用，即从BM向脾脏迁移或从脾切除小鼠的BM至血液。

therapeutic value in aplastic anemia patients with defects in the HM. BM fibroblast growth can also be stimulated by the presence of megakaryocytes and the evolution of myelofibrosis is often linked with abnormal megakaryocytopoiesis, and IL-11 has been shown to modulate megakaryocyte-dependent myelofibrosis. IL-11 with other cytokines has been shown to mobilize primitive hematopoietic stem/progenitor cells both in vitro and in vivo. Treatment with IL-11 and SCF can enhance mobilization of long-term repopulating cells from the BM to the spleen and from the BM to the blood of splenectomized mice.

2. IL−11对非造血系统的作用

2. Non-hematopoietic effects of IL-11

（1）对上皮细胞的作用。肺泡和支气管上皮细胞会产生大量的IL-11。炎症性细胞因子、呼吸道合胞病毒（respiratory syncytial virus，RSV）和油酸等均可上调IL-11的水平，说明IL-11和IL-11Rα对肺部炎症可能起重要作用。肠胃道上皮中亦表达IL-11和IL-11R。体外研究显示，IL-11可直接与胃肠道上皮细胞相互作用，并且可逆性地抑制肠腺管干细胞系（LEC-6和EC-18）的增殖。因此，IL-11可能参与肠胃（GI）上皮细胞正常生长的调控，其作用可能由于延长G_1-S期的移行，此时也伴有视网膜细胞瘤敏感基因产物的低磷酸化的堆积。另外，IL-11可增强大鼠GI对铁的吸收，而此作用与红细胞的生成改变无关。

(1) Effects of IL-11 on epithelial cells: Alveolar and bronchial epithelial cells produce large amounts of IL-11. The up-regulation of IL-11 production by inflammatory cytokines, respiratory syncytial virus (RSV), and retinoic acid (RA) suggests that IL-11 may play an important role in pulmonary inflammation. IL-11 and IL-11R are also expressed in epithelial cells of the gastrointestinal (GI) tract. In vitro studies showed that IL-11 directly interacted with GI epithelial cells and reversibly inhibited proliferation of the intestinal crypt stem cell lines (IEC-6 and IEC-18). Thus, IL-11 may be involved in the normal growth control of GI epithelial cells. IL-11–induced decrease in proliferation of these cells may be due to prolongation of the G_1-S phase transition which is also associated with accumulation of the hypophosphorylated form of the retinoblastoma susceptibility gene product. In addition, IL-11 had been found to enhance GI absorption of iron in rats, which did not appear to bc related to changes in erythropoiesis.

（2）破骨细胞生成。在骨髓细胞培养中或骨髓细胞与颅骨细胞的共培养中联合使用IL-11、1α-25-二羟基维生素D3［1α-25（OH）$_2$D$_3$］和甲状旁腺素（parathyroid hormone，PTH），可刺激破骨细胞的产生并抑制骨结节的形成。成骨细胞是破骨细胞调控吸收的重要影响因子，IL-11介

(2) Osteoclastogenesis: IL-11 in combination with lα, 25-dihydroxyvitamin D3 [1 α,25(OH)$_2$D$_3$] and parathyroid hormone (PTH) had been shown to stimulate osteoclast development and inhibited bone nodule formation in BM cultures and cocultures of BM with calvaria cells. Osteoblasts are important regulators of osteoclast-mediated bone resorption. The requirement of the presence of osteoblastic cells in IL-11–induced osteoclast development suggests that the effect of IL-11 may be mediated through the

导的破骨细胞发育需要成骨细胞的参与，提示 IL-11 的调节作用可能是通过成骨细胞衍生出的其他刺激因子所产生的。初生成骨细胞表达 IL-11Rα 和 gp130 mRNA，IL-1、PTH 和 1α-25（OH）$_2$D$_3$ 可上调 gp130 mRNA 的表达量。成熟的破骨细胞也表达 IL-11Rα mRNA。这些研究说明，IL-11 是骨代谢中重要的成骨细胞衍生旁分泌调节因子，并且骨生成和骨再吸收细胞都是 IL-11 潜在的作用靶点。

stimulation of other factors derived from osteoblastic cells. Primary osteoblasts express both IL-11R α and gp130 mRNA, and gp130 mRNA can be up-regulated by IL-1, PTH, and 1 α,25(OH)$_2$D$_3$. Mature osteoclasts also express IL-11R α mRNA. These studies suggest that IL-11 is an important osteoblast-derived paracrine regulator of bone metabolism and that both bone-forming and bone-resorbing cells are potential targets of IL-11.

（3）参与神经发育作用。有研究发现 IL-11 mRNA 表达于海马神经细胞、脊髓神经的运动神经细胞和交感神经细胞中。外源性 IL-11 可刺激海马神经祖细胞（H19-7）的增生，此作用具有剂量依赖性。另外，研究证实 IL-11 和其他几种造血生长因子是小鼠胎儿海马神经祖细胞的分化因子。肺部上皮下的部位有大量的神经末梢分布，而肺泡和支管上皮细胞可产生 IL-11，进一步说明 IL-11 是感觉和运动神经元重要的存活因子。

(3) Neurogenesis: Studies had showed that IL-11 mRNA was expressed in hippocampal neuronal cells and in motor and sympathetic neurons of the spinal cord. Exogenous IL-11 stimulates the proliferation of hippocampal neuronal progenitor cells (H19-7) in a dose-dependent fashion. In addition, it had been shown that IL-11 and several other hematopoietic growth factors were differentiation factors for murine fetal hippocampal neuronal progenitors (MK31). The production of IL-11 by alveolar and bronchial epithelial cells further suggest that IL-11 is an important survival factor for sensory and motor neurons because the subepithelial space of lung is rich in nervous innervation.

（4）其他作用。IL-11 也参与非造血活动如抑制脂肪的形成，诱导发热反应和调节细胞外基质（ECM）代谢，这些功能可能对结缔组织具有保护作用，也可能与肝纤维化和肝硬化的发病机理有关。在体外细胞培养中发现 IL-11 可降低炎症性细胞因子的表达，尤其是单核细胞和巨噬细胞释放的肿瘤坏死因子 α（tumor necrosis factor，TNF-α）。

(4) Other effects: IL-11 had also been shown to have other nonhematopoietic activities such as inhibition of adipogenesis, induction of a febrile response, and modulation of extra cellular matrix (ECM) metabolism, which may have a protective effect on connective tissues or could be involved in the pathogenesis of liver fibrosis and cirrhosis. In vitro cell culture, IL-11 appeared to reduce pro-inflammatory cytokine expression, particularly the release of tumor necrosis factor- α (TNF- α) by monocytes/macrophages.

1.2.3 生物学活性测定

用于 IL-11 生物活性的常规测定方法，比较成熟的有 B9-11 细胞株 / MTT 比色法，其测定原理及其步骤

1.2.3 IL-11 Biological Activation Identification

The conventional testing method of IL-11 biological activity is B9-11 cell line/MTT colorimetric assay, and the measurement principle and steps are as follows.

简述如下。

1. 原理　IL-11的效价测定采用B9-11细胞株/MTT比色法。适当浓度的IL-11，对B9-11细胞具有剂量依赖性促增殖作用。MTT在活细胞的线粒体中可以定量地被还原为MTT Fornazan，通过比色法测定MTT Fornazan的量可以直接表示B9-11细胞的生长状态。制备一系列IL-11稀释溶液，按50%最大效应点的稀释倍数可以折算为待检样品中IL-11的效价，测定结果需用IL-11效价测定用参考品校正。

1. Principle　Using B9-11 cell line/MTT colorimetric assay to identify the potency of IL-11. Appropriate concentration of IL-11 promotes B9-11 cells'proliferation in a dose-dependent fashion. The insoluble Fomazan produced by the living cells in response to MTT dye can be quantitatively reverted to the numbers of B9-11 cells. So to prepare a series of IL-11 diluted solution, and the point 50% maximum dilution effect was converted into the potency of IL-11 in unknown sample. The determined results should be corrected by potency of IL-11 in standard sample.

2. 试剂

（1）RPMI-1640培养液，置4℃保存，设为A液。

（2）95% RPMI-1640培养液,5% FBS，设为B液。

（3）B液添加5ng/ml的IL-11（终浓度），即为C液。

（4）B9-11细胞株：用C液在37℃，5%CO_2条件下培养。

（5）PBS缓冲液。取8g NaCl、0.2g KCl、1.44g Na_2HPO_4、0.24g KH_2PO_4用蒸馏水配成1000ml的溶液。经121℃、15min高压灭菌，保存于室温。

（6）MTT溶液。用PBS缓冲液配成5.0mg/ml的溶液，经0.22μm滤膜过滤除菌，置于玻璃瓶中，4℃条件下避光保存。每次试验前检查，发现蓝色沉淀者应重新配置。

（7）DMSO裂解液。

（8）IL-11标准品、IL-11样品。

2. Reagent

(1) RPMI-1640 culture solution, conserving at 4℃. (A solution)

(2) 95% RPMI-1640 culture solution combined with 5% FBS. (B solution)

(3) 5 ng/ml IL-11 was added into B solution (final concentration). (C solution)

(4) B9-11 cell line: These cells were cultured in C solution under the condition of 37℃ and 5% CO_2.

(5) PBS buffer solution: It is prepared by distilled water with 8g NaCl, 0.2g KCl, 1.44g $NaHPO_4$, and 0.24g KH_2PO_4. To sterilize it under 121 ℃ for 15 minutes then store it under room temperature.

(6) MTT solution: To dilute PBS buffer solution into 5.0 mg/ml, and degerm it by filter membrane. To store it under 4℃ in glass bottle avoiding light. Make sure there is no blue sediment before use it.

(7) DMSO lysis buffer.

(8) IL-11 standard sample and IL-11 unknown sample.

3. 仪器　超净工作台，CO_2培养箱，酶标仪，移液器，显微镜，血球计数器，离心机，Eppendorf管，取液

3. Apparatus　Aseptic bench, CO_2 incubator, ELISA, pipettor, hemacytometer, centrifugal machine, Eppendorf pipe, pipette tips, 96 well cell culture plates

器吸头，96孔细胞培养板，储液槽。以上器材必须洁净、干燥和无菌。

4. 操作步骤　第（1）至第（6）步骤均应置于无菌条件下进行。

（1）制备细胞悬液。取对数生长期B9-11细胞，用A液洗3遍，再用B液调整至1.4×10^5个/ml的细胞悬液待用。

（2）制备样品溶液。取1支标准品按说明书配成标准样品溶液。取规定数量的待检样品按说明书配成待检样品溶液。

（3）制备样本溶液。用B液将标准样品溶液和待检样品溶液稀释至500IU/ml，每步稀释不得超过10倍。按以上预稀释程序制备的溶液称为样本溶液。

（4）制备样本梯度。于96孔细胞培养板中，第2～11列各孔加入150μl B液，在A6、A7各孔中每孔加入50μl标准样本溶液，在A2/A3、A4/A5、A8/A9、A10/A11各孔中每孔加入50μl各待检样本溶液，每个待检样品做2个复孔。自A行至H行做四倍稀释，每孔留150μl余液。

（5）加样。将（4）制备的实验样本梯度按位置对应关系移入96孔培养板，每孔50μl。将（1）制备的细胞悬液移入96孔培养板。37℃、5% CO_2条件下培养48～72h。

（6）加入MTT溶液并培养。每孔加入20μl MTT溶液，37℃、5%CO_2条件下培养4～6h。

（7）加入裂解液并测定结果。96孔培养板离心去上清，加入裂解溶液100μl，混匀，测定波长500nm，记录测定结果。

and accumulater. These materials must be clean, dry and sterile.

4. Operation steps　It needs to operate under sterile condition from first step to sixth step.

(1) Preparation of cell suspension: Using the A solution to wash B9-11 cells when they are at the period of logarithmic phase, then using the B solution to modulate the solution into cell suspension of 1.4×10^5/ml.

(2) Preparation of specimen solution: To prepare the standard specimen solution using one piece of standard specimen and prepare the unknown specimen solution using specified amount of unknown specimen solution according to the instruction book.

(3) Preparation of sample solution: Using the B solution to dilute the standard specimen solution and unknown specimen solution into 500IU/ml respectively. Dilution factor shouldn't be more than 10 times in every diluted step.

(4) Preparation of sample gradient: In the 96 well cell culture plates, to add 150μl B solution into each well from the second line to the eleventh line, then add 50μl standard sample solution into each well of A6 and A7 rows, and add 50μl unknown sample solution into each well of A2, A3, A4, A5, A8, A9, A10, and A11 rows. Each unknown sample has two repeated wells. 4 times dilution the previous solution from A row to H row, then keep 150μl in each well.

(5) Loading sample: To add 50 μl sample gradient into 96 well cell plates according to the position-match relation, then add 50 μl cell suspension into another 96 well cell plates. To culture them under the condition of 37℃ and 5% CO_2 for 48-72 hours.

(6) Adding MTT solution: To add 20 μl MTT solution into each well from step 5 then culture them under the condition of 37℃ and 5% CO_2 for 4-6 hours.

(7) Adding DMSO lysis buffer and measurement: To centrifuge the solutions from 96 well cell plates respectively then remove the supernates, and add DMSO lysis buffer into each well, and mix the solutions. Then measured with an ELISA reader at the wavelength of 500 nm and record the results.

5. 结果处理　对各实验样品的各实验点OD值、预稀释倍数、样本梯度等数据采用计算机程序进行处理，分别计算各实验样品的半效稀释倍数，即从样本溶液至相当于标准品50%最大效应点的稀释倍数，并按下式计算实验结果：

$$P_u=P_s\times\frac{D_{preu}}{D_{pres}}\times\frac{D_{hpu}}{D_{hps}}$$

式中，P_u为待见样品效价；P_s为标准品效价；D_{preu}为待检样品预稀释倍数；D_{pres}为标准品预稀释倍数；D_{hpu}为待检样品半效稀释倍数；D_{hps}为标准品半效稀释倍数。

5. Results analysis Using computer program manipulation to analyze the data of experimental samples including OD value, pre-dilution factor, sample gradient, *et al.* then calculate the half-potency dilution factor of each experimental sample respectively (i.e. equivalently to the dilution factor of the point of 50% maximum dilution effect of the standard solution). To calculate experimental results according the following equation.

$$P_u=P_s\times\frac{D_{preu}}{D_{pres}}\times\frac{D_{hpu}}{D_{hps}}$$

P_u: Potency of unknown sample, P_s: Potency of standard sample, D_{preu}: Pre-dilution factor of unknown sample, D_{pres}: Pre-dilution factor of standard sample, D_{hpu}: Half-potency dilution factor of unknown sample, D_{hps}: Half-potency dilution factor of standard sample.

1.2.4 IL-11药物及生产工艺

目前市场上可购买的增加血小板数量的生长因子有IL-11。利用基因工程技术，在大肠杆菌中表达重组人IL-11，提取得到IL-11的衍生物，其活性与天然人IL-11相比有很大的提高。产品的冻干制剂经国家药品监督管理局批准获得临床批文，属I类新生物制品。大概的生产流程是从菌种库得到种子液进行制备，发酵、收菌、碎菌后，制备融合蛋白粗制品，然后制备融合蛋白精制品；接着用精制品制备重组IL-11的粗制品，得到精制品后进行稀释、分装和冻干，最后得到成品，并加以检定（图5-4）。该生产工艺和制剂的优点有：①得到纯度>95%的重组人IL-11；②产物重组人IL-11的比活性高，比活性>5×10^{6}U/mg；③生产工艺简单，产物收率高，生产成本低。

1.2.4 IL-11 Medicine and Production Technology

Currently, IL-11 is commercially available as a growth factor for increasing platelet counts. Derivative of IL-11 was extracted from *E.coli* by recombinant human interleukin 11 (rhIL-11) genetic techniques. The activation of this derivative is higher than that of human nature IL-11. The freeze-dry preparation of this product was permitted for clinical application by State Drug Administration, and this product belongs to the novel biopharmaceuticals of type I. The general production process is included the preparation of seed solution, fermentation, collection of bacillus, crude product of fusion protein, refined product of fusion protein, crude product of rhIL-11, refined product of rhIL-11, dilution of the refined product, sub-package, lyophilization, and verification of the final product (Figure 5-4). This productive technology and preparation have some virtues: ① the purity of rhIL-11 > 95%, ② specific activity of rhIL-11 > 5 × 10^{6}U/mg, ③ simple production technology, higher yield of product, and lower cost of production.

1997年美国食品与药物管理局（FDA）批准美国GI公司研制的重组人IL-11上市（商品名：

Recombinant human IL-11 named Neumega which manufactured by American GI company was permitted for commercial use in 1997 by FDA, and it was used for the prevention and treatment

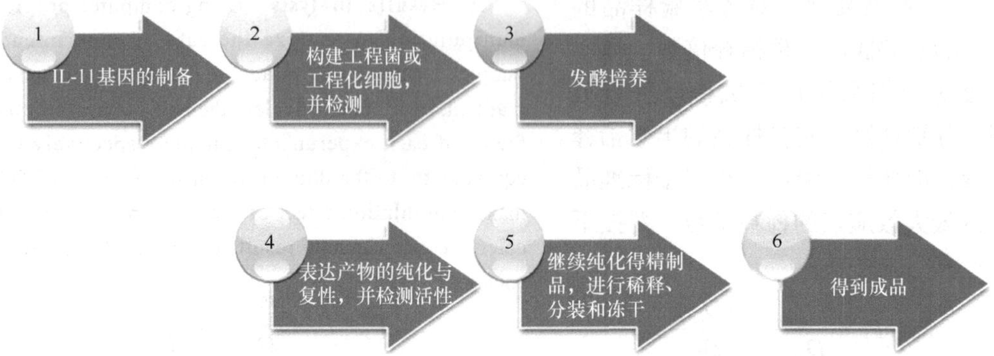

图 5-4　IL-11 的生产工艺流程

Neumega），用于预防和治疗放化疗所引起的血小板减少症。国内有许多同类产品（表 5-2），以下是这类产品的基本信息。

of thrombocytopenia caused by radiation and chemotherapy. Many congeneric products have been used in China (Table 5-2); the basic information about this kind of products as follows.

表 5-2　中国 SFDA 的部分重组人 IL-11 产品注册批准信息

批准文号（国药准字）	S20030015 S20030014	S20030017 S20030016	S20030034	S20030077	S20050046
产品名称	注射用重组人白介素 -11	注射用重组人白介素 -11	注射用重组人白介素 -11	注射用重组人白介素 -11	注射用重组人白介素 -11
商品名	迈格尔	巨和粒	依星	吉巨芬	欣美格
生产单位	北京双鹭药业股份有限公司	齐鲁制药厂	成都地奥九泓制药	杭州九源基因工程有限公司	上海中信国健药业有限公司
批准日期	2003-03-17	2003-03-18	2003-04-24	2003-09-18	2005-04-26
产品类别	生物制品	生物制品	生物制品	生物制品	生物制品
规格	1200×10^4U（1.5mg）/ 支 600×10^4U（0.75mg）/ 支	1.2×10^7U（1.5mg）/ 支 2.4×10^7U（3mg）/ 支	2400×10^4U（3mg）/ 支	3mg/ 支	4mg/ 支
剂型	冻干制剂	注射剂	冻干粉针	冻干粉针	针剂

【性状】本品系无菌冻干制剂，类白色疏松体，加注射用水后能迅速溶解。

[Traits] This product is sterile lyophilized preparation, white likeness loose body, and dissolves rapidly after adding water for injection.

【主要成分】主药重组人 IL-11，辅料甘氨酸、氯化钠、磷酸氢二钠。

[Bases] Primary drug is rhIL-11, and the adjuvant includes glycine, sodium chloride and disodium hydrogen phosphate.

【药理毒理】

[Pharmacology and toxicology]

（1）药理作用。本品是应用基因

(1) Pharmacological action: This product is a type

重组技术生产的一种促血小板生长因子，可直接刺激造血干细胞和巨核祖细胞的增殖，诱导巨核细胞的成熟分化，增加体内血小板的生成，从而提高血液血小板计数，而血小板功能无明显改变。

of pro-platelet growth factor manufactured by gene recombination technique. It stimulates the proliferation of hemopoietic stem cells and megakaryocytic progenitor cells, induces mature differentiation of megakaryocyte, and enhances growth of platelet. Platelet count will increase while the functions of platelet have no significant changes.

（2）毒理研究。

(2) Toxicological research.

① 安全性药理。在各种动物模型上，观察到本品的一些非造血系统效用，包括控制肠上皮细胞生长、抑制脂质形成、诱导急性期反应蛋白合成、抑制巨噬细胞分泌促炎症细胞因子、抑制破骨活动及神经发生等。

① Safety pharmacology: In various animal models, some non-hematopoietic effects of this product include regulation of growth of intestinal epithelial cells, inhibition of formation of lipid, induction of synthetization of acute responsive protein, inhibition of secretion of pro-inflammatory factors by macrophages, and inhibition of osteoclastogenesis and neurogenesis.

② 重复给药毒性。非灵长类动物连续给药 2～13 周，未见关节囊和肌腱纤维化及骨膜增厚现象，但长期用药患者是否会有类似改变目前尚不清楚。

② Toxicity of repeated dose: After administration on non-human primates for 2-13 weeks, there was no indication of joint capsule fibrosis, tendon fibrosis and periosteal thickening. However, it is not clear yet whether patients have the similar consequences with long time administration.

③ 遗传毒性。体外研究显示本品无致突变作用。

③ Toxicity of heredity: There are on mutational effects in vitro.

④ 生殖毒性。在正常人用剂量的 2～20 倍时，可见大鼠的发情期延长，但即使剂量高达每天 1000μg/kg，对大鼠的生育力也未见影响。家兔在人用剂量的 0.02～2 倍时（每天不少于 1μg/kg），可见母体毒性。大鼠试验中可见母鼠一过性呼吸困难，早期胚胎死亡率增加。家兔可见摄食和排泄量减少、体重减轻、流产、胚胎及胎仔死亡率增加。

④ Reproductive toxicity: Oestrous period of rat had been prolonged when the dose was 2-20 times of the dose of normal human using. But the dose of 1000μg/kg/d showed no influences on rats'fertility. Maternal toxicity had been observed on rabbit when the dose was 0.02-2 times of the dose of normal human using (no less than 1μg/kg/d). Results of animal models had shown that temporary dyspnea on female rats, increased mortality of early embryo, reduced ingestion and excretion, decreased body weight and induced abortion on rabbits.

⑤ 致癌性。尚未进行本品潜在致癌性研究。体外研究显示，本品对一系列来源于不同患者的肿瘤细胞尚未表现出生长刺激作用。

⑤ Carcinogenicity: There are no reports about potential carcinogenicity of this product. In vitro, this product shows irritant proliferative effects on cancer cells from different patients.

【药代动力学】文献报道：重组人 IL-11 50μg/kg 单剂量皮下注射给药，血浆药物浓度的峰值（C_{max}）为

[Pharmacokinetics] Researches had shown that after administration of rhIL-11 50μg/kg by subcutaneous injection, the peak value of blood plasma

（17.4±5.4）μg/ml，T_{max} 为（3.2±2.4）h，终末半衰期为（6.9±1.7）h。25μg/kg 单剂量皮下注射或静脉注射，结果提示男性和女性健康受试者在药代动力学参数上没有差别。未观察到药物体内蓄积或消除率降低的现象。在大鼠模型中，放射标记的重组人 IL-11 给药后从血浆中很快被消除并分布到一些血液灌流量大的组织器官，肾脏是主要的药物清除途径。但尿液中以原型排泄的重组人 IL-11 量很少，提示药物在排泄前经过代谢处理。

drug concentration (C_{max}) was 17.4 ± 5.4 μg/ml; T_{max} was 3.2 ± 2.4 h, and the terminal half-life was 6.9 ± 1.7 h. Administration of rhIL-11 25μg/kg by subcutaneous injection or intravenous injection revealed no difference of pharmacokinetic parameter between male and female subjects. There was no report about drug accumulation in vivo and decreased elimination rate. In rat model, after administration radio-labeled rhIL-11 was eliminated in blood plasma then spread into organs of huge blood flow. Kidney is the major pathway of drug elimination. However traces of rhIL-11 is little in urine which suggests that this drug had been metabolized before being excreted.

【适应证】用于实体瘤、非髓性白血病化疗后Ⅲ、Ⅳ度血小板减少症的治疗；实体瘤及非髓性白血病患者，前一疗程化疗后发生Ⅲ / Ⅳ度血小板减少症（即血小板数≤5.0×10^9 个 /L）者，下一疗程化疗前使用本品，以减少病人因血小板减少引起的出血和对血小板输注的依赖性。同时有白细胞减少症的病人必要时可合并使用重组人粒细胞集落刺激因子（rhG-CSF）。

[Indication] chemotherapy induced Ⅲ / Ⅳ degree thrombocytopenia (platelet count ≤5.0 × 10^9/L) before next chemotherapy or after treating solid tumors and non-myeloid leukemia. Reduce patients' hemorrhage caused by thrombocytopenia and requirements of platelet transfusion. Patients with leukopenia use it in combination with rhG-CSF in the case of necessity.

【用法用量】根据本品临床研究结果，推荐本品应用剂量为 50μg/kg，于化疗结束后 24～48h 开始或发生血小板减少症后皮下注射（以 1ml 注射用水稀释），每日一次，疗程一般 7～14d，血小板计数恢复后应及时停药。

[Usage] According to the research results from clinical trials, it is better to use 50μg/kg/day (subcutaneous injection, diluted by water for injection) 24-48 hours after chemotherapy or occurring thrombocytopenia. Commonly, 7-14 days is a treatment phase; dosing should be stopped when platelet count returns to normal level.

【不良反应】重组人 IL-11 的大部分不良反应均为轻至中度，且停药后均能迅速消退。约有 10% 的临床病人在观察期间有一些不良事件出现，主要包括水肿以及发热等。

[Untoward effect] Most of the untoward effects are mild or moderate, and abolish rapidly without medication. About 10% patients from clinical trials have side effect events including edema, fever and so on.

【药物相互作用】未发现使用重组人 IL-11 的同时使用 G-CSF 对二者疗效产生任何不良影响。目前尚未

[Drug interaction] So far, there are no untoward interactions between rhIL-11 and G-CSF had been reported. Drug interactions between rhIL-11 and other drugs have not been assessed yet. According to

对重组人IL-11与其他一些药物之间的相互作用进行评价，根据已有的体外和动物试验数据，重组人IL-11与P450药酶的一些已知底物之间不会有相互作用。

experiment data of in vitro and animal models, there are no interactions between rhIL-11 and substrates of P450 enzyme.

【禁忌】同类产品国外曾发生严重过敏反应。因此，对IL-11及本品中其他成分过敏者禁用，对血液制品及大肠杆菌表达的其他生物制剂有过敏史者慎用。

[Taboo] Severe anaphylactic reactions had happened when using congeneric products abroad, so people who are allergic to IL-11 and related components should not use this product. People have allergic history to blood products and *E.coli* will use it with caution.

【注意事项】①本品应在化疗后24～48h开始使用，不宜在化疗前或化疗过程中使用。②使用本品过程中应定期检查血象（一般隔日一次），注意血小板数值的变化。在血小板升至100×10^9个/L时应及时停药。③器质性心脏病患者，尤其充血性心衰及房颤、房扑病史的患者慎用。④使用期间应注意毛细血管渗漏综合征的监测，如体重、浮肿、胸腹腔积液等。⑤该药仅供医嘱或在医生指导下使用。

[Notes] ① Product should be used by 24-48 hours after chemotherapy, it is inappropriate when using it before or during chemotherapy. ② To exam patients' hemogram and platelet count changes during dosing period (once/two days). Stop use it when platelet count returns to 100×10^9/L. ③ Patients with heart diseases, particularly congestive heart failure and atrial fibrillation should use the drug with caution. ④ Capillary leakage syndrome should be monitored during using period, such as weight, dropsy, hydrothorax and seroperitoneum. ⑤ Only use it when prescribed by a professional.

1.2.5 IL-11临床应用

许多恶性疾病的治疗都会用到化学疗法，然而化疗会引起骨髓抑制及与之相关的贫血、白细胞减少症和血小板减少症等疾病。红细胞生成素（EPO）、集落刺激因子（CSF）等生长因子的应用，可减轻贫血和粒细胞减少对患者的危害，但血小板减少症却是一个一直未解决的问题。虽然输入血小板可缓解血小板减少，但是这种作用是暂时的，同时还可能出现其他问题如感染、输血反应、抗原抗体反应。在寻找血小板生长因子的过程中，人们陆续发现了TPO、SCF、白介素（包括IL-1、IL-3、IL-6和IL-11）以及GM-CSF，其中IL-11显示出较好的效

1.2.5 IL-11 Clinical Trials

Therapies for malignant diseases often includes chemotherapy, while chemotherapy may induces bone marrow (BM) inhibition, anemia, leukocytopenia, and thrombocytopenia. Although several growth factors had been administrated, such as treatments of erythropoietin (EPO) and colony stimulating factor (CSF) reduce damage to patients of anemia and granulocytopenia, the thrombocytopenia has not been improved. Platelet transfusion can be used to cure thrombocytopenia, but this enhanced trend is just temporary, at the same time other problems such as infection, transfusion reaction, and antigen antibody reaction may occur. During the period of finding platelet growth factor, such as thrombopoietin (TPO), stem cell factor (SCF), interleukin-1, 3, 6, 11, and granulocyte-macrophage colony-stimulating factor (GM-CSF). Among these cytokines, IL-11 has the best therapeutic effect. Recombinant human interleukin-II（rhIL-11）had

果。人重组IL-11能够增加动物和人类血小板数量，它是唯一用于治疗由化学疗法导致的血小板减少症（CIT）的药物。

been shown to increase platelet counts in animals and human, and it is the only drug approved for using in chemotherapy induced thrombocytopenia (CIT).

近年来，国内外对IL-11的临床应用研究主要集中于有关血液类的疾病如血友病、骨骼增生异常综合征、再生障碍性贫血、白血病和血液恶性肿瘤等，也涉及其他有关血小板减少的疾病如乳腺癌、自发性脑内出血以及登革热等。

Nowadays, clinical trial studies about IL-11 focus on blood related diseased such as hemophilia, myelodysplastic syndrome, aplastic anemia, leukemia and hematologic malignancies. These studies also refer to thrombocytopenia associated diseased such as breast cancer, spontaneous intracerebral hemorrhage and dengue fever.

1. *血友病*　凝血因子是一种多聚体糖蛋白，调节血小板黏附降低血管性血友病。血管加压素是治疗血管性血友病的传统手段，但是由于诸多原因，它存在着局限，如快速抗药反应、使用不方便等，20%的病人对其无反应。人重组IL-11是130kDa的信号因子，具有造血和抗炎活性、增加凝血因子抗原，以及增强鼠和犬杂合子凝血因子的活性。为了验证重组IL-Ⅱ在人体内的生物活性和安全性，Ragni MV等选择非出血性中度血管性血友病患者为研究对象，实行二期前瞻性开放性试验。他们对9名患者每天给予重组IL-11皮下注射干预，连续7d。他们发现，凝血因子和Ⅷ因子水平逐渐升高，随后静脉注射血管加压素可使凝血因子和Ⅷ因子水平进一步升高；定量PCR检测血小板凝血因子mRNA表达，发现其表达量超过基线2～8倍。这些发现说明重组IL-11可能上调凝血因子mRNA的表达，IL-11与血管加压素的作用机理不同。14d后凝血因子和Ⅷ因子水平又会恢复到基础水平。

1. Hemophilia　Von Willebrand factor (VWF) is a multimeric glycoprotein that mediates platelet adhesion and is decreased in von Willebrand disease (VWD). Arginine vasopressin, the most common treatment for VWD, is limited by tachyphylaxis and inconvenience, and in 20% of the patients, unresponsiveness. Recombinant human interleukin-II (rhIL-11), a gp-130 signalling cytokine with haematopoietic and anti-inflammatory activity, increases VWF antigen and its activity in heterozygous $VWF^{+/-}$ mice and dogs. To determine the biological efficacy and safety of rhIL-11 in non-bleeding human subjects with mild VWD, Ragni MV *et al.* conducted a phase II prospective open-label trial of rhIL-11 by subcutaneously (s.c.), given daily for 7 days in nine subjects with mild VWD. They had found that VWF and factor Ⅷ (FⅧ) levels increased gradually and progressively after s.c. rhIL-11. Following intravenous arginine vasopressin, on day 7 there was a further boost in VWF and FⅧ levels, suggesting that the mechanism of rhIL-11 differs from that of arginine vasopressin. Platelet VWF mRNA expression measured by quantitative PCR increased from two to eight fold over baseline, suggesting that the mechanism of rhIL-11 effect may be upregulation of VWF mRNA. However, VWF and FⅧ levels returned to baseline by day 14.

2. *骨骼增生异常综合征*　骨髓

2. Myelodysplastic syndrome　Severe

增生异常综合征患者严重缺少血小板，具有出血的高风险。目前对其的治疗手段主要是给患者输入血小板。Montero AJ 等报道了骨髓增生异常综合征患者使用低剂量 IL-11［10mg/(kg·d)］后他们的临床反应和毒性反应。32 名患者中只有 9 名表现为血小板数量增加。按照世界卫生组织的规定，其中 5 名显示出较强的血小板反应，4 名反应较弱。低剂量 IL-11 也易耐受，没有出现 4 级毒性。他们的研究为长期使用 IL-11 提供了临床参考依据，低剂量的 IL-11 能使小部分骨髓增生异常综合征患者血小板数量增加，减少输入血小板的需求。

3. *再生障碍性贫血* 为评估重组 IL-11 治疗前再生障碍性贫血的临床功效，研究者给 6 名患者注射重组 IL-11，每天一次，连续 7～14d。在治疗后第 8、15、30 和 60 天进行血小板计数，第 15 天检测骨髓并与治疗前的情况作比较。结果显示 50% 的患者血小板数量显著增加，总疗效率达 83.3%，6 名患者骨髓中巨核细胞数量都增加，并且重组 IL-11 的副作用很小。因此在本次研究中重组 IL-11 对前再生障碍性贫血的疗效还是令人满意的，但是由于实验样本数量太小，还需进一步大样本量研究。

4. *白血病* IL-11 与 IL-3、GM-CSF 和 SCF 协同作用可刺激人早期白血病细胞、骨髓性白血病细胞系、成巨核细胞系、红白血病细胞系增生，并能刺激白血病细胞集落形成。IL-11 mRNA 表达于白血病细胞，使用 IL-11 反义寡核苷酸技术可抑制白血病细胞的生长，说明 IL-11 可能是

thrombocytopenia results in myelodysplastic syndrome (MDS) at risk of serious hemorrhage. Currently, therapeutic options are limited to platelet transfusions. Montero AJ *et al.* reported the results of a phase Ⅱ trial to the clinical response and toxicity data for low-dose IL-11 (10 microg/kg/day) in patients with MDS. In that study, there were nine of 32 assessable patients (28%) demonstrated increases in their platelet counts after treatment. Of these, five were considered major platelet responses (15%), as defined by World Health Organization criteria, and four patients had minor platelet responses (13%). Low-dose IL-11 was well tolerated, with no observed grade 4 toxicities. Montero AJ's study provided additional clinical evidence that chronic administration of IL-11, at low doses, can raise platelet counts and reduce platelet transfusion requirements in a subset of patients with MDS.

3. Aplastic anemia Another Chinese study was to evaluate the clinical efficacy of recombinant human interleukin 11 in the treatment of pre-aplastic anemia, six patients with pre-six patients with pre-aplastic anemia were injected with rhIL-11 of 6 million units once a day during 7-14 days. Blood platelet counts were taken on day 8, 15, 30 and 60 after the treatment, and bone marrow examination was performed on day 15 as compared with those before treatment. The results showed that platelet counts in 3 out of 6 patients increased remarkably (50%), the total efficacy rate was 83.3%, and the amount of megakaryocyte in bone marrow of all six patients increased, the side effect of the rhIL-11 treatment was mild. In conclusion, the efficacy of recombinant human interleukin-II in the treatment of thrombocytopenia patients with pre-aplastic anemia was satisfactory. As the number of the cases was too small to conclude, further exploration is needed to accumulate more applications.

4. Leukemia IL-11 acts as a synergistic factor with IL-3, GM-CSF, and SCF to stimulate proliferation of human primary leukemia cells, myeloid leukemia cell lines, megakaryoblastic cell lines, and erythroleukemic cell lines and to stimulate leukemic blast colony formation. IL-11 mRNA expression in leukemic cells and inhibition of leukemic cell growth by IL-11 antisense oligonucleotides suggest that IL-11 may take its function as an autocrine growth factor in leukemic cell lines.

白血病细胞系中的自分泌生长因子。

5. *血液恶性肿瘤* Zhang XL 等研究和评估低剂量 IL-11 对伴有血液恶性肿瘤的血小板减少症患者经过化疗之后的临床治疗作用。他们发现伴有血液恶性肿瘤的血小板减少症患者在化疗后使用 IL-11，不仅可以显著地提高其血小板的数量，缩短血小板恢复时间，而且只有轻微的副作用。

6. *乳腺癌* 女性晚期乳腺癌临床Ⅰ期的治疗中，使用大剂量的化学疗法并在化疗前和 4 个周期之后对患者使用 IL-11，其使用量逐渐增加［直到 100mg/（kg·d）］。服用重组 IL-11 剂量高于 25mg/（kg·d）的患者在第一和第二个化疗周期后其血小板减少的情况有所减轻。外周血血小板数量增加与 IL-11 刺激血小板产生和巨核细胞成熟有关，也证实了骨髓巨核细胞集落生成单位增多和巨核细胞数量增多。但在此临床试验中，化疗导致的白细胞减少和中性粒细胞减少并未因使用 IL-11 治疗而改善。

7. *自发性脑内出血* 有研究显示，大脑组织受伤后会释放 IL-11 入外周血。Fang HY 等人测定了自发性脑内出血患者血浆和脑脊液中 IL-11 的含量，建立了 IL-11 水平与生存者、脑水肿、血肿以及脑积液患者之间的关系。他们发现 IL-11 与自发性脑内出血患者和脑积液患者的死亡率高度相关，因此认为 IL-11 水平可作为临床上判定自发性脑内出血的重要指标。

然而并非所有的临床实验都是令人满意的，虽然 IL-11 对治疗血小板减少症有疗效，但是仍存在着不可忽

5. Hematologic malignancies Zhang XL *et al.* investigated and evaluated the clinical therapeutic effect of low-dose interleukin-11 treatment of thrombocytopenia in patients with malignant hematologic diseases after chemotherapy. They reported that administration of interleukin-11 in thrombocytopenic patients with hematologic malignancies after chemotherapy may not only remarkably enhanced platelet counts and shortened the recovery time of thrombocytopenia, but also had only mild side effects.

6. Breast cancer In the initial phase I trial, women with advanced-stage breast cancer undergoing high-dose chemotherapy were treated with increasing doses of IL-11 (up to 100 mg/kg/d) both before therapy and after each of four cycles of combined chemotherapy. Patients receiving rhIL-11 at doses > 25 mg/kg/d showed attenuated postchemotherapy thrombocytopenia after the first and second cycles. The increased number of peripheral platelet counts was associated with both stimulation of platelet production and megakaryocyte maturation, as evidenced by increased numbers of BM colony-forming unit-megakaryocyte (CFUMK) and megakaryocyte. However, IL-11 treatment in this trial had no significant effect on leucopenia or neutropenia due to chemotherapy.

7. Spontaneous intracerebral hemorrhage Studies have confirmed that IL-11 would be released into the peripheral blood from damaged brain tissue. Fang HY *et al.* determined the IL-11 levels in patients'plasma and cerebral spinal fluid (CSF) with spontaneous intracerebral hemorrhage (ICH) and correlated IL-11 with survival, edema of the brain, volume of hematoma, and hydrocephalus. They found that IL-11 was highly associated with mortality caused by spontaneous ICH and correlated with the hydrocephalus occurring after ICH onset. So it is believed that IL-11 is a useful clinical marker for spontaneous ICH patients.

However, not all the results of clinical trials are satisfactory. While IL-11 plays a pivotal role in treating thrombocytopenia, but due to its inevitable side effects, its clinical use has been limited. Some studies have adverse

略的副作用，从而限制了它的应用。有的研究结论相反，有的研究仍需进一步探讨。

conclusions and some studies still need to be further explored.

在自体骨髓移植术中，重组 IL-11 不但未能减少血小板输入，同时还会出现水肿和心血管系统的不良反应。最近，Jung Y 等人采用定点突变技术来提高人重组 IL-11 的功效。他们发现 IL-11 突变蛋白具有更好的稳定性、更高的功效并且毒性更小。Schinkel C 等研究发现多处损伤后全身 IL-11 水平升高并且抗炎反应也会在创伤后立即发生，然而外伤后 IL-11 水平升高到底起什么作用仍有待进一步研究。登革热的主要临床表现之一就是血小板数量减少。而 Cardier JE 等研究发现，登革热患者血小板缺乏只与血清促血小板生成素的水平相关，而与 IL-11 的水平无关。

After autologous bone marrow transplantation, rhIL-11 treatment didn't reduce the requirements of platelet transfusion, even induced edema and untoward effect of cardiovascular system. Recently, Jung Y *et al.* had used site-directed mutagenesis technique improving efficacy of rhIL-11. They found that interleukin-II mutein (mIL-11) had the improved stability, enhanced efficacy, and less toxicity. The study of Schinkel *et al.* had shown that elevated levels of systemic IL-11 after multiple injuries, and the anti-inflammatory responses to major injury occurred immediately after trauma. But the role of a posttraumatic increase of IL-11 still has to be further analysis. Thrombocytopenia is one of the main clinical features of dengue fever (DF). However, the study of Cardier JE *et al.* had shown that the thrombocytopenia in dengue disease was associated with changes in the serum levels of thrombopoietin (TPO), but not IL-11.

（编者 邓林红，蒋雪梅）

(Compiled by Linhong Deng & Xuemei Jiang)

2 干 扰 素

2 Interferon

2.1 干扰素概述

2.1 Introduction of Interferon

1957 年 Isaacs 和 Indenmann 首次发现，利用灭活的流感病毒作用于鸡胚绒毛尿囊膜后，可使细胞产生一种可溶性物质，这种物质可干扰其他病毒的复制，将这一物质命名为干扰素（interferon，IFN）。1980 年，国际干扰素命名大会对干扰素的定义为：一种在同种细胞上具有广谱抗病毒活性的蛋白质。干

In 1957, the British virologist Alick Isaacs and the Swiss researcher Jean Lindenmann noticed that an interference effect caused by heat-inactivated influenza virus on the chicken egg chorioallantoic membrane. The effect could make the cells to produce a kind of soluble substances which could interfer the replication of other viruses. In 1980, International Congress of interferon nomenclature defined the soluble substance as interferon (INFs) that had a broad spectrum antiviral activity in the same kind of cells. Interferon was the first multifunctional cytokine that was discovered, cloned

扰素是最早被发现的、第一个被克隆且第一个用于临床治疗疾病的多功能细胞因子。它在生物中普遍存在，在正常个体的脾脏、肝脏、肾脏、外周淋巴细胞和骨髓中都可以检出，又可以被多重因素诱发分泌，如各种病毒、双链 RNA、细胞内繁殖的微生物、细菌产物、多聚体等，其中以病毒和双链 RNA 的诱导最为有效。各种类型的干扰素具有抗病毒、抑制细胞增殖和免疫调节等一系列生物活性，成为多种疾病的有效治疗药物。由于 20 世纪 80 年代以来基因工程技术的迅猛发展，使得干扰素的大规模、低成本生产成为可能。重组人干扰素 α1b（rhIFN-α1b）是我国药品监督管理部门批准上市的第一个基因工程药物。目前世界上已有 20 多个国家致力于干扰素的生产、临床及其作用机理的研究。

and used for clinical treatment of disease. Prevalently, interferon existes in many organisms such as spleen, liver, kidney, peripheral lymphocytes and bone marrow of a normal individual. At same time, the secretion of interferon can be induced by multiple factors, such as a variety of viruses, double-stranded RNA, the propagation of intracellular microorganisms, bacterial products, poly body and etcetera, in which viruses and double-stranded RNA were the most effectively interferon induced factors. Various types of interferon with a series of biological activity of antiviral, inhibition of cell proliferation and immune regulation should be an effective therapy for many diseases. Since the 1980s, the rapid development of genetic engineering techniques made it possible to realize large-scale and low-cost production of the interferon. Recombinant human interferon α1b (rhIFN- α1b) was the first genetically engineered drug that had been approved for marketing by Drug Supervision and Management Departments of China. More than 20 countries in the world dedicated to the product, clinical studies and the mechanistic research of interferon.

2.2 干扰素类型

干扰素是由多种细胞产生的具有广泛的抗病毒、抗肿瘤和免疫调节作用的可溶性糖蛋白。干扰素在整体上不是均一的分子，根据干扰素的产生细胞、受体和活性等综合因素将其分为两种类型：Ⅰ型和Ⅱ型。Ⅰ型干扰素包括普遍存在的 IFN-α、IFN-β、IFN-ω 和 IFN-κ 亚类，Ⅱ型干扰素仅有 IFN-γ，见表 5-3。根据特定型内氨基酸或组成的不同，又将其分为不同的亚型，包括 23 种 IFNα、1 种 IFNβ、1 种 IFNγ，如 IFN-α1、IFN-α2、IFN-α3 等亚型；此外还有亚亚型，如 IFN-α1a、IFN-α1b、IFN-

2.2 Classification of Interferon

The IFNs are a large family of multifunctional secreted proteins involved in antiviral defense, antitumor activities and immune properties. The IFNs are classified into two distinct types (type Ⅰ IFN and type Ⅱ IFN) according to their cellular origin, receptors and activity. The type Ⅰ IFNs include the ubiquitous IFN-α, IFN-β, and IFN-ωsubtypes, and the species-restricted IFN-κ. The only type Ⅱ IFN is IFN-γ (Table 5-3). Several subtypes are also being identified for each type of interferon. These types are divided by different amino acids. Type Ⅰ IFNs include 23 kinds of IFN-α, 1 IFN-β and 1 IFN-γ. Different types of interferon have different clinical application, and the same subtypes of interferon have similar clinical effect (Table 5-4). Type Ⅰ and Ⅱ IFNs differ in their primary amino-acid sequences, evolutionary relationships, receptor cross reactivity, sites of production, and inducibility. Nonetheless, α2a、IFN-α2b 等。不同型别干扰素的临床应用不同，而相同型别，特别是亚型接近的干扰素不仅适应证相同，临床效果也相似，见表 5-4。Ⅰ型和Ⅱ型干扰素在一级氨基酸序列、进化关系、受体交叉反应、产生部位和可诱导性上均不同。但是，Ⅰ型和Ⅱ型干扰素都诱导应答基因的重叠部分和激活相似的信号级联反应。

type Ⅰ and Ⅱ IFNs induce overlapping sets of responsive genes and activate similar signaling cascades.

表 5-3 人干扰素分子家族

家族	人染色体	基因数	氨基酸数	同源性
α	9	13	165～166	75%～85%
β	9	1	166	29%
γ	12	1	173	50%
ω	9	>6	172	30%
κ	9	1	207	30%

表 5-4 不同类型的干扰素临床应用

干扰素型别	适应证
IFN-α1a	病毒性疾病
IFN-α2a	病毒性疾病，肿瘤
IFN-α2b	病毒性疾病，肿瘤
IFN-β	多发性硬化症
IFN-γ	类风湿性关节炎

2.2.1 Ⅰ型干扰素

Ⅰ型干扰素又称为抗病毒干扰素，其生物活性以抗病毒为主。Ⅰ型干扰素基因从普通原始祖细胞进化而来，排列在人第 9 号染色体的短臂上。IFN-α 和 IFN-β 是Ⅰ型干扰素的两种主要形式。IFN-α 主要由白细胞产生，为多基因产物，含有至少 14 种不同基因编码的蛋白质，各成分之间氨基酸顺序的同源性为 70% 以上，成熟的 IFN-α 的分子量约 19kDa。IFN-β 是单基因产物，主要由成纤维细胞和白细胞以外的其他细胞产生，分子量约 20kDa，与 IFN-α 的同源性在氨基酸水平上仅为 29%，在核苷酸水平上约为 45%。IFN-α 和 IFN-β 的受体为同一种分子，其基因位于第 21 号染色体上，在几乎所有类型的有核细胞表面表

2.2.1 Type I IFNs

Type Ⅰ IFNs called anti-viral interferon, are also known primarily for their ability to make cells resistant to viral infections. Type Ⅰ IFN genes have evolved from a common primordial progenitor and are arrayed on the short arm of chromosome 9 in humans. Interferon-alpha, -beta, which have relatively similar amino acid sequences, are classified as type Ⅰ IFNs. IFN-α proteins are produced by leukocytes, there are at least 14 distinct human IFN- α proteins. Individual molcculcs of IFN- α generally show in excess of 70% amino acid homology with each other. Mature IFN-α has a molecular weight of 19 kD. IFN-β is a single copy gene produced by fibroblast and epithelial cells with a molecular weight of 20 kD. Comparison of the sequences of IFN-α and IFN-β has demonstrated approximately 45% homology of the encoding nucleotides and 29% homology of subsequently translated amino acids. IFN- α and -β share the same receptor, coded for by chromosome 21. Both receptors are present on most cell types. Most of Type Ⅰ IFNs are stable at pH2.

达，因此Ⅰ型干扰素作用范围十分广泛。多数Ⅰ型干扰素对酸稳定，在pH 2.0时不被破坏。

Ⅰ型干扰素是身体防御抵抗病毒、外源细胞（包括肿瘤细胞）、细菌和其他病原体第一道防线的组成部分。细胞能响应病毒、细菌和其他细胞因子而产生大量的干扰素，并依据诱导物和细胞类型产生准确的干扰素亚型。与获得性免疫应答不同，Ⅰ型干扰素应答是非专一性的并且非常迅速，其峰值通常出现在最初感染的几小时内。Ⅰ型干扰素也能激活免疫系统成分、提高T细胞毒性、上调主要组织相容性复合物（major histompatibility complex，MHC）Ⅰ类抗原。

Type Ⅰ IFNs are parts of the body's first line of defense against viruses, foreign cells (including tumor cells), microbes, and other pathogens. Cells are capable of producing substantial quantities of IFNs in response to virus, bacteria, and other cytokines. The exact IFN subtype produced depends on the inducer and the cell type. Unlike acquired immune responses, the type Ⅰ IFN response is nonspecific and very rapid, usually peaking within several hours of initial infection. Type Ⅰ IFNs can also activate components of the immune system, enhance T-cell toxicity, and up-regulate class I major histocompatibility complex (MHC) antigens.

1. IFN-α　IFN-α基因家族包含14个人类基因，这些基因编码13个不同的蛋白质，由165或166个氨基酸残基组成，在氨基酸序列上有15%～25%的差异，每一种蛋白都具有不同侧重的抗病毒和抗增殖活性。IFN-α的不同亚型具有细胞作用专一性，例如，IFN-α1对B细胞瘤有较大的抗增殖活性，而IFN-α2则对其他肿瘤作用更特异。当用人细胞进行分析时，IFN-α1的抗病毒活性要低于IFN-α2；然而，IFN-α1对牛、猫和其他灵长类动物的细胞则有更强的抗病毒活性。另外，IFN-α7可作为自然杀伤（NK）细胞刺激活性的拮抗药。

1. IFN-α　There are 14 human genes that comprise the IFN- α family. These genes generally encode mature polypeptides with either 165 or 166 amino acids and have a 15% to 25% variation in primary amino-acid sequence. Thirteen separate proteins are expressed from these 14 genes, each with a different profile of antiviral and antiproliferative activity. Evidence for cell specificity of IFN-α subtypes have been reported. For example, IFN-α 1 has greater antiproliferative effects on B-cell neoplasms, whereas IFN-α2 is more specific for other malignancies. When assayed on human cells, IFN-α1 has less antiviral activity than IFN-α2. Nevertheless, IFN-α1 has potent antiviral activity in bovine-, feline-, and primate-derived cells. Furthermore IFN-α7 can act as an antagonist of natural killer (NK) cell stimulatory activity.

世界范围内注册用于抗肿瘤的重组干扰素是IFN-α2。临床上使用的两种类型重组IFN-α2是IFN-α2a（Roferon A）　和IFN-α2b（Intron

The recombinant IFN licensed internationally for use in oncology is IFN-α2. The two recombinant types of IFN-α2 used clinically are IFN-α2a (Roferon A) and IFN-α2b (IntronA). They differ by a single amino-acid substitution at residue 23-lysine in

A）。两者结构上有一个氨基酸的差别，即IFN-α2a第23位是Lys，而IFN-α 2b是Arg。IFN-α另外一个重组亚型IFN-α1已在我国用于临床。随机的临床试验显示IFN-α1和IFN-α2在刺激NK细胞毒性以及提高细胞内2-5A合成酶活性上是等效的。然而，接受IFN-α1治疗的病人与接受IFN-α2治疗的病人相比，前者较少发生发热、疲劳和胃肠道不适等症状。

IFN- α2a and arginine in IFN-α2b. A second major interferon-α termed "IFN α1" is a recombinant subtype used clinically in China. A randomized clinical trial has found IFN-α1 and IFN-α2 to be equipotent in stimulating NK cell cytotoxicity as well as in increasing intracellular 2-5A synthetase activity. Patients receiving IFN-α1, however, experienced less fever, fatigue, and gastrointestinal disturbance when compared to patients receiving IFN-α2.

在所有IFN-α亚型的序列和它们的活性都研究清楚后，可以进一步设计它们的融合体以改进天然亚型的活性，也可以通过制造融合体从根本上改变IFN的性质。例如，IFN-α1的前61个氨基酸残基与IFN-α2的后104个氨基酸残基组成的融合体对鼠细胞表现出高度抗病毒活性；而另一个由IFN-α1的前91个氨基酸残基与IFN-α4的后72个氨基酸残基组成的融合体虽没有抗病毒活性，但抗小鼠肿瘤的活性却提高了，而这一性质是它们母体分子所不具备的。此外，对融合分子的结构-功能分析已阐明IFN-α分子的N末端对它们的生物活性至关重要。

After the sequences of all IFN- α subtype and their activities were deduced, recombinant hybrids were designed to improve the activity of natural subtypes. It is possible to radically change the properties of an IFN by making hybrids. For example, a hybrid made from the first 61 amino-acid residues of IFN- α1 and the last 104 amino-acid residues of IFN- α2 exhibited high antiviral activity on mouse cells, while another hybrid produced from the first 91 amino acid residues of IFN-α1 and the last 72 amino-acid residues of IFN-α4 had no antiviral effect but displayed enhanced antitumor activity in mice, a property neither of the parent molecules exhibit. Structure-function analysis of hybrid molecules has indicated that the NH_2 terminus portion of the IFN- α molecule is important for its biological activity.

据报道，IFN-α半衰期是4～8h，注射24h后，血清中极少或无法测得IFN-α，为了维持有效浓度需要每日注射。自20世纪90年代早期，就有研发PEG型IFN-α的尝试。最初，5kDa的PEG被连接到IFN-α 2a上，但是，经Ⅱ期临床早期证实其疗效比IFN- α2a差而被迫终止试验。最近，人们研发出了另外两个PEG-IFN：40kDa的PEG交联到IFN-α2a上

Since the reported half-life of IFN- α range from 4 to 8 hours, and little or no IFN- α is detected in serum 24 hours following administration, daily injections of IFN- α are required to achieve sustained efficacy. Since the early 1990s attempts have been made to develop a PEG form of IFN-α. Initially 5kDa PEG was linked to IFN-α2a, but it was abandoned when early phase Ⅱ trials demonstrated inferior efficacy when compared to IFN- α2a. Recently two PEG-IFNs have been developed, 40 kDa PEG was conjugated to IFN- α2a (Pegasys: Hoffman-LaRoche) and a 20kDa PEG was

（Pegasys: Hoffman-LaRoche）和 20kDa 的 PEG 交联到 IFN-α2b 上（PEG- Intron: Schering）。

conjugated to IFN- α2b (PEG-Intron: Schering).

2. IFN-β　人 IFN-β 是由大多数体细胞包括成纤维细胞和上皮细胞表达的一种 IFN 亚型。人 IFN-β 基因是单拷贝基因。天然的 IFN-β 是由 166 个氨基酸组成的蛋白质，在其第 80 位氨基酸残基有 N 连接的单糖基化，然而，该糖基化对 IFN-β 多数生物活性而言不是必需的。

2. IFN-β Human IFN-β is the predominant subtype expressed by most somatic cells, including fibroblasts and epithelial cells. IFN-β gene is a single copy gene in humans. Natural IFN-β is a 166 amino-acid protein with a single N-linked glycosylation at amino-acid residue 80. This glycosylation is not necessary for most of its biological activity.

IFN-β1a（Rebif：Serono 和 Avonex：Biogen）是一种重组的商业化使用的 IFN-β，产自中国仓鼠卵巢（CHO）细胞。IFN-β1b（Betaseron: Berlex）也是商业化的重组产品，是一种由细菌产生的由 Ser 取代了第 17 位的 Cys 的合成突变体。IFN-β1a 和 IFN-β1b 被美国 FDA 批准用于治疗多发性硬化症，二者生物活性相当。

A recombinant IFN-β, IFN-β1a (Rebif: Serono and Avonex: Biogen) is produced for commercial use in Chinese hamster ovary (CHO) cells. A synthetic mutant produced in bacteria has a substitution of serine for cysteine at amino acid 17, yielding IFN-β1b (Betaseron: Berlex). Both IFN-β1a and IFN-β1b are approved by the FDA for treatment of multiple sclerosis (MS) and have shown comparable biological activity.

尽管 IFN-α 和 IFN-β 有大约 30% 氨基酸的同源性，但 IFN-α 与 IFN-β 的抗血清不发生交叉反应，有趣的是两者都竞争相同的细胞受体。体内研究证实与 IFN-α2 相比，IFN-β 对不同的非造血细胞系和成纤维细胞有更强的抗增殖效果。然而，对淋巴衍生细胞系或其他淋巴祖细胞，IFN-α 则具有更强的抗增殖效果。临床上，IFN-α 对治疗多发性硬化症患者无效。

Although IFN- α and IFN-β have approximately 30% amino acid homology, antisera to IFN-α and IFN-β do not cross-react. Interestingly both compete for the same cellular receptor. *In vitro* studies demonstrate greater antiproliferative effects of IFN on different nonhematopoietic cell lines and fibroblasts compared to IFN-α2. However, on lymphoidderived cell lines or other myeloid progenitor cells, IFN- α displayed greater antiproliferative effects. Clinically IFN- α is largely ineffective in treating patients with MS.

3. 其他 Ⅰ 型 IFN（IFN-ω、IFN-τ 和 IFN-κ）　IFN-ω 约一亿年前由 IFN-α 进化而来，可能是原型的 IFN-α 扩充前的形式。与 IFN-α 一样，IFN-ω 基因在反刍动物被复制

3. Other Type Ⅰ IFNs (IFN-ω, IFN-τ and IFN-κ) IFN-ω evolved from IFN- α approximately 100 million years ago, probably before amplification of prototypic IFN-α species. Unlike IFN- α, IFN-ω protein is 172 amino-acid residues long and is capable of eliciting similar biological responses *in vitro* as compared to other type Ⅰ IFNs.

成高拷贝数。和IFN-α不同，IFN-ω蛋白质有172个氨基酸，它能在体外激发与其他Ⅰ型干扰素相似的生物学效应。

最先在反刍动物中鉴定的IFN-τ基因是IFN-ω复制的结果。IFN-τ由早期的胎盘组织（如胚胎滋养层）分泌，因而推测其功能是引起妊娠。它也具有其他Ⅰ型干扰素的生物活性，包括抗病毒和免疫调节。与IFN-τ一样，成熟的IFN-τ蛋白也是由172个氨基酸组成，它与普通的Ⅰ型受体相互作用。重组的绵羊IFN-τ已被报道对治疗多发性硬化症和AIDS病毒感染有效。已知人的IFN-τ没有绵羊蛋白的潜在免疫原性，具有治疗价值，但至今尚未发现人的IFN-τ基因。

The IFN-τ gene first was identified in ruminants as a result of duplication of IFN-ω. IFN-τ is secreted by early placenta tissue (i.e., trophoblasts) where it is proposed to function in establishing pregnancy. It also possesses biological activities of other type Ⅰ IFNs, including antiviral and immunomodulation. Like IFN-ω, mature IFN-τ protein is 172 amino acids in length and interacts with the common type Ⅰ receptor. A recombinant ovine IFN-τ has been reported effective in treatment of multiple melanoma and AIDS virus. Clearly, a huIFN-τ, lacking potential immunogenicity of ovine protein, would be of therapeutic value; however, no such human gene has yet been identified.

IFN-κ是与其他Ⅰ型干扰素蛋白有30%同源性的一种新颖的干扰素，在表皮角化细胞中被发现。IFN-κ基因定位在9号染色体上，其蛋白含有207个氨基酸。当病毒感染或与双链RNA、IFN-γ及IFN-β接触时，角化细胞将诱导产生IFN-κ。IFN-κ利用与其他Ⅰ型干扰素相同的受体和信号途径来产生细胞保护以抵抗病毒感染。

A novel IFN (IFN-κ) with 30% homology to other type Ⅰ proteins has been identified in epidermal keratinocytes. IFN-κ gene is located on chromosome 9, and its protein consists of 207 amino acids. It is induced in keratinocytes upon viral infection, exposure to double stranded RNA, IFN-γ, or IFN-β. It imparts cellular protection against viral infections, utilizing the same receptors and signaling pathways as other type Ⅰ IFNs.

2.2.2 Ⅱ型干扰素

Ⅱ型干扰素又称免疫干扰素或IFN-γ，由T细胞产生，主要参与免疫调节，是体内重要的免疫调节因子。IFN-γ基因的进化似乎出现在较近的年代，与Ⅰ型干扰素基因不同，IFN-γ是更典型的真核基因，含有4个外显子和3个内含子（IFN-α和IFN-β基因没有内含子），其基因定位在12号

2.2.2 Type Ⅱ IFNs (IFN-γ)

Type Ⅱ IFNs, also called immune interferon or IFN-γ. IFN-γ is largely produced by T cells. This interferon is known for its ability to regulate overall immune system functioning. The type Ⅱ *IFN* gene evolved from a distinct progenitor that seems to have appeared in more recent times. Unlike type Ⅰ *IFN* genes, *IFN-γ* is more typical of eukaryotic genes in tha.t it consists of four exons and three introns (*IFN-α* and *IFN-β* have no introns in their respective gene transcripts). It is located on chromosome 12.

染色体上。成熟蛋白质含有143个氨基酸，有糖基化，与Ⅰ型干扰素有非常少的同源性。IFN-γ诱导巨噬细胞表达细胞因子（IL-12、TNF-α）、MHCⅠ类和MHCⅡ类分子和Fc受体，从而调节宿主防御系统。

IFN-γ与Ⅰ型干扰素几乎在所有方面均有不同：IFN-γ是只有一种活性形式的蛋白质，由一条分子量为18kDa的多肽链进行不同程度的糖基化修饰而成；IFN-γ的基因只有一个，位于人类第12号染色体上；IFN-γ的受体与Ⅰ型干扰素的受体无关，其基因位于第6号染色体上，但也同样表达在多数有核细胞表面；IFN-γ对酸不稳定，在pH 2.0时极易破坏，利用此特性可以很容易的将其与Ⅰ型干扰素区分开来。

各类干扰素的生物学特性见表5-5。

The mature protein is 143 amino acids in length, glycosylated, and shares very little homology with the type Ⅰ IFNs. IFN-γ induces macrophages to express cytokines (IL-12, TNF-a), class Ⅰ and Ⅱ MHC molecules, and Fc receptors, thereby modulating the host defense system. Therefore IFN-γ is often described as an immunomodulatory cytokine and has no direct antiviral-inducing activity.

IFN-γ has different characters from other interferons. IFN-γ has only one biologically active form. It is a glycosylated polypeptide chain with molecular weight of 18kDa. In addition to their structural makeup, type Ⅰ and type Ⅱ IFNs have other differences. IFN-γ is a single copy gene located on human chromosome 12. IFN-γ binds to a receptor which is separate and different from the receptor for IFN- α and IFN-β. The binding chain of the IFN-γ receptor is located on human chromosome 6, which also lies on the surface of most cells. IFN-γ is instable at acidic pH 2. Using this feature can easily separate type Ⅰ interferon and IFN-γ.

Biological characteristics of the interferon are shown in Table 5-5.

表5-5　干扰素的生物学特性

干扰素	氨基酸残基数量	分子质量/ku	糖基化	主要来源	主要生物学活性
α（多种）	通常166	16～27	部分	淋巴细胞、单核细胞、巨噬细胞	提供靶细胞抗病毒活性；抑制多种细胞的增殖；调节Ⅰ类MHC抗原的表达
β	166	20	否	成纤维细胞、上皮细胞	与IFN-α相似
γ	143	40～70	是	T淋巴细胞、自然杀伤细胞	调节免疫与炎性反应的各阶段；抗病毒活性和抗增殖活性较弱
ω	172	20	是	白细胞为主	与其他Ⅰ型干扰素相似
τ	172	19	否	滋养层细胞	怀孕早期维持黄体；具有其他Ⅰ型干扰素的主要生物学特性

注：IFN-τ的数据来源于反刍动物IFN，其他IFN的数据来源于人类IFN。

2.3　干扰素的生物学作用

干扰素具有抗病毒繁殖、抗细胞分裂增殖及调节机体免疫三大基本功能。Ⅰ型干扰素的抗病毒作用较强，而Ⅱ型干扰素则有较强的抑制肿瘤细

2.3 The Biological Effects of Interferon

Interferons (IFNs) are the proteins made and released by host cells in response to the presence of pathogens such as viruses, bacteria, or parasites or tumor cells. IFN Ⅰ has more strong antiviral activity than IFN Ⅱ. IFN Ⅱ also has other functions that they can activate immune cells

胞增殖和免疫调节作用。

2.3.1 抗病毒作用

人们首先是在抗病毒的研究中发现了干扰素，虽然干扰素的作用远远超过抗病毒的范围，但目前仍以干扰素诱导细胞产生抗病毒活力作为干扰素的主要特性，而且干扰素的活力测定，一般也是根据其抗病毒作用。

I 型干扰素具有广谱的抗病毒活性，对多种病毒如DNA病毒和RNA病毒均有抑制作用，表现在病毒增殖量减少，细胞损伤程度降低。干扰素的抗病毒活性较高，1mg纯干扰素有2×10^8～2×10^9IU，每10个分子就可以使1个细胞产生抗病毒状态。干扰素对病毒增殖的抑制作用与其浓度成正比，其抗病毒作用是抑制病毒的增殖，而不是将病毒杀灭，一旦从培养细胞中除去干扰素，病毒还会再度增殖起来，只有用足够浓度的干扰素反复处理，才能从培养细胞中将病毒消除。

2.3.2 免疫调节作用

免疫系统细胞在受到抗原、促有丝分裂原或肿瘤细胞等刺激时，或在自发情况下可以产生干扰素，如产生IFN-γ及少量的IFN-α和IFN-β。机体内具免疫功能的各种细胞，如粒细胞、单核细胞、巨噬细胞、T淋巴细胞、B淋巴细胞、自然杀伤细胞（NK细胞）等，不论在体内或体外受干扰素作用后，其特定的功能都会发生变化，从而发挥干扰素的免疫调节作用。干扰素可使淋巴细胞、单核-巨噬细胞等的表面受体和表面抗原发生改变，这些

and increase recognition of infection or tumor cells.

2.3.1 Antiviral Affect of Interferon

Interferon is the name given to a group of proteins known primarily for their role in inhibiting viral infections and in stimulating the entire immune system to fight disease. Research has also shown that these proteins play numerous roles in regulating many kinds of cell functions. However, we just evaluate the potency of the interferon through measuring the ability of inhibiting the virus.

The interferon can inhibit every single- or double-stranded DNA- or RNA- viruses. The most impressive characteristic of interferons to us, however, is their very great biological activity. The specific activity of alpha and beta interferons ranged from 2×10^8 to 2×10^9 units /mg pure protein. For the interferons this means that one units contains as little as 2×10^7molecules. Since ten molecules of interferon can protect one cells. This is lower than the effective concentration of many other highly active biological substances. The antiviral activity is proportional to the concentration of interferon.

2.3.2 Effects on the immune system

Interferons (IFNs) , such as IFN-γ and a small amount of IFN-α and IFN-β, are mainly released by immune system cells in response to the presence of antigens, mitogens as well as tumor cells, or produced spontaneously. The specific functions of all types of immune cells could change under the effects of interferons in vitro and in vivo, such as granulocytes, monocytes, macrophages, T lymphocytes, B lymphocytes and natural killer cells (NK cells). Invariably, interferons play roles in modulating immune responses. Furthermore, interferons alter the surface antigen and receptor of lymphocytes, macrophages, etc. These membrane molecules are important in many immune functions. Interferons, therefore, regulate immune function through changing these immune cell surface molecules.

膜分子在许多免疫功能上是重要的，干扰素可能通过诱导免疫细胞膜分子的改变而调节免疫功能。

干扰素的免疫调节作用主要表现在：①干扰素对Fc受体和HLA-A、HLA-B等位点的作用，促进抗体依赖的杀伤作用和特异性杀伤T细胞的细胞毒作用，从而扩大免疫效应。②干扰素主要通过IFN-γ改变细胞表面DR（Ia）抗原的表达，影响抗原的输送和识别，从而调节巨噬细胞、T细胞、B细胞间的相互关系以影响免疫应答。组织中只要有少量IFN-γ，就可能作用于有DR抗原存在的细胞之间的关系。

The main immunomodulatory effects of interferon: ① the interferon can promote the cytotoxicity of antibody-dependent cells and specific T cell, in addition to expand the immune response through the effecting with the Fc receptors, HLA-A and HLA-B. ② interferon influence the delivery and recognition system of antigens through changing DR (Ia) antigen expression on cell surface by IFN-γ , to regulate the relationship among the macrophage, T cells and B cells. As long as there is a small amount of IFN-γ, it will have some effects to existing DR antigens cells.

2.3.3 抗肿瘤作用

干扰素的两个性质——抗病毒和抗肿瘤活性——使它们成为极具吸引力的治疗疾病的候选药物。IFN-α2作为单一制剂在临床上对毛细胞白血病、AIDS相关的卡波西肉瘤、慢性髓性白血病（CML）、B细胞淋巴瘤和T细胞淋巴瘤、黑色素瘤、多发性硬化症和肾细胞癌具有良好的治疗活性。这些发现使得干扰素成为第一个能够提高肿瘤患者生存率的人体蛋白质，在世界范围内获准上市（表5-6）。

2.3.3 Antitumor Effect of Interferon

Two properties of IFNs—potent antiviral and antitumor activities—make them very attractive candidates for the treatment of disease. The degree of activity and improvement of quality of life of patients with hairy-cell leukemia led to the first licensed approval of IFN in the United States. Clinically beneficial therapeutic activity of IFN-a2 as a single agent has been demonstrated in hairy-cell leukemia, AIDS-related Kaposi's sarcoma, chronic myelogenous leukemia (CML), B- and T-cell lymphomas, melanoma, multiple myeloma, and renal cell carcinoma. These findings have led to regulatory approvals around the world, with IFN representing the first human protein demonstrated to increase survival of cancer patients (Table 5-6).

表5-6 商业销售的人干扰素制品

干扰素类型	商品名	FDA批准适应证
IFN-α2b	Intron-A（Schering）	黑色素瘤，丙型肝炎，CML，毛细胞白血病，AIDS相关卡波西肉瘤，皮肤T细胞淋巴瘤
IFN-α2a	Roferon-A（Roche）	慢性丙型肝炎
IFN-β1a	Avonex（Biogen Inc.）；Rebif（Serono）	多发性硬化症
IFN-β1b	Betaseron（Berlex）	多发性硬化症
IFN-γ1b	Actimmune（InterMune Pharmaceutical，Inc.）	延缓严重恶性骨硬化症患者疾病发展
IFN-α con-1	Infergen（Amgen Inc.）	慢性丙型肝炎

干扰素抗肿瘤机制主要表现在：①抗增殖和凋亡作用。干扰素调控基因表达、调节细胞表面蛋白质表达、激活与细胞生长相关的酶，因此干扰素参与肿瘤细胞增殖速度、凋亡以及功能活性的改变。②增强免疫效应细胞作用效果。干扰素是多效性的细胞调节物，推测其抗肿瘤作用是直接作用于肿瘤细胞的功能性区域或抗原成分，或间接作用于与肿瘤细胞相互作用的免疫细胞群体。干扰素能增强所有类型能消除肿瘤靶细胞的免疫效应细胞（病毒T细胞、自然杀伤细胞和单核细胞）的作用效果。③血管生成抑制作用。随着干扰素的治疗，肿瘤内皮细胞表现出微血管创伤和凝结坏死。IFN-α会在干扰素敏感细胞中直接下调血管生成蛋白bFGF的表达，以降低肿瘤细胞的生长。

Anti-tumor mechanism of interferon: ① Antiproliferative and apoptotic effects. IFNs regulate gene expression, modulate expression of proteins on the cell surface, and activate enzymes that modulate cellular growth. On a cellular basis, these effects translate into alterations of the state of differentiation, rate of proliferation, apoptosis, and functional activity of cancer cells. ② Ehance the effect of the immune effector cells. IFNs are pleiotropic cellular modulators. Antitumor effects are postulated to result from either a direct effect on functional capacity or antigenic composition of tumor cells, or from an indirect effect on the modulation of immune effector cell populations that interact with tumor cells. IFNs augment the effectiveness of all immune effector cell types (cytotoxic T cells, natural killer cells, and monocytes) that have the potential to eliminate tumor target cells. ③ Angiogenesis Inhibition. Following treatment with IFN, tumor endothelial cells exhibit microvascular injury and a pattern of coagulation necrosis. Systemic administration of IFN-a reduces tumor cell growth in IFN-sensitive cells by directly regulating expression of the angiogenic protein, bFGF.

2.4 干扰素的生产工艺

2.4 The Interferon Preparation Process

最初，因为受到各种条件的限制，干扰素的制备来源主要是人体细胞。如α干扰素是用血白细胞或者淋巴细胞制备，β干扰素用人成纤维母细胞制备，γ干扰素用类淋巴母细胞（Namalva）制备。但是传统的干扰素制备方法有很多不足之处：①首先是成本高。由于细胞在短期内产生少量的干扰素，所以大规模生产需要大量的细胞，血源的来源比较困难。②其次是组分纯度低、活性低。一种细胞在诱生剂作用下产生的为多种干扰素的混合物。用病毒或者多聚核苷酸处理白细胞、类淋巴母细胞、成纤维母细胞，得

Initially, the main sources of interferon preparation are human cells, for example, IFN-α is originally isolated from white blood cells or lymphocytes, when it comes to IFN-β and IFN-γ, the primary cells of origin are Fibroblasts and Namalva respectively. However, there are many inadequacies of the traditional interferon preparation methods. ① Firstly, the high manufacturing cost is a troublesome problem. Since large-scale production requires a large number of cells in order to produce an amount of interferon in a short term, the blood cells become scarce. ② Secondly, the low purity and activity level in components becomes inevitable barriers in interferon preparation process. One kind of cells produces a mixture of a variety of interferon under the effect of inducer. All the products of the white blood cells, lymphoblastoid cells (Namalva) and fibroblasts that treated with virus or polynucleotide are mixtures. The productions,

到的均为混合物，需要进行纯化、活性检定后才能使用。③再次是生产的不稳定性。不同厂家、不同批号干扰素有很大的差别，必须有选择地使用。以上原因造成了干扰素的使用受限。用细胞培养制备干扰素是第一代产品，后来科学家从基因工程方面进行研究，研制出了干扰素第二代产品。在我国，基因工程 IFN-α 已于 1989 年转入中试，1990 年获得生产许可证而进入工业化生产阶段。

therefore, need purification and activity test before using. ③ Last but not least, the instability of production is a fatal disadvantage. The interferons from different manufacturers and different batches are so different that they must be used selectively. All the limitations that have been mentioned above restrain the using of traditional interferon preparation methods. Interferons obtained from cell culture preparation were the first generation products, and later scientists developed the second generation of interferons which were produced from the genetic engineering. In China, genetically engineered IFN- α was transferred to the pilot in 1989, and obtained the production license in 1990 and then entered in the production stage of industrialization.

基因工程生产干扰素最早采用的是原核表达系统，以大肠杆菌的应用最为广泛。真核表达体系中，CHO 细胞已成为干扰素表达生产最重要的系统。两种主要的表达系统在生产干扰素方面具有各自不同的特点。

Prokaryotic expression system was firstly used in interferon expression and the *E. coli* had been the most commonly used host, while CHO cells were the most important production system for interferon expression in eukaryotic expression system. The two main expression systems have different characteristics in the production of interferon.

（1）大肠杆菌遗传背景清楚，具有周期短、高密度、易操作、使用安全等优点，是干扰素基因的首选表达系统，能够在较短时间内获得干扰素表达产物，且所需成本相对较低。但用大肠杆菌生产的蛋白质药物缺乏糖基化而导致在人体内易被降解，因此它的药效低。此外，它还存在易产生内毒素和包含体的问题。

(1) *E. coli* has so many advantages, such as a clearer genetic background, a shorter growth cycle, a high density, easier operation and safer usage. Therefore, the *E. coli* was the preferred interferon expression system, in which the interferon expression products were obtained in a short time, and the costs were relatively low. However, the protein drugs of the *E. coli* production were susceptible to be degraded in the human body because of the lack of glycosylation, which had a bad treatment for many diseases. There were also some problems in *E. coli* expression system, such as endotoxins and inclusion bodies.

（2）CHO 细胞具有转录后的糖基化修饰准确、产物胞外分泌、高效扩增和表达等特点。CHO 细胞是目前重组糖基化蛋白生产的首选体系。在 CHO 细胞中分泌表达的干扰素其抗原性、免疫原性和功能与天然蛋白质最接近，生物学性质最接近天然状态。但该法表达率低、周期长、

(2) CHO had many characteristics. For example, the high-level expression of the recombinant protein could be achieved, and the proteins secreted into the fermentation medium were properly folded, and etc. CHO cells were the preferred system for recombinant glycosylated protein production. In most cases, glycosylation and other post-translational modifications occurred in a more or less "human-like" manner, while minor differences may exist. Nevertheless, the development of a stable cell line that expressed the therapeutic protein at a high level might take a long term, and the manufacturing cost was

成本高。细胞收液中干扰素含量为0.2～200mg/L。

下面以大肠杆菌表达系统为例，介绍基因工程生产干扰素 α2b 的生产工艺。

干扰素 α2b 由 165 个氨基酸组成，有 4 个 Cys，1 位 Cys 与 98 位 Cys、29 位 Cys 与 138 位 Cys 之间形成二硫键，后者对其生物活性有很重要的作用，相对分子量约为 19 219。其等电点范围 5～6，在 pH 2.5 溶液中稳定，对热也稳定，对蛋白酶敏感。干扰素 α2b 制备工艺路线如图 5-5 所示。

high. The content of interferon in cell liquid collection in CHO expression system was 0.2～200mg/L.

Let's introduce the INF-α2b preparation process in *E. coli* expression system.

IFN-α2b is a 165 amino-acid protein with 4 Cys residues, there are disulfide bonds between Cys-1 and Cys-98, Cys-29 and Cys-138, and the later plays a very important role in IFN-α2b's biological activity. Molecular weight of IFN- α 2b is about 19 219. The isoelectric point is between 5 and 6 with stability in pH2.5 solution. IFN- α 2b is characterized with thermal stability and protease-sensitivity (Figure 5-5).

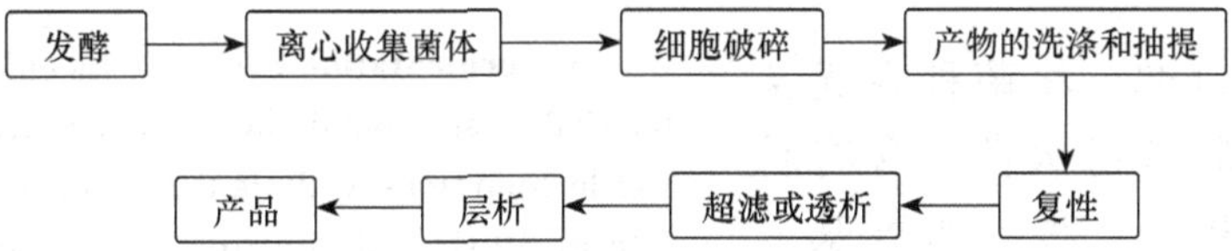

图 5-5　干扰素 α2b 制备工艺路线

2.4.1　发酵

1. 基因工程菌的组建　实验表明 12S 的 mRNA 所合成的干扰素抗病毒活性最高，因此首先将能产生 IFN-α2b 的白细胞中 mRNA 分级分离，收集得到 12S 的 mRNA。将 12S 的 mRNA 反转录取得 cDNA，将此 DNA 克隆到含有四环素和氨苄青霉素的抗性基因的质粒 PBR322 中，转化到大肠杆菌受体中，最终筛选出抗四环素但对氨苄青霉素敏感的细胞克隆，经过多轮筛选获得产生干扰素的 cDNA。最后将干扰素 DNA 克隆入表达载体，转化大肠杆菌 K12，获得干扰素基因工程菌。

2.4.1　Fermentation

1. Construction of genetically engineered bacteria Experiments show that the interferon synthesized by 12S mRNA has the highest antiviral activity. Interferon 12S mRNA is extracted from white blood cells stimulated to produce IFN- α 2b and a complementary DNA (cDNA) copy is made by reverse transcriptase. The 12S cDNA is attached to a plasmid vector PBR322 containing tetracycline and ampicillin resistance gene which will naturally transform an appropriate strain of bacteria, *E. coli*. The cell clones with tetracycline-rejective but ampicillin-sensitive are filtered out, in which the INF cDNA is obtained after several rounds of screening. Finally, the INF cDNA was cloned into expression vectors and transformed into the host cells, *E. coli* K12. The interferon genetically engineered bacteria was constructed.

2. 发酵培养液组成　培养基是影响基因工程菌生长及产物合成的主要因素之一。常用的氮源有酵母粉、蛋白胨、酪蛋白氨基酸、玉米浆和硝酸铵、硫酸铵等。常用的碳源有葡萄糖、甘油、乳糖、甘露糖、果糖等。另外，培养基中还常加一些无机盐、微量元素、维生素、生物素等。

2. Composition of fermented medium The medium is one of the main factors that influent growth of genetically engineered bacteria and product synthesis. Nitrogen sources commonly come from yeast extract, peptone, casein amino acids, corn steep liquor, ammonium nitrate and ammonium sulfate and etc. Carbon sources commonly come from glucose, glycerol, lactose, mannose, fructose and etc. In addition, some inorganic salts, trace elements, vitamins and biotin are often added to the medium.

基因工程生产干扰素 α2b 种子培养液组成：含氨苄青霉素的 LB 液体培养基（含胰化蛋白胨、酵母粉、氯化钠）；发酵培养液组成：蛋白胨、酵母粉、$Na_2HPO_4 \cdot 12H_2O$、KH_2PO_4、NH_4Cl、NaCl、$MgSO_4$、$CaCl_2$ 和少量防泡剂。

In the INF-α2b preparation process, seed medium of genetically engineered bacteria is LB liquid medium which is composed of tryptone peptone, yeast extract, sodium chloride and ampicillin. Fermentation medium is composed of peptone, yeast extract, $Na_2HPO_4 \cdot 12H_2O$, KH_2PO_4, NH_4Cl, NaCl, $MgSO_4$, $CaCl_2$ and a small amount of anti-foaming agent.

3. 发酵过程　将菌种在无菌条件下接入装有已灭菌、含有种子培养液的摇瓶。种子培养基一般采用 LB 培养基。30℃摇床培养 10～12h，培养过程中控制 pH、温度和溶氧值，当吸光度达到规定值，无菌转入装有已灭菌培养液的发酵罐中。控制发酵参数（搅拌转速 500r/min，通气量 10L/min，溶氧 50%）于 30℃发酵 8h，42℃诱导 2～3h，当吸光度达到规定值，将发酵罐降温，放罐。发酵结束后离心除去上清液，收集固体菌体。发酵流程图如图 5-6 所示。

3. Fermentation process Inoculate the strains in the shake flask containing sterilized seed medium under sterile conditions. Culture the strains in shaker for 10 ~ 12h at 30℃, and control the pH, temperature and dissolved oxygen during the culture process. Transfer strains into the fermenter which contains sterilized fermentation medium, when the absorbance reaches to the typical value. Control fermentation parameters, such as agitation speed 500r/min, ventilation 10L/min, and dissolved oxygen 50%. Ferment for 8h at 30℃ and induce for 3h at 42℃ . Cool the fermenter and harvest the products, when the absorbance reaches to the typical value. Centrifugate to remove the supernatant and collect the dried bacterias, when fermentation finishs(Figure 5-6).

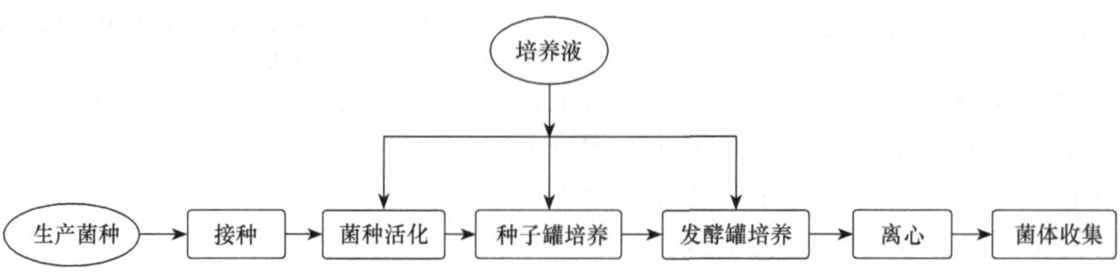

图 5-6　发酵流程

4. 发酵过程中的影响因素

（1）热诱导对发酵的影响。常用的温度诱导型启动子的启动温度是 42℃，而菌种发酵的最佳温度为 30℃，因此发酵分为两个阶段：①前期温度控制在 30℃，此时细菌细胞正常合成阻遏蛋白，使干扰素表达受到阻遏而有利于菌体细胞的生长和重组质粒稳定性。②发酵后期升温至 42℃，经热诱导后，阻遏蛋白失活，干扰素开始表达合成。由于表达产物过量的形成以及其细胞内生理条件的限制，干扰素在细胞内形成包含体，它们会对细胞的生长和质粒的稳定性产生一定的影响。因此，升温的时间选择要合理，一般选择在对数生长期中后期，这样不仅能提高整个细胞密度，又能提高干扰素的表达量。

（2）pH 对发酵的影响。发酵过程中，pH 可以抑制或者激活菌体中某些酶的活性，也会影响细菌细胞膜电荷携带的状况，菌体携带电荷的改变会使膜的渗透性发生改变，从而有可能影响微生物的营养吸收及代谢产物的分泌。pH 对大肠杆菌生长和干扰素表达的影响不同，大肠杆菌的生长最佳 pH 一般是 6.8，而干扰素的最佳表达 pH 往往是 6.0 左右，因此发酵采用两段式培养工艺，前期着重于优化工程菌最佳的生长条件，后期着重于优化干扰素表达的条件。发酵前期，pH 控制在 6.8 左右，当热诱导开始时，关闭碱泵，由于细胞自身代谢的结果，pH 逐步下降；当降到 6.1 时，重新启动碱泵，避免环境 pH 的变化对细胞生

4. Influential factors affecting the fermentation process

(1) Thermo-inducible fermentation　The start temperature of commonly used temperature-inducible promoters is 42℃ while the optimum temperature of bacterial fermentation is 30℃ . Therefore, fermentation can be divided into two phases: ① when the temperature is controlled at 30℃ in the early stage, the bacterial cells synthesize repressor proteins normally and inhibit the expression of interferons which benefit the growth of bacterial cells and the stability of recombinant plasmid; ② At fermentation anaphase, the temperature was up to 42℃. After thermal induction, the repressor protein was inactivated and the synthesis of interferon was started. Due to the overproduction of the expressed product and the limitation of its physiological conditions on the cells, interferons form inclusion bodies in the cells, which can influence cell growth and stabilize plasmids. Up-shifting the culture temperature of cell incubation, generally at the middle and late logarithmic stage of growth, increases the density of cells and improves the production of interferons.

(2) Effects of pH on fermentation　The effects of pH on fermentation are complex beacause the pH can not only inhibit or activate the activity of some enzymes in bacteria but also impact the status of charges of bacterial cell membrane which will effect on the membrane permeability change to affect microbial nutrition absorption and secretion of metabolites. *E. coli* thrives when the pH is a neutral to slightly acid -pH 6.8, while the expression of the interferon is rising to well at pH 6.0. In that case, we choose the two-stage cultivation process. The earlier stage focuses on optimization of growing conditions of the engineered bacteria, the later one focuses on the conditions to optimize interferon expression. In detail, pH is controlled at 6.8 in fermentation prophase. Once interferon is induced, we will close the base pump, and the pH can be decreased substantially by bacterial metabolism. The pH of the medium must be monitored and be adjusted by base addition to maintain a constant pH 6.

长和代谢的不利影响。

（3）泡沫的形成与控制。在发酵培养液内含有各种易产生泡沫的蛋白质，它们与通气搅拌所产生的小气泡混和，会在发酵中产生一定数量的泡沫。随着菌体繁殖，使整个发酵过程形成过多和稳定持久的泡沫。气泡内的空气有隔热作用，造成培养及灭菌不彻底。泡沫多时会从排气口甚至从轴封中溢出，容易造成染菌。一般应通过机械搅拌和加入少量表面活性剂来消除泡沫。

(3) Control foam and bubble in fermentation The proteins which are contained in the fermentation medium can produce bubbles easily. In fermentation process, these proteins will mix with small bubbles to produce a certain amount of foam. Along with the bacteria propagation, there tends to form the excessive, stable and lasting foam in the entire fermentation process. The air bubbles will cause inadequate culture and poor disinfection. Foam will overflow from the exhaust port of the bioreactor, which causes contamination of the fermented medium. Generally, we can eliminate foam and bubbles through stirring and adding a small amount of surfactant.

（4）溶解氧浓度对发酵的影响。一般基因工程菌发酵要求培养液中有足够的溶解氧，为了提高反应器的供氧能力，往往需要增快搅拌转速，增加空气流量，或通入纯氧。氧气不足的一般表现为乙酸、乙醇等发酵副产物的积累，同时造成菌体生长、生产速率下降。因此可以根据菌种生长和产物表达的不同需要，在发酵的不同时期给予不同的溶解氧浓度，这对提高干扰素的发酵水平非常有利。

(4) Dissolved oxygen concentration in fermentation Optimal growth of genetically engineered bacteria usually requires large amounts of dissolved oxygen. Increasing the agitation speed and air flow, or passing pure oxygen into fermentors will increase the oxygen capacity of the reactor. The lack of oxygen is usually manifested as the accumulation of any toxic metabolites, such as acetic acid and ethanol, which inhibites the cells growth and decreases the IFN products. The dissolved oxygen level of the medium is changing according to different demands of the bacteria growth and product expression in the different periods, which will be very favorable to the production of recombinant INF protein.

2.4.2 产物的提取和纯化

表达系统以可溶性蛋白形式表达干扰素，分离纯化工艺流程如图 5-7 所示。

2.4.2 Extraction and Purification of the Products

Flow chart for producing soluble recombinant interferon protein without denaturing is as below (Figure 5-7).

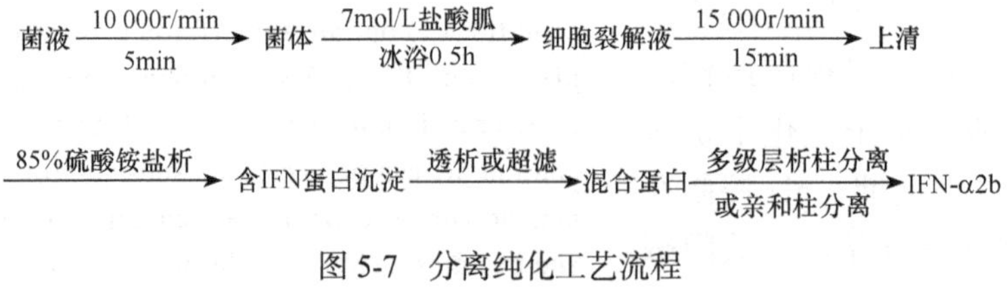

图 5-7 分离纯化工艺流程

1. 细胞破碎及产物的抽提 将收集到的湿菌体用 pH 7.5、20mmol/L

1. Cell disruption and protein extraction The collected wet cells are washed with pH 7.5, 20 mmol/L

磷酸缓冲液洗涤并重悬，然后进行菌体裂解。常用的细菌裂解方法是冰浴超声裂解法，往往会加入裂解液，其中含有 EDTA、SDS、7mol/L 盐酸胍、溶菌酶或者尿素等物质，冰浴裂解 30min。15 000r/min 离心 15min，取上清液加入硫酸铵至85%饱和度后，离心收集蛋白质沉淀，沉淀重悬于缓冲液中，经透析后得干扰素粗液。

of phosphate buffer and resuspended in lysate containing EDTA, SDS, 7mol/L of guanidine hydrochloride, lysozyme or urea. A commonly used method for bacterial lysis is ultrasound accompanied with freezing lysis processing. Cells are disrupted by sonication using sonicator pulsed separated by 30 min incuations on ice. The extract is centrifuged 15 min at 15 000 r/min. The protein content of the supernatant is salting –out after adding ammonium sulfate into the supernatant until the salt concentration reaches 85% saturation, Collect the protein precipitate by centrifugation, then resuspend the protein in buffer. After dialysis, we can get crude interferon solution.

2. *产物的分离纯化* 经上述方法所得到的混合蛋白可以通过常规的离子交换分离得到 INF-α2b，也可直接以亲和分离的模式得到更加纯净的 INF- α2b。

2. Separation and purification of interferon We can get pure INF- α2b from crude products by applying conventional ion exchange or affinity separation mode.

（1）离子交换柱层析法。采用 DEAE Sepharose（FF）柱，柱型 XK30/50。用 pH 7.5，10mmol/L Tris-HCl 液平衡后，将上述透析后的样品上样，流速为 20ml/min。上样结束后，以含不同浓度 NaCl 的 pH 7.5、10mmol/L Tris-HCl 液洗脱，收集干扰素活性峰。选用 Sephacry 100/50XK 层析柱，S.100 填料，柱体积 1700ml，用 pH 5.0、0.05mol/L PBS 液平衡后，将离子交换柱层析收集的干扰素活性峰样品经凝胶过滤，上样量为柱床体积的 5%，再收集干扰素活性峰，合并样品进行质量检定。

(1) Ion exchange chromatography DEAE Sepharose (FF) column (XK30/50) is applied, which is balanced with pH 7.5, 10mmol/L of Tris.HCl. After completion of dialysis, load the crude interferon on the column with flow rate 20ml/min. Afterwards, elute the sample with pH7.5 10mmol/L of Tris.HCl containing different concentrations of NaCl, then collect interferon activity peak. Select Sephacryl00/50XK column packed with S.100, whose volume is 1700ml. Balance the column with pH 5.0,0.05mol/L of PBS. The sample collected by ion exchange column chromatography is purified by gel-filtration chromatography. The loading volume accounts for 5% of the column bed volume. Re-collect the interferon activity peak and combine the samples for quality testing.

（2）亲和柱层析法。亲和柱色谱分离一般可用抗 INF-α 单克隆抗体结合至合适的载体上，然后将混合蛋白过该柱使之得到分离。所使用的柱材料可以这样制备：将 Sepharose 4B 与纯化的抗 IFN-α2b 单克隆抗体混合振摇，室温放置 4h 后低速离心，并用 0.1mol/L $NaHCO_3$（pH 8.5）

(2) Affinity column chromatography In affinity column chromatography, monoclonal antibodies against INF- α are bound to a suitable carrier and crude interferons applied on this affinity column are purified. The affinity column can be prepared like this: mix and shake Sepharose 4B and the purified anti-IFN- α 2b monoclonal antibody. Centrifuge the mixture after place it at room temperature for 4h, Wash away the unbound antibodies with 0.1 mol/L of $NaHCO_3$(pH 8.5) buffer. An equal volume of ethanolamine is added

的缓冲液洗去未结合的抗体，加入等体积的乙醇胺并振摇4h以上，将残存的活性基团加以封闭。用pH 4.0的醋酸盐缓冲液及pH 8.0的Tris-HCl缓冲液交替洗涤，并用pH 7.5的0.1mol/L Tris-HCl加以平衡后装柱。用pH 7.2的0.001mo1/L PBS洗柱，将含有IFN的混合液上样。用PBS反复洗柱，至流出液的吸收回到基线。用pH 2.5的0.1mol/L Gly-HCl洗脱并分段收集，并对各部分浓度进行测定。

into the mixture, which is shaking for more than 4h to block the residual active groups. Wash the mixture with pH 4.0 of acetate buffer solution and pH 8.0 of Tris-HCL buffer alternately. After balanced with pH 7.5 0.1 mol/L Tris-HC1, the mixture is packed into a column. Wash the column with pH 7.2 0.001mo1/L of PBS and load the IFN-α 2b-containing culture on the affinity column. The column should be washed with PBS repeatedly until the absorption of the effluent returns to the baseline. Elute the IFN-α2b sample with pH 2.5, 0.1mol/L of Gly-HC1 and collect the sample in fraction. Finally, determine the concentration of each collection.

若基因工程菌表达干扰素以不溶性蛋白-包含体的形式存在，则在菌体破碎提取包含体后，需经历包含体的溶解及复性过程，具体如下：菌体在裂解液中冰浴裂解30min后，离心收集沉淀，洗涤，即得包含体。沉淀加入8mol/L盐酸胍（含5mmol/L DDT），冰浴条件下抽提2h，再将抽提液迅速用pH 9.0、0.15mol/L硼酸复性液100倍稀释，4℃放置2h，于pH 7.5、10mmol/L Tris-HCl液透析20～24h，即得干扰素复性液。进一步分离纯化方法同上。

There will be dissolution and refolding process after cell disruption and extraction when interferons expressed by genetically engineered bacteria exist in the form of inclusion bodies. Specific steps as follows: The engineered bacterias are lysised in an ice bath for 30 min. The inclusion bodies are obtained after centrifuging to collect precipitation. Add 8mol/L of guanidine hydrochloride containing 5 mmol/L of DDT into the precipitation and extract it in ice bath for 2h. The extraction is quickly diluted 100 times with renatured buffer (pH 9.0, 0.15mol/L of boric acid), then the extraction is placed at 4℃ for 2h. The solution is dialysised against pH 7.5, 10mmol/L of Tris. HCl for 20-24h. Finally, we get the renatured mixture of interferons which is separated and purified furtherly as above mentioned.

2.5 干扰素质量控制标准和要求

2.5 Quality Control and Safety Standards of Interferon

2.5.1 半成品检定

1. 干扰素效价测定　用WISH/VSV（人羊膜细胞/水疱性口炎病毒）系统，可采用细胞病变抑制法，用国家参考品校准为国际单位（IU）。现将测定方法叙述如下。

（1）试验材料（所用试剂均需分析纯或与指定产品相当）。

① MEM培养液。按说明书配制，经除菌过滤，置于玻璃或塑料瓶

2.5.1 Semi-finished Products Test

1. Determination of interferon activity　Interferon activity is assayed by the inhibition of the cytopathic effect (CPE) of vesicular stomatitis virus (VSV) on the human amnion cell line WISH. The national reference preparation is standardized against the human IFN international reference.

The method of determination is described as follows:

(1) Material (All reagents shall be of analytical grade or equivalent to the reference preparation)

① MEM medium　Prepare and sterilize the MEM medium according to the instructions, and preserve it in glass or plastic bottle at 4℃. The validity should not

中，4℃保存。使用期限不得超过产品标示有效期。

exceed the expiration date.

② 牛血清。应符合《中国生物制品主要原辅材料质控标准》中牛血清的有关要求。

② Fetal Bovine serum　The bovine serum should be consistent with the standards against *Quality Control Standards For the Main Raw Material of Biological Products of China*.

③ 完全培养液。MEM 培养液添加 10% 牛血清。

③ Complete medium　MEM medium with 10% of fetal bovine serum (FBS).

④ 测定培养液。MEM 培养液添加 7% 牛血清。

④ Assay medium　MEM medium with 7% of FBS.

⑤ 攻毒培养液。MEM 培养液添加 3% 牛血清。

⑤ Challenge medium　MEM medium with 3% of FBS.

⑥ 消化液。取 0.5g 胰酶、0.2g 乙二胺四乙酸二钠、8g 氯化钠、0.2g 氯化钾、1.152g 磷酸氢二钠、0.2g 磷酸二氢钾，用蒸馏水配成 1000ml 的溶液，经 121℃、15min 高压灭菌。

⑥ Trypsin solution　Dissolve 0.5g of trypsin, 0.2g of EDTA, 8g of NaCl, 0.2g of KCl, 1.152g of Na_2HPO_4, 0.2g of KH_2PO_4 with distilled water to a final volume of 1000ml, then autoclave at 121℃ for 15min.

⑦ 染色液。取 50mg 结晶紫加入 20ml 乙醇溶解，用蒸馏水定容至 100ml。

⑦ Staining solution　Dissolve 50mg of crystal violet with 20ml of ethanol. The distilled water is added to achieve a final volume of 100ml.

⑧ 脱色液。由 50% 乙醇、50% 蒸馏水、0.1% 乙酸（ml/ml。）混匀而成。

⑧ Destaining solution　A solution containing 50% ethanol, 50% distilled water, 0.1% acetic acid (ml/ml).

⑨ 标准品。干扰素效价测定用国家标准品。

⑨ Standard preparation　The national standard preparation is used for determination of interferon activity.

⑩ PBS。取 8g 氯化钠、0.2g 氯化钾、1.44g 磷酸氢二钠、0.24g 磷酸二氢钾，用蒸馏水配成 1000ml 的溶液，经 121℃、15min 高压灭菌。

⑩ PBS　Dissolve 8g of NaCl, 0.2g of KCl, 1.44g of Na_2HPO_4, 0.24g of KH_2PO_4 with distilled water to a final volume of 1000ml, then autoclave at 121℃ for 15min.

⑪ WISH 细胞。WISH 细胞在培养基中呈单层、贴壁生长，每周 2 次 1∶4 消化传代，于完全培养基中生长。

⑪ WISH cells (Human amnion cells)　The cells are maintained in monolayer cultures in complete medium, which are passaged 2 times per week. A sub- cultivation ratio of 1∶4 is recommended.

⑫ VSV。于－70℃保存。

⑫ VSV (vesicular stomatitis virus)　The virus is stored frozen at －70℃ .

（2）试验步骤。①～⑦项各步骤应于无菌条件下进行。

(2) Methods (Each step should be carried out under sterile conditions)

① 铺板。弃去 WISH 细胞培养瓶中的培养液，用 PBS 洗 2 次后消化和收集细胞，用完全培养液配成 2.5×10^4～3.5×10^4 个细胞 /ml 的细

① Plank the cells　Confluent monllayers of WISH cells in flasks is rinsed with PBS 2 times and incubated with the trypsin solution. The Cells in the trypsin solution is collected and replated at 2.5×10^4-3.5×10^4 cells /well in

胞悬液，接种于96孔细胞培养板中，每孔100μl。37℃、5%CO_2条件下培养4～6h。

② 制备标准溶液。取1支标准品按说明书溶解后，用测定培养液稀释至1000IU/ml。

③ 制备样品溶液。将待检样品按说明书溶解后，用测定培养液稀释至1000^3IU/ml。

④ 96孔板稀释。于96孔细胞培养板中每孔加入150μl测定培养液。在A6、A7各孔中每孔加入50μl标准溶液，在A2、A3，A4、A5，A8、A9，A10、A11各孔中每孔加入50μl各待检样品溶液，每个待检样品做3个复孔。自A行取50μl至H行做4倍稀释，每孔也留150μl余液。

⑤ 加样。取①项制备的细胞培养板。将④项制备的溶液移入该细胞培养板，每孔100μl。37℃、5%CO_2条件下培养18～24h。

⑥ 制备病毒液。攻毒剂量为100TC ID_{50}，取保存的VSV用攻毒培养液稀释至工作浓度。

⑦ 攻毒。取⑤项制备的细胞培养板，弃去上清。加入病毒液，每孔100μl。37℃、5%CO_2条件下培养24h（镜检标准溶液的50%病变点在D或E行）。

⑧ 染色。弃去上清，每孔加入50μl染色液，室温放30min。

⑨ 脱色。流水小心冲去染色液，吸干残留水分。每孔加入100μl脱色液，室温放置3～5min。

⑩ 比色。测定波长570nm，参比波长630nm，记录测定结果。

⑪ 结果计算。采用计算机程序

96 well plates which are incubated in the incubator for four to six hours at 37 ℃ and 5% CO_2.

② Standard preparation solution Dissolve the standard preparation according to the instructions, then dilute the solution to 10^3 IU/ml with assay medium.

③ Sample preparation solution Dissolve the samples according to the instructions, then dilute the solution to 10^3 IU/ml with assay medium.

④ Dilution in 96 well plate Put 150μl assay medium into each hole of 96 well plate, then put 50μl standard preparation solution into A6 and A7. Put 50μl each sample preparation solution into A2, A3, A4, A5, A8, A9, A10 and A11. Each sample preparation is added to triplicate wells. Four fold dilutions of the samples are made by transferring 50μl each sample preparation from row A to row H. At last, leave 150μl solution in each pore.

⑤ Transfer samples solution Decant the medium from the 96 well plate of ① .Transfer the solution preparated according to item ④ into the 96 well plate of ①, 100μl of the solution was added to each well. The plate is incubated at 37℃ in 5% CO_2 for 18-24 hours.

⑥ Preparation of virus dilutions Dilute the saved VSV with challenge medium to a suspension containing VSV in concentration 100TC ID50.

⑦ Challenge assay Decant the solution from the 96 well plate of ⑤ . Add 100μl virus dilutions into the 96 well plate of ① each well. The plate is incubated at 37 ℃ in 5% CO_2 for 24 hours (The 50% endpoints are determined by visual examination of the monolayers with a microscope at row D or E).

⑧ Staining process The medium from each well is decanted, and cell monolayers are stained with 50μl staining solution at room temperature for 30min.

⑨ Destaining process Wash off the staining solution with running tap water carefully and remove the residual drops. Add 100μl of destaining solution to each hole, and dry the plate at room temperature for 3-5min.

⑩ Colorimetric analysis The plate reader reads the absorbance at 570nm with a reference wavelength of 630nm. Record the results of the determination.

⑪ Calculation of results Data are processed by

或直线回归计算法进行处理。分别计算各试验样品的半效稀释倍数（即从样品溶液至相当于标准品 50% 最大效应点的稀释倍数），并按下式计算试验结果。

the computer program or a linear regression. The semi-effective dilution factors of each test sample is calculated.

The formula is as follows:

Sample titer = Standard preparation titer × (Sample predilution factors/Standard pre-dilution factors Sample) × (Sample semi-effective dilution factors/Standard semi-effective dilution factors)

$$待检样品效价=标准品效价\times\frac{待检样品预稀释倍数}{标准品预稀释倍数}\times\frac{待检样品半效稀释倍数}{标准品半效稀释倍数}$$

2. 蛋白含量测定　Lowry 法（Folin- 酚显色反应）是国内外常用的测定微量蛋白质的经典方法，常用来测定干扰素中的蛋白质含量。本部分内容以中国药品生物制品检定所提供的标准蛋白作标准。

2. Determination of interferon concentration Lowry method (Folin-phenol color reaction) is a classic method commonly used for the determination of microprotein. The Lowry procedure is most commonly employed for the determination of interferon concentration using standard protein provided by The National Institute for the Control of Pharmaceutical and Biological Products as a standard.

3. 比活性　比活性（specific activity）是指每毫克蛋白中所含干扰素的活性单位，即干扰素效价的国际单位与蛋白含量的毫克数之比。国际上规定，能保护 50% 细胞免受病毒攻击的浓度即为一个干扰素活性单位。

3. Determination of interferon specific activity Specific activity is defined as the amount of interferon activity units per mg of protein. In other words, the specific activity is derived by comparing the international units of interferon to the mg of protein content. In international regulations, an interferon activity unit is the concentration of which the interferon can protect 50% of the cells against virus attacks.

随着干扰素的逐步纯化，其比活性逐步升高，可借以鉴定干扰素的纯化程度和纯化步骤的正确性。其检定步骤以及标准具体参照《中国生物制品规程》，应不低于 1.0×10^{7}IU/mg 蛋白。

Along with the purification of interferon, the specific activity is increasing gradually. We can identify the degree of purification and the correctness of purification steps according to the specific activity. The specific methods and standars should be carried out according to *Procedures for China Biologic Products*. The specific activity should not below 1.0×10^{7}IU/mg protein.

4. 纯度测定　用非还原型 SDS-PAGE 法，加样量应不低于 5μg（银染法）或 10μg（考马斯亮蓝 R250 染色法）。经扫描仪扫描，纯度应不低于 95.0%；用高效液相色谱法，波长 280nm 检测，应呈一个吸收峰，或主峰应不低于总面积的 95.0%。

4. Determination of interferon purity The non-reduced SDS-PAGE is introduced and the loading quantity of protein sample should not be less than 5μg (silver staining) or 10μg (Coomassie brilliant blue R250 staining).The purity of interferon should not be less than 95.0%, when the results of electrophoresis are obtained by scanning. As for HPLC 280nm, there would be a single absorption peak or the main peak area should not be less than 95.0% of the total area.

5. 相对分子质量测定　用还原型 SDS-PAGE 法，加样量应不低于

5. Determination of interferon relative molecular mass The reduced SDS-PAGE is introduced and the

1μg，同时用已知相对分子质量的蛋白标准系列作对照，以迁移率为横坐标，相对分子质量的对数为纵坐标作图，计算相对分子质量，再与干扰素理论值比较，其误差不得高于10%。

6. 外源性DNA残留量测定 用固相斑点杂交法，以地高辛标记的核酸探针法或经国家药品鉴定机构认可的其他敏感方法测定。每剂量中残余外源性DNA应低于100pg。

7. IgG残留量测定 如采用单克隆抗体亲和层析法纯化，应进行本项检定。采用双抗体夹心酶联免疫法测定。IgG含量应不高于100ng/剂量。

8. 宿主菌蛋白残留量测定 用酶联免疫法测定。宿主菌蛋白残留量应不高于总蛋白质的0.1%。

9. 残余抗生素活性 凡在菌种传代中曾经使用过抗生素者，均应测定相应抗生素活性，半成品中不应有残余抗生素活性存在。

10. 细菌内毒素含量测定 采用凝胶法检查细菌内毒素，即用鲎试剂与细菌内毒素产生凝结反应的机理，以判断供试品中细菌内毒素的限量是否符合规定。具体步骤按2000版规程通则《生物制品细菌内毒素试验规程》测定。细菌内毒素含量应不高于10EU/300 000 IU。

11. 等电点测定 用等电点聚焦电泳法测定，批与批之间等电点应完全相同。

12. 紫外光谱扫描 检查半成品的光谱吸收值，用全自动扫描紫外分

loading quantity of protein sample should not be less than 1μg. The protein standard molecular weight marker should be used in the electrophoresis as a control. Make the standard curve with the migration rate as the abscissa and the logarithm of relative molecular weight as vertical axis. Calculate the relative molecular mass of interferon, and then compare it with interferon theoretical value. The inaccuracy should not be higher than 10%.

6. Determination of residual exogenous DNA Residues of exogenous DNA is determined by the method of solid-phase dot blot hybridization with digoxigenin-labeled nucleic acid probe or other sensitive method recognized by the State Drug accreditation bodies. Residual exogenous DNA per dose should be lower than 100pg.

7. Determination of residual IgG This test should be carried out when interferon is purified by monoclonal antibody affinity chromatography. Double antibody sandwich enzyme-linked immunosorbent assay is introduced. Content of IgG in interferon preparation should not be higher than 100ng per dose.

8. Determination of residues of protein of host cell Enzyme-linked immunosorbent assay is introduced. Residues of protein of host cell should not be higher than 0.1% of the total protein.

9. Determination of residual antibiotic activity If antibiotics are used in the bacteria passages, the determination of the antibiotic activity should be carried out. There should not have residual antibiotic activity in semi-finished products.

10. Determination of bacterial endotoxin The limulus amebocyte lysate reagent and gel method are used in the Bacterial Endotoxins Test to judge whether the bacterial endotoxin in the sample is to be standard-complian. The specific steps should be carried out in accordance with *Procedures for China Biologic Products—Testing Procedures of Bacterial Endotoxin For Biologic Products*. The bacterial endotoxin should not be higher than 10EU/300 000 IU.

11. Determination of isoelectric point Isoelectric focusing electrophoresis (IEF) is used for the determination of the isoelectric point. Isoelectric point among different batches should be identical.

12. UV scanning the products The spectral absorption value of the semi-finished products should be

光光度计观察紫外光范围内光谱图，最大吸收峰波长应为278nm±3nm，批与批之间应一致。

13. 肽图测定 肽图谱分析是基因工程多肽药物质量控制的重要手段。肽图谱对每一种蛋白质来说是特征的、专一的。通过肽图分析可以鉴别干扰素，并检测各批产品的蛋白质一级结构的一致性。肽图测定采用胰蛋白酶裂解反相HPLC法。其测定结果应符合干扰素的图形，批与批之间肽图应一致。

2.5.2 成品检定

1. 物理性质 冻干制品外观多为白色或者黄色疏松体，加注射水迅速溶解，肉眼观察呈澄清透明液体。

2. 鉴别试验 用中和实验法，干扰素抗病毒活性能被抗干扰素血清中和。或用免疫印染法，应为阳性。

3. 水分测定 用卡式法，应低于3%。

4. 无菌试验 基因工程药物不得含有杂菌，干扰素制品采用直接接种法进行无菌试验，结果以无杂菌生长判为合格。若无菌试验发现杂菌生长，可复试。复试样品量应加倍，若复试仍有杂菌生长，该制品判为不合格。具体步骤按《生物制品无菌试验规程》A项进行。

5. 热原质实验 本试验将一定剂量的干扰素供试品静脉注入家兔，在规定期间内观察家兔体温升高情况，以判定干扰素供试品中所含热原质的限度是否符合规定。具体方法按《中华人民共和国药典》(2015年，第三部)进行。

6. 效价测定 同2.5.1半成品检

determined. Observe the spectra in the UV range with automatic scanning spectrophotometer. The max wavelength of absorption peak should be 278 nm ± 3 nm. The results among different batches should be identical.

13. Determination of peptide mapping of interferon The peptide mapping analysis is an important method for the quality control of genetically engineered peptide drug. Peptide mapping is characterized and specific for each protein. With peptide mapping analysis, we can not only identify interferons but also detect the consistency and primary structure of protein among different batches. Peptide mapping is determined by Trypsin-degradation and Reverse-Phase HPLC. The results should be consistent with the graphics of interferon, and the peptide mappings of different batches should be in compliance with each other.

2.5.2 Finished Products Test

1. Physical properties Freeze-dried interferon products mostly have an appearance of white or yellow loose body. It can be dissolved with injection water rapidly. The solution is clear and transparent on visual inspection.

2. Identification test If neutralization test is introduced, the antiviral activity of interferons would be neutralized by anti-interferon serum. On the other hand, the result should be positive if immunohistochemical staining method is used.

3. Determination of moisture content Carle Fischer's method for the determination of moisture is introduced. The moisture concent should be less than 3%.

4. Sterility test The genetically engineered drugs shouldn't have bacterias. With direct inoculation method, there should be no bacteria growing in sterility test. If not, re-test the sample with double volume. If there still have bacterias growth, the products is considered unqualified. The specific steps should be carried out in accordance with *Procedures for China Biologic Products item A — Sterility Test Procedures For Biologic Products*.

5. Pyrogen test After an intravenous injection of a certain dose of interferons, observe the fervescence of rabbit body within a prescribed period to determine the limitation of the pyrogens contained in interferon test sample. The specific methods should be carried out according to *Pharmacopoeia of The People's Republic of China*. 2015. Volume Ⅲ.

6. Interferon products activity determination Same

定，效价不低于标示量。

7. 安全试验　用豚鼠或者白鼠作为实验对象，进行腹侧皮下注射或者腹腔注射，观察若干天，如均健存，每只体重增加者判为合格。如不符合上述要求，则进行复试再次判定。具体按《中华人民共和国药典》（2015 年，第三部）相关要求进行。

基因工程干扰素 α2b 半成品及成品质量标准见表 5-7 及表 5-8。

as in Section 2.5.1 Semi-finished Products Test.

7. Safty test The guinea pigs or mice are observed for a number of days after ventral subcutaneous or intraperitoneal injecting interferon. If the test objects is healthy and gaining weight, the products are judged to be qualified. Otherwise, re-test and determine again. The specific methods should be carried out according to *Pharmacopoeia of The People's Republic of China*. 2015. Volume Ⅲ.

Quality standards for recombinant interferon- α2b are shown in Table 5-7 and Table 5-8 respectively.

表 5-7　干扰素 α2b 半成品质量标准

测定项目	标准	方法
外观	澄清透明溶液	肉眼
分子量	18 500～20 000	SDS-PAGE 法
肽图	4～6 个	反向 HPLC 法（化学裂解）
等电点	5.5～6.8	等电聚焦电泳
纯度	＞98%	HPLC 法
紫外最大吸收峰波长（nm）	280±2	紫外分光光度法
N- 端氨基酸序列	与文献一致	氨基酸序列仪
外源性 DNA	＜100pg	点杂交法
残留鼠 IgG	100ng	酶标法
蛋白质含量	80～100μg	Folin- 酚法
比活性	1×10^{8}IU/mg	细胞毒抑制试验

表 5-8　干扰素 α2b 成品质量标准

测定项目	标准	方法
外观	白色疏松体，加注射用水迅速溶解，呈澄清透明液体	肉眼
水分含量	＜3%	卡氏法
热原试验	通过	中华人民共和国药典（2015 年，第三部）
无菌试验	通过	中华人民共和国药典（2015 年，第三部）
安全试验	通过	中华人民共和国药典（2015 年，第三部）
生物效价	超过标示量	细胞毒抑制试验

（编者　康　宁）

(Compiled by Ning Kang)

3 肿瘤坏死因子

3 Tumor Necrosis Factor (TNF)

3.1 TNF 概述

1975 年 Carswell 等发现接种卡介苗（BCG）的小鼠注射脂多糖（LPS）后，血清中含有一种能杀伤某些肿瘤细胞或使体内肿瘤组织发生血坏死的因子，称为肿瘤坏死因子（Tumor Necrosis Factor,TNF）。1985 年 Shalaby 把由单核 - 巨噬细胞产生的 TNF 命名为 TNF- α，也称恶病质素（Cachectin），把 T 淋巴细胞产生的淋巴毒素（Lymphotoxin,LT）命名为 TNF-β。

肿瘤坏死因子是指一种由活化的单核 - 巨噬细胞产生，能杀伤和抑制肿瘤细胞的细胞因子。根据其产生来源和结构不同，可分为 TNF- α 和 TNF-β 两类，前者由单核 - 巨噬细胞产生，后者由活化 T 细胞产生。二者具有相似的生物学活性，能促进中性粒细胞吞噬、抗感染、引起发热、诱导肝细胞急性期蛋白合成、促进髓样白血病细胞向巨噬细胞的分化、促进细胞增殖和分化，是重要的炎症介质，并同时参与某些自身免疫性疾病的病理过程。

3.1 Introduction of TNF

In 1975, one type of factor in mouse serum was discovered by Carswell that could kill some cancer cells or cause hemorrhagic necrosis of tumour organization in vivo. the mouse which had an inoculation against BCG, were injected to LPS, named it tumor necrosis factor (TNF). In 1985, Shalaby named the factor produced by the monocyte-macrophage cells TNF- α, also known as cachectin. Subsequently, a cytotoxic factor produced by lymphocytes (lymphotoxin, LT) was named as TNF-β.

Tumor necrosis factor is the factor that produced by monocyte-macrophage cells, and can kill or restrain the tumour cells. Based on the main source and the different structures, the tumor necrosis factor (TNF) family of cytokines consists of two types TNF- α and TNF-β. The former produced by monocyte-macrophage cells, and the latter produced by ATC. Both of them have the similar biological activities, such as promoting neutrophil to phagocytose, anti-infection, causing fever, inducing acute phase protein synthesis in liver cells, promoting the differentiation from myeloid leukemia cells to macrophage and promoting the cells' proliferation and differentiation, they also are the important inflammatory mediators, and participate in some of their own immune disease of pathological damage.

3.2 TNF 的分子结构和基因

TNF 是主要由含 212 个氨基酸的 II 型横跨膜蛋白组成的稳定同源三聚体。并在这种形式下发挥生物学效应。人的 TNF- α 基因与 TNF-β 基因，均位于第 6 对染色体上，由 4 个

3.2 Molecular Structures and Gene of TNF

TNF is primarily produced as a 212-amino acid-long type II transmembrane protein arranged in stable homo-trimers and in this form can play biological effect. The human TNF- α and -β genes map to chromosome 6, both contain four exons and three introns. Although both molecules bind the same receptor and induce very similar biological activities, they display limited sequence

外显子和3个内含子组成。尽管两分子均绑定在同一受体上，具有非常相似的生物活性，但是它们却显示有限的同源序列。

homology.

人TNF-α基因长约为2.76kb。人TNF-α前体由233个氨基酸残基组成，含76个氨基酸残基的信号肽。切除信号肽后成熟的TNF-α含157个氨基酸残基，为非糖基化形式，在第69位和101位两个半胱氨酸形成分子内二硫键。

The human TNF-a gene spans about 2.76 kilobases and the precursor is initially synthesized as a 233 amino acid residues, which contains about 76 amino acid residues signal peptide. Mature human TNF-α, after signal peptide excision,contains 157 amino acid residues , unglycosylation and an intramolecular disulfide bond between the 69-bit and 101 CYS.

人TNF-β分子由205个氨基酸残基组成，其中包含34氨基酸残基的信号肽。成熟的人TNF-β分子含171个氨基酸残基，分子量25kDa。人的TNF-β和TNF-α DNA序列同源性达56%，氨基酸序列同源性为36%，两者都有重要的生物学功能。

The human TNF-β molecular composes of 205 amino acid residues, including 34 amino acid residues signal peptide. Mature human TNF-β consists of 171 amino acid residues and has an apparent molecular weight of approximately 25-kilodalton (kDa). Between Hu TNF-β and Hu TNF- α, there're about 56% of DNA sequence homology, and about 36% amino acid sequence homology. TNF- α and -β play important biological fucntions.

3.3 TNF的受体

3.3 TNF-R

3.3.1 两类TNF受体（TNF-R）

TNF -R可分为两类：Ⅰ型TNF-R，即TNF-R55，分子量55kDa，CD120a，439个氨基酸残基，该类受体可能在溶细胞活性上起主要作用；Ⅱ型TNF-R，即TNF-R75，分子量75kDa，CD120b，426个氨基酸残基，该类受体可能与信号传递和T细胞增殖有关。两类TNF-R均包括胞膜外区、穿膜区和胞浆区三个部分，可能与TNF-α和TNF-β都结合。它们的胞膜外区有28%的同源性，而胞浆区无同源性，这可能与依赖不同介导的信号转导途径有关。

3.3.1 Classification of TNF-R

TNF receptor divided into two types. The type I receptor displays a molecular mass of 55 kDa, CD120a and 439 amino acid residues, is known as TNF-R55, it may play a leading role in the activities of dissolving cells; The type II receptor is 75 kDa, CD120b and 426 amino acid residues,is known as TNF-R55, it may relate to signal transmission and T cells proliferation. Both receptors consist of three parts: extracellular domain, transmembrane region and cytoplasmic domain, both receptors may bind both TNF- α and TNF-β. The extracellular domain exhibits 28% homology while the cytoplasmic domain is devoid of any homlogy, indicating the likely existence of distinct signalling mechanisms .

TNF-R属于神经生长因子受体（NGFR）家族。TNF-R存在于多种正常及肿瘤细胞表面，一般每个细胞

Both receptor types are members of the nerve growth factor receptor superfamily. They exist in the surface of normal and cancer cells, and each cell receptor has the number of 103-104. Different number of cell surface TNF

受体数目有103～104。不同细胞表面TNF-αR的数目和亲和力似乎与细胞对TNF-α的敏感性并不成正比。TNF-α与相应受体结合后的信号传递机理尚不清楚，可能与活化蛋白激酶C（PKC）催化受体蛋白磷酸化有关。

alpha R and the affinity of cells seem not parallel to the sensitivity of TNF-α. The signaling mechanism of TNF-α combined with the corresponding receptor is not clear, may concern that the activated protein kinase C (PKC) catalyzes the phosphorylation of receptor protein.

3.3.2　可溶性TNF-R

TNF结合蛋白（TNF-BP）是TNF-R的可溶形式，有sTNF-RⅠ（TNF-BPⅠ）和sTNF-RⅡ（TNF-BPⅡ）两种。一般认为sTNF-R具有限制TNF活性或稳定TNF的作用，同时在细胞因子网络中有重要的调节作用。1988年Seckiner发现发热患者尿中有TNF抑制物，分子量为33kDa。1989年Olsson在慢性肾功能不全患者血和尿中也发现有TNF-BP。TNF-BP可与TNF特异性结合，抑制TNF活性，如抑制其细胞毒活性和诱导IL-1产生，促进皮下接种Meth A肉毒的生长，可见于正常妊娠尿中。炎症、内毒素血症、脑膜炎双球菌感染、SLE、HIV感染、肾功能不全时以及肿瘤时可见其升高。可溶性TNF-R可有效地减轻佐剂性关节炎和败血症休克的病理表现。

3.3.2 sTNF-R

TNF binding protein (TNF-BP) is a soluble form of TNF-R, contains sTNF-RⅠ (TNF-BPⅠ) and sTNF-RⅡ (TNF-BPⅡ). TNF-R is generally believed that it has a role in limiting the activity and stability of TNF, and has an important regulatory role in the cytokine network. In 1988, Seckiner discovered that TNF inhibitor was in urine of fever patients, with the molecular weight of 33kDa. In 1989 TNF-BP had also been found in blood and urine of patients, with chronic renal insufficiency, by Olsson. TNF-BP can specifically bind TNF and inhibit TNF activity, such as inhibiting the cytotoxic activity and inducing the production of IL-1, can promote the subcutaneous inoculation of Meth A botulinum's growth, be found in normal pregnancy urine. Inflammation, endotoxemia, meningococcal infections, SLE, HIV infection, renal insufficiency, and tumor can be increased. Soluble TNF-R can effectively reduce the pathological changes in adjuvant arthritis, and septic shock.

3.4　TNF的生物学活性

TNF-α与TNF-β的生物学作用极为相似，这可能与分子结构的相似性和受体的同一性有关，但某些方面也有不同之处。TNF-α有时直接简化为TNF，对它的研究要比TNF-β多。

重组TNF-α的产生有助于研究该细胞因子在体外对细胞的影响以及在体内的全身影响。跟大多数的细胞因子一样，TNF-α对细胞类型有多种作用。这一细胞因子诱导的生物学反应

3.4 Biological Activities of TNF

The biological activities of TNF-α and TNF-β are very similar, which may be because the similarity of the molecular structure and identity of the relevant receptors. But there are also some differences in some biological effects. TNF-α, sometimes referred to simply as TNF, has been studied in significantly greater detail than TNF-β.

The availability of large quantities of recombinant TNF-α facilitates rigorous investigation of the effects of this cytokine on cells in vitro, as well as its systemic in vivo effects. Like most cytokines, TNF-α exhibits pleiotropic effect on various cell types. The major biological responses induced by this cytokine include:

有：①激活非特异性和特异性免疫的某些成分，尤其是对革兰氏阴性菌的反应；②诱导或调节炎症反应；③选择性杀伤多种肿瘤细胞；④介导多种病理反应，包括感染性休克、恶病质和厌食。

① Activation of certain elements of both non-specific and specific immunity, in particular in response to Gram-negative bacteria; ② Induction/regulation of inflammation; ③ Selective cytotoxic activity against a range of tumour cells; ④ Mediation of various pathological conditions, including septic shock, cachexia and anorexia.

由肿瘤坏死因子引起的生物效应的确切范围与多种因素有关，其中最重要的是 TNF-α 的产生水平。在低浓度时，TNF-α 在局部副分泌和自分泌，主要影响白血细胞和血管内皮细胞。在这种情况下，TNF-α 主要是对免疫和炎症进行调节。然而在某些情况下，可能产生大量 TNF-α，例如严重感染革兰氏阴性菌，在这种情况下，TNF-α 进入血液，进行内分泌。TNF-α 的全身影响是指对整个身体的影响，不仅仅是特定区域和器官，如严重休克。许多 TNF-α 的生物效应都是由干扰素 -γ 增强的。

The exact range of biological effects induced by TNF-α is dependent upon a number of factors, most notably the level at which TNF-α is produced. At low concentrations, TNF-α acts locally in a paracrine and autocrine manner, predominantly influencing white blood cells and endothelial cells. Under such circumstances, TNF-α's major activity relates to regulation of immunity and inflammation. In some situations, however, very large quantities of TNF-α may be produced (e.g. during severe Gram-negative bacterial infections). In such instances, TNF-α enters the blood stream and acts in an endocrine manner. Systemic effects of TNF-α (systemic means relating to the whole body, not just a specific area or organ), which include severe shock, are largely detrimental. Many of TNF-α's biological effects are augmented by interferon-γ.

3.4.1 免疫与炎症

在低浓度时，TNF-α 激活一系列介导特异性和非特异性免疫反应的特定白细胞。这些 TNF- α 的作用包括：①激活各种吞噬细胞，包括巨噬细胞、中性粒细胞和多形核白细胞；②增强嗜酸性粒细胞和与病原体相关的巨噬细胞毒性；③发挥抗病毒活性（类似于 I 类干扰素），以及增强 MHC- I 类分子在敏感细胞表面的表达；④增强依赖 IL-2 的 T 淋巴细胞的增殖。

3.4.1 Immunity and Inammation

At low concentrations, TNF-α activates a range of leukocytes which mediate selected elements of both specific and non-specific immunity. These TNF-α actions include: ① Activation of various phagocytic cells, including macrophages, neutrophils and polymorphonuclear leukocytes; ② Enhanced toxicity of eosinophils and macrophages towards pathogens; ③ Exerts antiviral activity somewhat similar to class I interferons, and increases surface expression of class I MHC molecules on sensitive cells; ④ Enhances proliferation of IL-2-dependent T lymphocytes.

TNF-α 的另一个重要作用就是调节炎症反应，而事实上，这可能是其主要的正常生理作用。它促进了炎症的方法有多种，其中包括：①促进中性粒细胞和嗜酸性粒细胞等炎症白细胞的活化；②诱导表达多种黏附分

TNF-α also plays a prominent role in mediating the inflammatory response and, indeed, this may be its major normal physiological role. It promotes inflammation by a number of means, including: ① Promoting activation of neutrophils, eosinophils and other inflammatory leukocytes; ② Induction of expression of various adhesion molecules on the

子的在血管内皮细胞表面的表达；③它们是中性粒细胞、单核细胞和淋巴细胞停留的位点，能促进这些细胞在炎症部位的聚集；④有趋化作用，特别是对单核细胞和多形核白细胞；⑤通过促进对血管内皮细胞骨架的重组来增强血管的通透；⑥诱导合成各种依赖脂质的炎症介质，这些介质包括一些前列腺素、血小板活化因子（PAF）、巨噬细胞和其他细胞，其中很多能促进维持血管扩张和增加血管的通透性；⑦诱导前炎性细胞因子（如 IL -1 和 IL -8）的合成。

因此，TNF-α 影响着免疫和炎症的各个方面。例如通过服用抗 TNF-α 抗体来阻碍其活性已被证明将降低机体遏制和消灭病原体的能力。

3.4.2 杀伤或抑制肿瘤细胞

TNF 在体内外均能杀死某些肿瘤细胞（溶解作用），或抑制增殖作用（抑制作用）。肿瘤细胞株对 TNF-α 敏感性有很大的差异，TNF-α 对极少数肿瘤细胞甚至有刺激作用。用放线菌素 D、丝裂霉素 C、放线菌酮等处理肿瘤细胞（如小鼠成纤维细胞株 L929）可明显增强 TNF- α 杀伤肿瘤细胞的活性。体内肿瘤对 TNF- α 的反应也有很大的差异，与其体外细胞株对 TNF- α 的敏感性并不成正比。同一细胞系可能有敏感株和抵抗株如 L929-S 和 L929-R。此外，靶细胞内源性 TNF 的表达可能会使细胞有抵抗外源性 TNF 的细胞毒性作用，因此，通过诱导或抑制内源性 TNF 的表达可改变细胞对外源性 TNF 的敏感性。巨噬细胞膜结合型 TNF 可能参与对靶细胞的杀伤作用。

surface of vascular endothelial cells; ③ These acts as docking sites for neutrophils, monocytes and lymphocytes, facilitating their accumulation at local sites of inflammation; ④ Displays chemotactic effects, especially for monocytes and polymorphonuclear leukocytes; ⑤ Enhances vascular leakness by promoting a reorganization of the cytoskeleton of endothelial cells; ⑥ Induction of synthesis of various lipid-based inflammatory mediators, including some prostaglandins and platelet-activating factor (PAF), macrophages and other cells. Many of these promote sustained vasodilation and increased vascular leakage; ⑦ Induction of synthesis of pro-inflammatory cytokines, such as IL-1 and IL-8.

TNF-α, therefore, promotes various aspects of immunity and inflammation. Blockage of its activity, e.g. by administration of anti-TNF-α antibodies, has been shown to compromise the body′ s ability to contain and destroy pathogens.

3.4.2 Kill/inhibit Tumor Cells

TNF can kill certain tumor cells (cytolytic action), or inhibition of proliferation (cytostatic action) no matter in vivo or in vitro. The sensitivity of tumor cell lines to TNF- α is very different, and TNF-α even stimulates few tumor cells. Processing tumor cells (such as the mouse fibroblast cell line L929) with actinomycin D, mitomycin C, cycloheximide and so on can significantly enhanced TNF- α activity in killing tumor cells. In vivo the responses of tumor to TNF- α are also very different, and their sensitivities of cell lines to TNF- α are not parallel in vitro.The same cell line may have susceptible and resistant strains such as the L929-S and L929-R.In addition, the expression of endogenous TNF of target cells may cause the cells against exogenous TNF cytotoxicity, Therefore, by inducing or inhibiting the expression of endogenous TNF can chang the sensitivity of cells to exogenous TNF.Macrophage membrane-bound TNF may be involved in the killing effect on target cells.

TNF 杀伤肿瘤的机理还不十分清楚，与补体或穿孔素杀伤细胞相比，TNF 杀伤细胞没有穿孔现象，而且杀伤过程相对比较缓慢。TNF 杀伤肿瘤组织细胞可能与以下机理有关。

The mechanism of TNF anti-tumor is not very clear. Compared with complement or perforin NK cells, TNF killer cells do not perforated, and the destruction process is relatively slow.TNF killing tumor cells may be related to the following mechanisms.

（1）直接杀伤或抑制作用。TNF 与相应受体结合后向细胞内移，被靶细胞溶酶体摄取导致溶酶体稳定性降低，各种酶外泄，引起细胞溶解。也有人认为 TNF 激活磷脂酶 A2，释放超氧化物而引起 DNA 断裂，磷脂酶 A2 抑制剂可降低 TNF 的抗病效应。TNF 可改变靶细胞糖代谢，使细胞内 pH 降低，导致细胞死亡。

(1) Direct destruction/inhibition After binding corresponding receptor, TNF moves to the cells, uptaken by target cells lysosomal led to decreased lysosomal stability, leakage of kinds of enzymes and cell lysis.Also thought that TNF activates phospholipase A2, releases superoxide and causes DNA breakage. phospholipase A2 inhibitors can reduce the resistance effect of TNF. TNF can change the glucose metabolism of target cells, decrease intracellular pH and cause cell death.

（2）TNF 对机体免疫功能的调节作用，促进 T 细胞及其他杀伤细胞对肿瘤细胞的杀伤。

(2) Promoting T cells and other killer cells on tumor cell killing by regulating role of TNF on the immune function.

（3）TNF 作用于血管内皮细胞，损伤内皮细胞或导致血管功能紊乱，使血管损伤和血栓形成，造成肿瘤组织的局部血流阻断而发生出血、缺氧坏死。

(3) TNF effect on vascular endothelial cells, damage vascular endothelial cells or cause dysfunction of the vascular, lead vascular injury and thrombosis ,and result in bleeding, hypoxic necrosis by tumor partial blood flow occlusion.

3.4.3 提高中性粒细胞的吞噬能力

增加过氧化物阴离子产生，增强 ADCC 功能，刺激细胞脱颗粒和分泌髓过氧化物酶。TNF 先与内皮细胞培养可使其增加 MHC-Ⅰ类抗原和 ICAM-1 的表达，IL-1、GM-CSF 和 IL-8 的分泌，并促进中性粒细胞黏附到内皮细胞上，从而刺激机体局部炎症反应，TNF-α 的这种诱导作用要比 TNF-β 强。TNF 刺激单核细胞和巨噬细胞分泌 IL-1，并调节 NHC-Ⅱ类抗原的表达。

3.4.3 Enhance the Phagocytic Activity of Neutrophils

Increase the produce of peroxide anion, enhance ADCC function, stimulate cell degranulation and secrete myeloperoxidase. TNF can be pre-cultured with endothelial cells to increase the expression of MHC-Ⅰ antigen and ICAM-1, and the secretion of IL-1, GM-CSF and IL-8, and promote neutrophil adhesion to endothelial cells, thus stimulate the body part inflammatory response. The induction of TNF-α is stronger than that of TNF-β. TNF stimulates monocytes and macrophages to secrete IL-1, and adjusts the expression of NHC-Ⅱ antigen .

3.4.4 抗感染

如抑制疟原虫生长，抑制病毒复制（如Ⅱ型腺病毒、Ⅱ型疱疹病

3.4.4 Anti-infective

Such as the inhibition of parasite growth, inhibition of viral replication (such as adenovirus type Ⅱ, herpes

毒），抑制病毒蛋白合成、病毒颗粒的产生和感染性，并可杀伤病毒感染细胞。但是TNF抗病毒机理不十分清楚。

virus type Ⅱ), inhibition of viral protein synthesis, production of viral particles and infectious, and anti-virus infected cells. Antiviral mechanism of TNF is not very clear.

3.4.5 TNF是一种内源性热原

TNF引起发热，并诱导肝细胞急性期蛋白的合成。TNF引起发热可能是通过直接刺激下丘脑体温调节中枢和刺激巨噬细胞释放IL-1而引起，还可通过IL-1、TNF-α刺激其他细胞产生IL-6。

3.4.5 TNF is An Endogenous Pyrogens

TNF causes fever and induces the acute protein synthesis of liver cell. TNF causes fever probably by direct stimulating hypothalamic thermoregulation center and stimulating macrophages to release IL-1. IL-1 and TNF- α also can stimulate other cells to produce IL-6.

3.4.6 促进髓样白血病细胞向巨噬细胞分化

TNF可促进髓样白血病细胞ML-1、单核细胞白血病细胞U937、早幼粒白血病细胞HL60的分化，机理不清楚。TGF-β可抑制TNF-α多种生物学活性，但不抑制TNF-α对髓样白血病细胞分化的诱导作用，甚至还有协同效应。

3.4.6 Promoting the Differentiation of Myeloid Leukemia Cells to Macrophages

Such as promoting the differentiation of myeloid leukemia cells ML-1, monocytic leukemia cells U937, promyelocytic leukemia cell HL60,but the mechanism is unclear. TGF-β can inhibit kinds of biological activities of TNF-α, but does not inhibit the induction of TNF- α on the differentiation of myeloid leukemia cells, which even has synergistic effects.

3.4.7 促进细胞增殖和分化

TNF促进T细胞NHC-Ⅰ类抗原表达，增强依赖IL-2和T细胞的胸腺细胞的增殖能力，促进IL-2、CSF和IFN-γ等淋巴因子产生，增强有丝分裂原或外来抗原刺激B细胞的增殖和Ig的分泌。TNF-α对某些肿瘤细胞具有类似生长因子的作用，并协同EGF、PDGF和胰岛素的促增殖作用，促进EGF受体表达。TNF也可促进*c-myc*和*c-fos*等与细胞增殖密切相关原癌基因的表达，引起细胞周期由G_0期向G_1期转变。

3.4.7 Promoting Cell Proliferation and Differentiation

TNF promotes the expression of T cell and NHC-Ⅰ antigen, enhances the proliferation of thymocyte that depends on IL-2 and T cell, promotes the production of IL-2, CSF, IFN-γ and other lymphokines, enhances B cell proliferation in the stimulation of mitogen or foreign antigen and the secretion of Ig. TNF- α has growth factor-like effect for some tumor.Together with the effect on the proliferation of EGF, PDGF and insulin, TNF- α promotes the expression of EGF receptor. TNF can also promote the expression of *c-myc*, *c-fos* and other oncogene closely related to cell proliferation, causing cell cycle transition from G_0 to G_1 phase.

IL-1、IFN-γ和GM-CSF对TNF的生物学功能有明显的增强作用，可能与增加细胞TNF受体的表达有关。

IL-1, IFN-γ and GM-CSF significantly enhanced the biological effects of TNF, this may because of the increased expression of TNF receptors. An anti-TNF- α

已经报道了一种抗 TNF- α 单克隆抗体，它可模拟 TNF- α 的某些生物学作用，这种现象在其他因子中还尚未见到。

monoclonal antibody was reported, it can simulate certain biological effect of TNF- α, but this phenomenon had yet to be seen in other factors.

3.5 TNF 的治疗作用

应用 TNF 在治疗肿瘤等方面已经开始Ⅱ期临床试验，也可与 IL-2 联合治疗肿瘤。目前认为全身用药的疗效不及局部用药，后者如病灶内注射，局部浓度高且副作用也较轻。近年来已采用 TNF 基因治疗开始对黑素瘤等肿瘤进行临床验证。值得重视的是 TNF 又与临床某些疾病的发生有关。

目前，对 TNF 临床研究的焦点主要集中在当 TNF 过度表达诱导阴性临床效果时中和其生物效应。TNF 与数十种疾病（表 5-9）的不良反应有关。服用抗 TNF 单克隆抗体或者可溶 TNF 受体应该能帮助缓解这些疾病的许多症状。

3.5 TNF: Therapeutic Aspects

Application of TNF in the treatment of cancer had been starting phase Ⅱ clinical trials, and TNF can treat cancer associated with IL-2. Now, experts think that the effect of systemic administration is less than that of Topical administration. In latter situation, if intralesional injection, local concentration is higher with less side effects. TNF gene therapy has been used in the clinical validation of melanoma and other tumors in recent years. In addition, It is remarkable that TNF has something to do with clinical occurrence of certain diseases.

Most clinical interest in TNF now centres on neutralizing its biological effects in situations where over expression of TNF induces negative clinical effects. TNF has been firmly implicated in mediating many of the adverse effects associated with dozens of diseases (Table 5-9). Administration of anti-TNF monoclonal antibodies or soluble forms of the TNF receptor should help reduce the severity of many of the symptoms of these diseases.

表 5-9 TNF 诱导的一些症状

疾病	TNF 诱导的症状	病理
癌症	恶病质	刺激某些肿瘤的生长
感染性休克	血管渗漏，组织坏死，低血压	激活血液凝固
类风湿关节炎	组织炎症	可能破坏关节
多发性硬化	炎症	
糖尿病	胰岛细胞死亡	诱导抗胰岛素性

3.5.1 感染性休克

目前认为革兰氏阴性杆菌或脑膜炎球菌引起的弥漫性血管内凝血、中毒性休克是由于细菌内毒素刺激机体产生过量 TNF- α，引起发热，心脏、肾上腺严重损害，呼吸循环衰竭，甚

3.5.1 Septic Shock

Disseminated intravascular coagulation and septic shock caused by gram-negative bacteria or meningococcus are because of excessive production of TNF- α stimulated by bacterial endotoxin,causing fever, serious damage of heart and adrenal, respiratory circulatory failure, and even death, where the TNF

至引起死亡，其TNF水平与病死率正相关。其发病机理可能是TNF刺激内皮细胞，导致炎症、组织损伤和凝血。TNF也是急性肝坏死的重要因素。病毒性暴发型肝衰竭外周血细胞诱生TNF，IL-1活性升高，且与病情程度相关。目前有关TNF介导内毒素性休克的机理还不很清楚。有人认为TNF能促进前凝血酶原活性物质生成，抑制内皮细胞凝血酶调节毒素休克。TNF抗体（抗血清或单克隆抗体）在小鼠、家兔和狒狒体内均有效地阻止致死性内毒素休克的发生。应用抗TNF mAb治疗脓毒症和化脓性休克已进入Ⅲ期临床试验，抗TNF嵌合抗体治疗细菌性感染也已开始Ⅰ期临床试验。

levels and mortality are related. Its pathogenesis may be that TNF stimulates endothelial cells, leading to inflammation, tissue damage and blood coagulation. TNF is also an important factor in acute hepatic necrosis. Peripheral blood cells of fulminant hepatic failure of viral induces TNF, leading IL-1 activity increased, which in the correlation with the severity of disease. Currently, the mechanism of TNF-mediated endotoxin shock is not very clear. Some think that TNF can promote the production of Former prothrombin active substances, and inhibit the toxic shock of endothelial cell thrombomodulin.TNF antibodies (anti-serum or monoclonal antibody) are effectively prevented the occurrence of lethal endotoxin shock no matter in mice, in rabbits and in baboons in vivo. Treating sepsis and septic shock with anti-TNF mAb has entered Phase Ⅲ clinical trials. In addition, treating bacterial infections with anti-TNF chimeric antibody has also begun to phase Ⅰ clinical trial.

3.5.2 恶病质

TNF-α又称恶病质素（cachectin），可诱发机体发生恶病质。

3.5.2 Cachexia

TNF-α, is also known as cachectin, by which the body can be induced to have cachexia.

3.5.3 TNF与病毒复制的关系

TNF还具有类似于IFN抗病毒作用，阻止病毒蛋白质的早期合成，从而抑制病毒的复制，并与TNF-α和TNF-γ协同抗病毒。另一方面，TNF诱导*HIV-Ⅰ*基因在T细胞中表达。TNF和HIV感染的$CD4^+$细胞中活化或诱导NF-κB，NF-κB结合于HIV的长末端重复序列（LTR）的增强子部位，活化*HIV*基因，可能与艾滋病发病有关。艾滋病患者单核细胞TNF-α产生增加，血清中TNF-α水平升高。

3.5.3 TNF and Viral Replication

TNF also has similar anti-viral activity of IFN, prevents the synthesis of virus early protein, thereby inhibits virus replication, and has synergistic antiviral activity with TNF-α and TNF-γ.On the other hand, *HIV- Ⅰ* gene expresses in T cells induced by TNF.$CD4^+$ cells infected by TNF and HIV activates or induces NF-κB.NF-κB binds to the enhancer site of HIV long terminal repeat (LTR), activating *HIV* genes, which may be related to the pathogenesis of AIDS. The production of TNF- α increases in mononuclear cells of AIDS patients, so do the TNF-α serum levels.

3.6 TNF技术发展现状和展望

3.6 Status and Prospects of TNF

3.6.1 TNF-α为主导的联合用药隔离性器官灌流法

3.6.1 TNF-α Dominated Combination Isolation Organ Perfusion Method

1. 隔离性肢体灌流法（ILP） 该法是由Lienard和Eggermont率先

1. Isolation limb perfusion (ILP) ILP is first started by Lienard and Eggermont. Firstly, tie the roots of physical cycle with a tourniquet to separate the

创导的。方法就是用止血带先绑住肢体循环部分的根部，把肢体的血液从全身循环中分隔开来，再用叠套法把肢体血管连接到体外的心肺机中，加入 TNF-α 为主导的联合化疗药物美法伦或γ-干扰素循环灌流在肢体血液中达 1h 左右。结果表明，用这种方法治疗后的瘤部周围的 TNF-α 的峰值浓度可达到全身给药途径时的 50 倍，且无明显的副作用，治疗效果显著。选择周围血管丰富的肿瘤对这种治疗会更加敏感。如肢体性软组织肉瘤和骨瘤，其治疗效果达到了 90% 和 70%，替代了截肢，提高了病人的生活质量。目前，这种方法还用在治疗转移性黑色素瘤，效果达到了 85%。

body's blood; secondly from the systemic circulation, and connect blood vessels to the body of the heart-lung machine in vitro by fold sets method; thirdly, add TNF-α combined with Melphalan or interferon-γ in the blood perfusion of the limb about 1h.The results show that the TNF-α peak concentration around the tumor by this method is 50 times more than that by systemic administration3.In addition this has no significant side effects, and the effect is remarkable. This treatment will be more sensitive if tumors rich in vascular are chose. Such as limb soft tissue sarcoma and bone tumor and the effect was 90% and 70%.This method improved patient quality of life avoiding amputation. At present, the effect in the treatment of metastatic melanoma by this method reached 85%.

2. 隔离性肝脏灌流法　肝脏是肿瘤高发性的器官，用 TNF-α 为主导联合用药进行隔离性肝脏灌流法（IHP）目前已经在临床研究的Ⅱ期。有报道对转移性肝脏肿瘤治疗共有 60 例：术后死亡 6 例（大部分死于由于灌流引起的血栓），CR：2 例，PR：34 例，MR：6 例，有效率为 60%。具体方法：把肝区循环从全身循环系统中隔离出来，用心肺机体外循环肝区，再用 TNF-α 联合美法伦或者阿霉素对肝脏进行灌流。副作用为灌流 18～24h 后肝的 GOT、GPT 可逆性升高。停止处理 30 d 恢复正常。除了肝脏之外，IHP 法还可以用在肺、肾等器官。

2. Isolation hepar perfusion (IHP) The liver is a high incidence of tumor organ. Using TNF-α dominated combination drug by IHP is now in phase Ⅱ clinical studies. It was reported that the treatments of metastatic liver tumor totally have 60 cases: 6 cases of postoperative death (mostly due to thrombosis as a result of perfusion).CR: 2 cases; PR: 34 cases; MR: 6 cases; effective rate was 60%.Method: to isolate the liver from the systemic circulation system loop, using heart-lung machine bypass liver area in vitro, and then perfuse liver with TNF-α combined with Melphalan or interferon-γ.Side effect was that the reversibility of GOT, GPT increased in18-24h after perfusion. It became to normal in 30 days after stopping processing. Except the liver, IHP method can also be used in the lung, kidney and other organs.

3. 低氧性隔离肝脏灌流法　隔离性肝脏灌流法（IHP ）带来的最大问题是剖腹手术给患者带来的压力，影响了生活质量。Eggermont

3. Isolation hypoxic hepar perfusion(IHHP) The biggest problem caused by is the stress for patients after laparotomy which affected the quality of life.IHHP first created by Eggermont will be the direction of the clinical treatment of tumors by TNF- α in the future.

首创的低氧性隔离肝脏灌流法（IHHP）将会是今后 TNF-α 用于临床治疗实体肿瘤的方向。IHHP 法主要是通过血管造影术将气囊管插入肝区相应的血管，再用 TNF-α 加化疗药物对肝肿瘤部分作灌流。IHHP 法的优势在于：无需剖腹手术，提高了病人的生活质量；方法简便，对临床基地的要求也较低，而且可以重复操作。

According to the IHHP, first thing is to insert the balloon tube into the corresponding liver vessels mainly through angiography, and then perfuse liver tumor areas with TNF-α and chemotherapy drugs. No laparotomy is the advantage of IHHP, and it improves life quality of patients. The method is simple, has lower clinical requirements, and can be repeated.

3.6.2　TNF-α 的变构体

用“随机突变法”和“定位突变法”把 TNF-α 的 157 个氨基酸上的每一段甚至每一个氨基酸通过缺失和更换的手段来检测它在功能上的作用。TNF-α 活性三聚体的间沟是由近 N 端和 C 端的极性基团及中部的疏水基团所组成，而这些亚单位之间的沟是 TNF-α 与其受体结合的部位。实验表明，C 端上最后一个氨基酸 Leu 改变成 Phe、Met 或者 Glu，其活性能提高 18 倍左右，因为这种改变使疏水性的 TNF-α 亚单位之间的结构变得更紧凑，从而提高了 TNF-α 同受体结合的效率。另外，缺失 1～7 个 N 端的氨基酸，增加了 TNF-α 的碱性度，提高了和受体的亲和力，活性增加了 5 倍。不仅如此，N 端的 8、9、10 和 17 个氨基酸变成碱性氨基酸，使某些对野生型 TNF-α 不敏感的肿瘤细胞变得敏感了，扩大了 TNF-α 的疗效范围，而且还增强了其抑制肿瘤的活性 .

3.6.2　The Isomers of TNF- α

Test the roles of every section or even each amino acid of 157 amino acids of TNF-α through deletion and replacement by random mutation method and positioning mutations method. The grooves of TNF-α active trimer is made of the polar groups (between the N-and C-side) and hydrophobic groups (in the middle), and those grooves between these subunits is sites of TNF- α binding to its receptor. Experiments show that, if Leu, the last amino acid of C-side, was changed to Phe, met or Glu, its activity could increase about 18 times, because the changes have led the structure among hydrophobic TNF-α subunits become more compact, thus the efficiency of TNF-α binding to the receptors increased. In addition, a lack of one to seven N-amino acids increased TNF-α 's basicity, increased affinity with receptors, and increased its activity by 5 times. Moreover, if the 8,9,10 and 17 N-amino acids are changed into basic amino acids, some wild-type tumor cells which are not sensitive to TNF-α become sensitive, the scope of the efficacy of TNF-α is expanded, and also its activity of inhibiting tumor is enhanced.

细胞水平和动物试验上的实验数据显示了这些变构体不仅提高了同受体结合的能力，还增加了其摧毁肿瘤

Cellular and animal studies show that those isomers not only improve improved the ability to combine with receptor, but also increase the ability to destroy vessels a than round the tumor .In addition, the side effects caused

周围血管上皮组织的能力。另外，变构体引起的毒副作用也低至野生型TNF-α的1/18。

by isomers are 1/18 lower than that caused by wild-type TNF-α.

中国国内对于TNF-α变构体的研究起步虽然较晚，但是效率很高。国内学者现有研究TNF-α变构体有7家机构。目前，已有数家机构的TNF-α变构体已经进入临床研究阶段，而且疗效显著、进程顺利，相信不久TNF-α的变构体会在中国正式面市，造福于肿瘤患者。

Studying the TNF-α isomers in china started at a late time, but efficiency was high. There are already seven research institutions studying the TNF-α isomers. Currently, the productions of several institutions have been in the clinical research phase, the effect is significant, and the process goes well. It is believed that soon the TNF-α isomers would be officially available for the cancer patients in China.

（编者　彭方毅）

(Compiled by Fangyi Peng)

4 胰 岛 素

4 Insulin

胰岛素是由胰岛β细胞产生的多肽激素。它是调节血糖水平的重要激素，正常情况下，胰岛素能维持血糖水平于一定的小范围内。

Insulin is a polypeptide hormone produced by the β cells of the pancreatic islets. It plays a central role in regulating blood glucose levels, generally keeping it within narrow defined limits in healthy individuals.

胰岛素是由Banting和Best在1921年发现的。在这之后，从牛和猪胰脏中提取胰岛素的生产工艺很快发展起来。1921—1980年，人们主要致力于提高胰岛素的纯度以及配制不同时效的制品，从而改进对血糖的调控；并且通过优化提取和加工条件以及运用层析方法来改进纯化的效果，降低蛋白质杂质的浓度。重组DNA技术的出现可使人们无限制地获得人胰岛素，这样就消除了给病人提供胰岛素的来源限制问题。现在将改进的纯化方法与重组DNA技术结合在一起，可以生产出纯度大于98%的极纯胰岛素。

Insulin was discovered by Banting and Best in 1921. Soon afterward, manufacturing processes were developed to extract the insulin from porcine and bovine pancreas. From 1921 to 1980, efforts were directed at both increasing the purity of the insulin and providing different formulations for altering time-action for improved glucose control. Purification was improved by optimizing the extraction and processing conditions and by implementing chromatographic processes to reduce the levels of both general protein impurities. The evolution of recombinant DNA technology led to the unlimited availability of human insulin, which has eliminated issues with sourcing constraints while providing the patient with a natural exogenous source of insulin. Combining the improved purification methodologies and recombinant DNA technology, manufacturers of insulin are now able to provide the purest human insulin (>98%) ever made.

4.1　胰岛素概述

4.1.1　胰岛素分子结构

在 1921 年，胰岛素被看做一种抗糖尿病因子，第二年便开始在临床上应用。其完整的氨基酸序列是在 1951 年确定的。尽管成熟的胰岛素是二聚体结构，但最初合成的单链多肽前体——前胰岛素原，是含有 108 个氨基酸残基的多肽。它在氨基末端有一个 23 个氨基酸残基的信号肽，引导多肽通过内质网膜。信号肽在内质网膜上由特异性的肽酶切除，成为前胰岛素。前胰岛素从内质网转运到高尔基体，形成含有前胰岛素的囊泡。这些囊泡是 β 细胞中胰岛素的贮存形式。

前胰岛素在囊泡中通过蛋白质水解形成成熟的胰岛素与一个 34 个氨基酸残基的连接肽（C 肽，图 5-8）。胰岛素在胰岛 β 细胞的囊泡中以锌 - 胰岛素六聚体的形式存在，包含 6 个由 2 个锌原子固定的

4.1 Introduction of Insulin

4.1.1 The Molecular Structure of Insulin

Insulin was first identified as an anti-diabetic factor in 1921, and was introduced clinically the following year. Its complete amino acid sequence was determined in 1951.Although mature insulin is a dimeric structure, it is synthesized as a single polypeptide precursor, preproinsulin. This 108 amino acid polypeptide contains a 23 amino acid signal sequence at its amino terminal end. This guides it through the endoplasmic reticulum membrane, where the signal sequence is removed by a specific peptidase. Proinsulin-containing vesicles bud off from the endoplasmic reticulum and fuse with the Golgi apparatus. As they move away from the Golgi, they become non-coated secretory vesicles. These vesicles serve as a storage form of insulin in the β cell.

Proinsulin is proteolytically processed in the coated secretory granules, yielding mature insulin and a 34 amino acid connecting peptide (C peptide, Figure 5-8). The insulin is stored in the form of a characteristic zinc–insulin hexamer in secretory granules, consisting of six molecules of insulin stabilized by two zinc atoms. Elevated levels of blood glucose, or other appropriate signals, cause the vesicles to fuse with the plasma membrane, thereby

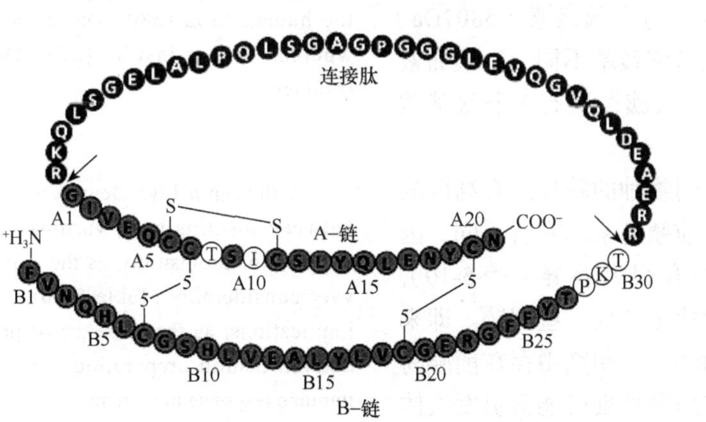

图 5-8　人前胰岛素（胰岛素和 C 肽）的一级结构

图中单个字母表示一种氨基酸，被蛋白内切酶切除的连接肽用黑色圆圈表示。胰岛素的氨基酸用灰色和白色圆圈表示，其中白色圆圈中的字母表示商品化的胰岛素中不保守的氨基酸

胰岛素分子。当血糖水平升高，或者其他信号使囊泡与细胞膜融合，囊泡将其中的内容物通过胞外分泌释放到血液。

releasing their contents into the blood via the process of exocytosis.

成熟的胰岛素有两条多肽链，含有 51 个氨基酸残基，通过两个链间二硫键连接。A 链含有 21 个氨基酸残基；B 链较大，含有 30 个氨基酸残基。胰岛素分子的净电荷源自 4 个谷氨酸残基、4 个酪氨酸残基、两个 α- 羧基、两个 α- 氨基、两个组氨酸残基、一个赖氨酸残基以及一个精氨酸残基。胰岛素在变性状态下的等电点（pI）为 5.3，因此在中性条件下胰岛素分子带负电荷。

Mature insulin consists of two polypeptide chains connected by two interchain disulphide linkages. The A-chain contains 21 amino acids, whereas the larger B-chain is composed of 30 residues. The net charge on the insulin molecule results from the four glutamic acid residues, four tyrosine residues, two α -carboxyl groups, two α-amino groups, two histidine residues, a lysine, and an arginine residue. Insulin has an isoelectric point (pI) of 5.3 in the denatured state; thus, the insulin molecule is negatively charged at neutral pH.

胰岛素的一个重要的内在性质是能够缔合成二聚体和高级聚集体。二聚体的驱动力是 B 链 C 端疏水相互作用力。胰岛素在各种二价金属离子存在的条件下缔合成分散的六聚体。在生理上，胰岛素以含锌六聚体的形式储存在胰腺 β 细胞。

Another important intrinsic property of insulin is its ability to readily associate into dimers and higher-order aggregates. The driving force for dimerization appears to be the formation of favorable hydrophobic interactions at the C-terminus of the B-chain. Insulin can associate into discrete hexameric complexes in the presence of various divalent metalions. Physiologically, insulin is stored as a zinc-containing hexamer in the β-celis of the pancreas.

不同物种的胰岛素基本结构相同，但氨基酸序列略有差异。猪胰岛素（5777Da）与人胰岛素（5807Da）仅有一个氨基酸残基不同，牛胰岛素（5733Da）与人胰岛素有 3 个氨基酸残基不同。

Insulins from various species conform to this basic structure, while varying slightly in their amino acid sequence. Porcine insulin (5777 Da) varies from the human form (5807 Da) by a single amino acid, whereas bovine insulin (5733 Da) differs by three residues.

尽管不同物种的胰岛素有高度的同源性，但前胰岛素却有所不同，因为 C 肽序列有很大差异（表 5-10），这样导致治疗上产生一些问题。即来源于动物的胰岛素制剂中存在前胰岛素时，这些前胰岛素可能会引发人体的免疫反应。

Although a high degree of homology is evident between insulins from various species, the same is not true for proinsulins, as the C peptide sequence can vary considerably (Table 5-10). This has therapeutic implications, as the presence of proinsulin in animal-derived insulin preparations can potentially elicit an immune response in humans.

表 5-10 不同物种前胰岛素结构序列差异

物种	A^{8}	A^{10}	A^{21}	B^{28}	B^{29}	B^{30}	B^{31}	B^{32}
人	Thr	Ile	Asn	Pro	Lys	Thr	—	—
猪	Thr	Ile	Asn	Pro	Lys	Ala	—	—
牛	Ala	Val	Asn	Pro	Lys	Ala	—	—
赖脯人胰岛素	Thr	Ile	Asn	Lys	Pro	Thr	—	—
天冬胰岛素	Thr	Ile	Asn	Asp	Lys	Thr	—	—
甘精胰岛素	Thr	Ile	Gly	Pro	Lys	Thr	Arg	Arg

4.1.2 胰岛素受体与信号转导

胰岛素受体是一个四聚体的膜糖蛋白，含有两条735个氨基酸残基的α链和两条620个氨基酸残基的β链，这些链通过二硫键连接（图5-9）。α链完全位于细胞膜外侧，含有富含半胱氨酸的胰岛素结合域。

4.1.2 The Insulin Receptor and Signal Transduction

The insulin receptor is a tetrameric integral membrane glycoprotein consisting of two 735 amino acid α -chains and two 620 amino acid β-chains. These are held together by disulphide linkages (Figure 5-9). The α -chain resides entirely on the extracellular side of the plasma membrane and contains the cysteine-rich insulin-binding doma.

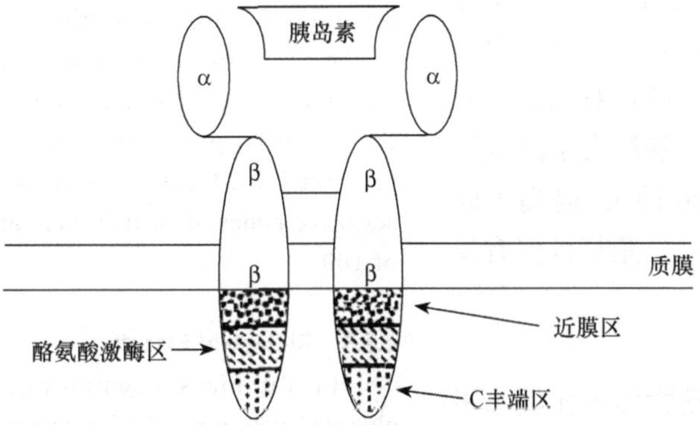

图 5-9 胰岛素受体结构

每条β链都由3个结构域组成：胞外区、穿膜区和具有酪氨酸激酶活性的胞内区。在胰岛素缺乏时，酪氨酸激酶活性很弱。完整的α亚单位能抑制β亚单位的内在激酶活性，阻止β亚单位与胰岛素的结合。α亚单位构象改变才能解除这种抑制作用。

Each β-subunit is composed of three regions; the extracellular domain, the transmembrane domain and a large cytoplasmic domain that displays tyrosine kinase activity. In the absence of insulin, tyrosine kinase activity is very weak. It is believed that the intact α-subunit exerts a negative influence on the endogenous kinase of the β-subunit and that binding of insulin, by causing a conformational shift in α-subunits, relieves this negative influence.

β 亚单位的胞内区含有 3 个亚结构域：①近膜区，识别或结合细胞内的底物分子；②酪氨酸激酶区，受体激活后发挥酪氨酸激酶活性；③ C 末端区，它的确切功能尚不清楚，但位点突变研究提示它能够促进胰岛素的促有丝分裂作用。

胰岛素细胞内的信号转导机制十分复杂。在过去的 10 年中，这方面取得了重大进展。胰岛素与其受体结合后，促进酪氨酸激酶区的 3 个特异的酪氨酸残基自身磷酸化，这种磷酸化又促进整个 β 亚单位构象改变，暴露出 ATP 结合位点与底物结合位点，从而激活酪氨酸激酶活性。酪氨酸激酶活化后，可使特异性的细胞内底物磷酸化，触发两条以上不同的信号转导通路：MAP 激酶通路的激活最终激发胰岛素的促有丝分裂作用；PI-3 激酶通路的激活则介导了胰岛素对新陈代谢的调节作用；通过对胰岛素敏感基因的转录调控（胰岛素敏感基因的种类大约超过 100 种），提高了胰岛素的多种作用效果，尤其是促有丝分裂作用的效果。

4.1.3 生物学活性

胰岛素是调节动物体内血糖水平的重要激素。正常情况下，胰岛素能够维持血糖水平在一定的小范围内（3.5～8.0 mmol/L）。胰岛素对蛋白质与脂类代谢也有重要影响，还具有一定的促有丝分裂活性。在体外试验中，高浓度的胰岛素促有丝分裂活性非常明显。

机体的许多细胞都表达有胰岛素受体，但胰岛素最重要的靶位是骨骼肌纤维、肝细胞与脂肪细胞。在这

The cytoplasmic domain of the β-subunit displays three distinct sub-domains: ① the juxtamembrane domain, implicated in recognition/binding of intracellular substrate molecules; ② the tyrosine kinase domain, which (upon receptor activation) displays tyrosine kinase activity; and ③ the C-terminal domain, whose exact function is less clear although site-directed mutagenesis studies implicate it promoting insulin' s mitogenic effects.

The molecular mechanisms central to insulin signal transduction are complex and have yet to be fully elucidated. However, considerable progress in this regard has been made over the last decade. Binding of insulin to its receptor promotes the autophosphorylation of three specific tyrosine residues in the tyrosine kinase domain. This in turn promotes an alteration in the conformational state of the entire β-subunit, unmasking ATP binding sites as well as substrate docking sites and activating its tyrosine kinase activity. Depending upon which specific intracellular substrates are then phosphorylated, at least two different signal transduction pathways are initiated: activation of the MAP kinase pathway is ultimately responsible for triggering insulin's mitogenic effects; activation of the PI-3 kinase pathway apparently mediates the majority of insulin's metabolic effects. Many of these effects, particularly the mitogenic effects, are promoted via transcriptional regulation of insulin-sensitive genes, of which there are probably in excess of 100.

4.1.3 Biological Effects

Insulin plays a central role in regulating blood glucose levels, generally keeping it within narrow defined limits (3.5-8.0 mmol/l) in normal animals. It also has a profound effect on the metabolism of proteins and lipids and displays some mitogenic activity. The latter is particularly evident in vitro studies and at high insulin concentrations.

Although many cells in the body express the insulin receptor, its most important targets are skeletal muscle fibres, hepatocytes and adipocytes, where it often antagonizes the effects of glucagon. The most

些部位，胰岛素与胰高血糖素的作用相互拮抗。刺激胰腺分泌胰岛素的最有效因素是血液葡萄糖水平的升高。胰岛素通过几种途径对葡萄糖及其他营养成分的吸收发挥着重要的代谢调节作用：①促进葡萄糖转运并转运氨基酸、K^+与其他分子至细胞内，从而降低这些物质在血液中的浓度；②促进细胞内的生物合成（合成代谢）通路，例如糖原的合成，这样有助于营养素在细胞内转化成贮存形式；③抑制分解代谢通路，如抑制糖原分解；④促进DNA和蛋白质合成。

总的来说，胰岛素通过诱导分解与合成代谢主要通路的数种关键调节酶的脱磷酸化而调节新陈代谢，它能抑制分解代谢通路，促进合成代谢通路。胰岛素的这些调节作用常与其他激素对抗，尤其是胰高血糖素和肾上腺素。当血糖浓度降低时，胰岛素水平降低，胰高血糖素的作用（主要是分解代谢）更为显著。胰岛素还通过改变多种基因的转录而发挥调节作用，这些基因多数是编码新陈代谢中的酶，以及细胞膜上的葡萄糖转运蛋白。

4.1.4　胰岛素产品

胰岛素产品根据浓度、起效和持续时间、纯度和来源而不同。胰岛素可分为速效、中效、长效制品（表5-11）及它们的混合物。速效制品包括普通、含锌结晶品及含锌快速胰岛素，平均起效时间为0.5～1h，有效作用时间6～8h（普通、锌胰岛素）或12～16h。中效制品和锌胰岛素悬浮液的起效时间约为2h，有效作用时间为24～48h。长效制品（低

potent known stimulus of pancreatic insulin release is an increase in blood glucose levels. Insulin orchestrates a suitable metabolic response to the absorption of glucose and other nutrients in a number of ways: ① it stimulates glucose transport (and transport of amino acids, K+ ions and other nutrients) into cells, thus reducing their blood concentration; ② it stimulates intracellular biosynthetic (anabolic) pathways, such as glycogen synthesis, which helps to convert the nutrients into a storage form in the cells; ③ it inhibits catabolic pathways, such as glycogenolysis; ④ it stimulates DNA and protein synthesis.

In general, insulin achieves such metabolic control by inducing the dephosphorylation of several key regulatory enzymes in mainline catabolic or anabolic pathways. This inhibits the former and stimulates the latter pathway types. These effects are often opposed by other hormones, notably glucagon and adrenaline. Thus, when blood glucose concentrations decrease, insulin levels decrease and the (largely catabolic) effects of glucagon become more prominent. Insulin also induces its characteristic effects by altering the level of transcription of various genes, many of which code for metabolic enzymes. Another gene upregulated by insulin is that of the integral membrane glucose transporter.

4.1.4 Insulin Products

Insulin products differ in concentration, onset and duration, purity, and species of origin. Insulins are divided into rapid-acting preparations, intermediate-acting preparations, long-acting preparations(Table 5-11), or a mixture of both. Rapid onset preparations include regular, crystalline zinc, and prompt insulin zinc,and have an average onset of 0.5 to 1 hour and a duration of effect of 6 to 8 hours (regular, crystalline zinc) or 12 to 16 hours. Intermediate-acting preparations and insulin zinc suspension have an approximate onset of about 2 hours and a duration of 24 to 48 hours. Long-

精锌悬浮液、特慢锌胰岛素悬浮液）平均起效时间为4～6h，作用时间超过36 h。

acting preparations (protamine zinc suspension and extended insulin zinc suspension ultralente) have a general onset of 4 to 6 hours and a duration of more than 36 hours.

表 5-11 速效、中效和长效胰岛素的一些药动学参数

种类	起效（用药后）/h	活性高峰（用药后）/h	持续时间 /h
速效	0.5～1	2～5	6～8
中效	2	4～12	24～48
长效	4～6	10～20	超过 36

在健康机体，血液中的胰岛素水平是根据血糖水平的变化而进行上调或下调。一般情况下，胰岛素是以很低的基础水平持续释放入血，一旦血糖水平升高，就会引起胰岛素水平的迅速提高。如饭后约1h，胰岛素的分泌达到高峰，然后2h内回落至基础水平。胰岛素的治疗均是仿照机体的内在调控进行的。

In healthy individuals, the blood insulin level is continuously up- or down-regulated as appropriate for the blood glucose levels at any given instant. Insulin is typically secreted continuously into the bloodstream at low basal levels, with rapid increases evident in response to elevated blood sugar levels. Insulin secretion usually peaks approximately 1 h after a meal, falling off to base levels once again within the following 2 h. Current Insulin preparation reproduces normal serum insulin baseline levels.

速（短）效胰岛素在用药后，血液胰岛素水平升高较为迅速。通常采用皮下注射，偶尔也采用肌内注射方式。速效胰岛素在注射后1.5～2h内血糖下降到达谷底，效果可维持5h。这种胰岛素制剂需要在饭前1h左右注射，并且病人不能改变他们的吃饭时间。

Fast (short)-acting insulins are those preparations that yield an elevated blood insulin concentration relatively quickly after their administration, which is usually by s.c. or, less commonly, by i.m. injection. An injection of fast-acting insulin will not produce a plasma hormone peak for 1.5-2 h post-injection, and levels then remain elevated for up to 5 h. Such traditional animal or human insulin preparations must thus be administered 1 h before eating—and the patient must not subsequently alter his/her planned mealtime.

正常血浆浓度的胰岛素（约1×10^{-9}mol/L）是以单体形式存在于溶液中。任何胰岛素直接注入血液中后半衰期仅为数分钟。

Insulin, at typical normal plasma concentrations (approximately 1×10^{-9}mol/L) exists in true solution as a monomer. Any insulin injected directly into the bloodstream exhibits a half-life of only a few minutes.

目前可溶性的胰岛素制剂（即速效胰岛素）中的胰岛素浓度很高（约1×10^{-3}mol/L）。在这样高的浓度时，可溶性胰岛素是以单体、二聚体、三聚体、锌-胰岛素六聚体等的混合体

The concentration of insulin present in soluble insulin preparations (i.e. fast-acting insulins), is much higher (approximately 1×10^{-3}mol/L). At this concentration, the soluble insulin exists as a mixture of monomer, dimer, tetramer and zinc-insulin hexamer. These insulin complexes have to dissociate

的形式存在。这些胰岛素复合体必须分离才能从注射位点进入血液，这就使激素的作用起效减慢。

in order to be absorbed from the injection site into the blood, which slows down the onset of hormone action.

为了延长胰岛素的作用时间，可溶性的胰岛素被配制成胰岛素混悬液。缓效胰岛素从注射位点进入血液循环的时间较为缓慢。其特点是起效慢，但作用时间长。缓效胰岛素的制备通过以下两种方法：①加入锌以促进锌-胰岛素晶体生长，晶体分离的时间较长，因此从注射点进入血液所需的时间也延长；②加入能够结合胰岛素且使胰岛素缓慢释放的蛋白质。常用的是鱼精蛋白，它存在于数种鱼的精子中，是一种与核苷酸结合的碱性蛋白质。根据胰岛素-鱼精蛋白的质量比，所得的长效胰岛素为鱼精蛋白-锌-胰岛素或者鱼精蛋白锌胰岛素。为了模仿正常机体的胰岛素释放规律，将速效胰岛素和长效胰岛素混合起来就形成了双相胰岛素。常见的胰岛素产品术语见表 5-12。

In order to prolong the duration of insulin action, soluble insulin may be formulated to generate insulin suspensions. Slow-acting insulins, on the other hand, enter the circulation much more slowly from the depot (injection) site. This is characterized by a slower onset of action, but one of longer duration. This is generally achieved in one of two ways: ① Addition of zinc in order to promote zinc–insulin crystal growth (which take longer to disassociate and, hence, longer to leak into the blood stream from the injection depot site); ② Addition of a protein to which the insulin will complex, and from which the insulin will only be slowly released. The proteins normally used are protamines, basic polypeptides naturally found in association with nucleic acid in the sperm of several species of fish. Depending on the relative molar ratios of insulin:protamine used, the resulting long-acting insulins generated are termed protamine–Zn–insulin or isophane insulin. Biphasic insulins include mixtures of short- and long-acting insulins, which attempt to mimic normal insulin rhythms in the body. The glossary terms for insulin products is seen in Table 5-12.

表 5-12 胰岛素产品术语表

术语	释义
常规胰岛素	仅通过结晶纯化的牛或猪胰岛素
单峰胰岛素	经过进一步凝胶过滤纯化的牛或猪胰岛素
高纯胰岛素	经过凝胶过滤及离子交换层析纯化的胰岛素
人胰岛素（emp）	猪胰岛素经过酶修正而成的人序列的胰岛素
人胰岛素（crb）	重组人胰岛素，其 A、B 链分别在细菌内表达
人胰岛素（prb）	重组人胰岛素，其前体在细菌内表达
人胰岛素（pyr）	在酵母内生产的重组人胰岛素
胰岛素 B.P./Eur.P./USP	牛或猪来源的胰岛素
人胰岛素（USP）	猪胰岛素经过酶修饰或通过重组 DNA 技术生产的人序列的胰岛素
可溶性胰岛素或正规胰岛素或未修饰胰岛素	任何来源的胰岛素溶液，未经任何方法设计 / 改变以延长其作用时间
鱼精蛋白锌胰岛素	复合了过量鱼精蛋白的胰岛素，其作用时间延长
等量鱼精蛋白锌胰岛素	等物质的量的鱼精蛋白与胰岛素复合的制剂，其作用时间延长
胰岛素 - 锌悬液	与锌共同结晶生成的胰岛素悬液，其作用时间延长
双相胰岛素	速效与长效胰岛素制剂的混合物

4.1.5 传统方法生产胰岛素

最初使用的胰岛素产品基本上是天然胰腺提取物。这种产品的治疗价值不明确，经常会发生副作用（存在大量杂质）。由于胰岛素需要频繁注射，使得不良反应非常严重。后来使用酸 - 乙醇沉淀的方法，使得胰岛素的纯度得到了适当的提高，在一定程度上减少了副作用的发生与严重程度。

胰岛素第一次结晶是在 1926 年，但由于当时对于促进晶体生长的因素还不甚了解，结晶效果不稳定。大约 10 年后，研究人员发现在天然提取物中加入锌能够促进可再生的结晶（加入锌可生成特征性的菱形晶体，基本结晶单位是胰岛素六聚体，2 个锌原子使结构稳定）。

20 世纪 30 年代中期，天然猪或牛胰岛提取物通过结晶制备成临床用胰岛素。通常采用重复结晶的步骤以提高产品纯度。这种制剂称为常规胰岛素。常规胰岛素制剂中仍存在少量其他杂质，例如胰高血糖素、生长激素抑制素、胰腺多肽与有血管活性的肠多肽。

杂质的存在不仅会以多种方式影响胰岛素产品的安全性与功效，而且许多杂质对人体具有免疫原性。这对胰岛素来说是非常不利的，因为胰岛素与大多数生物制剂不同，一般而言它需要终生每天用药。即使是很弱的免疫抗原，频繁用药最终也会触发强烈的免疫反应。猪胰岛素与人胰岛素仅有一个氨基酸残基的差异（人胰岛素 B 链的第 30 个残基是苏氨酸，猪胰岛素是丙氨酸），对人体几乎没有免疫原性。牛胰岛素与人胰岛素有 3 个氨

4.1.5 Traditional Method for Insulin Preparation

Insulin preparations used initially were little more than crude pancreatic extracts. The therapeutic value of such products was marginal, as severe adverse reactions were commonplace (due to the presence of impurities). This was made worse by the frequency of injections required. The introduction of an acid–alcohol precipitation step yielded insulin preparations of moderate purity, thus partially overcoming the range and severity of side effects noted.

Although insulin was first crystallized in 1926, the factors promoting crystal growth were poorly understood and yielded inconsistent results. It was almost 10 years later when researchers discovered that the addition of zinc to a crude extract promoted reproducible crystallization (zinc addition yields a characteristic rhombohedral crystal, the basic crystal unit being the insulin hexamer, stabilized by the two zinc atoms).

By the mid-1930's, commercial insulin was being prepared by crystallization from crude porcine or bovine extracts. The crystallized preparation was generally subjected to a recrystallization step in order to further increase the product's purity. Such preparations are termed conventional insulins. Additional impurities, such as glucagon, somatostatin, pancreatic polypeptide and vasoactive intestinal polypeptide, are present in most conventional insulin preparations at lower levels.

The presence of such contaminants can impact upon product safety and efficacy in a number of ways, and many of the contaminants are immunogenic in man. This is particularly significant in the case of insulin as, unlike most other biopharmaceuticals, it usually must be administered daily for life. The frequency of administration will ultimately trigger a strong immunological response, even against weak immunogens. Porcine insulin, differing from human insulin by only a single amino acid (residue 30 of the B-chain; threonine in humans, alanine in pigs), is essentially nonimmunogenic in man. Bovine insulin, differing from the human form by three amino acids, does elicit an immunological response in humans. Additonally, many of the insulin contaminants (including proinsulin) from animals are immunogenic.

基酸残基的不同，会引发人体的免疫反应。另外，许多动物来源的胰岛素污染物（包括前胰岛素）具有免疫原性。

1. 酸醇提取减压浓缩法生产胰岛素　胰岛素产品的制备方法根据原料和过程的不同，分为胰脏提取法、生物技术法、化学合成法。动物来源的胰岛素以提取法为主，比较成熟且被普遍采用的是酸醇提取减压浓缩法（图 5-10），原料多采用猪的新鲜胰。

1. Acid-alcohol Extract and Reduced Pressure Condense for Insulin Products The insulin products can be achieved by pancreas extract, biological technique and chemical synthesis due to different raw materials and preparation processes. The insulins derived from animals mainly are prepared by extract, in that acid-alcohol extract and reduced pressure condense have been well-eatablished and widely adopted (Figure 5-10), and most of the raw materials come from porcine fresh pancreas.

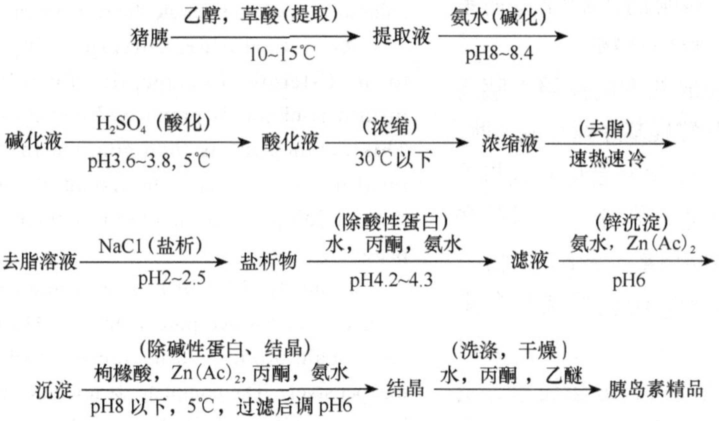

图 5-10　传统的酸醇提取减压浓缩法生产胰岛素的技术路线

传统的酸醇提取减压浓缩法的原理是根据胰岛素在约 70% 乙醇的酸性水溶液中易溶解，且较稳定。首先选用 pH 约 2.5 的酸醇作提取溶剂，通过调节 pH，沉淀胰脏中碱性和酸性蛋白杂质。然后在减压低温下蒸发除去乙醇，速热速冷除去油脂杂质，再用氯化钠盐析，即得胰岛素粗品。将粗品溶解，在适当浓度的丙酮 - 水溶液中调 pH 去除杂蛋白，并利用锌盐在一定的 pH 溶液中重结晶，即得精制的胰岛素。

The principle of traditional acid-alcohol extract and reduced pressure condense for insulin is that insulin is easy to be resolved and keep stable in about 70% alcohol of acidulous water solution. The first step of extraction is to use acid-alcohol of about pH 2.5 as solvent, and then the basic and acid protein impurities of pancreas are precipitated by pH adjust. The second step is to remove the impurities by centrifugation and alcohol by reduced pressure and low temperature, and the lipid impurities are removed by rapid hot and cold, then the crude product of insulin is got by NaCl salting out. Furthermore, crude product of insulin is dissolved into the certain concentration of acetone-water solution and the impurities are removed by pH adjust. As a result, the refined insulin is achieved by using Zincic salts recrystallization in certain pH solution.

2. 转化法生产人胰岛素　如前所述，猪胰岛素与人胰岛素仅有一个

2. Enzymatic Conversion of Porcine Insulin As previously described, porcine insulin differs from human

氨基酸残基的差异（B30 残基），在 20 世纪 70 年代早期，发现了一种可使猪胰岛素转化为人胰岛素的方法。这种方法是用酶和化学方法处理猪胰岛素的产物。

首先用 pH7.5 的无糜蛋白酶的胰酶消化猪胰岛素不低于 45min。这样就选择性地切断胰岛素 B 链 22 位精氨酸残基与 23 位甘氨酸残基之间的肽键，结果得到一段缩短了的胰岛素分子与一段 B 链末端 8 肽（可被胰酶进一步水解）。胰酶消化液经过葡聚糖凝胶过滤分离反应产物。

通过化学合成的方法按照人胰岛素 B 链 C 端氨基酸残基序列合成 8 肽。然后通过成熟的化学方法将 8 肽与缩短的胰岛素分子连接起来，这样就产生了半合成的人胰岛素（emp）产物。

一般来说，10g 纯化的猪胰岛素可制备 7.0g 人胰岛素。这种方法可以大量生产人胰岛素，方法简单，费用低廉。尽管完全化学合成胰岛素在技术上是可能的，但从经济上考虑不太合适。

4.2 基因工程生产胰岛素

1982 年，在美国、德国、英国与荷兰，通过重组 DNA 技术生产的人胰岛素首次获得许可应用于临床，它也是第一个准许用于人体治疗的重组 DNA 技术产品。这种产品有几个优点：①供应充分；②消除了由于动物胰岛组织中存在病原体而导致疾病意外传播的危险；③经济实用。

从一只猪的胰腺所纯化得到的纯胰岛素的量可以供一位糖尿病患者使用 3 天。胰腺组织的供给需要大量动

insulin by only one amino acid (the B30 residue). In the early 1970s, a method was developed by which porcine insulin could be converted into a preparation of identical amino acid sequence to that of human insulin. This method utilizes a combination of enzymatic and chemical treatment of the porcine product.

The porcine insulin is first subjected to digestion with chymotrypsin-free trypsin at pH 7.5 for in excess of 45 min. This results in the selective cleavage of the peptide bond linking arginine 22 and glycine 23 in the insulin B-chain. The reaction products are a shortened insulin molecule and the B-chain terminal octapeptide (which can be further proteolysed by trypsin). The tryptic digest is then applied to a Sephadex G-75 gel filtration column in order to separate the reaction products.

An octapeptide, corresponding in sequence to the C-terminal octapeptide found in the human insulin B-chain, is prepared by chemical synthesis. This octapeptide is then coupled to the shortened insulin using a well-established chemical method, thus yielding a semi-synthetic human insulin (emp) product.

Typically, 7.0 g of human insulin can be prepared from 10 g of purified porcine insulin. The process allows large-scale production of human insulin simply and inexpensively. The complete de novo synthesis of human insulin, while technically possible, is economically unattractive.

4.2 Production of Human Insulin by Recombinant DNA Technology

Human insulin produced by recombinant DNA technology was first approved for general medical use in 1982, initially in the USA, West Germany, the UK and The Netherlands. As such, it was the first product of recombinant DNA technology to be approved for therapeutic use in humans. Production of insulin by recombinant means displays several advantages, including: ① reliability of supply; ② elimination of the risk of accidental transmission of disease, due to the presence of pathogens in animal pancreatic tissue; ③ economically attractive.

The quantity of purified insulin obtained from the pancreas of one pig satisfies the requirements of one diabetic for 3 days. The supply of pancreatic tissue is dependent upon the demand for meat, which does not

物组织，难以满足全世界发病率逐步升高的糖尿病患者的需求。重组DNA技术能够保证充足的胰岛素供给量。

4.2.1　基因工程生产人胰岛素的技术路线

基因工程科学家最初采用的方法是分别将编码胰岛素A链与B链的核苷酸序列插入两个不同的大肠杆菌细胞中。在发酵罐中分别大规模培养，层析纯化表达的胰岛素链。然后在适宜氧化条件下将A链与B链共同孵育，促进链间二硫键的形成，这样就形成了人胰岛素，这种方法称为AB链合成法。

另一种方法是在大肠杆菌工程细胞中插入编码人前胰岛素的核苷酸序列，然后纯化所表达的前胰岛素，在体外通过蛋白水解酶切去C肽。这种方法仅需要一次发酵与纯化，所以使用得更为广泛。

一系列研究结论性地表明重组胰岛素在结构与功能上与天然人胰岛素均相同：序列分析证明其氨基酸序列正确；高效液相色谱（HPLC）分析也呈现单峰；分析分子三级结构的圆二色性研究与X射线晶体学分析也都证明重组胰岛素的空间结构与天然人胰岛素相同；对胰岛素的放射受体分析、放射免疫分析与传统的生物检测（检测激素诱导兔低血糖反应的能力）均证明了同样的结果。在获得临床应用许可前，人的临床追踪也发现重组胰岛素在控制血糖水平方面与以前的胰岛素产品同样有效。现在重组人胰岛素已使用大规模的发酵罐生产。

necessarily correlate with the increasing worldwide incidence of diabetes. Recombinant DNA technology provides an obvious way to ensure future adequate supply of insulin.

4.2.1 The Technological Scheme for Recombinant Human Insulin

The initial approach is that entailed inserting the nucleotide sequence coding for the insulin A- and B-chains into two different *E. coli* cells (both *E. coli* strain K12). These cells were then cultured separately in large-scale fermentation vessels, with subsequent hromatographic purification of the insulin chains produced. The A- and B-chains are then incubated together under appropriate oxidizing conditions in order to promote interchain disulphide bond formation, forming human insulin (crb). This method is called as A-B-chain-synthesis.

An alternative method entails inserting a nucleotide sequence coding for human proinsulin into recombinant *E. coli*. This is followed by purification of the expressed proinsulin and subsequent proteolytic excision of the C peptide in vitro. This approach has become more popular, largely due to the requirement for a single fermentation and subsequent purification scheme.

A number of studies have shown conclusively that the recombinant insulins are chemically and functionally identical to native human insulin. Sequencing data confirms the expected amino acid sequence. HPLC analysis also yields identical chromatograms. Circular dichromism studies, which evaluate tertiary structure, also yield identical patterns, as does X-ray crystallographic analysis. Radioreceptor- and radioimmuno-assays yield identical results, as does the traditional bioassay for insulin (measure of the hormone's ability to induce a hypoglycaemic response in rabbits). Prior to its approval for medical use, human clinical trials also found the recombinant insulin to be as effective as previously available products in terms of controlling hyperglycaemia. Recombinant human insulins are now routinely produced in large volume of fermentation vessels and several such products have gained marketing approval.

4.2.2 基因工程生产人胰岛素的纯化

尽管重组人胰岛素与天然胰岛素一样，但任何来源于大肠杆菌的杂质都具有潜在的高免疫原性。重组人胰岛素必须严格纯化，纯化过程包括数个层析步骤（通常是凝胶层析与离子交换，附加根据分子疏水性不同而分离分子的疏水作用层析或反相层析），见图 5-11。

4.2.2 Purity of Recombinant Human Insulin

While the recombinant product is identical to native insulin, any impurities present will be *E. coli*-derived and, hence, potentially highly immunogenic. Stringent purification of the recombinant product must thus be undertaken. This entails several chromatographic steps (often gel filtration and ion-exchange, along with additional steps which exploit differences in molecular hydrophobicity, e.g. hydrophobic interaction chromatography or reverse-phase chromatography) (Figure 5-11).

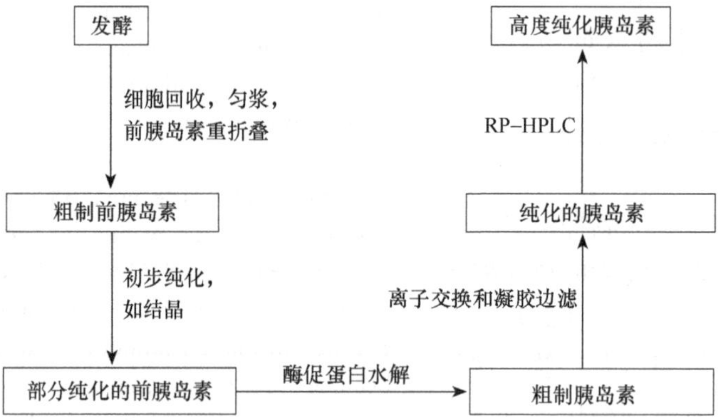

图 5-11 人胰岛素的纯化步骤

基因工程生产人胰岛素的过程中使用了反相高效液相色谱（RP-HPLC）。使用容量 80L 或更大的 C8 或 C18 的 RP-HPLC 柱，一次纯化就可以承载 1200g 的胰岛素。使用酸性（常为乙酸）流动相分离（即流动相的 pH 值要低于胰岛素的 pI5.3，以保持胰岛素充分溶解）。通常是用水溶性酸性流动相将胰岛素载于柱上，然后用乙腈梯度洗脱（一般 15%～30% 的乙腈能够把胰岛素洗脱下来）。

The reverse-phase HPLC (RP-HPLC) step has been introduced into production of human insulin prb. The C8 or C18 RP-HPLC column used displays an internal volume of 80 L or more, and up to 1200 g of insulin may be loaded during a single purification run. Separation is achieved using an acidic (often acetic acid-based) mobile phase (i.e. set at a pH value sufficiently below the insulin pI value of 5.3 in order to keep it fully in solution). The insulin is usually loaded in the water-rich acidic mobile phase, followed by gradient elution using acetonitrile (insulin typically elutes at 15%-30% acetonitrile).

由于载于柱上的胰岛素的纯度已经比较高（约 92%），经过这个最终纯化步骤后，产品的纯度可达约 99%。柱上的胰岛素具有 95% 的活

While the starting material loaded onto the column is fairly pure (about 92%), this step yields a final product of approximately 99% purity. Over 95% of the insulin activity loaded onto the column can be recovered. A single column run takes in the

性。一次纯化过程需要 1h。

RP-HPLC 的精细纯化步骤不仅能够去除大肠杆菌来源的杂质，还能有效分离天然胰岛素产品中的修饰胰岛素。这种胰岛素制剂中残留的极微量的杂质不会引起使用者任何明显的免疫反应。

4.2.3 基因工程胰岛素产品

重组 DNA 技术不仅便于利用微生物系统生产人胰岛素，而且还便于生产修饰氨基酸序列的胰岛素。产生这种基因工程胰岛素类似物的主要目的在于：①鉴定胰岛素的不同药物代谢动力学特征，如速效胰岛素或长效胰岛素；②制备强效型胰岛素（受体亲和力更高的胰岛素）。一些胰岛素类似物结构上的巧妙设计，将为我们进行生物药物的研发提供启示。

在胰岛素分子中，与胰岛素受体相互作用的氨基酸残基已经确定，它们是 A1、A5、A19、A21、B10、B16 和 B23～25。在这些位点上由氨基酸取代的类似物有的已经生产，例如用组氨酸代替 B10 的谷氨酸，可使胰岛素的体外活性增加 5 倍。其他氨基酸残基替代也会产生更高特异性活性的胰岛素类似物。

生产速效胰岛素的焦点集中在治疗剂量下，不会形成二聚体或更大聚合体的胰岛素。形成胰岛素二聚体或多聚体需要胰岛素单体分子间 B8、B9、B12～13、B16 与 B23～28 的位点接触。因此在这些位点由不同氨基酸残基替代的胰岛素类似物也有生产。通常采用的方法是插入带电或体积较大的氨基酸，促进各胰岛素单体间的电荷排斥或造成空间阻碍。有几种这

order of 1 h.

The RP-HPLC polishing step not only removes *E. coli* -derived impurities, but also effectively separates modified insulin derivatives from the native insulin product. The resultant extremely low levels of impurities remaining in these insulin preparations fail to elicit any significant immunological response in diabetic recipients.

4.2.3 Recombinant Insulin Preparations

Recombinant DNA technology facilitates not only production of human insulin in microbial systems, but also facilitates generation of insulins of modified amino acid sequences. The major aims of generating such engineered insulin analogues include: ① identification of insulins with altered pharmacokinetic properties, such as faster-acting or slower-acting insulins; ② identification of super-potent insulin forms (insulins with higher receptor affinities). The primary structures of some insulin analogues are skillful arranged, which may provide ideas for study and development of our biopharmaceuticals.

The insulin amino acid residues which interact with the insulin receptor have been identified (A1, A5, A19, A21, B10, B16, B23–25), and a number of analogues containing amino acid substitutions at several of these points have been manufactured, e.g. conversion of histidine to glutamate at the B10 position yields an analogue displaying five-fold higher activity in vitro. The higher specific insulin analogues may be produced by other amino acid substitutions.

Attempts to generate faster-acting insulins have centered upon developing analogues which do not dimerize or form higher polymers at therapeutic dose concentrations. The contact points between individual insulin molecules in insulin dimers/oligomers include amino acids at positions B8, B9, B12–13, B16 and B23–28. Thus, analogues with various substitutions at these positions have been generated. The approach adopted generally entails insertion of charged or bulky amino acids, in order to promote charge repulsion or steric hindrance between individual insulin monomers. Several

样的类似物从注射位点进入血液的速度比天然可溶（速效）胰岛素快得多。这样修饰的胰岛素可以在进餐时注射，而不必提前 1h。有几种这样的速效基因工程胰岛素已经在临床上应用。胰岛素 Lispro 是第一个进入市场的速效基因工程胰岛素类似物。

1. 重组人胰岛素 Humulin　Humulin 通过基因工程插入人的胰岛素基因，在实验室大肠杆菌或酵母中生产。所有人胰岛素制品均为 100U/ml。Humulin 有 6 种不同的制品，包括普通型 R、中效制品 NPH（N）、锌胰岛素混悬液 Lente（L）、Humulin50/50 和 Humulin70/30。这些制剂用于皮下给药。

Humulin50/50 含 50% 人低精蛋白锌胰岛素悬浮液和 50%Humulin，它是中效胰岛素与快速起效的普通胰岛素的结合，注射后作用时间可长至 24h。Humulin70/30 含 70% 人低精蛋白锌胰岛素悬浮液和 30%Humulin，它也是中效胰岛素与快速起效的普通胰岛素的结合，注射后作用时间可长至 24h。Humulin70/30、R 型 和 N 型均是预装在盒形和笔形容器中。

Humulin L 是无菌、无定型结晶悬浮液重组人锌胰岛素。该药属中效，作用起效慢，比普通作用时间长（长至 24h）。Humulin N 是无菌结晶鱼精蛋白锌人胰岛素。该药也是中效，作用起始慢，比普通作用时间长（长至 24h）。Humulin R 是锌胰岛素结晶体的无菌水溶液，没有添加任何辅料以改变其作用速度和时间。与其他胰岛素相比，它作用快，活力滞留时间较短（4～12h）。与 Humulin R

are absorbed from the site of injection into the bloodstream far more quickly than native soluble (fast-acting) insulin. Such modified insulins could thus be injected at mealtimes, rather than 1 h before, and several such fast-acting engineered insulins have now been approved for medical use. Insulin Lispro was the first such engineered short-acting insulin to come to market.

1. Humulin　Humulin is synthesized in laboratory strains of *Escherichia coli* bacteria and *Saccharomyces cerevisiae* (bakers'yeast) that has been genetically altered by the addition of the gene and recombinant DNA technology for human insulin. All of human insulin preparation are 100U/ml. Humulin is available in six formulations—Regular (R), NPH(N), Lente (L), Humulin 50/50, and Humulin 70/30. They are for subcutaneous injection only.

Humulin 50/50 is a sterile suspension containing a mixture of 50% Human Insulin Isophane Suspension and 50% Human Insulin (rDNA origin).It is an intermediate-acting insulin combined with the more rapid onset of action of regular insulin. The duration of activity may last up to 24 hours following injection. Humulin 70/30 is a mixture of 70% Human Insulin Isophane Suspension and 30% Human Insulin (rDNA origin). It is also an intermediate-acting insulin combined with the more rapid onset of action of regular insulin. The duration of activity may last up to 24 hours following injection. Humulin 70/30 is also available in the form of a prefilled cartridge and pen. Humulin R and N are also available in cartridges and Humalin N is also available in a pen.

Humulin L is a sterile, amorphous, and crystalline suspension of recombinant human insulin with zinc providing an intermediate-acting insulin with a slower onset and a longer duration of activity (up to 24 hours) than regular insulin. Humulin N is a sterile, crystalline suspension of human insulin with protamine and zinc providing an intermediate-acting insulin with a slower onset of action and a longer duration of activity (up to 24 hours) than that of regular insulin. Humulin R consists of zinc-insulin crystals dissolved in a sterile solution. Humulin R has had nothing added to change the speed or length of its action. It takes effect rapidly and has a relatively short duration of activity

一样，Humulin R（U-500）为锌胰岛素结晶体的无菌水溶液，但是含量为500 U/ml，而不是通常的100 U/ml。与其他胰岛素相比，Humulin R（U-500）作用快，活力滞留时间较长（长达24h）。与普通胰岛素或中效胰岛素比，Humulin U是锌胰岛素结晶体的无菌水溶液，起效慢，作用时间较长（长达28h），但强度较低。

胰岛素的注射使病人葡萄糖水平接近正常。对新诊断的Ⅰ型糖尿病患者，胰岛素的起始剂量为每天每千克体重0.6～0.75U。对于新诊断的Ⅱ型糖尿病患者，开始的剂量一般为10U胰岛素，每周增加5~10U。另外，需要注意的是，胰岛素的任何改变，包括强度、厂商、类型（即普通、NPH或者Lente）、物种（牛、猪、牛-猪或人）及制备过程（重组或动物来源的），这些都会影响胰岛素的剂量。

2. 胰岛素Lispro　胰岛素Lispro是第一个获得市场许可的重组速效胰岛素类似物。胰岛素Lispro的氨基酸残基序列与天然人胰岛素相似，唯一的不同在于胰岛素B链28、29位的脯氨酸-赖氨酸（Pro-Lys）顺序倒置。这一简单的改变，显著地降低了治疗剂量浓度的胰岛素分子之间自联的倾向。胰岛素Lispro二聚值是未经修饰的天然人胰岛素的1/300。从结构上来看，这种氨基酸序列倒置破坏了对自联很关键的链间疏水作用。

胰岛素Lispro这种序列改变的基本原理是基于对胰岛素样生长因子1（IGF-1）的研究。IGF-1的结构与前胰岛素非常相像，IGF-1的A、B链结构域的氨基酸残基有50%与胰岛素A、B

(4–12 hours) as compared with other insulins. Humulin R (U-500), like Humulin R, consists of zincinsulin crystals dissolved in a sterile solution but contains 500 units/ml rather than 100 units/ml. Humulin R (U-500) takes effect rapidly but has a relatively long duration of activity following a single dose (up to 24 hours) as compared with other Regular insulins. It is available by prescription only. Humulin U is a sterile crystalline suspension of human insulin with zinc providing a slower onset and a longer and less intense duration of activity (up to 28 hours) than regular insulin or the intermediate-acting insulins (NPH and Lente).

Insulin dosage should be titrated to achieve near-normal levels of glucose, while avoiding hypoglycemia. For newly diagnosed type Ⅰ diabetes mellitus, the initial daily dosage of insulin is 0.6 to 0.75 Units/kg. For patients with type Ⅱ diabetes mellitus, the starting dosage of insulin is usually 10 units and may be increased weekly in increments of 5 to 10 units.The label for all products that contain human insulin (recombinant DNA origin) warns: human insulin products differ from animal-source insulins because they are structurally identical to the insulin produced by your body's pancreas and because of their unique manufacturing process. Any change of insulin should be made cautiously and only under medical supervision. Changes in strength, manufacturer, type (e.g., regular, NPH, lente), species (beef, pork, beef-pork, human), or method of manufacture (rDNA versus animal-source insulin) may result in the need for a change in dosage.

2. Insulin Lispro Insulin Lispro was the first recombinant fast-acting insulin analogue to gain marketing approval. It displays an amino acid sequence identical to native human insulin, with one alteration—an inversion of the natural proline–lysine sequence found at positions 28 and 29 of the insulin B-chain. This simple alteration significantly decreased the propensity of individual insulin molecules to self-associate when stored at therapeutic dose concentrations. The dimerization constant for Insulin Lispro is 300 times lower than that exhibited by unmodified human insulin. Structurally, this appears to occur as the change in sequence disrupts the formation of inter-chain hydrophobic interactions critical to self-association.

Insulin Lispro was developed by scientists at Eli Lilly (which, along with Novo Nordisk, are the world's largest producers of therapeutic insulins). The rationale underlying the sequence alteration was rooted in studies, not of insulin, but of insulin-like growth factor-1 (IGF-1). The latter

链相应位置的氨基酸残基一致。与胰岛素相比，IGF-1 分子的自联倾向非常小。早期的序列研究揭示胰岛素脯氨酸 B28（ProB28）、赖氨酸 B29（LysB29）的顺序在 IGF-1 分子上是颠倒过来的。这提示了如果这种顺序与分子自联倾向相关，那颠倒这两个氨基酸残基的顺序就会降低胰岛素的自联作用。直接的试验证实了这个假设的正确性。

displays a strong structural resemblance to proinsulin, with up to 50% of amino acid residues within the IGF-1 A- and B-domains being identical to those found in comparable positions in the insulin A- and B-chains. When compared to insulin, IGF-1 molecules display a significantly decreased propensity to self-associate. Sequencing studies earlier revealed that the ProlineB28–LysineB29 sequence characteristic of insulin is reversed in IGF-1. It was suggested that, if this sequence difference was responsible for the differences in self-association propensity, then inversion of the Proline B28 – Lysine B29 sequence in insulin would result in its decreased self-association. Direct experimentation proved this hypothesis accurate.

胰岛素 Lispro 的商业生产过程与天然重组人胰岛素的前胰岛素路线非常相似。在大肠杆菌中表达编码 LysB28-ProB29 前胰岛素的合成基因。随后发酵与收获前胰岛素，用胰酶和羧肽酶 B 处理前胰岛素，蛋白水解基因工程的胰岛素分子。然后经过一系列高效层析步骤将胰岛素纯化成单一组分。最终产品的配方中还含有 m- 甲苯酚（保存剂与稳定剂）、氧化锌（稳定剂）、丙三醇（张力修饰成分）与磷酸盐缓冲液。商业产品在 2～8℃条件下的保质期为 2 年。

Insulin Lispro is manufactured commercially in a manner quite similar to the proinsulin route used to manufacture native recombinant human insulin. A synthetic gene coding for LysB28–ProB29 proinsulin is expressed in *E. coli*. Following fermentation and recovery, the proinsulin is treated with trypsin and carboxypeptidase B, resulting in the proteolytic excision of the engineered insulin molecule. It is then purified to homogeneity by a number of high-resolution chromatographic steps. The final product formulation also contains m-cresol (preservative and stabilizer), zinc oxide (stabilizer), glycerol (tonicity modifier) and a phosphate-based buffer. The commercial product has a shelf-life of 2 years when stored at 2–8℃ .

3. *胰岛素 Aspart*　胰岛素 Aspart 是第二个批准临床使用的速效基因工程人胰岛素类似物。它与天然人胰岛素的不同在于由天冬氨酸取代了 B 链 28 位的脯氨酸残基。这种单个氨基酸残基的取代降低了胰岛素分子自联的倾向，确保胰岛素一经注射就能从注射位点立即进入血液循环。由于 Aspart 起效快，作用时间短，因此它通常与长效和中效胰岛素一起使用。

3. Insulin Aspart　Insulin Aspart is a second fast-acting engineered human insulin analogue now approved for general medical use. It differs from native human insulin in that the prolineB28 residue has been replaced by aspartic acid. This single amino acid substitution also decreases the propensity of individual molecules to self-associate, ensuring that they begin to enter the bloodstream from the site of injection immediately upon administration. Because Aspart has a more rapid onset and a shorter duration of action than human regular insulin, it should normally be used in regimens together with intermediate or long-acting insulin.

4. *长效胰岛素类似物*　研究注重生产长效胰岛素类似物。目前使用的锌 - 胰岛素混悬液，或鱼精蛋白 - 锌 - 胰岛素混悬液，其血浆半衰期为 20～25h。选择性的氨基酸替代会使胰

4. Longer-acting Insulin Analogues　A number of studies have also focused upon the generation of longer-acting insulin analogues. The currently used Zn–insulin suspensions, or protamine–Zn–insulin suspensions, generally display a plasma half-life of 20-25 h. Selected amino acid substitutions have

岛素类似物，甚至是可溶性的胰岛素类似物的血浆半衰期长达35h。

于2000年批准入市的一种长效胰岛素类似物的商业名为Optisulin或Lantus。这种基因工程胰岛素的国际非专利名称（inn）是Insulin glargine。它与天然胰岛素的不同在于：胰岛素A链C端的天冬氨酸由甘氨酸取代，且B链的C端加入了两个精氨酸残基。其最终效果是使胰岛素分子的等电点（pI）从5.4提高到近7.0。这种基因工程胰岛素是通过前胰岛素路线在大肠杆菌工程株K12宿主菌中表达的。纯化产物的pH调节到4.0，在此pH条件下，Optisulin能够完全溶解。

Optisulin经皮下注射后，pH增高至接近中性（生理条件下pH约为7.4），因此在皮下沉淀。它的再次溶解过程非常缓慢，释放到血液中所需的时间也大为延长。每天注射1次即可维持令人满意的基础血液胰岛素水平，并且用药24h后在注射位点仍可检测到胰岛素分子。

基因工程胰岛素类似物的生产带来了几个关系到产品安全性和药效的重要问题。而天然蛋白质氨基酸序列的改变可能会使基因工程产品出现免疫原性。由于胰岛素基本上需要终生每天注射，因此，对胰岛素来说，这个问题非常突出。另外，结构的改变可能会对药物代谢动力学或药效造成意想不到的影响。然而，对于这些已经批准应用的类似物的临床前、特别是临床评估证明它们是安全有效的。引入的序列变化相对较小，似乎不会引起免疫反应。幸运的是，临床证明这些序列变化既没有影响胰岛素分子

generated insulins which, even in soluble form, exhibit plasma half-lives of up to 35 h.

Optisulin or Lantus are the trade names given to one such analogue that gained general marketing approval in 2000. The international non-proprietary name (inn) for this engineered molecule is "insulin glargine". It differs from native human insulin in that the Cterminal aspargine residue of the A-chain has been replaced by a glycine residue and the B-chain has been elongated (again from its C-terminus) by two arginine residues. The overall effect is to increase the molecule's isoelectric point from 5.4 to a value approaching 7.0. The engineered insulin is expressed in a recombinant *E. coli* K12 host strain and is produced via the proinsulin route, as previously described. The purified product is formulated at pH 4.0, a pH value at which it is fully soluble.

Upon s.c. injection of Optisulin, the insulin experiences an increase in pH towards more neutral values (about 7.4) and, consequently, appears to precipitate in the subcutaneous tissue. It resolubilizes very slowly and hence a greatly prolonged duration of release into the bloodstream is noted. Consequently, a single daily injection supports the maintenance of acceptable basal blood insulin levels, and insulin molecules are still detected at the site of injection more than 24 h after administration.

The generation of engineered insulin analogues raises several important issues relating to product safety and efficacy. Alteration of a native protein's amino acid sequence could render the engineered product immunogenic. Such an effect would be particularly significant in the case of insulin, as the product is generally administered daily for life. In addition, alteration in structure could have unintended (in addition to intended) influences upon pharmacokinetic and/or pharmacodynamic characteristics of the drug. Preclinical and, in particular, clinical evaluations undertaken upon the analogues thus far approved, however, have confirmed their safety as well as efficacy. The sequence changes introduced are relatively minor and do not seem to elicit an immunological response. Fortuitously, neither has the alterations made affected the ability of the insulin molecule to interact with the insulin receptor and trigger the resultant characteristic biological responses.

与胰岛素受体之间的相互作用，又没有影响任何生理反应。

4.2.4 重组人胰岛素的质量标准

重组人胰岛素为重组DNA技术生产的由51个氨基酸残基组成的蛋白质，宿主细胞为大肠杆菌。按干燥品计算，含重组人胰岛素（包括A_{21}脱氮人胰岛素）应为95.0%～105.0%。每单位重组人胰岛素相当于0.0347mg。

1. 性状　本品为白色或类白色的结晶性粉末。本品在水、乙醇和乙醚中几乎不溶，在稀盐酸和稀氢氧化钠溶液中易溶。

2. 鉴别

（1）在效价测定项下记录的色谱图中，供试品主峰的保留时间应与重组人胰岛素对照品峰的保留时间一致。

（2）取本品适量，用0.1%三氟醋酸溶液制成每毫升中含10mg的溶液，取20μl，加0.2mol/L三羟甲基氨基甲烷-盐酸缓冲液（pH7.3）20μl、0.1% V8酶溶液20μl与水140μl，混匀，置37℃水浴中2h后，加磷酸3μl，作为供试品溶液；另取重组人胰岛素对照品适量，同法制备，作为对照品溶液。照效价测定项下的方法，以0.2mol/L硫酸盐缓冲液（pH2.3）-乙腈（90∶10）为流动相A，乙腈-水（50∶50）为流动相B，进行梯度洗脱。取对照品溶液和供试品溶液各25μl，分别注入液相色谱仪，记录色谱图，供试品的肽图谱应与对照品的肽图谱一致。

3. 锌含量的测定　精密称取本品适量（约相当于锌0.1mg），置10ml量瓶中，用0.01mol/L盐酸溶液溶解并稀释至刻度，作为供试品

4.2.4 Quality Standard of Recombinant Human Insulin

Recombinant human insulin is a protein having 51 amino acid residues and produced in *E.coli* as host cell by a method based on recombinant DNA technology. The content of human insulin $C_{257}H_{283}N_{65}O_{77}S_6$ plus A_{21} desamido human insulin is not less than 95.0% and not more than 105.0%, calculated on the dried basis. By convention, for the purpose of labeling insulin preparations, 0.0347 mg of human insulin is equivalent to 1 IU of insulin.

1. Description　A white or almost white crystalline power. Practically insoluble in water, ethanol or ether, freely soluble in dilute hydrochloric acid solution or dilute sodium hydroxide solution.

2. Identification

(1) The retention time of principal peak of the substance being examined in the chromatogram obtained in the Assay is identical with that of the principal peak of recombinant human insulin RS.

(2) Dissolve a quantity substance being examined in 0.1% trifluoroacetic acid solution of 10 mg per ml, measure 20 μl of above solution, add 20 μl of 0.2mol/L trometamol-hydrochloric acid BS (pH 7.3), 20 μl of 0.1% V8 enzyme solution and 140 μl of water, mix well, warm in a water bath at 37C for 2 hours, then add 3 μl of phosphate acid as test solution. Repeat the operation, using a quantity of recombinant human insulin CRS instead of the substance being examined, as reference solution. Carry out the method for Assay, using a mixture of 0.2 mol/L sulfate BS (pH 2.3)-acetonitrile (90∶10) as mobile phase A, and a mixture of acetonitrile-water (50∶50) as mobile phase B. Inject separately 25 μl the reference solution and the test solution into the column, and record the chromatogram. The peptide map of the test solution is identical with that of the reference solution.

3. Zinc　Dissolve a quantity substance being examined equivalent to about 0.1 mg of zinc, accurately weighed, in 0.01 mol/L hydrochloric acid solution in a 10 ml volumetric flask, dilute to volume and mix well as a test solution and zinc standard solution (each ml

溶液；取供试品溶液与标准锌溶液（精密称取硫酸锌44mg，置100ml量瓶中，加水溶解并稀释至刻度，摇匀，精密量取10ml，置100ml量瓶中，加水稀释至刻度，摇匀，每毫升相当于10μg的锌）各1.6ml，分别置10ml量瓶中，各加硼酸-氯化钠缓冲液（pH 9.0）2ml与新制的锌试剂溶液1ml，加水稀释至刻度，摇匀，放置30min，照分光光度法，在620nm的波长处分别测定吸收度，计算。含锌量不得超过1.0%。

4. 氮的测定 取本品约10mg，精密称定，照氮测定法测定，按干燥品计算，含氮量应为14.5%～16.5%。

5. 干燥失重 取本品0.2g，在105℃干燥至恒重，减失重量不得超过10.0%。

6. 炽灼残渣 取本品约0.4g，依法检查，遗留残渣不得超过2.0%。

7. 无菌检查 取本品，加适量的灭菌0.01mol/L盐酸溶液制成每1ml中含2mg的溶液，用薄膜过滤法处理后，依法检查，应符合规定。

8. 细菌内毒素检查 取本品，依法检查，每毫克重组人胰岛素中含内毒素的量应小于10EU。

9. 含量测定 照高效液相色谱法测定。色谱条件与系统适用性试验：用十八烷基硅烷键合硅胶为填充剂（5～10μm）；0.2mol/L硫酸盐缓冲液-乙腈（74∶26）为流动相；柱温为40℃；检测波长为214nm。取系统适用性试验用溶液20μl注入液相色谱仪，胰岛素主峰和A_{21}脱氨胰岛素峰之间的分离度应不小于1.8，拖尾因子应不大于1.8。测定法：取

of the solution is equivalent to 10 μg of Zn) separately to two 10 ml volumetric flasks, to each flask add 2 ml of boric acid-sodium chloride BS (pH 9.0) and 1 ml of freshly-prepared zinc test solution, dilute with water to volume, and mix well. Allow to stand for 3 minutes, measure the absorbances of the two resulting solutions at 620 nm. Calculate the content of zinc, not more than 1.0%.

4. Nitrogen Carry out the method for determination of nitrogen, not less than 14.5% and not more than 16.5% of N, calculate on the dried basis.

5. Loss on drying When dried to constant weight at 105℃, loses not more than 10.0% of its weight using 0.2 g.

6. Residue on ignition Not more than 2.0%, using 0.4 g.

7. Microbial limits The total bacterial count does not exceed 300 per g, the test being performed on a portion of about 0.2 g, accurately weighed.

8. Bacterial endotoxin Carry out the test for bacterial endotoxin, less than 10 EU per mg of recombinant human insulin.

9. Assay Carry out the method for high performance liquid chromatography, using a column packed with octadecylsilane bonded silica gel (5-10 μm) and a mixture of 0.2 mol/L sulfate BS as the mobile phase. Column temperature is 40℃ and detection wavelength is 214 nm. Inject 20 μl of solution for system suitability test into the column, the resolution factor between the peaks of insulin and A_{21}-desamino-insulin is not less than 1.8, the tailing factor is not more than 1.8. Procedure: Dissolve a quantity, accurately weighed, in 0.01 mol/L hydrochloric acid solution to produce a solution of 10 Units per ml. Inject 20 μl into

本品适量，精密称定，用0.01mol/L盐酸溶液制成每毫升中含10.0单位的溶液（临用新配）。取20μl注入液相色谱仪，记录色谱图；另取重组人胰岛素对照品适量，同法测定。按外标法以峰面积计算。

10. 贮藏　遮光，密闭，在−15℃以下保存。

11. 制剂　①重组人胰岛素注射液；②精蛋白重组人胰岛素注射液。

the column and record the chromatogram. Repeat the operation, using recombinant human insulin RS instead of the substance being examined. Calculate the content of recombinant human insulin with respet to the peak area obtained in the chromatogram by the external standard method.

10. Storage Preserve in well closed containers, stored below −15℃ and protected from light.

11. Preparation ① Recombinant human insulin injection; ② Isophane protamine recombinant human insulin injection.

4.3 胰岛素的临床应用与给药方法

4.3.1 胰岛素的临床应用

胰岛素的适应证是Ⅰ型糖尿病病人和口服其他降血糖药不明显的Ⅱ型糖尿病病人。

几千年前，人们就已经认识并记录了糖尿病的症状，包括口渴、多尿、体重减轻；直到最近，糖尿病仍有一定的死亡率。最常见的糖尿病是由于部分或完全缺乏胰岛素引发的生理病理反应。糖尿病有两种类型：胰岛素依赖性糖尿病（IDDM）和非胰岛素依赖性糖尿病（NIDDM）。IDDM，又称Ⅰ型糖尿病，通常在孩童时期或青少年时期发病。发病的直接原因是由于自身免疫破坏了产生胰岛素的胰腺β细胞而导致的，因此其特征是血液中没有或几乎不存在胰岛素，即使血糖水平升高，血液胰岛素的含量也不会增加。对于有糖尿病遗传倾向的个体，当T细胞介导的自身免疫损伤产生胰岛素的β细胞时就会引发IDDM。这可能是由于这些个体的β胰岛细胞表面具有异常抗原所

4.3 Therapeutic Uses of Insulin and Means of Administration

4.3.1 Therapeutic Uses of Insulin

Insulin is indicated for all patients with type Ⅰ diabetes mellitus and for patients with type Ⅱ diabetes mellitus who are inadequately controlled by oral medication.

The symptoms of diabetes have been recognized and recorded for thousands of years. They generally consist of excessive thirst and urination, loss of weight and, until recently, certain death. Diabetes mellitus, the most common form of diabetes, is caused by the partial or complete absence of insulin-triggered biological responses. Two forms of diabetes mellitus exist: insulin dependent diabetes mellitus (IDDM) and non-insulin dependent diabetes mellitus (NIDDM). IDDM, also known as type I diabetes, normally occurs during childhood or teens. It is caused by an autoimmune-mediated destruction of the insulin-producing pancreatic β cells. It is thus characterized by the absence or near-absence of insulin in the blood, even at elevated blood glucose levels. In genetically predisposed individuals, IDDM is caused by T cell-mediated autoimmune destruction of the insulin-producing bpancreatic islet cells. This is probably due to the expression of antigen on the surface of the β cells in such individuals. In the case of NIDDM (maturity onset or type Ⅱ diabetes mellitus), insulin is present in the blood at normal (or even elevated) levels, but fails to promote any of its characteristic effects. A number of factors can contribute to such insulin resistance, including: ① reduced numbers of insulin receptors on sensitive cells; ② reduced receptor affinity

致。对于NIDDM（成人型或Ⅱ型糖尿病），血液中胰岛素水平正常（甚至增高），但胰岛素不能发挥正常的调节作用。许多因素可能造成这种胰岛素抗性，其中包括：①敏感细胞上的胰岛素受体减少；②胰岛素与受体的亲和力降低；③胰岛素完全/部分不能结合受体而引起细胞无反应。

for insulin; ③ total/partial failure of insulin binding to initiate an intracellular response.

体内若胰岛素不足，可导致细胞膜转运葡萄糖的速率下降，出现高血糖症；细胞内酶系统催化葡萄糖转化为糖原的速率下降，而蛋白质转化为糖的速率异常提高。注射的胰岛素可刺激糖代谢，并协助葡萄糖转运到心肌、骨骼肌和脂肪组织中，并将葡萄糖转化为糖原，这样血糖的浓度和糖基化血红蛋白的浓度才能接近正常水平。

Insulin deficiency results in marked reduction in the rate of transport of glucose across cell membranes, resulting in hyperglycemia; there is also a reduction in the enzyme system that catalyzes the conversion of glucose to glycogen, and an abnormally high rate of conversion of protein to glucose. Insulin stimulated endogenously or administered exogenously stimulates carbohydrate metabolism and facilitates transfer of glucose into cardiac muscle, skeletal muscle, and adipose tissue and converts glucose to glycogen.

胰岛素在肝脏和肾脏中代谢。胰岛素作用时间的开始、峰效应和滞留时间主要决定于胰岛素产品本身以及给药的部位。肾小球的过滤速率（GFR）在10～50ml/min的患者，给药剂量应降低25%。GFR小于10ml/min的患者，剂量应降低50%。胰岛素主要分布在细胞外，表观分布容积0.15L/kg。

Insulin is metabolized in the liver and kidney. The onset of action, peak effect, and duration of action depend primarily on the insulin product and site of administration. For patients with a glomerular filtration rate (GFR) between 10 and 50 ml/min, a dosage reduction of approximately 25% may be indicated; a dosage reduction of approximately 50% may be indicated in patients with a GFR less than 10 ml/min. Insulin is distributed primarily in extracellular fluids and has an apparent volume of distribution of 0.15 liter/kg.

Ⅰ型糖尿病人必须接受胰岛素的治疗才能生存。IDDM可以通过注射外源胰岛素制剂而控制病情，通常是定期皮下注射。病人应很好地自我监测血糖浓度。大多数Ⅱ型糖尿病病人不需要立即使用胰岛素治疗。静脉胰岛素给药用于治疗有酮酸症的糖尿病患者。糖尿病患者有微脉管（如视网膜病、肾病和神

All patients with type Ⅰ diabetes mellitus require insulin therapy to survive. IDDM may be controlled by parenteral administration of exogenous insulin preparations, usually by regular s.c. injection. For optimal control, the patient should perform self-monitoring of blood glucose. The majority of patients diagnosed with type Ⅱ diabetes mellitus do not require immediate treatment with insulin. Intravenous regular insulin is recommended for treating patients with diabetic ketoacidosis. Patients with diabetes mellitus are at risk for microvascular (i.e. retinopathy, nephropathy, and neuropathy) and macrovascular (i.e. myocardial infarction)

经性疾病）和大脉管（如心肌梗死）并发症的危险，导致并发症或死亡。糖尿病控制和并发症实验表明，加强胰岛素的治疗，能够使糖基化血红蛋白的浓度接近正常，降低Ⅰ型糖尿病患者的微脉管并发症。高纯度的重组人胰岛素的使用降低了Ⅰ型糖尿病患者的局部反应，如胰岛素过敏、免疫抵抗和脂肪萎缩症的发生率。

complications that lead to morbidity and mortality. The Diabetes Control and Complications Trial demonstrated that intensive insulin therapy, designed to achieve a near-normal glycosylated hemoglobin, reduced microvascular complications in patients with type Ⅰ diabetes. The increased use of human recombinant insulin and highly purified insulin has led to a decline in the incidence of local reactions, insulin allergy, immune resistance, and lipoatrophy in patients with type Ⅰ diabetes.

4.3.2 胰岛素的给药方法

胰岛素制剂可通过肌肉、皮下、口腔或肺部用药。还有一种用药方法是利用注射系统持续为患者输入胰岛素，能够更为近似地模拟血液中胰岛素水平的自然变化。这种方法中最简单的设计称为开环系统。它包括一个通过导管自动皮下注射可溶性胰岛素的注射泵。血糖水平是手动监控，并据此调整注射速率。

尽管这种方法效果明显，但在临床上并不常用，主要是因为会引起许多并发症，包括：①注射部位容易形成脓肿或蜂窝织炎；②注射泵可能会出现故障；③导管堵塞；④对系统组分产生过敏反应；⑤需要手动监控血糖水平。

这种方法中更加复杂的设计是闭环系统（也常称为人工胰腺）。它不仅包括泵和注射装置，还包括一个整体的葡萄糖感受器和计算机，用以分析血糖数据并将流速做相应的调整。这种系统的应用价值仍有待评定。

注射泵已经在某种程度上朝着模拟血液胰岛素水平的自然调控走近了一步，而移植合成胰岛素的胰腺细胞

4.3.2 Means of Insulin Administration

Insulin preparations can be administrated by i.m., s.c., even delivered via the oral or pulmonary route. An additional approach, which may mimic more closely the normal changes in blood insulin levels, entails the use of infusion systems which constantly deliver insulin to the patient. The simplest design in this regard is termed an open loop system. This consists of an infusion pump which automatically infuses soluble insulin subcutaneously, via a catheter. Blood glucose levels are monitored manually and the infusion rate is programmed accordingly.

While the potential of such systems is obvious, they have not become popular in practice, mainly due to complications which can potentially arise, including: ① abscess formation or development of cellulitis at the site of injection; ② possible pump malfunction; ③ obstruction; ④ hypersensitivity reactions to components of the system; ⑤ requirement for manual blood glucose monitoring.

The closed-loop system (often termed the artificial pancreas) is essentially a more sophisticated version of the system described above. It consists not only of a pump and infusion device, but also an integral glucose sensor and computer, which analyses the blood glucose data obtained and adjusts the flow-rate accordingly. The true potential of such systems remains to be assessed.

While infusion pumps can go some way towards mimicking normal control of blood insulin levels, transplantation of insulin-producing pancreatic cells should effectively “cure”the diabetic patient. This

将会有效地“治愈”糖尿病。目前已经有患者移植了合成胰岛素的胰腺细胞，结果鼓舞人心。

approach has been adopted thus far with patients, with encouraging results.

在20世纪70年代，用先天品系小鼠做的实验证明了这种方法的可行性。给小鼠注射药物如链脲霉素破坏其胰腺β细胞，同一品系的某些小鼠捐出合成胰岛素的胰岛细胞，植入这些患糖尿病的小鼠体内。胰岛移植后，糖尿病动物血糖水平可持久地恢复正常。更鼓舞人心的是，这种治疗方法能够预防破坏肾和眼功能的糖尿病并发症（人糖尿病会导致肾衰和部分失明）。

Initial experiments in the 1970s using inbred strains of rats illustrated the feasibility of this approach. Insulin-producing pancreatic islet cells donated by one set of rats were transplanted into other rats of the same strain, first made diabetic by injection with drugs such as streptozotocin, which destroy the pancreatic β cells. Such islet grafts permanently returned blood glucose levels in the diabetic animals to normal values. Even more encouraging, this treatment prevented development of diabetic-associated complications of kidney and eye function (which, in human diabetics, can lead to kidney failure and partial blindness).

移植技术很简单，从门静脉注入合成胰岛素的胰岛细胞，这些细胞聚集在静脉的小分支血管中。在此，它们持续地依据血糖水平释放胰岛素。通常移植400 000～800 000个胰岛β细胞。大多数情况下，这种调节功能最多能够保持3年，有时候这些移植的胰岛细胞不能完全控制血糖水平，因此，有时还需要附加注射胰岛素。

The technique simply entails injecting the insulin-producing islets into the portal vein. These cells subsequently lodge in smaller vessels branching from the vein. Here, they can constantly monitor blood glucose levels and secrete insulin accordingly. About 400 000–800 000 islet β cells are usually transplanted. In most instances, these function for up to 3 years, although they often fail to control blood glucose levels fully, i.e. supplemental insulin injections are sometimes required.

植入人体的胰岛细胞来源于去世的人体捐献者。接受者的免疫系统会产生反应，这些细胞最多只能暂时发挥治疗作用，继发的免疫反应会迅速破坏这些外来细胞。迄今在人体做的实验利用的是接受肾移植的糖尿病患者，因为他们已经接受了免疫抑制治疗。可以预见，这种方法无法广泛应用的最大障碍在于在移植的同时，需要进行免疫抑制。

The islet cells transplanted into humans are obtained from pancreatic tissue of deceased human donors. Implantation of these cells in recipients displaying a competent immune system would, at best, be of transient therapeutic benefit. The ensuing immune response would quickly destroy the foreign cells. Studies conducted thus far in humans have utilized diabetic patients who have received kidney transplants, as these are already subject to immunosuppressive therapy. However, a major stumbling block to the widespread adoption of this therapeutic approach is, predictably, the requirement to induce concurrent immune suppression.

（编者　付爱玲）

(Compiled by Ailing Fu)

5 促红细胞生成素

5 Erythropoietin (EPO)

5.1 EPO 概述

1906 年，Carnot 和 DeFlandre 发现，将大量失血造成贫血的兔的血浆注入到正常兔子体内，后者出现红细胞增多现象。他们由此假设动物体内存在一种体液因子并称之为生血素。此后，Hjort（1936 年）、Krumdieck（1943 年）和 Erslev（1953 年）的报告中均证明了一种能够控制红细胞生成体液因子的存在，1948 年 Bonsdorff 和 Jalavisto 将其命名为促红细胞生成素（erythropoietin，简称 EPO）。在 1950 年，Reissman 证明组织缺氧会刺激 EPO 的产生。后来发现，哺乳动物体内绝大部分 EPO 是由肾脏产生的。

EPO 是第一个被克隆的造血生长激素。1977 年，Miyake 等人从严重贫血患者的尿液中分离提取和纯化了 EPO。这种尿蛋白（即 EPO）的氨基酸序列被鉴定后，使得合成用于分离和克隆 EPO 基因的 DNA 探针成为可能。中国仓鼠卵巢（CHO）细胞转染了结合表达载体的 EPO 基因后，能够产生具有生物学活性的重组人 EPO（recombinant human erythropoietin，简称 rhEPO），此后，rhEPO 开始在临床上得到应用。EPO 是最主要的红细胞生成调节因子，同时能够促进幼稚红细胞的存活、增殖和分化。

20 世纪 80 年代，Adamson 等在美国西北肾学中心进行了安进（Amgen）公司的合成 EPO——Epogen 的临床研究，实验取得了成功。1985 年，Lin

5.1 Introduction of EPO

In 1906, Carnot and DeFlandre demonstrated the occurrence of a retuculocytosis in normal rabbits injected with plasma from rabbits made anemic by bleeding and thus postulated a humoral factor they called hemopoietin. The reports by Hjort in 1936, Krumdieck in 1943, and Erslev in 1953 confirmed the existence of a humoral factor that controls erythropoiesis, and in 1948 Bonsdorff and Jalavisto named it erythropoietin (abbreviated, EPO). In 1950, Reissman demonstrated that hypoxia stimulated the production of erythropoietin. In the following years the kidney of mammals was found mostly to be the site of production of erythropoietin.

EPO was the first hematopoietic growth hormone to be cloned. In 1977, Miyake *et al.* isolated and purified EPO from the urine of severely anemic patients. The identification of the amino acid sequence of the pure urinary protein (i.e. EPO) enabled it to synthesize EPO DNA probes for isolation and cloning of the EPO gene. Chinese hamster ovary (CHO) cells transfected with the EPO gene linked to an expression vector were shown to produce biologically active recombinant human erythropoietin (abbreviated, rhEPO), and hence, rhEPO became available for clinical use. EPO is the primary regulator of erythropoiesis, and promotes the survival, proliferation and differentiation of erythroid progenitor cells.

In the 1980s, Adamson *et al.* conducted a clinical trial at U.S. Northwest Kidney Centers for a synthetic form of the hormone, Epogen produced by Amgen. The trial was successful. In 1985, Lin *et al.* isolated the human EPO gene from a genomic phage library

等从噬菌体基因组文库中分离得到了人EPO基因，并使其可用于研究和生产。研究表明哺乳动物体内编码EPO的基因在体内外均具有生物学活性，这就打开了生产用于治疗贫血病人的rhEPO的大门。1989年，Epogen获得FDA的批准上市，适应证为多种贫血。由于其市场表现不俗，因此目前仍在使用。

and were able to characterize it for research and production. Their research demonstrated that the gene for EPO encoded the production of EPO in mammalian cells that is biologically active *in vitro* and *in vivo*. This opened up the door for the industrial production of rhEPO for treating anemia patients. In 1989, U.S. Food and Drug Administration approved the hormone, called Epogen whose indication was many kinds of anemia. Because of the excellent market performance, Epogen remains in use now.

5.2 EPO的基因、受体和分子结构

5.2 EPO Gene, Receptor and Structure

5.2.1 EPO的基因

EPO基因在生物物种间高度保守。人类和猪、羊、小鼠和大鼠的EPO基因序列中有80%～82%的氨基酸是相同的。编码人类EPO的基因包括一个5.4 kb的单拷贝区域，位于染色体7q11～q22上。它包括5个外显子和4个内含子，编码一个193个氨基酸的激素原。在内含子Ⅰ，5和3侧翼区在生物物种间也是高度保守的，参与EPO基因的表达调控。3.9 kb的5侧翼序列含有一个识别位点（供各类细胞活素类因子、金属效应元件、糖皮质激素反应元件和3，5-环磷酸腺苷反应元件共同使用）及一个结合位点（供蛋白质AP1、核因子kB、核转录因子Sp1共同使用）。这三个侧翼区包括氮调（氧传感）共有序列、组织特异性调控元件以及一个AP1和Sp1的结合位点。

5.2.1 EPO Gene

The EPO gene is highly conserved between species. There is a 80%-82% amino acid identity between the human sequence and that of pig, sheep, mouse and rat EPO genes. The gene encoding human EPO contained a single-copy in a 5.4 kb region and is located on chromosome 7q11-q22. It contains five exons and four introns coding for the prohormone of 193 amino acids. Intron Ⅰ, the 5 and 3 flanking regions are also highly conserved between species and are involved in the regulation of the EPO gene expression. The 3.9 kb of the 5 flanking sequence contains consensus recognition sites for various cytokines, metal-response elements, glucocorticoid-response elements, and adenosine 3, 5-cyclic monophosphatase-response elements, as well as binding sites for protein AP1, nuclear factor (NF) kB, and Sp1 transcription factors. The 3 flanking region includes nitrogen-regulatory/ oxygen sensing consensus sequences, tissue-specific regulatory elements, and binding sites for AP1 and Sp1.

5.2.2. EPO的受体

人EPO受体由507个氨基酸残基组成，包含具有24个氨基酸残基的前导序列，其糖基化后分子量为66 kDa（表5-13）。

5.2.2 EPO Receptor

Human EPO receptor consists of 507 amino acid residues (Table 5-13), containing the leader sequence of 24 amino acid residues. Its molecular weight is about 66 kDa when glycosylated .

表 5-13　人类 EPO 及其受体的性质

项目	EPO	EPO 受体	项目	EPO	EPO 受体
蛋白			基因		
前体蛋白氨基酸数	193		基因大小（kb）	5.4	
成熟蛋白氨基酸数	165	507	外显子	5	8
			内含子	4	7
分子量（kDa）	30.4	66	染色体基因定位	7q11～q22	19p
N- 糖基化位点	3	0			
二硫键个数	2	2			
细胞质内酪氨酸		8	细胞来源	肾间质细胞，肝细胞（脑、胎盘、睾丸、脾、肺）	红系集落形成单位，红系暴增式集落形成单位，巨噬细胞，胎盘组织，内皮细胞，神经细胞

5.2.3　EPO 的分子结构

人 EPO 是一种以 193 个氨基酸的多肽激素原形式合成的糖蛋白。它是在初级转录产物合成基础上，在细胞内切除 27 个 N- 末端氨基酸的前导序列和 1 个 C- 末端精酰胺残基而形成的。这种成熟的糖蛋白激素分子由多肽和糖基构成，多肽链部分共有 165 个氨基酸，分子量 18.4 kDa（图 5-12）。通过广泛结合大约 40% 的糖类进行糖基化，最后总分子量约为 30.4 kDa。其内部有三个四天线型 N- 连接［天冬氨酸（Asp）24、38 和 83］和一个小型 O- 连接［丝氨酸（Ser）126］的酸性寡糖支链，这对 EPO 正确加工和运输、保持分子稳定性和在机体内的生物学活性都是必需的。糖类部分包括果糖（fructose）、甘露糖（mannose）、N- 乙酰氨基葡萄糖（N-acetylglucosamine）、N- 乙酰氨基半乳糖（N-acetylgalactosamine）和唾液酸（sialic acid）。4 个半胱氨酸残基由二硫键（disulfide bond）连接（7-161 和 29-33），对于 EPO 保持体内生物学

5.2.3 EPO Structure

Human EPO is a glycoprotein synthesized as a 193 amino acid polypeptide. A leader sequence of 27 amino acids from the N-terminus and the arginyl residue from the C-terminus are cleaved intracellularly from the primary translation product. The mature glycoprotein hormone consists of 165 amino acids, and the EPO peptide has a molecular mass of 18.4 kDa (Figure 5-12). It is extensively glycosylated with a carbohydrate moiety of approximately 40%, resulting in a total mass of 30.4 kDa. There are three tetraantennary N-linked (Asp 24, 38 and 83) and one small O-linked (Ser 126) acidic oligosaccharide side chains, required for the correct processing and export of the hormone, molecular stability, and for its full *in vivo* activity. The carbohydrate contains fructose, mannose, N-acetylglucosamine, N-acetylgalactosamine and sialic acid. Four cysteines residues linked by disulfide bonds between cysteine 7-161 and 29-33 are essential for full *in vivo* biological activity. Despite little similarity in the amino acid sequence (i.e. primary structure), EPO has a tertiary conformation analogous to that of growth hormone, prolactin and interleukin-6.

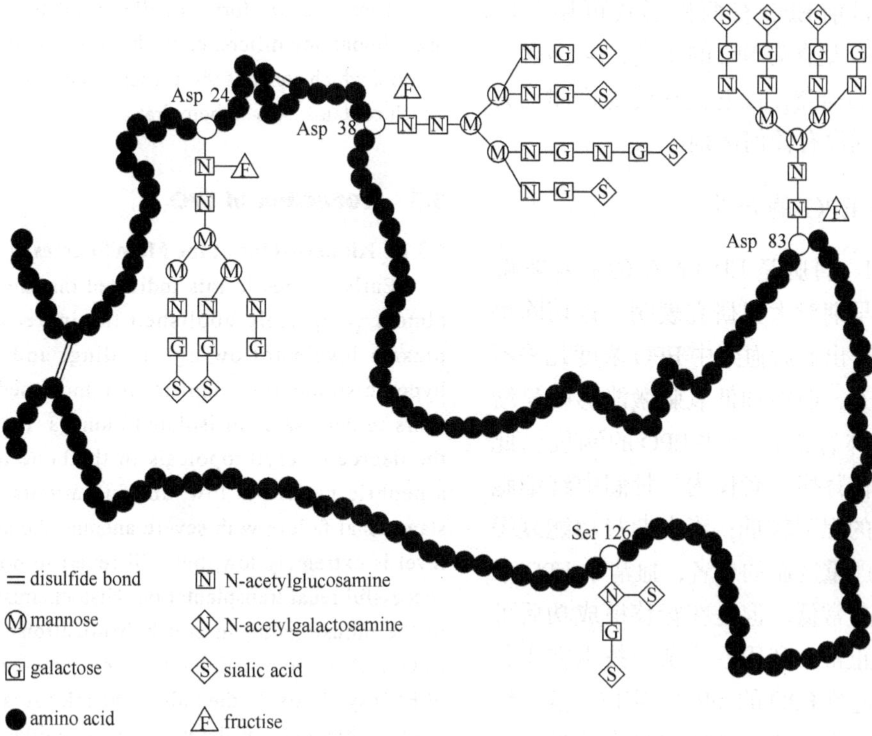

图 5-12　人 EPO 的典型结构

活性也是必需的。尽管 EPO 的氨基酸序列（即一级结构）与生长素、催乳素、白细胞介素 -6 相似度较低，但是三级构象却较为相似。

通过圆二色谱、紫外吸收光谱和荧光光谱等方法对 CHO 细胞中表达的 rhEPO 与人尿中提取的 EPO 进行比较，发现二者的氨基酸序列、两个二硫键位置、糖基化位点和二级结构均没有明显的区别。然而，rhEPO 比人尿 EPO 的体内活性更高。起初，对于低压冻干人尿 EPO 的活性，其 1 个国际单位（IU）被定义为与 5μg 钴所引起红细胞生成的效果相同时所需的 EPO 的量；但是现在由于 DNA 重组技术的出现，EPO 的单位定义已发生了变化，现在定义每毫克完全糖基化的 EPO 为 130 000 IU。

Recombinant human EPO expressed in CHO cells has been compared with human urinary EPO by circular dichroism, UV light absorbance, and fluorescence spectroscopy, without identifying any significant difference in amino acid sequence, position of their two disulfide bridges, glycosylation position, and secondary structure. However, rhEPO has a higher *in vivo* activity than urinary erythropoietin. Initially, a lyophilized human urinary erythropoietin with a specific activity of 1 international unit (IU) was defined as the activity which elicits the same erythropoiesis-stimulating effect as 5 μg cobalt, but has been replaced now by a DNA-derived human EPO standard with a specific activity of 130 000 IU per milligram fully glycosylated protein.

目前EPO有两种形式可用于临床：α-型含有9%的糖类，β-型含有24%的糖类，而它们在药物代谢动力学方面没有任何区别。

There are two forms of EPO available for clinical use without any difference in the pharmacokinetics: the α-form, which contains 9% carbohydrate, and the β-form, which contains 24% carbohydrate.

5.3 EPO的产生

5.3 Production of EPO

5.3.1 肾脏是EPO产生的主要来源

早期对大鼠研究表明，肾切除手术使得出血后血浆中EPO浓度几乎不再上升，而钴和低氧刺激能够明显增加离体肾灌注液中EPO的滴度。此外，在肾病患者体内，骨髓中红细胞生成的程度较低；终末期肾功能衰竭合并严重贫血的患者，血清中EPO水平也非常低，但是在肾移植成功后可恢复正常。利用原位杂交技术在小鼠体内进行EPO的mRNA定位，这一组织化学手段证明肾是EPO合成的主要部位。在许多物种中，肾内的mRNA在间质细胞（interstitial cell）中表达，这些间质细胞位于有成纤维细胞样特征的肾小管周围的毛细血管床中。细胞内没有EPO成分的储备，而贫血可导致间质细胞数量、EPO的mRNA和血清中EPO浓度的指数级增长。在患有肾腺癌和红细胞增多症的病人体内，肿瘤细胞亦可产生EPO。

5.3.1 Kidney—One of the Main Sources of EPO

Early studies in rats indicated that nephrectomy almost completely abolished the increase in EPO plasma levels following bleeding, and cobalt or hypoxic stimulation significantly increased the EPO titers in perfusates of isolated kidneys. Furthermore, the degree of erythropoiesis in the bone marrow of a nephric patient is low, and in patients with end-stage renal failure with severe anemia, the serum EPO level is extremely low, but will revert to normal after successful renal transplantation. Histochemistry studies in the mouse using in situ hybridization to localize EPO mRNA confirmed that the kidney is a primary site of EPO synthesis. In the kidney the mRNA is expressed in interstitial cells of the peritubular capillary bed with fibroblast-like characteristics in various species. There are no intracellular stores of EPO, and anemia leads to an exponential increase in the number of interstitial cells in parallel with the exponential increase in EPO mRNA and serum EPO. In patients with renal adenocarcinoma and polycythemia, their tumoral cells produce EPO.

5.3.2 产生EPO的另一重要器官是肝脏

成人体内另一个产生EPO的主要器官是肝脏，提供血浆中循环EPO的10%～15%。但是一旦发生慢性肾衰竭，肝脏产生的EPO并不能补偿肾脏所减少生成的EPO。在肝脏中肝细胞是产生mRNA的主要细胞种类。转基因小鼠体内原位杂交技术证明，EPO的mRNA由环绕在中央静脉周围的上

5.3.2 Liver—The Other Important Organ Producing EPO

The other major site of EPO production in adults is the liver, contributing to approximately 10%-15% to circulating plasma levels. However, hepatic production cannot compensate for loss of the kidneys in case of chronic renal failure. The hepatocytes are the major cell type producing EPO mRNA in the liver. Using in situ hybridization in transgenic mice, epithelial cells surrounding the central veins have been identified to express EPO mRNA. Those cells are supposed to share some

皮细胞表达。目前推测这些细胞与肾脏中产生EPO的细胞具有某些共同特性。另一方面，尽管近来的研究证实在肾脏中有早期的EPO产物，但在胎儿时期肝脏却是产生EPO的主要部位，且EPO生产部位是在怀孕过程中逐渐由肝脏转变到肾脏中的。在大脑、胎盘、睾丸、脾脏以及肺中，EPO的mRNA也都有极少量的表达。

characteristics with the EPO producing cells in the kidney. On the other hand, the liver appears to be the major site of EPO production during fetal life with a shift from the liver to the kidney during gestation, although more recent studies also demonstrated an early production in the kidneys. To a minimal degree expression of EPO mRNA has also been found in the brain, placenta, testes, spleen and lung.

5.4 EPO的生物学作用

5.4.1 红细胞生成之外的促红细胞生成素

EPO是一种特异性作用于红细胞前体细胞的生长因子，对其他细胞系几乎没有作用。但EPO不仅仅是“促红细胞生成素”，可能发挥着更广泛的生物学作用。已经检测到EPO受体存在于神经细胞内，且神经元和星形胶质细胞在含氧量低的情况下有产生EPO的能力。此外，EPO可能在大脑发育中发挥作用。在动物模型中，EPO预处理降低了神经损伤，而给予溶解的EPO受体加重了组织损伤，有人据此提出EPO通过一种特殊的受体-介导的传输机制穿过血-脑屏障。此外，对大脑中动脉闭塞的大鼠全身用药后，EPO极大地减少了梗死的体积。EPO诱导的神经蛋白可能是通过激活特殊的转录因子抑制程序性细胞死亡的结果。因此，EPO可能提供新的中风疗法。

EPO是诱导机体骨髓产生红细胞的激素。虽然EPO主要是由肾脏在感应到血液含氧量下降时产生的，但是总量的10%～15%来自肝脏。一次剂量的EPO作用持续大约1d，但是在约5d后，当被诱导的红细胞增殖

5.4 Bioactive Effects of EPO

5.4.1 EPO Beyond Erythropoiesis

EPO is an erythroid precursors-specific growth factor with almost no effects on other cell lines. It is not just an “EPO” but may play a broader biologic role. The presence of EPO receptors in neural cells and the ability of neurons and astrocytes to produce EPO under hypoxic conditions have been detected. Furthermore, EPO might play a role in brain development. Pretreatment with EPO attenuated neural damage in animal models, but administration of soluble EPO receptor exacerbated tissue damage. It has been suggested that EPO is able to cross the blood-brain barrier via a specific receptor-mediated transport mechanism. Furthermore, systemic administration of EPO after middle-cerebral artery occlusion in rats dramatically reduces the volume of infarction. The EPO induced neuroprotection maybe a result of the inhibition of programmed cell death via activation of specific transcription factors. Thus, EPO might provide new stroke treatment.

EPO is the hormone responsible for inducing red blood cell production by the body's bone marrow. EPO is primarily produced by the kidney when a drop in blood oxygen level is perceived, though 10%-15% of the total EPO produced comes from the liver. A dose of EPO lasts about a day but its effect is seen approximately 5 days later when the red cell proliferation it has induced is mature enough for

到足够成熟而能进入循环时，它的效果还可以观察到。

release into the circulation.

EPO是一种蛋白质，1983年人们第一次阐明了它的氨基酸序列。1985年，利用重组DNA技术制造的商品化人EPO产品问世，用于治疗透析中的贫血症（低红细胞计数）和癌症病人。不久，人们对天然EPO进行了一些合成上的改进，将EPO与糖结合（称为“糖基化”）减缓了其从体内被清除的速度，使其作用能维持得更久。糖基化的EPO有三种形式：α型（在兽药中最常用的形式）、β型（与α型EPO的临床功效相似）和darbepoetin（糖基化程度非常高而且在体内维持得最久）。

EPO is a protein and its amino acid sequence was first mapped out in 1983. By 1985, human EPO was being manufactured commercially using recombinant DNA technology and used for the treatment of anemia (low red blood cell count) in human dialysis and cancer patients. Before long, some synthetic changes improving upon the natural hormone were being added. Binding of EPO with sugars (called “glycosylation”) slows the clearance of EPO from the body thus allowing the hormone to last longer. Glycosylated EPO comes in 3 forms: “α”(the most commonly used type in veterinary medicine), “β”(of similar clinical efficacy to “α”), and darbepoetin (which is particularly heavily glycosylated and lasts the longest).

虽然rhEPO在犬和猫的体内作用比较可靠，但是它与犬和猫自身产生的EPO不一样，并且一些患病宠物的免疫系统能将人的蛋白质当作外源物质识别出来。接触后产生的抗体不仅破坏人的EPO，同样也破坏宠物自身的EPO。

While rhEPO works reliably in dogs and cats, it is not quite the same protein as what a dog or cat would make for itself and in some patients the immune system of the pet can recognize the human protein as foreign. Antibodies may develop in response to its exposure which not only destroy the human EPO but the pet’s natural EPO as well.

在动物中，EPO有一个主要的作用，即治疗慢性肾病引起的贫血。因为大多数EPO是由肾脏产生的，因此受损的肾脏不能产生正常数量的EPO从而导致贫血。EPO注射非常有效且简单易行，在家即可操作。贫血的缓解会使食欲增加、精力更旺盛及生活质量更高。

In animals, EPO has one major use: the treatment of anemia due to chronic renal disease. Since most EPO is produced by the kidney it should make sense that a damaged kidney cannot produce normal amounts of EPO and anemia results. EPO injections are very effective and easy to administer by owners at home. Resolution of anemia leads to better appetite, more energy and higher life quality.

对于任何骨髓不能充分应答的贫血病人使用EPO具有诱人的前景。在大多数骨髓不能应答的情况下，最终结果是多种因素共同产生的，因为在大多数骨髓不能响应的情况下，正常的肾脏早已响应贫血

It is tempting to use EPO for any anemia to which the bone marrow is not adequately responding. Results are mixed with this practice as in most such cases where the marrow is not responsive, there are healthy kidneys producing large amounts of EPO already in response the anemia. Giving yet more EPO is unlikely to be beneficial as, for these cases, lack of marrow

症状而产生大量的EPO，因此在这种情况下，仍然给予更多的EPO是不可能有益的，因为骨髓刺激缺乏不是问题的所在，问题在于骨髓不能应答。

stimulation is not the problem; the problem is that the marrow cannot respond.

5.4.2 EPO的正规用途

EPO适用于治疗由慢性肾脏疾病、脊髓发育不良、癌症的化疗或放疗及其他危重疾病（如心衰）引起的贫血。

5.4.2 Official Uses of EPO

EPO is available as a therapeutic agent in treating anemia resulting from chronic kidney disease and myelodysplasia, from the treatment of cancer (chemotherapy and radiation), and from other critical illnesses (heart failure).

5.5 EPO的使用方法

5.5 How This Medication Is Used

EPO最早经皮下注射给药，每周3次并结合口服补铁剂。最初每周进行红细胞计数的测量并据此对剂量进行调整。大多数病人在治疗的前四周内红细胞计数达到正常水平，在这之后许多病人通过每周注射即可维持正常水平。

EPO is given by subcutaneous injection initially three times a week in conjunction with an oral iron supplement. The red cell count is measured weekly at first and the dose is modified accordingly. Most patients achieve a normal red blood cell count within the first 4 weeks of therapy. Many patients can be maintained on weekly injections after that.

监测红细胞计数最普遍的方法是通过一个称为“血细胞比容”的测试进行，这种测试方法可以在医院内简便地进行操作而无需将样品送到外面的实验室检测。测试得到的数值反映了红细胞所吸收血容量的比例。犬的正常值是35%～60%（取决于年龄和性别），猫的正常值是29%～50%。由于有可能产生针对人源性EPO的抗体，因此重要的是，刚开始表现出贫血症状时不要使用EPO，直到真正地确诊后再开始使用。此外，要注意肾病病人的贫血除EPO缺乏还有很多原因。

The most common method of monitoring red blood cell count is via a test called a “packed cell volume” or “hematocrit” as these tests are readily run in the hospital without sending samples to an outside laboratory. The value obtained reflects the percentage of blood volume taken up by red blood cells. Normal for dogs is approximately 35%-60% (depending on age and gender). Normal for cats is 29%-50%. Because of the potential for antibody production against human origin EPO, is important to withhold use until it is truly warranted rather than beginning it at the first sign of anemia. It is important to note that there is actually quite a bit more to anemia in kidney patients than lack of EPO.

更多的问题尚在研究之中：①肾毒素的累积很可能导致胃肠道溃疡和出血；②红细胞在高浓度循环的肾毒素存在时会失去正常的寿命；③大量

Multiple issues are at work: ① the build-up of renal toxins is very ulcerating to the GI tract and bleeding may result; ② red blood cells do not have normal life spans in the presence of high levels of circulating renal toxins; ③ many renal toxins directly suppress the bone

肾毒素的存在直接抑制骨髓的制造血细胞能力；④肾脏 EPO 的产生不足。

marrow's ability to produce red blood cells; ④ poor production of EPO by the kidney.

5.6 EPO 的推荐剂量

5.6 Recommended Dosage of EPO

EPO 是澄清无色的液体，必须冷藏保存，通过静脉注射或直接皮下注射给药，用于治疗贫血病人有若干种治疗方案。

EPO is a clear, colorless liquid that must be kept refrigerated. It is administered as an intravenous injection or an injection directly underneath the skin. There are several dosing schedules used to treat patients with anemia.

1. *用于治疗癌症相关贫血* EPO 按照每千克体重一定单位进行给药，起始剂量为按每千克体重 150IU，每周给药 3 次，之后可升高至按每千克体重 300IU，每周给药 3 次。

1. To Treat Cancer-Related Anemia EPO is dosed in units per kilogram of body weight, starting at 150 units per kilograms of body weight administered three times per week. This dosage can be increased to 300 units per kilogram of body weight three times per week.

医生往往根据药物类型、品牌、给药方式及病人个人状况选择药物剂量。EPO 可在医生办公室或病人家里进行皮下注射。虽然 2004 年的一项研究表明 EPO 某种形式可以安全地每两周给药 1 次，使其使用更为方便，但比较典型的频率是每周注射 1 次。

Physicians select dosage amount and frequency based on the drug type, brand, method of administration, and the individual patient's situation. EPO may be administered at a physician office or at home subcutaneously. Typically, it has been administered once a week, though in 2004, a study revealed that a form of the drug could safely be dosed every two weeks for some cancer patients, making it more convenient.

2. *用于治疗肾衰病人* EPO 的起始剂量是按每千克体重 50～100IU，每周给药 3 次。这一剂量将根据血液及病人反应情况进行调整。

2. To Treat Patients With Renal Failure EPO starting dose is 50-100 units per kilogram of body weight three times a week. This would be adjusted based on blood work and patient response.

3. *用于正在使用叠氮胸苷的艾滋病人* EPO 的起始剂量是按每千克体重 100IU，每周给药 3 次，共计 8 周。这一剂量将根据血液数量及病人反应情况进行调整。

3. To Treat Aids Patients on the Drug Zid-ovudine EPO starting dose is 100 units per kilogram of body weight three times per week for 8 weeks. This would be adjusted based on blood counts and patient response.

4. *用于术前病人* EPO 的起始剂量是术前 10d 开始、手术当天及术后 4d，每千克体重 300IU。或使用替代方案：术前 3 周，每周一次按每千克体重 600IU，然后手术当天使用第 4 种剂量（按每千克体重 300IU）。

4. To Treat Patients Prior to Surgery EPO starting dose is 300 units per kilogram of body weight per day for 10 days prior to surgery, the day of surgery, and four days after surgery. An alternate schedule is EPO 600 units per kilogram of body weight administered once weekly beginning three weeks before surgery, then a fourth dose on the day of surgery (300 units per kilogram of body weight).

手术病人在开始注射EPO时，需要摄入铁置换剂。

Surgery patients need to take iron replacement with the start of EPO injections.

5.7 使用EPO的注意事项

（1）在使用EPO前，需对血细胞数量进行监测，在使用过程中，应进行规律性监测，以便于医生决定病人是否适应这一治疗以及剂量是否需要调整。

（2）在使用EPO时血压也同样需要监测。病人的血压若不在控制范围之内，则不应使用EPO。

（3）为提高EPO疗效，病人可能会被要求同时口服补铁片。

（4）不推荐患有癌症（如白血病）的病人使用EPO。

（5）曾有对EPO或白蛋白过敏史的病人应事先告知医生。

（6）可能或准备怀孕的病人应在接受EPO之前告知医生。

5.7 Precaution When Using EPO

(1) Blood counts will be monitored before receiving EPO and regularly while on the drug EPO. This allows the doctor to determine if patients are candidates for this treatment and if the dose the patient is receiving needs to be increased or decreased.

(2) Blood pressure should also be monitored regularly while on EPO. Patients who have high blood pressure that is not under control should not use EPO.

(3) Patients may be instructed to take oral iron tablets while on EPO to increase the drug' s effectiveness.

(4) It is not recommended to give EPO to patients who have cancer, such as leukemias, arising from their bone marrow.

(5) Patients with a known previous allergic reaction to EPO or the drug albumin should tell their doctor.

(6) Patients who may be pregnant or trying to become pregnant should tell their doctor before receiving EPO.

5.8 EPO的副作用及与其他药物的相互作用

5.8.1 EPO的副作用

需要考虑的主要不良反应是产生抗体。抗体是针对体内的外源蛋白质产生的。问题是产生的抗体不仅对抗外源（人体）的EPO，也对抗患者本身的EPO。这会导致更严重的贫血，而这种贫血只能用输血来治疗。进行规律性监测以检查红细胞计数的突然降低是至关重要的，因为这是问题刚刚出现的标记，而处于稳定期时可以停止用药。已有人提出darbepoetin（Aranesp®）比其他重组EPO导致这种反应的可能性小，因为它与人体内EPO差异最大。

5.8 Side Effects of EPO and Its Interactions with Other Drugs

5.8.1 Side Effects of EPO

The main adverse event to be concerned about is antibody production. Antibodies are generated in response to an alien protein in the body. The problem is that antibodies that are generated not just against the alien (human) EPO but also against the patient's natural EPO. This makes for a profound anemia which can only be treated with blood transfusions. Regular monitoring to check for a sudden drop in red cell count, is crucial as this is the sign that a problem is incipient and the product can be stopped while there is still time. It has been proposed that darbepoetin (Aranesp ®) is less likely to cause this reaction than other recombinant EPOs since it differs from the human product the most.

一项研究发现，由于产生针对EPO的抗体，用人EPO治疗90d以上后，2/3的犬发展成为顽固性贫血；用人EPO治疗180d以上后，5/7的猫也患上了顽固性贫血。而通常报道的统计数值是有30%发展成顽固性贫血的发生率。

In one study 2 out of 3 dogs treated for more than 90 days with human EPO and 5 out of 7 cats treated for more than 180 days with human EPO developed refractory anemia due to anti-EPO antibodies. A more commonly reported statistic is a 30% incidence in development of refractory anemia.

在中止EPO治疗后，抗体在2～12个月内逐渐减少，红细胞计数恢复到治疗前的水平，在此期间可能需要输血维持病人存活。

After discontinuing EPO, antibodies wane over 2-12 months and the red cell count returns to its pre-treatment level. Blood transfusions may be needed to keep the patient alive during this time.

一些使用EPO的病人发展成高血压。这被认为是当红细胞计数过高而血液变得异常黏稠时产生的。可通过血压药物治疗来控制并且可以调整EPO的剂量来减少红细胞计数。

Some patients on EPO develop high blood pressure. It is thought that this occurs when the red cell count gets too high and the blood becomes abnormally viscous. Blood pressure medications can be used to control this and the EPO dose can be modified to reduce the red cell count.

如果EPO用于升高血红蛋白超过13.0 g/dL，则有引起肾病病人的有害心血管并发症的风险。

EPO is associated with an increased risk of adverse cardiovascular complications in patients with kidney disease if it is used to increase hemoglobin levels above 13.0 g/ dL.

早期用于治疗早产儿贫血时，EPO往往和早产儿视网膜病的发病风险有关，使得人们更加担忧EPO的生血作用会加重视网膜病变。不过，由于贫血本身也会增大视网膜疾病的风险，因此EPO治疗引起的视网膜病变可以简单地视为偶然事件并可归咎于贫血导致的视网膜病变。

Early treatment with EPO correlated with an increase in the risk of Retinopathy of prematurity in premature infants who had anemia of prematurity, raising concern that the angiogenic actions of EPO may exacerbate retinopathy. However, since anemia itself increases the risk of retinopathy, the correlation with EPO treatment may simply be incidental and reflect that anemia induced retinopathy.

5.8.2 EPO与其他药物的相互作用

在给病人注射EPO的时候，同时给予铁剂是非常重要的，因为只有这样，骨髓才有所有必需的原料以生成红细胞。含铁的多维素也是经常使用的，因为可以提供生成红细胞时需要的B族维生素。用于人类治疗时，给予肉毒碱能改善某些个体对EPO的反应作用。

5.8.2 Interactions with Other Drugs

It is very important to administer an iron supplement in conjunction with the EPO injections so that the bone marrow will have all the necessary ‘supplies’ with which to make red blood cells. A multivitamin with iron is typically used as this will also provide the B vitamins needed in red cell production as well. In humans, carnitine administration improves the response to EPO in some individuals.

5.9 EPO 与体育作弊

EPO 属于国际奥委会规定的违禁药物。相对来说，EPO 是最近才进入为了追求胜利光环而造假的体育领域的。在医疗上，EPO 可用于治疗特定的贫血（如有长期的肾脏衰竭而引起的贫血）。从逻辑上来说，既然 EPO 能加速红细胞的生成，它就也能增加细胞携带运输氧气的能力。这个事实就引起了运动领域的注意。

血液回输技术是为了提高运动员的表现，而人为地增加机体内红细胞的数量的做法。过去，这个过程是通过输血来完成的。运动员将“捐献”出一份血而储存起来，3 个星期以后，等机体已经完全补充了损失的血液后，再把抽出来的血液重新输入到机体内。这种情况仅仅发生在大赛前，这样可以为运动员有效地增加一份“额外”的血。这使得在耐力运动（如自行车、划船、长跑、竞走、越野滑雪、冬季两项及铁人三项等）中的运动员由于增加了额外的携氧能力而表现得更好。这种做法现在已经被法律禁止。不仅仅是因为它使得比赛不公平，同时也是因为这项操作具有危险性。

EPO 使得血液回输技术有了一个全新的方式。不需要烦琐地输血，只需要将 EPO 注射到人体内来增加循环红细胞总量。直到 2000 年，准确测试仍是一个难题，因为在实验室合成的重组人 EPO 实际上和人体自然产生的 EPO 形式是完全一样的，而且也没有严格的规定人体内正常

5.9 Cheating with EPO

EPO is in the illicit medication list of the International Olympic Committee (IOC). EPO is a relatively recent entry into the deceitful pursuit of glory. Medically, EPO is used to treat certain forms of anemia (e.g. due to chronic kidney failure). Logically, since EPO accelerates erythrocyte production it also increases oxygen carrying capacity. This fact did not long escape notice of the athletic community.

Blood doping is the process of artificially increasing the amount of red blood cells in the body in an attempt to improve athletic performance. In the past this was accomplished by transfusion. The athlete would “donate” a unit of blood into storage and then 3 weeks later, after the body had completely replaced the blood loss, transfuse the unit back into the body. This would occur just before a big race, effectively giving the athlete an “extra” unit of blood. This enables performance improvements in endurance sports such as cycling, rowing, distance running, race walking, cross country skiing, biathlon, and triathlons. because of the extra oxygen carrying capacity. The practice has been outlawed. Not just because it is unfair but because of the dangers involved.

EPO has put a whole new spin on blood doping. No need for messy transfusions, just shoot up with EPO to increase your circulating erythrocyte mass. Until 2000, accurate testing has been difficult because the recombinant human EPO made in the lab is virtually identical to the naturally occurring form and there are no firmly established normal ranges for EPO in the body. The only previously available route to curtail cheating for sports governing bodies was to ban an athlete if the hematocrit level was too high (e.g.,

的 EPO 范围。目前，唯一可行的确定一个运动员在比赛中有没有作弊的途径就是看运动员的血细胞比容是否过高（如果超过 50% 就视为作弊）。因此，在过去的 10～15 年里一些运动员选择作弊，因为只要他们控制血细胞比容低于 50%，被查出来的可能性就很小。

above 50%). Thus, over the past 10 - 15 years some athletes chose to cheat because, as long as they kept their hematocrit levels below 50%, there seemed little risk of getting caught.

幸运的是，测试技术的进步有望使作弊的趋势得以遏制。现在有一种精确的尿检方法能区分正常的 EPO 和合成 EPO。在 2004 年雅典奥林匹克运动会中，这是唯一检验运动员是否使用了 EPO 的标准技术。这项技术的可靠性帮助解释了一些运动员随后被取消比赛资格的原因。

Fortunately, testing technology has now caught up and promises to stem the tide of abuse. There is now an accurate urine test that can detect the differences between normal and synthetic EPO. This test is now the standard and was the sole means to detect for EPO use in the 2004 Athens Olympic Games. The reliability of this test helps explain the cascade of athletes who have been caught and, subsequently, banned from competition.

2002 年盐湖城冬季奥林匹克运动会上，加利福尼亚大学奥林匹克分析实验室创始人及时任主任——医学博士 Don Catlin，第一次报道了体育竞技中使用 EPO 的一种形式 α-darbepoetin。同年开始，美国体育局对 EPO 的检测就有了尿检或称“直接”测试；而从 2006 年起，奥运会的 EPO 检测包含了血检和尿检。在 2008 年北京奥运会中第一例的兴奋剂检查呈阳性的西班牙自行车选手玛丽亚·莫里诺，就是被发现使用了可增加耐力的 EPO 而被纪律委员会取消参赛资格的。2010 年，专业自行车手 Floyd Landis 承认在其职业生涯的大部分时间中使用了增强运动功能的药物，包括 EPO。

In 2002, at the Winter Olympic Games in Salt Lake City, Don Catlin, MD, the founder and then-director of the UCLA Olympic Analytical Lab, reported α-darbepoetin, a form of EPO, for the first time in sports. Since 2002, EPO tests done by U.S. sports authorities have consisted of only a urine or “direct” test. From 2006, EPO tests at the Olympics were conducted on both blood and urine. In 2008, Spanish cyclist Maria Moreno became the first athlete to be expelled from the Beijing Games by the disciplinary commission after returning a positive drug test for EPO which was believed to boost endurance. In 2010, Floyd Landis admitted using performance-enhancing drugs, including EPO, throughout the majority of his career as a professional rider.

5.10 使用 EPO 的危险性

5.10 Why EPO Is Dangerous

之所以输入 EPO 及血液回输技术具有危险性是因为这样会增加血液

The reason why EPO and transfusion blood doping is dangerous is because of increased blood viscosity. Basically, whole blood consists of red

黏度。基本上来说，全血包括红细胞和血浆（水、蛋白等）。红细胞在全血中所占到的比例被称为血细胞比容。一个低的血细胞比容就意味着较稀（淡）的血液，而一个高的血细胞比容则意味着较浓（稠）的血液。血细胞比容超过某个数值就会引起毛细血管的拥堵甚至是堵塞。如果发生在大脑，则会引起脑中风。如果发生在心脏，则会引起心脏病发作。不幸的是，这在使用了EPO的多名运动员身上发生了。

blood cells and plasma (water, proteins, etc.). The percentage of whole blood that is occupied by the red blood cells is referred to as, the hematocrit. A low hematocrit means dilute (thin) blood, and a high hematocrit mean concentrated (thick) blood. Above a certain hematocrit level whole blood can sludge and clog capillaries. If this happens in the brain it results in a stroke. In the heart, a heart attack. Unfortunately, this has happened to several elite athletes who have used EPO.

对于长期锻炼的运动员来说，EPO的危害更为严重。一个耐力良好的运动员比一个久坐的人的抗脱水能力要好。机体通过几种方式来完成这个过程，但是一个最重要的因素是能在安静的时候“留住”更多的水。循环的全血是这个过程完成的场所，在整个过程当中发挥着一个“水库”的作用。在一个耐力要求高的运动过程中，当液体流失时，水从血流中转移出来（这样，血细胞比容就上升了）。如果一个人已经用人为的手段（注射EPO）提高了血细胞比容，那么很快就会出现这个问题——血管堵塞。

EPO use is especially dangerous to athletes who exercise over prolonged periods. A well-conditioned endurance athlete is more dehydration resistant than a sedentary individual. The body accomplishes this by several methods, but one key component is to “hold on” to more water at rest. Circulating whole blood is one location in which this occurs and, thus, can function as a water reservoir. During demanding exercise, as fluid losses mount, water is shifted out of the blood stream (hematocrit rises). If one is already starting with an artificially elevated hematocrit then you can begin to see the problem—it is a short trip to the critical “sludge zone.”

EPO的另外的危险包括在睡觉时猝死（在过去的15年当中，这使得大约18名职业自行车运动员失去了生命）和机体内形成直接对抗EPO的抗体；在后一种情况中，机体由于反复对抗EPO的注射，就会发展出现贫血症。

Additional dangers of EPO include sudden death during sleep, which has killed approximately 18 pro cyclists in the past fifteen years, and the development of antibodies directed against EPO. In this later circumstance the individual develops anemia as a result of the body’s reaction against repeated EPO injections.

（编者　潘　扬，张　芳）

(Compiled by Yang Pan & Fang Zhang)

6 激素制剂

6 Hormons

激素是机体产生的一组最重要的调节分子。内分泌激素的定义是机体的特定腺体合成并释放的一种物质，经过与远程敏感细胞内或细胞表面的受体相互作用而使靶细胞发生变化。临床上所有用于治疗的蛋白激素都是内分泌激素，例如胰岛素、胰高血糖素、生长激素、促性腺激素与降钙素等。

Hormones are amongst the most important groups of regulatory molecules produced by the body. The endocrine hormones are defined as a substance synthesized and released from a specific gland in the body which, by interacting with a receptor present in or on a distant sensitive cell, brought about a change in that target cell. All of the hormones used therapeutically in clinic fit into this grouping, such as insulin, glucagon, growth hormone, gonadotrophins and Calcitonin.

6.1 激素药物的制备方法

治疗用蛋白激素的发展历程可分为三个阶段：前重组DNA时期、重组DNA时期和未来的合理设计蛋白激素的时期。不同的时期激素的制备方法不同。前重组DNA时期是通过动物体内提取获得激素药物；重组DNA时期是通过基因工程技术，使用微生物发酵获得激素产品，这开辟了治疗和利用激素家族的新领域；设计蛋白激素的时期是通过合理设计激素的结构，进行氨基酸序列改变，由此将导致其药代动力学和生物学活性改变。

6.1 The Preparation Methods of Therapeutic Hormones

The progress of therapeutic hormones can be divided into three stages: the prerecombinant DNA era, the recombinant DNA era and rational design of protein hormones. The different era has the differently representive preparation methods of therapeutic hormones: hormones originally are isolated from animal sources (traditional method) during prerecombinant DNA era, protein hormones are produced through genetically engineering technology during the recombinant DNA era, which has allowed greater access to treatment and the exploration of new uses for familiar hormones. The development of designer hormones represents the next step with alterations in amino-acid sequence resulting in changes in pharmacokinetics and differences in biological activity.

6.1.1 动物体内提取激素药物及其应用限制

25年前，从动物和人体中提纯的肽和蛋白类激素已经应用于临床（表5-14）。在临床中，一些小片段的人工合成肽也有生物活性。如果某种动物蛋白的氨基酸序列（与人的相近）在人体内有生物活性，那么可以从屠宰的食用动物中提取。常见的例子如胰岛素（猪和人的胰岛素只相差

6.1.1 Traditional Method of Hormones Preparations and Application Limits

Twenty-five years ago, the peptide and protein hormones used clinically were purified from animal or human sources (Table 5-14). In a few cases, where a smaller fragment had biological activity, synthetic forms were also available. If the amino acid sequence of the animal protein was such that the protein had biological activity in humans, its purification from animals slaughtered for food was possible. Common examples were insulin (where only one amino acid is different between porcine and human insulin and three amino acids difference between bovine and human insulin),

1个氨基酸，牛和人的胰岛素只相差3个氨基酸）、胰高血糖素、甲状腺旁素（PTH）。尽管这些生物制品有效，但常有其他动物蛋白污染，因此使用这些制品会产生相应的抗体。有的时候，高效价抗体与这些蛋白的结合能力也很高，导致药物失活。

glucagon, and parathyroid hormone (PTH). Although effective, these preparations were often contaminated with other animal proteins, and their use could lead to the development of antibodies. In some cases, where high antibody titers were accompanied by high binding capacities, the drug could become ineffective.

表 5-14 一些最初从动物体内分离的激素药物

产品	适应证	来源
胰岛素	糖尿病	猪 / 牛的胰腺组织
胰高血糖素	用于纠正胰岛素诱导的低血糖	猪 / 牛的胰腺组织
人生长激素	侏儒症	人脑垂体
促卵泡激素	治疗不孕不育	绝经妇女的尿液
人绒毛膜促性腺激素	治疗不孕不育	怀孕妇女的尿液
其他（促性腺激素释放激素、催产素、抗利尿激素）	多种	大多来源于脑垂体腺

传统治疗用的胰高血糖素是从牛或猪胰腺组织中经过层析纯化而成（牛、猪与人的胰高血糖素的结构是相同的，这样就不会直接造成免疫系统并发症）。商业用胰高血糖素制剂的配方中还有乳糖和氯化钠，冻成干粉状态；给患者皮下注射或静脉注射，每次0.5～1.0IU（0.5～1.0mg 干粉）。

Traditionally, glucagon preparations utilized therapeutically are chromatographically purified from bovine or porcine pancreatic tissue (the structure of bovine, porcine and human glucagon is identical, thus eliminating the possibility of direct immunological complications). Such commercial preparations are generally formulated with lactose and sodium chloride and sold in freeze-dried form; 0.5-1.0 units of glucagon (approximately 0.5-1.0 mg freeze-dried hormone) are administered to the patient by s.c. or i.m. injection.

人的蛋白激素与牛或猪的激素的序列有差异，因此应用于临床的激素最好直接来源于人体组织。生长激素（Somatotropin/growth hormone，GH）要从死者的垂体中提取，来源少，因此限制了它的应用。1985年，由于传染性物质污染了来自垂体的HGH，引起了克雅病（Creutzfeldt-Jakob disease, CJD），从此终止了从尸体的垂体中提取HGH。在这一年，一位接受HGH治疗长达15年的年轻人死于CJD，调查人员认为是感染的垂体提取物导致了他的感染（朊病毒诱

For protein hormones where there was less sequence homology with the equivalent bovine or porcine protein, supplies for clinical use were derived directly from human tissues. Somatotropin or growth hormone (GH) was purified from cadaver pituitaries. The limitations on GH use were due to its scarcity. In 1985, concerns about contamination of human derived pituitary HGH with the infectious agent causing Creutzfeldt-Jakob disease (CJD) resulted in the discontinuation of its use in children with growth failure due to lack of endogenous GH secretion. In this year, a young man who had received HGH therapy some 15 years previously died from CJD, which, investigators concluded, he had contracted from infected pituitary extract. At least an additional 12 CJD cases suspected of being caused in the same way have

发 CJD）。随后又出现了至少 12 例可能是相同途径感染 CJD 的病例报道。幸运的是，当时已经开发了几种重组 HGH（rHGH）制剂。现在临床上使用的所有 HGH 制剂都是重组来源。由于现在已经没有必要节省来源稀少的 HGH 供治疗儿童使用，医生可以对缺乏生长激素的成年人研究 GH 的疗效。这些研究表明 GH 改善了人体的组成（降低了脂肪组织，增加了肌肉组织），并且降低了患心血管病的危险。

subsequently been documented. Fortunately, several recombinant HGH (rHGH) preparations were coming onstream at that time, and now all HGH preparations used clinically are derived from recombinant sources. Because it is no longer necessary to conserve scarce supplies of HGH for use in children, physicians are able to study GH effects in adults with growth hormone deficiency. These studies have shown that GH improves body composition (decreases fat mass and increases lean body mass) and decreases risk factors for cardiovascular disease.

其他人源的多肽激素包括人绒毛膜促性腺激素、从胎盘中提取的黄体激素（LH）类似物以及更年期促性腺激素（HMG）从尿中提取的 LH 和促卵泡激素（FSH）的混合物，这些激素来源充足，足够临床应用。虽然当时已知从死者垂体中提取的蛋白有可能被传染性病原体污染，但人们仍一直致力于通过提纯促性腺激素改进产品的效果。随着重组 DNA 技术的出现，这些工作随之结束。目前，重组的 LH、FSH 已用于诊断、治疗隐睾症以及体外受精。

Other peptide hormones purified from human sources include human chorionic gonadotropin, the luteinizing (LH) analogue isolated from human placenta, and human menopausal gonadotropins, a mixture of LH and follicle stimulating hormone (FSH) isolated from urine. The supply of these hormones was sufficient for clinical application. Even before the recognition that proteins purified from cadaver pituitaries could be contaminated with infectious agents, efforts were being made to purify the gonadotropins to increase potency. These efforts came to an end with the intro even before the recognition that proteins purified from cadaver pituitaries could be contaminated with infectious agents, efforts were being made to purify the gonadotropins to increase potency. These efforts came to an end with the introduction of recombinant DNA technology. Today recombinant LH and FSH are available for diagnostic use, the treatment of cryptorchidism, and *in vitro* fertilization.

一些小肽如促性腺激素释放激素类似物阿拉瑞林（Alarelin，一种修饰的 9 肽激素）、生长激素抑制素（一种 14 肽激素）可以化学合成，但是大分子的合成常常产量低，并且会被相似结构的肽污染。因此，采用基因重组的方法是获得这些大分子蛋白质激素的合理和有效的途径。

While small peptides, such as Alarelin and somatostatin, a modified nonapeptide and tetradecapeptide respectively, could be synthesized chemically, synthesis of larger molecules resulted in low yields of desired peptide, usually contaminated with peptides of similar structure. Therefore, it's a reasonable and effective method for protein hormones production through recombinant DNA technology.

6.1.2 基因重组技术获得激素产品

随着重组 DNA 技术制备蛋白质方法的出现，蛋白质激素的生产情

6.1.2 Recombinant DNA Technology for Hormones Preparations

The situation changed with development of methods for synthesizing proteins by recombinant

况得到了彻底的改观。现在市场上大量的蛋白质激素类药物都是利用各种不同的重组表达系统通过遗传工程的方法制造的。尽管可供利用的蛋白质表达系统有多种，但目前为止进入市场的大多数蛋白质都是通过重组大肠杆菌或重组哺乳动物细胞制造的。这些重组系统一般是通过将目的蛋白的基因或 cDNA 转入经过鉴定的细胞株中。这些细胞株包括大肠杆菌和中国仓鼠卵巢细胞；基因或 cDNA 的转运通过合适的表达质粒实现（表 5-15）。

DNA techniques. The bulk of biopharmaceuticals currently on the market are produced by genetic engineering using various recombinant expression systems. While a wide range of potential protein production systems are available, most of the recombinant proteins that have gained marketing approval to date are produced either in recombinant *Eschericia coli* (*E. coli*) or in recombinant mammalian cell lines. Such recombinant systems are invariably constructed by the introduction of a gene or cDNA coding for the protein of interest into a well characterized strain of the chosen producer cell. Examples include *E. coli* and Chinese hamster ovary (CHO). Gene/cDNA transfer is normally achieved by using an appropriate expression plasmid, or other standard gene-manipulating techniques (Table 5-15).

表 5-15　市场上蛋白质激素药物的表达系统

产品	来源	产品	来源
胰岛素	大肠杆菌	人绒毛膜促性腺激素	大肠杆菌
胰高血糖素	大肠杆菌	其他（促性腺激素释放激素、催产素、抗利尿激素）	大肠杆菌或中国仓鼠卵巢细胞
人生长激素	大肠杆菌		
促卵泡激素	中国仓鼠卵巢细胞		

1979 年，Goeddel 等发表了两篇文章，描述了将人的胰岛素基因和生长激素的基因导入大肠杆菌中，然后表达相应的蛋白质，并且应用于临床实验。1982 年，胰岛素成为第一个临床用重组蛋白激素在美国上市。

In 1979 Goeddel *et al.* published two papers describing the introduction of the genes for human insulin and human GH into *E.coli*. This was followed by expression of the respective proteins, which were then taken into clinical trials. In 1982 insulin became the first recombinant protein hormone marketed for clinical use in the United States.

1. 基因工程生产人生长激素　人生长激素（HGH，somatotrophin）是垂体前叶合成的一种多肽激素。它能促进正常机体生长、泌乳并影响细胞代谢的多个方面。

1. Recombinant Human Growth Hormone　Human growth hormone (HGH, somatotrophin) is a polypeptide hormone synthesized in the anterior pituitary. It promotes normal body growth and lactation and influences various aspects of cellular metabolism.

（1）HGH 分子结构。成熟的 HGH 含有 191 个氨基酸残基，分子质量为 22 kDa，含有两个特征性的链内二硫键。HGH 的 mRNA 也可经过另一种方式的剪接，形成缩短的

(1) Molecular Structure of HGH　Mature HGH contains 191 amino acid residues and displays a molecular mass of 22kDa. It also contains two characteristic intra-chain disulphide linkages. HGH mRNA can also undergo alternate splicing, yielding a shortened GH molecule (20kDa), which appears to display biological activities

GH 分子（20kDa），其生理活性与 22 kDa 的分子没有明显不同。

indistinguishable from the 22 kDa species.

HGH 的种族特异性非常明显。只有从其他灵长类动物体内分离到的 GH 在人体内才有生理活性（这就排除了牛 / 猪生长激素应用于人体治疗的可能性）。

HGH displays significant, although not absolute, species specificity. GHs isolated from other primates are the only preparations biologically active in humans (this precluded the earlier use of bovine/porcine preparations for medical use in humans).

GH 通过与特异性细胞表面受体结合而诱导其特征性的生理效应。人的 GH 受体是一个含 620 个氨基酸残基的单链穿膜多肽。序列分析表明它属于造血因子受体超家族（该家族包括 IL、GM-CSM 和 EPO 等受体）。X 射线结晶学分析表明 GH 同时与两个受体分子的胞外区结合，促进两个受体聚合并磷酸化，导致细胞内物质磷酸化和蛋白激酶 C 激活。

GH induces its characteristic biological effects by binding to a specific cell surface receptor. The human receptor is a single chain 620 amino acid transmembrane polypeptide. Sequence analysis indicates it is a member of the haemopoietic receptor superfamily (which includes receptors for several ILs, GM-CSF and EPO). X-ray crystallographic analysis shows that GH binds simultaneously to the extracellular domains of two receptor molecules, effectively promoting receptor dimerization and inducing autophosphorylation, as well as phosphorylation of additional cellular substances and protein kinase C activation.

（2）重组人生长激素的制备。20 世纪 80 年代早期，rHGH 首先在大肠杆菌中生产。最初的重组制剂与天然人生长激素的不同仅在于多了一个蛋氨酸（因为在基因的开始部位插入了 AUG 启动子），但这个额外的甲硫氨酸并不影响产品的生物活性。临床试验表明，重组 HGH 提高了人体内胰岛素样生长因子 - I 的水平以及生长素缺乏儿童的生长速率。重组 HGH 的下一个重要步骤是生产不带额外甲硫氨酸的分子。1987 年，Chang 等构建了一个含大肠杆菌热稳定肠毒素 Ⅱ 基因的分泌信号的编码区的杂合基因。当培养含有该杂合基因的细胞时，90% 的 HGH 分泌到细胞质中，这也简化了纯化步骤。

(2) The Preparation of Recombinant HGH

Recombinant HGH was first produced in *E. coli* in the early 1980s. The initial recombinant preparations differed from the native human hormone only in that they contained an extra methionine residue (due to the AUG start codon inserted at the beginning of the gene). The additional methionine did not affect the biological activity of the product, and clinical trials were soon underway. Recombinant GH increased levels of insulin-like growth factor-I (IGF-I) in adults and increased growth rates in GH-deficient (GHD) children. The next step in the saga of recombinant HGH was production of a molecule that did not have the extra methionine. In 1987, Chang *et al*. prepared a hybrid gene that included the secretion signal-coding region of *E. coli* heat-stable enterotoxin II. When cells containing this hybrid were grown in culture, 90% of the HGH produced was secreted into the periplasm. This also simplified the purification process.

胰蛋白酶多肽图谱、氨基酸序

In vitro analysis, including tryptic peptide

列分析与比较免疫检测等体外分析方法都表明重组产品与天然HGH分子相同。人体临床试验也证明重组HGH引导的生理反应与天然激素相同。基因工程科学家首次纯化rHGH的路线图见图5-13。工程级纯化方法与实验室纯化方法大致相同。

mapping, amino acid analysis and comparative immunoassays, show the native and recombinant form of the molecule to be identical. Clinical trials in humans have also confirmed that the recombinant version promotes identical biological responses to the native hormone. rHGH was first purified (on a lab scale) by Genentech scientists using the strategy outlined in Figure 5-13. A somewhat similar strategy is likely used in its process-scale purification.

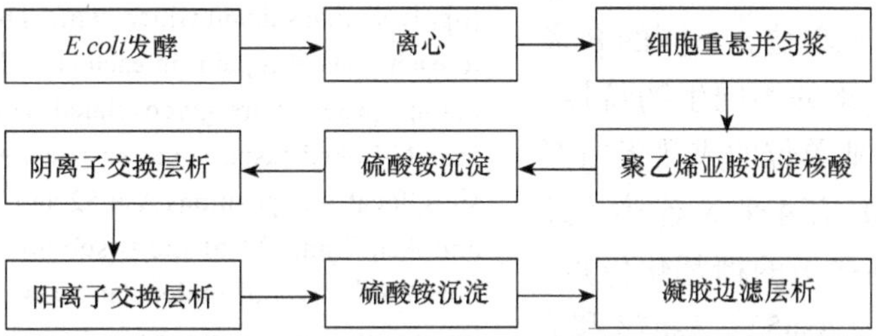

图5-13 在大肠杆菌中生产人生长激素

2. 基因工程生产促性腺激素

促性腺激素是以性腺为主要靶标的激素家族（表5-16）。它们直接或间接调节生殖功能，在某些情况下，还调节第二性征的发育。在这个激素家族中，任何成员的分泌不足都会影响生殖功能，这可以通过补给外源激素进行治疗。大多数促性腺激素是由垂体合成的，某些由生殖器官及其相关组织合成。

2. Recombinant Gonadotrophins The gonadotrophins are a family of hormones for which the gonads represent their primary target (Table 5-16). They directly and indirectly regulate reproductive function and, in some cases, the development of secondary sexual characteristics. Insufficient endogenous production of any member of this family will adversely affect reproductive function, which can be treated by administration of an exogenous preparation of the hormone in question. Most gonadotrophins are synthesized by the pituitary, although some are made by reproductive and associated tissues.

表5-16 促性腺激素的合成部位和主要生物学作用

促性腺激素	合成位点	主要作用
促卵泡激素（FSH）	垂体	刺激卵泡生长（雌性）、促进精子发生（雄性）
黄体激素（LH）	垂体	诱导排卵（雌性）、合成睾丸激素（雄性）
绒毛膜促性腺激素（CG）		维持妊娠雌性的黄体
抑制素	性腺	抑制FSH合成，可能是肿瘤抑制剂
刺激素	性腺	刺激FSH合成

（1）促性腺激素的分类。促性腺激素主要包括促卵泡激素（FSH）、黄体激素（LH）和人绒毛膜促性腺激素（HCG）、抑制素和刺激素等。

(1) Classes of Gonadotrophins Gonadotrophins is composed of follicle stimulating hormone (FSH), luteinizing hormone (LH) and human chorionic gonadotrophin (hCG), inhibins and activins.

促卵泡激素（FSH）与黄体激素（LH）在男性尤其是女性中，在其生殖功能的发育与维持方面发挥着关键作用。怀孕的女性分泌人绒毛膜促性腺激素（HCG），HCG 的作用是在妊娠早期维持胚胎发育的支持系统。

Follicle stimulating hormone (FSH) and luteinizing hormone (LH) play critical roles in the development and maintenance of male and particularly female reproductive function. Human chorionic gonadotrophin (hCG), produced by pregnant women, plays a central role in maintaining support systems for the developing embryo during early pregnancy.

这三种激素都是异源二聚体，含有一个相同的 α 多肽亚单位和一个各不相同的 β 多肽亚单位，不同的 β 多肽亚单位造成激素的不同生物特性。每种激素的两个亚单位的成熟蛋白都是糖基化的。FSH 有 4 个 N 连接（天冬酰胺或 Asn 连接）的糖基化位点，位于 α 亚单位的 Asn52、Asn78 和 β 亚单位的 Asn7、Asn24。糖类占了整个激素分子量的 30%。从结构上看，这种附加的寡糖实质上是异质性的，尤其是唾液酸基团和硫酸基团存在很大不同。这就是促性腺激素电荷异质性的结构基础。

All three are heterodimeric hormones, containing an identical α-polypeptide subunit and a unique β-polypeptide subunit which confers biological specificity to each gonadotrophin. In each case, both subunits of the mature proteins are glycosylated. Human FSH displays four N-linked (asparagine or Asn-linked) glycosylation sites, located at positions Asn 52 and 78 of the α -subunit and Asn 7 and 24 of the β-subunit. Some 30% of the hormone's overall molecular mass is accounted for by its carbohydrate component. Structurally, the attached oligosaccharides are heterogenous in nature, varying in particular in terms of the content of sialic acid residues and sulphate groups present. This represents the structural basis of the charge heterogeneity characteristic of this (and other) gonadotrophins.

寡糖组分对促性腺激素的生物合成、分泌、血浆半衰期和效力起直接且关键的作用。α 亚单位连接的糖组分对二聚体的装配和稳定、激素的分泌和信号转导起重要作用。β 亚单位连接的糖组分在二聚体的装配和激素的分泌方面也发挥作用，但在激素从血液循环中清除过程中其作用更为突出。

The oligosaccharide components play a direct and central role in the biosynthesis, secretion, serum half-life and potency of the gonadotrophins. The sugar components attached to the α-subunits play an important role in dimer assembly and stability, as well as hormone secretion and possibly signal transduction. The sugars associated with the β-subunit, while contributing to dimer assembly and secretion; appear to play a more prominent role in clearance of the hormone from circulation.

基因重组方法生产促性腺激素时，糖基的作用就具有重要意义。目前，已经利用重组系统（动物细胞系）生产了几种临床使用的产品。尽管重组产物的糖基与天然激素的有所不同，但这些细微差异对临床应用没有不良影响。

The functional effects of glycosylation take on added significance in the context of producing gonadotrophins by recombinant means. Several are now produced for clinical application in recombinant (animal cell line) systems. While the glycosylation patterns observed on the recombinant molecules can vary somewhat in composition from those associated with the native hormone, these slight differences bear no negative influence upon their clinical applicability.

（2）重组促性腺激素的制备。目

(2) The Preparation of Recombinant Gonadotrophins

Gonadotrophins are now also produced by recombinant

前促性腺激素可以通过重组 DNA 技术生产。几个物种的编码促性腺激素的基因或 cDNA 已经确定，并在不同重组宿主系统中表达，特别是哺乳动物细胞系。在中国仓鼠卵巢细胞（CHO）中生产的 rHFSH 在临床上证明有效。其氨基酸序列与人的分子一致，但糖的组分略有不同。给人应用这种重组制剂时，人体能够很好地接受，没有出现意想不到的副作用。它不会引起免疫反应，血浆半衰期与天然激素相同。对于促性腺激素不足导致的性腺机能障碍的女性，rHFSH 能够有效地促进其卵泡生长；对于患同种疾病的男性，rHFSH 也能有效地治疗。在欧洲，rHFSH 是第一批通过临床应用程序被批准用于临床的生物药品。

DNA technology. The genes or cDNAs coding for gonadotrophins from several species have been identified and expressed in various recombinant host systems, particularly mammalian cell lines. rHFSH produced in CHO cells has proved clinically effective. While exhibiting an amino acid sequence identical to the human molecule, its carbohydrate composition differs slightly. When administered to humans, the preparation is well tolerated and yields no unexpected side effects. It does not elicit an immunological response, and its plasma half-life is similar to the native hormone. rHFSH has proved efficacious in stimulating follicular growth in females suffering from hypogonadotrophic hypogonadism and is effective in the treatment of males suffering from similar conditions. rHFSH was amongst the first biopharmaceutical substances to be approved for general medical use in Europe via the centralized application procedure.

应用于临床的两种重组促性腺激素为 Ovitrelle 和 Luveris。Ovitrelle 是重组 HCG 产品的商品名称。这种产品是用编码 HCG α 亚单位和 β 亚单位的基因共转染 CHO 细胞系生产的。下游过程包括数种层析和超滤步骤，最终产品是冻干状态。该产品每瓶含有 285μg 活性 HCG，有效期为 2 年。使用前用水溶解后立即注射。

Two of the more recent such approvals are that of Ovitrelle and Luveris. Ovitrelle is the trade name given by Serono to its recombinant hCG-based product. The producer is an engineered CHO cell line that has been co-transfected with the genes coding for both the HCG α- and β-subunits. Downstream processing entails a combination of several chromatographic and ultrafiltration steps and the final product is presented in freeze-dried form. Each vial of product contains 285μg of active substance (HCG) and the product has been assigned a 2 year shelf-life. It is reconstituted with water for injections (WFI) immediately before use.

Luveris 是重组（人）黄体激素（rLH），与 FSH 联合使用刺激 LH 和 FSH 严重不足的女性卵泡发育。同 Ovitrelle 一样，Luveris 是通过编码 LH α 亚单位和 β 亚单位的基因共转染 CHO 细胞系而生产的。经过最初的超滤（浓缩）步骤后，利用数种层析步骤纯化激素。最后再经过超滤（除去病毒的安全步骤），将产品装入玻璃瓶中冷冻干燥。产品中除了活性物质，还含有磷酸盐缓冲液、蔗糖和

Luveris is a recombinant (human) lutcinizing hormone (rLH) used in conjunction with FSH to stimulate follicular development in women displaying severe LH and FSH deficiency. As in the case of Ovitrelle, the producing source is an engineered CHO cell line, co-transfected in this case with the genes coding for the LH α- and β-subunits. After an initial ultrafiltration (concentration) step, the hormone is purified using a combination of several chromatographic steps. After a final ultrafiltration step (viral removal safety step), the product is filled into glass vials and freeze-dried. In addition to the active substance, the product contains a phosphate buffer, sucrose and polysorbate 20 as excipients.

吐温 20 等赋形剂。

重组 DNA 技术也可用于修饰天然促性腺激素，产生某些功能特征增强的变体（如保质期延长、血浆半衰期延长、药效更强等）。例如，将 HCG 的 β 亚单位羧基末端肽融合到 FSH 的 β 亚单位羧基末端，形成的杂交结构具有与 FSH 相同的生理活性，但血浆半衰期比 FSH 长。

显然，FSH 与 LH 的重组产物便于生产纯的激素制剂，不会有第二种促性腺激素污染。这种制剂也有利于进一步明确每一种激素在生殖功能方面的确切作用。

3. 基因工程生产胰高血糖素　胰高血糖素也可用基因重组方法大量生产。这类重组胰高血糖素所用菌株为酿酒酵母。生产过程大致为，在上游工程（有氧发酵）结束后，将 pH 上调，溶解沉淀的产物（在 pH 3～9.5 的水相溶液中胰高血糖素是不溶的），通过离心去除酵母。通过一系列的进一步沉淀和高效层析步骤收获并纯化胰高血糖素。另外，也可用大肠杆菌工程菌株生产重组胰高血糖素。

4. 基因工程生产促甲状腺激素　在结构上，促甲状腺素（TSH）属于促性腺激素家族。但从功能上看，它的靶标是甲状腺。与其他促性腺激素一样，TSH 是一个异源二聚体糖蛋白，含有一个通用的 α 亚单位和独特的 β 亚单位。TSH 含有 118 个氨基酸残基，富含半胱氨酸，含有一个 N 链接糖基化位点（Asn23）。

商业重组 TSH 产品（商品名为 Thyrogen，国际性的非专利名称为促甲状腺素 α）是把携带编码 TSH α 亚单

Recombinant DNA technology also facilitates genetic modification of native gonadotrophins in the hope of generating variants displaying some enhanced functional characteristic (e.g. extended shelf-life, longer plasma half-life, greater potency, etc.), e.g. a carboxyl terminal peptide of the HCG β-subunit has been fused to the carboxyl terminus of the FSH β-subunit. The resulting hybrid structure displays identical biological activity to that of FSH, but with an extended plasma half-life.

The recombinant production of FSH and LH also obviously facilitates the generation of a pure preparation of any one gonadotrophin, without contamination by a second gonadotrophin. Such preparations continue to facilitate a greater understanding of the precise role each gonadotrophin plays in reproductive function.

3. Recombinant Glucagon Glucagon preparations produced via recombinant means have also become available. Such product was produced by using an engineered *Saccharomyces cerevisiae* strain. Upstream processing (aerobic batch-fed fermentation) is followed by an upward adjustment of media pH in order to dissolve precipitated product (glucagon is insoluble in aqueous-based media between pH 3-9.5). This facilitates subsequent removal of the yeast by centrifugation. Glucagon is then recovered and purified from the media by a series of further precipitation as well as high-resolution chromatographic steps. Recombinant glucagon product can be produced by using an engineered *E. coli* strain.

4. Recombinant Thyroid Stimulating Hormone Structurally, thyroid-stimulating hormone (TSH or thyrotrophin) is classified as a member of the gonadotrophin family, although functionally it targets the thyroid gland. As with other gonadotrophins, it is a heterodimeric glycoprotein displaying a common α-subunit and a unique β-subunit. It consists of 118 amino acids, is particularly rich in cysteine residues and contains one N-linked glycosylation site (Asn 23).

The commercial recombinant TSH product (trade name, Thyrogen; international nonproprietary name; thyrotrophin-α) is produced in a CHO cell line co-transfected with plasmids harbouring the DNA

位和β亚单位的DNA序列的质粒转染CHO细胞而生产的。这些细胞培养在动物细胞培养生物反应器中。收获并浓缩（超滤）后，TSH经过层析纯化，溶解在含甘露醇和氯化钠赋形剂的磷酸盐缓冲液中，最终浓度为0.9 mg/ml。经过过滤除菌后装入玻璃小瓶内，冷冻干燥。成品在2～8℃可以保存3年。

5. 基因工程生产鲑降钙素 所有物种合成的降钙素均为含有32个氨基酸残基的单链多肽。分子质量为3500Da。鲑的降钙素与人的降钙素有9个氨基酸残基不同，但比人体内的天然降钙素功能强100倍，因此备受关注。这是由于鲑降钙素与受体的亲和力大而且在体内更加不易降解，因此鲑降钙素也用于临床。传统的临床降钙素制剂是通过化学合成的，现在重组形式的降钙素也获得销售批准。重组降钙素是在大肠杆菌工程菌株中生产的。从结构上看，鲑的降钙素C端被酰胺化。在许多多肽激素中，都存在C端由氨基基团（$—CONH_2$）替代羧基基团的现象。这个氨基基团一般是维持激素生物活性/稳定性所必需的，但大肠杆菌不能进行转录后修饰。重组的降钙素的酰胺化是在体外，由同样是在工程CHO细胞系中生产的α-酰胺化酶催化完成的。

6.1.3 合理设计蛋白激素

重组胰岛素的出现本身并未对糖尿病患者疾病本身的控制产生重大改变。但是，控制高血糖可以防止糖尿病的某些长期的并发症（依据1993年糖尿病控制和并发症试验研究小组的结论），导致人们怀着更大的兴趣寻找理想的胰岛素摄取量。用过于严格的

sequences coding for the α - and β-TSH subunits. The cells are grown in batch harvest animal cell culture bioreactors. Following recovery and concentration (ultrafiltration), the TSH is chromatographically purified and formulated to a concentration of 0.9 mg/ml in phosphate buffer containing mannitol and sodium chloride as excipients. After sterile filtration and aseptic filling into glass vials, the product is freeze-dried. Finished product has been assigned a shelf-life of 3 years when stored at 2-8℃.

5. Recombinant Calcitonin Calcitonin produced by virtually all species is a single-chain, 32 amino acid residue polypeptide, displaying a molecular mass in the region of 3500 Da. Salmon calcitonin differs in sequence from the human hormone by nine amino acid residues. It is noteworthy, however, as it is approximately 100-fold more potent than the native hormone in humans. The higher potency appears due to both a greater affinity for the receptor and greater resistance to degradation *in vivo*. As such, salmon, as opposed to human calcitonin, is used clinically. Traditional clinical preparations were manufactured by direct chemical synthesis, although a recombinant form of the molecule has now gained marketing approval. The recombinant calcitonin is produced in an engineered *E. coli* strain. Structurally, salmon calcitonin displays C-terminal amidation. A C-terminal amide group ($—CONH_2$) replacing the usual carboxyl group is a characteristic feature of many polypeptide hormones. If present, it is usually required for full biological activity/stability. As *E. coli* cannot carry out post-translational modifications, the amidation of the recombinant calcitonin is carried out in vitro using an α -amidating enzyme which is itself produced by recombinant means in an engineered CHO cell line.

6.1.3 Rational Design of Protein Hormones

The introduction of recombinant insulin by itself did not lead to major changes in the management of patients with diabetes mellitus. However, demonstration of the importance of glycemic control in preventing certain long-term complications of diabetes (by The Diabetes Control and Complications Trial Research Group in 1993) resulted in greater interest in optimizing insulin regimens. Attempts at stricter control of blood glucose levels, however,

方法控制血糖水平，常常导致负面的结果，比如有低血糖症、体重增加的危险等。由于外源胰岛素的摄入出现的问题，这对婴儿和儿童的糖尿病患者的胰岛素使用提出挑战。父母在饭前给孩子使用胰岛素，结果却可能是发现孩子不吃饭，或不愿意吃饭。

also had undesirable consequences such as increased risk of hypoglycemia and weight gain. The problems associated with aggressive titration of insulin are particularly challenging in infants and young children with diabetes. Parents may give a dose of insulin before a meal only to discover that the toddler, as toddlers are wont to do, does not want to eat.

生理胰岛素的分泌有两种情形。胰岛素持续的基础分泌调节肝的糖异生。另外，膳食会引起胰岛素分泌出现高峰。胰岛素的替代治疗模拟这些因素，用中效或长效的胰岛素结合速效的胰岛素一起使用。但是，长效胰岛素从注射点的释放不稳定，通常胰岛素一般起作用滞后于饭后高血糖。因此注射胰岛素应在饭前。再者，实现良好控制血糖的尝试会导致胰岛素水平大大偏移以及生理性低血糖。

Physiologic insulin secretion has two elements. There is a continuous basal secretion of insulin that serves to regulate hepatic gluconeogenesis. In addition there are peaks of insulin secretion that occur upon eating. Efforts at insulin replacement therapy have attempted to mimic these elements by combining intermediate or longacting insulin, with more rapidly acting regular insulin. Unfortunately, release of longer-acting insulins from the injection site is not constant, and onset of action of regular insulin after injection lags behind postprandial hyperglycemia. Therefore insulin injection must precede a meal. Hence, attempts to achieve euglycemia result in wide excursions of insulin levels and sometimes precipitate clinical hypoglycemia.

新型的速效重组胰岛素是在胰岛素序列中引入突变，缩短了胰岛素从注射后到作用的时间，这样可以在膳食同时注射胰岛素。胰岛素 Lispro 和胰岛素 Aspart 都是速效胰岛素，可以在吃饭时给婴儿和儿童注射，以调节他们的实际食物吸收。后一种方法对治疗婴儿或儿童糖尿病患者特别重要，护理人员可以根据儿童的实际用餐量调整胰岛素的剂量。胰岛素 Glargine 是一种新的胰岛素类似物，可以解决速效胰岛素注射后出现的峰谷值问题，胰岛素水平如同分泌生理性胰岛素一样几乎保持恒定水平。

The introduction of new versions of rapid-acting recombinant insulin prepared by introducing modifications in the insulin sequence has decreased the time from injection to insulin action, allowing for an insulin injection at the same time as a meal. Insulin Lispro and insulin Aspart are rapidly acting forms of the hormone that can be given at mealtime or even after a meal. The latter approach is especially important in managing an infant or young child with diabetes as the caregiver can adjust the insulin dose to the child's actual food intake. A new analogue, insulin Glargine, appears to solve the problem of peaks and troughs of insulin action following rapid-acting rebombinant insulin injections; it provides nearly constant levels of insulin action, similar to physiologic insulin secretion.

6.2 激素制剂的质量控制标准和临床适应证

6.2 Quality Standard and Indications of Therapeutic Hormones

6.2.1 重组人生长激素

生长激素（GH）的基本作用是

6.2.1 Recombinant HGH

GH primarily displays an anabolic activity. It partially stimulates the growth of bone, muscle and

促进合成代谢，它能够直接刺激骨、肌肉与软骨细胞的生长。与肝细胞受体结合会促进胰岛素样生长因子（IGF-Ⅰ）的合成与释放，IGF-Ⅰ介导了大多数的GH生长促进活性，如骨与骨骼肌的生长。

cartilage cells directly. Binding of GH to its hepatic receptor results in the synthesis and release of insulin-like growth factor (IGF-Ⅰ), which mediates most of the GH's growth-promoting activities on bone and skeletal muscle.

在幼年机体生长活跃时期HGH分泌不足会导致垂体性矮小（对外源HGH有反应）。另一方面，在幼年机体生长活跃时期HGH分泌过量会导致巨人症。成年后HGH分泌过量会导致肢端肥大症，特征是手、足变大、面貌粗糙。

A deficiency in the secretion of HGH during the years of active body growth results in pituitary dwarfism (a condition responsive to exogenous HGH administration). On the other hand, overproduction of HGH during active body growth results in gigantism. HGH overproduction after primary body growth has occurred results in acromegaly, a condition characterized by enlarged hands and feet, as well as coarse features.

1. **重组人生长激素的产品质量标准** 重组人生长激素是由191个氨基酸残基组成的蛋白质，可加适量赋形剂或稳定剂。含重组人生长激素量应为标示量的90.0%～110.0%。本品由DNA重组技术生产，必须在生产过程中，用体内生物测定方法测定其生物效价，每毫克无水重组生长激素相当于3.0IU。每毫克蛋白中含重组人生长激素（$C_{990}H_{1528}N_{262}O_{300}S_7$）的量应不少于0.91mg。

1. Quality Standard of HGH Recombinant human growth hormone is a protein having 191 amino acid residues, mixed with suitable excipients or stabilizing agent. It contains not less than 90.0% and not more than 110.0% of the labeled amount of Recombinant human growth hormone. Recombinant human growth hormone is produced by a method based on recombinant DNA technology. It must be demonstrated that the process produces a product having a biological activity of equivalent to 3.0 Units per mg anhydrous Recombinant human growth hormone, using a suitable, validated vivo bioassay. The amount of Recombinant human growth hormone ($C_{990}H_{1528}N_{262}O_{300}S_7$) per 1 mg protein is not less than 0.91 mg.

重组人生长激素中残余宿主DNA和宿主细胞蛋白为与生产过程相关的、潜在的特异性杂质，必须在生产过程中严格控制，其限度应符合有关规定。

Host-cell and vector-derived DNA and host-cell derived proteins are potential specific impurities related to the manufacturing process and must be controlled strictly during the manufacturing process, the host cell derived proteins content of recombinant human growth hormone are determined by an appropriate and validated method.

（1）性状。本品为白色冻干粉末。

(1) Description A white lyophilized powder.

（2）鉴别。

(2) Identification

① 取本品适量，加0.05mol/L三羟甲基氨基甲烷缓冲液（pH7.5）制成每毫升含重组人生长激素2mg的溶液，作为供试品溶液；另取重组人生长激素或甲酰化重组人生长激素对照品适量，同法制备，作为

① Dissolve a quantity in 0.05mol/L trimetamol-hydrocholoric acid (pH 7.5) to produce a solution of 2 mg per 1 ml as a test solution. Repeat the operation using recombinant human growth hormone RS or formylate recombinant human growth factor as a reference solution (RS). Carry out the method for high performance liquid chromatography. The retention time of principal peak of the substance being examined

对照品溶液。照相关蛋白质检查项下的色谱条件试验，供试品溶液主峰的保留时间应与对照品溶液主峰的保留时间一致。

② 取重组人生长激素对照品，加鉴别①项下的缓冲液制成每毫升含2mg的溶液，取此液300μl、胰蛋白酶溶液20μl与鉴别①项下的缓冲液300μl，置37℃水浴4h，立即置－20℃终止反应，作为对照品溶液；取本品，同法制备，作为供试品溶液；照高效液相色谱法试验，以辛基硅烷键合硅胶为填充剂（5～10μm）；以0.1%三氟醋酸溶液为流动相A，含0.1%三氟醋酸的90%乙腈溶液为流动相B，进行梯度洗脱。流速为每分钟1.0 ml；柱温为35℃；检测波长为214 nm。取空白溶液、对照品溶液和供试品溶液各100μl，分别注入液相色谱仪，记录色谱图，扣除空白溶液峰供试品的肽图谱应与相应的重组人生长激素对照品的肽图谱一致。在相关蛋白质检查项下记录的色谱图中，供试品主峰的保留时间应与相应的对照品的保留时间一致。

（3）含量测定。按照分子排阻色谱法测定。以适合分离分子量为5 000～60 000球状蛋白的色谱用亲水硅胶为填充剂；以异丙醇－0.06 mol/L磷酸盐缓冲液（pH7.0，3：97）为流动相；流速为每分钟0.6ml；检测波长为214nm。取人生长激素单体与二聚体混合物对照品，用0.025mol/L的磷酸盐缓冲液（pH7.0）制成每毫升含1.0mg的溶液，取20μl注入液相色谱仪，重组人生长激素单体与二聚体的分离度应符合要求。测定时取

on the chromatogram obtained in the test for Related proteins is identical with that of the principal peak of recombinant human growth hormone RS or formylate recombinant human growth hormone RS in the chromatogram of the reference solution.

② Dissolve a quantity of Recombinant human growth hormone or formylate recombinant human growth hormone in buffer ① to produce a solution of 2 mg per 1ml, to 300 μl of the solution add 20 μl of trypsin solution and 300 μl buffer ①, mix well, warm in a 37℃ water bath for 4 hours, stop the reaction at －20℃, as a reference solution. Repeat the operation using instead of the substance being examined, as a test solution. Carry out the method for high performance liquid chromatography, using a column packed with octadecane bonded silica gel (5-10μm), a 0.1% solution of trifluoroacetic acid in water as mobile phase A and a 0.1% solution of trifluoroacetic acid in 90% acetonitrile as mobile phase B and performing the gradient elution program. Flow rate is 1.0 ml per minutes, column temperature is 35℃ and detective wavelength is 214 nm. The retention time of principal peak of the substance being examined on the chromatogram obtained in the test for Related proteins is identical with that of the principal peak of recombinant human growth hormone RS or formylate recombinant human growth hormone RS in the chromatogram of the reference solution.

(3) Assay Carry out the method for size exclusion chromatography, using a column packed with hydrophilic silica gel for chromatography suitable for separation of global protein with 5000~60000 daltons and 2-propanol-0.06 mol/L phosphate buffer solution (pH 7.0) (3:97) as the mobile phase. Flow rat is 0.6 ml per minute and detection wavelength is 214 nm. Inject 20 μl of solution for system suitability test into the column. The resolution between the peak of monomer of human growth hormone and dimmer should comply with the requirement. Procedure: Dissolve a quantity in 0.025 mol/L phosphate buffer solution (pH 7.0) to produce a solution of 1.0 mg per ml, inject 20 μl into the column. Repeat the operation, using Recombinant human growth hormone CRS instead of the substance being examined. Calculate the content of

本品，用0.025mol/L磷酸盐缓冲液（pH7.0）溶解并定量稀释制成每毫升含1.0mg的溶液，精密量取20μl注入色谱仪，记录色谱图；另取重组人生长激素对照品，同法测定。按外标法以峰面积计算，即得含量。

Recombinant human growth hormone with respect to the peak area obtained in the chromatogram by the external standard method.

（4）无菌检查。取本品，依膜过滤法进行检查。

(4) Sterility　Complies with the test for sterility (membrane filtration method).

（5）水分检测。水分应不超过10%。

(5) Water　No more than 10%.

（6）细菌内毒素检查。取本品，依法测定，每毫克含内毒素应小于5 EU。

(6) Bacterial endotoxin　Carry out the test for bacterial endotoxin, less than 5 EU per 1mg of Recombinant human growth hormone.

（7）贮藏。密闭，2～8℃保存。

(7) Storage　Preserve in tightly closed containers, stored at a temperature of 2℃ to 8℃ .

（8）制剂。注射用重组人生长激素。

(8) Preparation　Recombinant human growth hormone for injection.

2. 临床适应证　GH的治疗范围非常广泛。但是到目前为止，它在临床上主要用于治疗身材矮小。1958年，研究者首次利用从人体垂体中提取的HGH治疗垂体性矮小（即垂体分泌GH不足导致的身材矮小）。后来也用来治疗其他原因导致的身材矮小，疗效也很好，这些病症包括：①特纳综合征；②先天性身材矮小；③慢性肾衰竭。

2. Therapeutic uses of GH　GH has a potentially wide range of therapeutic uses. To date, however, its major application has been for the treatment of short stature. HGH extracted from human pituitary glands was first used to treat pituitary dwarfism (i.e. caused by sub-optimal pituitary GH secretion), in 1958. It has subsequently proved effective in the treatment of short stature caused by a variety of other conditions, including: ① Turner's syndrome; ② idiopathic short stature; ③ chronic renal failure.

（1）治疗垂体性矮小。许多研究都证明rHGH能够促进患垂体性矮小（典型的生长激素不足）的儿童长高。治疗需要每周肌内注射或皮下注射一次。疗程一般为6个月到2年，偶尔需要持续4年。

(1) Recombinant HGH (rHGH) and Pituitary Dwarfism　Various studies have confirmed that rHGH promotes increased linear growth rates in children suffering from pituitary dwarfism (classical growth hormone deficiency). Dosages are generally administered on a weekly basis by i.m. or s.c. injection. Duration of treatment typically varies from 6 months to 2 years, although on occasion administration has continued for up to 4 years.

一般来说注射生长激素后生长速率都有加快，但加快的程度不同，治疗开始的年龄、性别和基础生长速率都会影响rHGH的效果。在治疗的起始阶段，rHGH诱导的生长速

Increased growth rates are generally observed, although the extent varies with, for example, the recipient's age at onset of treatment, sex and baseline growth rates. rHGH-induced growth acceleration is most notable during the initial stages of treatment, with relative effect decreasing with time. In most cases, growth

率加快效果很明显，随着时间的推移，这种效果相对减弱。在大多数病例，生长激素的治疗会使患者身高长高数厘米。

hormone treatment ensures a final body height several centimetres greater than would otherwise be attained in recipients.

（2）先天性身材矮小与特纳综合征。在许多病例中，身材矮小的儿童较正常儿童生长缓慢的根本原因并不明显（先天性身材矮小），他们的内源性 GH 水平也在正常范围内（但可能它的波动的分泌模式有些改变）。然而，大量临床试验表明，应用 rHGH 能够加快先天性身材矮小儿童的生长速率。有几例持续 3 年的临床试验表明，应用 rHGH 后第 1 年的效果很显著，甚至第 3 年的效果也很明显，3 年内平均生长速率比预期增加了 3cm/ 年。

(2) Idiopathic Short Stature and Turner's Syndrome In many cases, a root cause for slower than normal growth rate in children of short stature is not immediately obvious (idiopathic short stature). Endogenous GH levels are often considered to be within a normal range (although there may be changes in its pusatile secretion patterns). A host of clinical trials have shown, however, that rHGH administration can increase the growth rate of many children with idiopathic short stature. Several trials lasting up to 3 years showed that, although the response was most dramatic during the first year, even during year three, mean growth rates were over 3 cm/year greater than expected.

特纳综合征是一种作用于女性的基因缺陷综合征（正常女性有两个 X 染色体；特纳综合征的患者仅携带一个 X 染色体）。这种患者不能生育，常出现发育缺陷、智力发育延迟、身材矮小。实际上，包括特纳综合征患者在内的临床试验证明，应用 GH 后患者生长速率加快，说明生长激素对这些疾病的治疗有效。

Turner's syndrome is a genetic defect that affects females (sufferers carry only one of the usual two X chromosomes). These individuals are infertile; often show developmental defects, mental retardation and short stature. Virtually all clinical trials involving Turner's syndrome patients confirm that administration of GH significantly increases growth velocity, indicating its therapeutic usefulness in these cases.

（3）HGH 的代谢作用。HGH 具有促进机体蛋白质合成和促进脂肪分解的双重作用，这就提示生长激素影响机体肌肉 / 脂肪比例。因此，人们将 HGH 用于治疗肥胖、烧伤以及对抗衰老。对正在节食的肥胖患者的临床研究提示，GH 治疗（通常 3～12 周）和仅靠节食相比，并不能促进脂肪快速减少。但在 rHGH 治疗期间不限制热量摄入的肥胖患者，可观察到脂肪分解。因此，这种激素对肥胖症

(3) Metabolic Effects of HGH The twin metabolic effects of HGH in promoting increased body protein synthesis and increased lipolytic activity suggest a role for the hormone in influencing body lean mass/fat composition. Attention in this regard has focused upon treating obesity and burns, as well as counteracting some of the effects of old age. Clinical studies in dieting obese people suggest that GH treatment (typically for 3-12 weeks) did not promote reduction of body fat levels any faster than in persons' dieting, but without treatment. A lipolytic effect was, however, observed in obese people who were not subject to caloric restriction during rHGH treatment. A future role

的治疗前景还不明确。

临床试验也证明rHGH可治疗严重烧伤，尤其是儿童烧伤。严重烧伤带来恐惧和情感创伤，激发神经系统与免疫系统介导的反应，称为应激反应。其特征为：①蛋白质分解；②肌肉量减少；③新陈代谢速率加快；④物质无效循环和脂肪分解；⑤体温增高。

rHGH治疗的目的主要在于减慢/阻止已经提高了的蛋白质分解水平。起初，对烧伤患者的临床试验表明，与对照（未经rHGH治疗）相比，应用GH能够减少蛋白质50%的损失量。随后对大面积烧伤（烧伤面积超过体表面积的50%）的儿童应用GH，伤口愈合加快，皮肤移植供区愈合尤其加快。这样，就使得短期内皮肤进一步移植成为可能，缩短了关闭烧伤创口所需的时间。在此基础上，未来GH对于烧伤护理还可能会发挥更大作用。

人体产生GH的水平受年龄影响。出生后GH水平最高，青春期会再次增高。40岁后GH分泌稳步下降，这种下降可能与肌肉、骨与皮肤的老化相关，这些都是衰老的原因。

近年来，为了评价GH对正在衰老的成年人的作用，研究者进行了几次临床试验。试验一般持续4～6个月，在一次试验中，机体肌肉量和皮肤厚度增加7%，脂肪减少14%。其他试验的结果没有这样显著。在老龄化的社会，要评估GH对抗衰老的效果，还需要更为详尽的临床试验和成本-效益分析。

（4）GH、泌乳与排卵。数年前，

for this hormone in treating obesity is, therefore, far from certain.

Clinical trials have also revealed a role for rHGH in the treatment of severe burns, particularly in children. The fear and emotional trauma (as well as physical damage) associated with receiving a severe burn triggers a neurological and immune-mediated response termed the stress response. This is characterized by: ① protein catabolism; ② loss of lean body mass; ③ increased metabolic rate; ④ futile substrate cycling and lipolysis; ⑤ elevated body temperature.

rHGH treatment is aimed largely at slowing/preventing elevated protein catabolism. Initial trials in burn patients showed that GH administration reduced protein loss by 50% compared to (untreated) controls. Subsequent GH studies in children with massive burning (over 50% of total body surface) revealed accelerated wound healing, particularly at the skin graft donor site. This, in turn, facilitated further skin grafting within shorter time periods, thus reducing the time to close the burn wound. On the basis of such results, GH may well play a future expanding role in burn care management.

The production of GH is age-modulated. The highest production levels are recorded immediately after birth, with a second increase noted at puberty. GH secretion decreases steadily after age 40, and this decline is likely linked to age-associated decreased muscle, bone and skin mass, all of which contribute to age-associated frailty.

In recent years, several pilot clinical trials, assessing the effects of GH administration to ageing adults, have been carried out. Typically, trial duration is 4-6 months. A 7% increase in lean body mass and skin thickness, along with a 14% drop in body fat, was observed in one trial, although results recorded in other trials were less striking. More detailed clinical trials and cost, benefit analysis must be carried out in order to fully assess the potential of GH to counteract some of the effects of ageing in the elderly population.

(4) GH, Lactation and Ovulation Recombinant

人们已经开始用重组牛 GH（牛生长激素，BST）促进奶牛泌乳（可提高泌乳量高达 20%）。近年来对猴的研究表明，GH 不仅可提高乳汁分泌量，同时也能使乳汁中脂肪含量略有提高。这些研究提示 GH 可能可以治疗人类泌乳障碍。给一小组哺乳女性应用 rHGH 治疗 1 周，乳汁分泌量有所增加（但不影响乳汁中营养成分的组成，包括脂肪水平）。

bovine GH (bovine somatotrophin, BST), has been used for a number of years to boost milk yield in dairy cattle (by up to 20%). More recent studies in monkeys showed that GH administration increased milk yields as well as promoting a slight increase in milk fat levels. This suggested that GH might be beneficial in the treatment of human lactation failure. Treatment of a small group of breast-feeding women with rHGH for 1 week has been shown to increase milk production (although it had no effect on the milk's nutritional composition, including fat levels).

GH 也影响卵巢生理，主要是通过 IGF-Ⅰ发挥作用。IGF-Ⅰ能够刺激培养的哺乳动物粒层细胞增殖、加强 FSH 对卵巢的作用。尽管 GH 似乎在促进排卵或生殖中不起主要作用，但它与促性腺激素和其他生殖激素具有协同作用。在一例由于内源性 FSH 分泌不足导致不能生育的女性病例中，用 rHGH 协同治疗能够减少诱导排卵所需的外源 FSH 的用量。

GH also appears to impact upon ovarian physiology, mainly although not exclusively through IGF-Ⅰ. IGF-Ⅰ stimulates replication of cultured mammalian granulosa cells and potentiates FSH action on the ovary. While GH appears not to play an essential role in ovulation or fertility, it seems to act synergistically with gonadotrophins and other reproductive hormones. One clinical trial, involving women subfertile due to the lack of endogenous FSH secretion, showed that co-treatment with rHGH decreased the quantity of exogenous FSH required to induce ovulation.

在临床医药领域，rHGH 的应用潜力还在不断增加。在大部分病例中，应用 GH 没有严重的副作用，这也促进了 GH 的应用。虽然 GH 促进身材矮小患者长高的效果确切，但如果在临床上用于其他方面，还需要更多令人信服的临床证据。

The range of potential applications of rHGH in clinical medicine continues to grow. The use of GH is also facilitated by the absence of any serious side effects in most instances. Although its efficacy in promoting growth in persons of short stature is beyond doubt, more convincing clinical evidence is required before its approved clinical applications are expanded further.

6.2.2 降钙素

降钙素是一种多肽激素，在调节血清钙离子（Ca^{2+}）和无机磷酸盐（PO_4^{3-}）水平方面发挥主要作用（协同副甲状腺激素，维生素 D 衍生物和 1，25- 二羟胆钙化醇）。成年人机体内约含有 2kg 钙，98% 存在于骨骼中。机体的磷含量约为 1kg，85% 也存在于骨骼中［骨的矿物成分主要是 $Ca_3(PO_4)_2$，它是机体钙与磷的贮

6.2.2 Calcitonin

Calcitonin is a polypeptide hormone which (along with parathyroid hormone and the vitamin D derivative, 1,25-dihydroxycholecalciferol) plays a central role in regulating serum ionized calcium (Ca^{2+}) and inorganic phosphate (Pi) levels. The adult human body contains up to 2 kg of calcium, of which 98% is present in the skeleton. Up to 85% of the 1 kg of phosphorus present in the body is also found in the skeleton (the so-called mineral fraction of bone is largely composed of $Ca_3(PO_4)_2$ which acts as a body reservoir for both calcium and phosphorus). Calcium concentrations in human serum

存器]。人体血清中的钙浓度约为0.1 mg/ml，变化较小（血清磷水平变化较大）。

approximate to 0.1 mg/ml and are regulated very tightly (serum phosphate levels are more variable).

降钙素的作用是降低血清钙和磷的水平，主要是通过抑制骨吸收实现的，降钙素也能够降低肾脏对钙和磷的吸收。降钙素受体主要存在于骨细胞（破骨细胞）和肾细胞，通过激活腺苷酸环化酶生成cAMP，cAMP在信号转导过程中发挥重要作用。

Calcitonin lowers serum Ca^{2+} and Pi levels, primarily by inhibiting the process of bone resorption, but also by decreasing resorption of Pi and Ca^{2+} in the kidney. Calcitonin receptors are predictably found primarily on bone cells (osteoclasts) and renal cells, and generation of cAMP via adenylate cyclase activation plays a prominent role in hormone signal transduction.

在大多数哺乳动物中，降钙素是由甲状腺的滤泡旁细胞合成的。在比哺乳动物低级的物种中，降钙素是由一种叫做后鳃体的特殊解剖结构合成的。

In most mammals, calcitonin is synthesized by specialized parafollicular cells in the thyroid. In sub-mammalian species, it is synthesized by specialized anatomical structures known as ultimobranchial bodies.

1. 重组鲑降钙素的质量标准 本品为化学合成的由32个氨基酸组成的多肽。按无水与无醋酸物计算，含鲑降钙素应为90.0%～105.0%，每毫克鲑降钙素相当于6000 IU。

1. Quality Standard of Calcitonin Calcitonin (Salmon) injection is a sterile solution of Calcitonin (salmon) in water for injection. It contains not less than 90.0% and not more than 115.0% of the labeled amount of calcitonin (salmon) ($C_{145}H_{240}N_{44}O_{48}S_2$)。

（1）性状。本品为白色或类白色粉末。本品在水中易溶。

(1) Description A white or almost white lyophilized powder. Freely soluble in water.

（2）鉴别。在含量测定项下记录的色谱图中，供试品溶液主峰的保留时间应与对照品溶液主峰的保留时间一致。

(2) Identification The retention time of the principal peak of calcitonin (salmon) is the substance being examined in the chromatogram obtained in the assay is identical with that of the principal peak of calcitonin (salmon) CRS in the chromatogram of the reference solution.

（3）含量测定。照高效液相色谱法测定。用十八烷基硅烷键合硅胶为填充剂；以0.402%四甲基氢氧化铵溶液-乙腈（9∶1）用磷酸调节pH至2.5为流动相A；以0.363%四甲基氢氧化铵溶液-乙腈（2∶3）用磷酸调节pH至2.5为流动相B，进行梯度洗脱，检测波长为220nm。取对照品溶液，置75℃加热15h，放冷，取200 μl注入液相色谱仪，降钙素C峰（溶剂峰与主峰间的最大峰即为降

(3) Assay Carry out the method for high performance liquid chromatography, using a column packed with octadecylsilane bonded silica gel, a mixture of 0.402% tetramethylammonium hydroxide pentahydrate solution acetonitrile (9∶1) (pH 2.5) as the mobile phase A, and a mixture of 0.363% tetramethylammonium hydroxide pentahydrate solution acetonitrile (2∶3) (pH 2.5) as the mobile phase B. Perform the gradient solution suing the gradient programme and detection wavelength is 220 nm. Here the references solution at 75℃ for 15 hours and cool. Inject 200 μl into the column, and the resolution factor between the peaks of calcitonin C (the largest peak to elute after the injection buffer salts and before the principal peak with a relative retention to that of calcitonin (salmon) of between 0.5 and 0.8) and calcitonin (salmon) is

钙素 C 峰，与主峰的相对保留时间为（0.5～0.8），与鲑降钙素峰的分离度应大于 3.0。精密量取本品 200 μl 注入液相色谱仪，记录色谱图；另取对照品适量，精密称定，加水制成每毫升中约含 10 μg 的溶液，同法测定。按外标法以峰面积计算，即得含量。

more than 3.0. Inject 200 μl of the injection being examined into the column and record the chromatogram. Dissolve a quantity of calcitonin (salmon) CRS, accurately weighed, in water to produce a reference solution containing about 10 μg per 1ml. Repeat the operation, using the references solution instead of the injection being examined. Calculate the content of $C_{145}H_{240}N_{44}O_{48}S_2$ with respect to the peak area obtained in the chromatogram by the external standard method.

（4）相关物质。按照含量测定项下的方法测定，按峰面积归一化法计算，单个杂质的峰面积不得大于总峰面积的 3.0%，各杂质峰面积的和不得大于总峰面积的 5.0%（峰面积小于主峰面积 0.1% 的峰除外）。

(4) Related substance　In the chromatogram obtained in the assay the area of any peak corresponding to calcitonin C is not greater than 7% and the sum of the areas of any peaks, other than the principal peak and the peak of calcitonin C is not greater than 5.0% of the total area of all the peaks by the peak area normalization method.

（5）水分检查。取本品，照水分测定法测定，含水分不得过 10.0%。

(5) Water　Carry out the test for aquametry, less than 10.0% of the calcitonin (salmon).

（6）醋酸和水分。醋酸和水分之和不得过 20%。

(6) Acetic acid and water　No more than 20% for total Acetic acid and water of the calcitonin (salmon).

（7）细菌内毒素。取本品，依法测定，每毫克含内毒素应小于 25EU（供注射用）。

(7) Bacterial endotoxin　Carry out the test for bacterial endotoxin, less than 0.5 EU per μg of the calcitonin (salmon) (for injection).

（8）贮藏。密封，遮光，2～8℃保存。

(8) Storage　Preserve in well closed containers, protected from light and stored at a temperature of 2℃ to 8℃.

（9）制剂。鲑降钙素注射液。

(9) Preparation　Calcitonin (Salmon) injection.

2. *降钙素的临床应用*　在临床上，降钙素主要用于治疗骨质疏松，也用于治疗某些恶性肿瘤和 Paget 病导致的高血钙。

2. Therapeutic uses of GH　Calcitonin is used clinically to treat osteoporosia, as well as hypercalcaemia associated with some forms of malignancy and Paget's disease.

6.2.3 促性腺激素

6.2.3 Recombinant Gonadotrophins

1. *临床应用*　促性腺激素的主要作用是维持生殖功能，因此它们对生育能力差和某些不能生育的情况具有潜在的治疗作用。促性腺激素还可用于诱导多种动物的过速排卵反应。

1. Medical Applications of Gonadotrophins　Because of their central role in maintaining reproductive function, the therapeutic potential of gonadotrophins in treating subfertility and some forms of infertility was obvious. Gonadotrophins are also used to induce a superovulatory response in various animal species.

用药的剂量要尽可能模拟生育期女性生殖周期的正常血清中促性腺激素水平。监测雌激素的合成情况或通过超声设备检测卵泡反应可以实现这

Dosage regimes attempt to mimic as closely as possible normal serum gonadotrophin profiles as they occur during the reproductive cycle of fertile females. This is achieved by monitoring the resultant oestrogen production, or by using ultrasonic equipment to monitor

一点。根据女性基础激素情况，计算用药的最佳剂量比较困难（治疗应该符合患者的生理需求）。用量过大就会导致多个卵泡发育，可能会造成多胎妊娠。

follicular response. Depending upon the basal hormonal status of the female, calculation of the optimal dosage levels can be tricky (treatments are tailored to meet the physiological requirements of individual patients). Over-dosage can, and does, result in multiple follicular development, with consequent risk of multiple pregnancy.

通常的用药方式是每天肌内注射促性腺激素，连续12d或更长时间，然后给予一剂HCG。或者每隔1d给予较大剂量的促生育素，连续3次用药，在最后一剂促生育素后1～2d给予HCG。

Treatment typically entails daily i.m. administration of gonadotrophin, often for 12 days or more, followed by a single dose of HCG. Alternatively, three equal larger doses of menotrophin may be administered on alternate days, followed by HCG administration 1-2 days after the final menotrophin dose.

促性腺激素也可用于协助生殖过程。给予治疗剂量的FSH，超过了个体卵泡的FSH需求极限，就会促进多个卵泡生长。这样就会有多个卵子形成，可以用于体外受精。这种技术常用于女性输卵管阻塞或者其他卵子与精子结合障碍的情况。女性经促性腺激素治疗后，收集形成的卵子，在体外与其伴侣的精子在培养基中共同孵育，直至胚泡形成，然后植入母亲的子宫。

Gonadotrophins are also used in assisted reproduction procedures. Here the aim is to administer therapeutic doses of FSH that exceed individual follicular FSH threshold requirements, thus stimulating multiple follicular growth. This, in turn, facilitates harvest of multiple eggs, which are then available for in vitro fertilization (IVF). This technique is often employed when a woman has a blocked fallopian tube, or some other impediment to normal fertilization of the egg by a sperm cell. After treatment, the resultant eggs are collected, incubated in vitro with her partner's sperm, incubated in culture media until the embryonic blastocyst is formed, and then implanted into the mother's uterus.

FSH和HCG也可用于治疗男性不育或类似情况。对于男性促性腺激素不足的性腺机能减退者，给予FSH或HCG能够促进精子形成，促进正常性功能。HCG对于治疗青春期前隐睾病（即睾丸不能完全从腹腔降落到阴囊）也有一定作用。HCG的刺激睾丸激素生成的活性也引起了某些运动员的注意，因此，国际奥林匹克委员会禁止运动员使用这种激素。

FSH and HCG also find application in the treatment of male subfertility or related conditions. Both are administered to males exhibiting hypogonadotrophic hypogonadism to stimulate sperm synthesis and normal sexual function. HCG has found limited application in the treatment of pre-pubertal cryptorchidism (a condition characterized by failure of the testes to descend fully into the scrotum from the abdomen). The ability of this hormone to stimulate testosterone production also caught the attention of some athletes and, as a result, the International Olympic Committee has banned its use.

在许多非性腺组织的细胞表面发现了LH/HCG受体，说明这些激素可能具有非性腺相关的生理功能。另外，肝、肾和肌肉组织的细胞缺乏这些受体，出生前的这些组织表

The LH/HCG cell surface receptor is found in a number of non-gonadal tissues, indicating that these hormones may exert physiologically relevant non-gonadal functions. In addition, while liver, kidney and muscle cells are devoid of such a receptor, it is expressed by a number of these tissues before birth, hinting at a potential

达这些受体，提示这些激素可能在发育时期具有某种作用。非性腺组织的受体水平明显低于性腺组织的受体水平。

developmental role. Receptor levels in non-gonadal tissues are generally much lower than in gonadal tissue.

研究表明，HCG可能具有一些妊娠/非妊娠相关的非性腺功能，将来可能具有临床应用价值。它似乎能够增强子宫血流量，加强母亲与胎儿之间的免疫抑制，这样，HCG可能对防治某些类型的流产有效。还有研究表明，将HCG注入子宫可以对抗前列腺素诱导的子宫收缩。因此，HCG可能对预防早产有效。HCG还具有抗乳腺癌的活性，例如，对啮齿动物的研究证明在给予动物不同化学致癌物前或后给予HCG，可以阻止乳腺肿瘤形成或生长。

Research indicates that HCG probably has a number of pregnancy/non-pregnancy-related non-gonadal functions that may give rise to future additional clinical applications. It appears to promote effects such as increased uterine blood flow and immunosuppression at the maternal–fetal interface. As such, HCG may prove potentially useful in preventing some types of pregnancy loss. Research has also indicated that infusion of HCG into the uterus can counteract prostaglandin-induced uterine contractions. As such, it may prove useful in preventing the onset of premature labour. HCG also exhibits anti-breast cancer activity. Studies in rodents, for example, illustrate that it prevents mammary tumor formation/growth when administered prior or subsequent to various chemical carcinogens.

重组抑制素和重组刺激素已有生产。用重组抑制素免疫不同动物能够增加动物的排卵速率，这可能是由于重组抑制素引发内源性抑制素的免疫中和作用。这种方法会被用于增强动物的生育能力。最近发现抑制素具有肿瘤抑制剂的作用，可能会对这种方法产生不利影响。

Recombinant inhibins and activins have been already produced. Immunization of various animal species with recombinant inhibin promotes increased ovulation rates, presumably due to ensuing immunoneutralization of endogenous inhibin. Such a strategy might be useful, e.g. in producing super-fertile animals. The recent discovery of inhibins' role as a tumor suppressor may militate against such an approach.

直接给予重组抑制素会抑制女性排卵（通过抑制FSH分泌发挥作用），给男性使用后能够抑制精子生成。虽然目前还没有文献报道这种避孕药的安全性或效果，但抑制素的这种抑制精子生成的作用提示了它可能会被用做男性避孕药。

Direct administration of recombinant inhibin inhibits ovulation in females (via inhibition of FSH secretion). It also inhibits spermatogenesis when administered to males. This last effect initially prompted interest in the use of inhibin as a potential male contraceptive, although no significant data on the safety or efficacy of such a treatment has yet been published.

2. 促性腺激素在兽医学中的应用 促性腺激素用于治疗动物的低生育力，以及广泛用于诱导有经济价值动物（马和牛）的超排卵反应。

2. Veterinary Applications of Gonadotrophins Gonadotrophins may be utilized to treat subfertility in animals and are routinely used to induce a superovulatory response in valuable animals, most notably valuable horses and cattle.

超排卵的理论和应用与促性腺激

The theory and practice of superovulation is quite

素辅助人体外受精相同。给予动物外源性的FSH，刺激动物多个卵泡同时发育。给予LH促进排卵，然后令动物交配，使释放的卵子受精。根据动物的个体差异和用药剂量，可以使动物产生0～50个胚胎，典型的胚胎数量是4～10个。收获胚胎（可以用手术方法，更常用的是非手术方法）后，在细胞培养基中培养很短的一段时间，然后将一个胚胎植入供者雌性动物，其余胚胎植入其他受体雌性动物，这些动物作为代理母亲，孕育这些后代直至生产。

这种技术最常用于价值较高的动物（如获奖的马，或产奶量高的奶牛），可以使其生育能力增强数倍。每个后代都继承生理母亲（或父亲）的基因，与孕育它们的代孕动物无关。

促性腺激素已常用于诱导排卵过速反应，包括猪FSH（pFSH）和猪LH（pLH）。pFSH是从屠宰场的猪的垂体提取的。天然垂体提取物通常用乙醇或盐沉淀，含FSH的沉淀物经过至少一种层析步骤纯化，最终产品常含有少量LH及其他垂体来源的蛋白质。

pLH也是从屠宰场的猪的垂体提取的。pFSH和pLH都是猪源性，而其靶受体基本上都是牛或马。这种给一种动物使用来源于另一个物种动物的产品的做法是可取的，这样将减少疾病传播的概率（很多病原体具有物种特异性）。

大多数超排卵疗法是给予动物pFSH，每日两次，连续4～5d。FSH的血浆半衰期较短，因此需要

similar to the use of gonadotrophins to assist in vitro fertilization procedures in humans. Exogenous FSH is administered to the animal such that multiple follicles develop simultaneously. After administration of LH to help promote ovulation, the animal is mated, thus fertilizing the released egg cells. Depending upon the specific animal and the superovulatory regime employed, anything between 0 and 50 viable embryos may be produced, although more typically the number is between 4 and 10. The embryos are then recovered from the animal (surgically or, more usually, non-surgically) and are often maintained in cell culture for a short period of time. A single embryo is then usually re-implanted into the donor female, while remaining embryos are implanted into other recipient animals, who act as surrogate mothers, carrying the offspring to term.

This technology is most often applied to valuable animals (e.g. prize-winning horses, or high milk-yield dairy cattle) in order to boost their effective reproductive capacity several-fold. Each of the offspring will inherit its genetic complement from the biological mother (and father), irrespective of which recipient animal carries it to term.

Gonadotrophins are usually utilized to induce a superovulatory response, and include porcine FSH (p-FSH) and porcine LH (p-LH). P-FSH is extracted from the pituitary glands of slaughterhouse pigs. The crude pituitary extract is usually subject to a precipitation step, using either ethanol or salts. The FSH-containing precipitate is normally subjected to at least one subsequent chromatographic step. The final product often contains some LH as well as low levels of additional pituitary-derived proteins.

P-LH is obtained, again, by its partial purification from the pituitary glands of slaughterhouse pigs. In both cases, a porcine source is utilized, as the target recipients are almost always cattle or horses. The use of a product derived from a species other than the intended recipient species is encouraged, as it helps minimize the danger of accidental transmission of disease via infected source material (many pathogens are species-specific).

Most superovulatory regimes that utilize p-FSH entail its administration to the recipient animal twice daily for 4-5 days. Regular injections are required, due

定期注射；皮下注射会延长每一剂量药物的半衰期。注射 FSH 4d 或 5d 后给予动物一剂 LH，能够促进卵泡成熟和排卵。

促性腺激素制剂的确切组分及用药方案均会影响药效。很多 pFSH 制剂中 FSH：LH 比例不同，也会影响结果。当 FSH 制剂中 LH 含量很低时，排卵过速反应最为稳定。

6.2.4 胰高血糖素

胰高血糖素的主要生理作用与胰岛素相反，尤其在调节代谢方面。胰高血糖素具有全面的分解代谢效果，刺激糖原、脂类与蛋白质分解，提高血糖水平。因此，胰高血糖素的主要生理作用是预防低血糖症。胰高血糖素在胰腺 α 细胞中合成后，便贮存于分泌小泡中，一旦血糖浓度降低，就经过胞吐作用释放到血液中去。

胰高血糖素的调节代谢（及其他）作用是通过与特异性细胞表面受体结合，激活腺苷酸激酶，然后促进 cAMP 依赖的蛋白激酶磷酸化，从而调节其活性。

低血糖症仍然是糖尿病患者注射胰岛素后最常见的并发症。其原因通常是：①过量注射胰岛素；②饭前注射胰岛素，但随后却没有用餐；③运动量增加。严重的低血糖症可能导致意识丧失甚至死亡。日服或者静脉注射葡萄糖可以治疗低血糖，但胰岛素诱导的低血糖症有时需要用胰高血糖素治疗。

在某些放射学检查中，胰高血糖素也可用于辅助诊断，如降低胃、小肠和大肠的运动以辅助检查（胰高血

to the relatively short halflife of FSH in serum; s.c. administration helps prolong the duration of effectiveness of each injection. The 4 or 5 days of treatment with FSH is followed by a single dose of LH, promoting final follicular maturation and ovulation.

The exact composition of the gonadotrophin preparations administered and the exact administration protocol also influences the outcome. The variability of FSH:LH ratios in many p-FSH preparations can affect the results obtained, with the most consistent superovulatory responses being observed when FSH preparations exhibiting low LH activity are used.

6.2.4 Recombinant Glucagon

The major biological actions of glucagon tend to oppose those of insulin, particularly with regard to regulation of metabolism. Glucagon has an overall catabolic effect, stimulating the breakdown of glycogen, lipid and protein. A prominent metabolic effect is to increase blood glucose levels. Indeed, the major physiological function of glucagon is to prevent hypoglycaemia. The hormone is stored in secretory vesicles, after synthesis in the pancreatic α cells, and released by exocytosis upon experiencing a drop in blood glucose concentration.

Glucagon initiates its metabolic (and other) effects by binding to a specific cell surface receptor, thus activating a membrane-bound adenylate kinase. This in turn, promotes activation of a cAMP-dependent protein kinase. The kinase phosphorylates key regulatory enzymes in carbohydrate metabolism, thereby modulating their activity.

Hypoglycaemia remains the most frequent complication of insulin administration to diabetics. It usually occurs due to ① administration of an excessive amount of insulin; ② administration of insulin prior to a mealtime, but with subsequent omission of the meal; or ③ due to increased physical activity. In severe cases, this can lead to loss of consciousness and even death. Although it may be treated by oral or i.v. administration of glucose, insulin-induced hypoglycaemia is sometimes treated by administration of glucagon.

Glucagon is also used medically as a diagnostic aid during certain radiological examinations of the stomach, small and large intestine where decreased intestinal motility is advantageous (the hormone has an inhibitory

糖素对消化道壁平滑肌的运动性有抑制作用)。

6.2.5 促甲状腺素

促甲状腺素(TSH)是由一种特别的垂体细胞合成的,促甲状腺素释放激素(TRH,一种下丘脑三肽激素)促进它的合成与释放。TSH通过与细胞表面的特异性受体结合而发挥作用,这些受体最初被发现位于甲状腺细胞表面,其他细胞也有这种受体。TSH与受体结合激活腺苷酸环化酶,导致细胞内cAMP水平提高。最终激发TSH的促甲状腺功能,包括促进甲状腺从血液中摄取碘、合成含碘激素甲状腺素(T_4)和三碘甲状腺氨酸(T_3)并将这些激素释放到血液中去,这些激素能够调节多种组织新陈代谢的许多方面。血浆T_4和T_3水平的升高也会通过负反馈机制抑制TSH的合成与释放。

TSH在医学上可用于甲状腺切除术后的患者检查甲状腺癌/甲状腺残余组织的辅助诊断。甲状腺癌相对较少见,主要发生于成人,尤其是女性。患甲状腺癌的首要治疗方案是切除所有或大部分甲状腺组织(甲状腺切除术)。然后进行甲状腺激素抑制疗法,服用足量的T_3或T_4,通过负反馈机制保持血清中TSH的低水平(抑制TSH的目的是防止TSH介导的对残余甲状腺癌细胞的刺激作用)。服用TSH和放射性碘还可以检测甲状腺癌的复发。这是由于TSH促进对放射性碘的摄取,通过适当的放射成像技术即可检测出来。

effect on the motility of the smooth muscle lining the walls of the gastrointestinal tract).

6.2.5 Recombinant Thyroid Stimulating Hormone

TSH is synthesized by a distinct pituitary cell type; the thyrotroph. Its synthesis and release is promoted by thyrotrophin releasing hormone (TRH, a hypothalamic tripeptide hormone). TSH exerts its characteristic effects by binding specific receptors found primarily, but not exclusively, on the surface of thyroid gland cells. Binding to the receptor activates adenylate cyclase, leading to increased intracellular cAMP levels. Ultimately, this triggers TSH's characteristic effects on thyroid function, including promoting iodine uptake from the blood, synthesis of the iodinecontaining thyroid hormones, thyroxine (T_4) and triiodothyronine (T_3) and release of these hormones into the blood, from where they regulate many aspects of general tissue metabolic activity. Elevated plasma levels of T_4 and T_3 also promote decreased TSH synthesis and release by a negative feedback mechanism.

TSH is approved for medical use as a diagnostic aid in the detection of thyroid cancer/thyroid remnants in post-thyroidectomy patients. Thyroid cancer is relatively rare, exhibiting the highest incidence in adults, particularly females. First-line treatment is surgical removal of all or most of the thyroid gland (thyroidectomy). This is followed by thyroid hormone suppression therapy, which entails administration of T_3 or T_4 at levels sufficient to maintain low serum TSH levels through the negative feedback mechanism mentioned earlier. TSH suppression is required in order to prevent the TSH-mediated stimulation of remnant thyroid cancer cells. The recurrence of active thyroid cancer can be detected by administration of TSH along with radioactive iodine. TSH promotes uptake of radioactivity, which can then be detected by appropriate radio-imaging techniques.

6.3 结语

激素制剂作为治疗药物在临床已经应用很长时间了，所有这些激素都是用来补充低于正常水平的内源性激素。自从胰岛素开始用于临床以来，已经挽救或延长了数百万糖尿病患者的生命；促性腺激素已经帮助数万甚至数十万不育者生育；生长激素提高了数千位身材矮小患者的生活质量。大多数激素在基因工程发明之前就已经应用于临床，而现在，在安全使用范围内，重组激素制剂更受欢迎。在未来的许多年里，激素仍将是临床上的重要药物。

（编者　付爱玲）

6.3 Conclusions

Several hormone preparations have a long history of use as therapeutic agents. In virtually all instances they are administered simply to compensate for lower than normal endogenous production of the hormone in question. Since it first became medically available, insulin has saved or prolonged the lives of millions of diabetics. Gonadotrophins have allowed tens, if not hundreds, of thousands of sub-fertile individuals to conceive. Growth hormone has improved the quality of life of thousands of people of short stature. Most such hormones were in medical use prior to the advent of genetic engineering. Recombinant hormonal preparations are now however gaining greater favor, mainly on safety grounds. Hormone therapy will remain a central therapeutic tool for clinicians for many years to come.

(Compiled by Ailing Fu)

参 考 文 献
References

陈慰峰. 2000. 医学免疫学. 3 版. 北京：人民卫生出版社

高建忠，祝卫东，黄玉帮. 2004. 干扰素的研究进展. 动物医学，21（1）：58-59

金伯泉. 2001. 细胞和分子免疫学. 2 版. 北京：科学出版社

刘占通，舒畅，刘金娥，等. 2006. 干扰素研究进展. 中国兽药杂志，40（10）：35-38

张林，戴留军，叶芳. 1996. 体内干扰素抗病毒机制初探. 河南医药信息，4（5）：53

张映. 2003. GFE1 多肽与新型肿瘤坏死因子融合基因的克隆、表达及表达产物的组织分布研究 . 中国人民解放军第四军医大学：13-15

张咏南. 1985. 干扰素入门. 合肥：安徽科学技术出版社

章崇杰. 2007. 医学免疫学纲要. 成都：四川大学出版社

R.J.Y. 霍，M. 吉巴尔地. 2006. 生物技术与生物药物——蛋白质药物与基因药物. 北京：化学工业出版社

Beutler B, Milsark I W, Cerami A C. 1985. Passive immunization against cachectin/tumor necrosis factor protects mice from lethal effect of endotoxin. Science; 229 (4716): 869-871

Carswell EA, Old LJ, Kassel RL, et al. 1975. An endotoxin-induced serum factor that causes necrosis of tumors. Proc Natl Acad Sci U S A，72(9):3666-3670

Cronmelin, D.J.A., Sindelar, et al. 2002. Pharmaceutical Biotechnology: An Introduction for Pharmacists and Pharmaceutical Scientists. Second Edition. Taylor & Francis Routledge

Gary Walsh. 2003. Biopharmaceuticals: Biochemistry and Biotechnology. Second Edition. Chichester: John Wiley & Sons, Ltd

Gresser I. 1981. Interferon 1981. Academic press

Gresser I. 1982. Interferon 1982. Academic press

Old LJ. 1985. Tumor necrosis factor (TNF).Science, 230 (4726): 630-632

Pennica D, Nedwin GE, Hayflick JS, et al. 1984. Human tumour necrosis factor: precursor structure, expression and homology to lymphotoxin. Nature. 312(5996): 724-729

Ruddle NH, Waksman BH. 1968. Cytotoxicity mediated by soluble antigen and lymphocytes in delayed hypersensitivity. 3. Analysis of mechanism. J Exp Med. 128(6): 1267-1279

第六章 抗体药物

Chapter 6 Antibody Pharmaceutics

1 抗 体

1 Antibody

抗体（antibody，Ab）是机体免疫细胞被抗原激活后，由分化成熟的终末B细胞-浆细胞合成、分泌的一类能与相应抗原特异性结合的具有免疫功能的球蛋白。早在19世纪后期，von Behring及其同事Kitasato就发现白喉或破伤风毒素免疫动物后可产生具有中和毒素作用的物质，称之为抗毒素（antitoxin），随后引入抗体一词来泛指抗毒素类物质。1937年Tiselius和Kabat用电泳方法将血清蛋白分为白蛋白、α1、α2、β及γ球蛋白等组分，并发现抗体活性存在于从α到γ的这一广泛区域，但主要存在于γ区，故相当长一段时间内，抗体又被称为γ球蛋白（丙种球蛋白）。1968年和1972年世界卫生组织和国际免疫学会联合会的专门委员会先后决定，将具有抗体活性或化学结构与抗体相似的球蛋白统一命名为免疫球蛋白（immunoglobulin，Ig）。免疫球蛋白可分为膜型（membrane immunoglobulin，mIg）和分泌型（secreted immunoglobulin，sIg）两种类型。前者作为B细胞膜上的抗原受体，识别抗原并启动B细胞的活化；后者由活化后的B细胞分泌，进入体液，通过各种生物学效应清除诱导其产生的抗原物质。

Antibodies, also known as immunoglobulins, are mainly synthesized and produced by plasma cells, and they are globulin which can bind specifically with antigens after immune activation. In the late 19th century, von Behring and his colleague Kitasato found that animals immunized with diphtheria or tetanus could produce "antitoxins", substances that can neutralize toxins, and later the term "antibody" was introduced to refer substances that anti toxins. In 1937, Tiselius and Kabat identified albumin, α1, α2, β and γ-globulin with serum protein electrophoresis, and found the antibody activity existed widely from the α to the γ domain, but mainly in the γ domain, so for a very long time, antibodies are also named γ globulins. In 1968 and 1972, the World Health Organization and International Union of Immunological Societies respectively determined to name globulins sharing similar activities or structures with antibodies as "immunoglobulins". Immunoglobulin (Ig) can occur either in a membrane-bound form or in a soluble form. The former binds on B cell membrane for antigen recognition and B cell activation, while the later enters the body fluids to activate a series of immune reactions to clear antigens.

1.1 免疫球蛋白的基本结构

X射线晶体衍射结构分析发现，免疫球蛋白由四肽链分子组成，各肽链间有数量不等的链间二硫键。结构上Ig可分为三个长度大致相同的片段，其中两个长度完全一致的片段位于分子的上方，通过一易弯曲的区域与主干连接，形成一Y形结构（图6-1），称为Ig单体，构成免疫球蛋白分子的基本单位。

1.1 The Basic Structure of Ig

Structural analysis of X-ray diffraction(XRD) showed that, the Ig consists of four polypeptide chains that are connected by different numbers of disulfide bonds. Structurally, an Ig could be divided into 3 fragments with similar length. Two fragments with an identical length are located at the upper side of the molecule, and are connected to the backbone by a pliable chain, forming a "Y"-shape molecule, which is named the "immunoglobulin monomer" and constitutes the basic functional unit of an antibody(Figure 6-1).

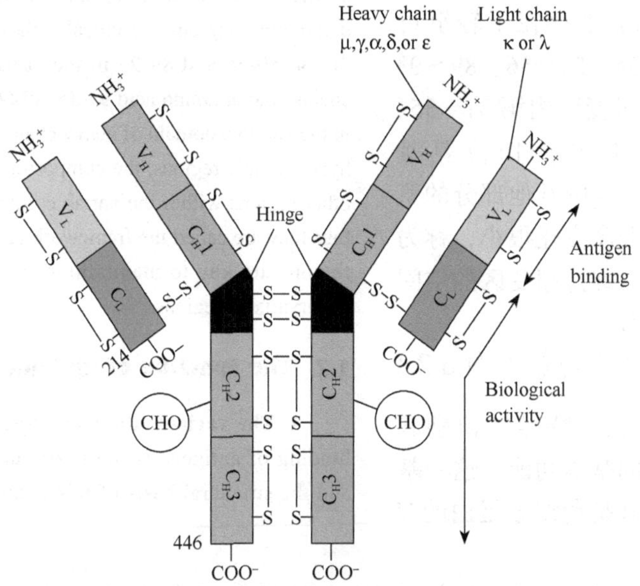

图6-1 免疫球蛋白的分子结构

1\. **重链和轻链** 免疫球蛋白分子中短的一对肽链称为轻链，由214个氨基酸残基组成，相对分子量约为25×10^3。长的一对肽链称为重链，由450～570个氨基酸残基组成，相对分子质量为50×10^3～70×10^3。

2\. **可变区与恒定区** 通过分析不同免疫球蛋白重链和轻链的氨基酸序列，发现重链和轻链靠近N端的

1. The heavy and light chains The light chain is the small polypeptide subunit of an Ig. It is composed of 214 amino acid residuals, and has a relative molecular weight of about 25×10^3. The heavy chain is the large polypeptide subunit of an antibody. It consists of 450-570 amino acid residuals, and has a relative molecular weight of 50 ×10^3-70 ×10^3.

2. The variable and constant domains By analyzing the amino acid sequences of heavy and light chains from different Ig, scientists found that the sequences of 110 amino acids locating near the

约110个氨基酸的序列变化很大，其他部分氨基酸序列则相对恒定。免疫球蛋白轻链和重链中靠近N端氨基酸序列变化较大的区域称为可变区（variableregion，V），分别占重链和轻链的1/4和1/2；而靠近C端氨基酸序列相对稳定的区域，称为恒定区（constant region，C区），分别占重链和轻链的3/4和1/2。

N terminal of both heavy and light chains are highly variable, while the arrangements of amino acids in other regions are relatively constant. Thus, in an antibody, the regions near the N terminal of the heavy and light chains are named as the variable domains, which occupy 1/4 of the heavy chain and 1/2 of the light chain, and constant regions near the C terminal are named as the constant domains, which occupy 3/4 of the heavy chain and 1/2 of the light chain.

3. *超变区和骨架区*　在可变区中，某些特定位置的氨基酸残基显示出更大的变异性，称为超变区或互补决定区。它们分别存在于轻链可变区内第26～34、50～56、89～93位氨基酸和重链可变区内第31～35、49～66、86～91和101～110位氨基酸的位置上。可变区中其他部分的氨基酸组成和排列顺序变化很小，称为骨架区。骨架区对维持超变区的空间结构具有重要作用。

3. The hypervariable region and the framework region　Within the variable domains, amino acid residuals in specific regions display even greater variability, known as the hypervariable regions or the complementarity determining regions. Specifically, they locate at amino acid 26-34, 50-56 and 89-93 in the variable domain of light chains, and at amino acid 31-35, 49-66, 86-91 and 101-110 in the variable domain of heavy chains. Comparing with the hypervariable regions, the composition and arrangement of other regions within the variable domain are less variable, thus they are called the framework regions. The framework regions are key to maintaining the spacial structure of hypervariable regions.

1.2　免疫球蛋白的功能（图6-2）

1.2　The Function of Ig(Figure 6-2)

1. *V区的功能*　特异性识别和结合抗原是抗体的基本功能。这一基本功能的结构基础是免疫球蛋白的V

1. The variable domain　Specific recognition and binding of antigens is the basic function of antibodies, and the structural basis of this is the variable domain of

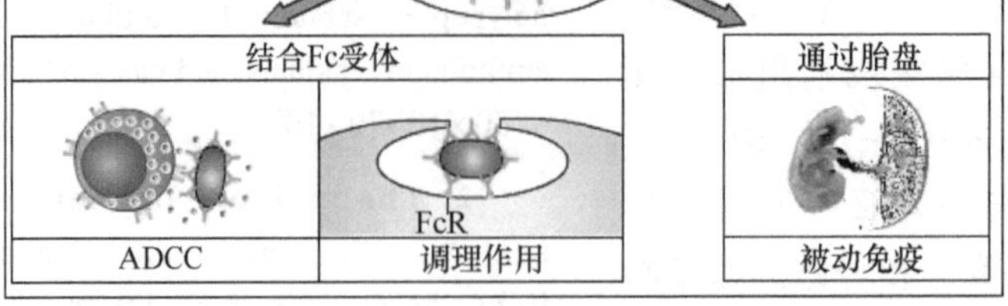

图6-2　免疫球蛋白的功能

区，特别是超变区组成的抗原结合部位。Ig结合抗原表位的个数称为抗原结合价，单体Ig为双价，分泌型Ig A为4价，五聚体IgM理论上为10价，但实际一般为5价。抗体与细菌抗原或病毒结合后，可中和细菌毒素、阻断病原微生物入侵和保护细胞免受损伤。

2. C区的功能

（1）激活补体。抗体（IgG1、IgG2、IgG3和IgM）与抗原结合后，可通过经典途径激活补体系统，产生多种效应功能；聚合的IgA、IgG4可通过旁路途径激活补体系统。

（2）结合细胞。抗体的Fc段可与多种细胞表面的Fc受体结合，介导一系列生物学功能。如调理作用，IgG的Fc段与巨噬细胞、中性粒细胞表面的IgG Fc受体结合，促进吞噬细胞对抗原的吞噬。

（3）抗体依赖的细胞介导的细胞毒作用（antibody-dependent cell-mediated cytotoxicity，ADCC）。表达Fc受体的杀伤细胞识别抗体的Fc段，通过释放介质直接杀伤被抗体结合的靶细胞。

（4）介导Ⅰ型超敏反应。IgE的Fc段与嗜碱性粒细胞、肥大细胞表面IgE Fc受体结合，参与Ⅰ型超敏反应的发生。

3. 穿过胎盘和黏膜 在人类，IgG是唯一能通过胎盘到达胎儿体内的免疫球蛋白，从而形成婴儿的天然免疫；IgA可通过呼吸道和消化道黏膜，是局部免疫的重要因素。

4. 各类免疫球蛋白的特性与功能 根据可变区抗原性质的不同，人们将重链分为γ、α、μ、δ、ε链

an Ig, particularly the antigen-binding sites formed by hypervariable regions. The number of antigen-binding epitopes on an Ig is called the antigenic valence. An Ig monomer has an antigenic valence of 2, a secretory Ig A has 4, and a pentameric IgM has 10 in theory, but often 5 in normal. After binding the bacteria or virus, antibodies could neutralize bacterial toxins, block the invasion of pathogenic microorganisms, and protect cells from damage.

2. The constant domain

(1) Complement activation After binding with antigens, antibodies (IgG1, IgG2, IgG3 and IgM) could activate the complement system via the classic pathway, which produces various effector functions, and the polymerized IgA and IgG4 could also activate complements through the alternative pathway.

(2) Cell binding The Fc regions of antibodies could bind the Fc receptors locating on the surface of many cells. Thus, the Fc regions could mediate a series of biological functions, such as the opsonization, where the Fc of IgG binds to its FcR on macrophages and neutrophils, thereby promoting the phagocytosis of antigens.

(3) Antibody-dependent cell-mediated cytotoxicity FcR-expressing killer cells could recognize the Fc of antibodies, thus to directly kill the antibody-bound target cells by releasing diverse cytokines.

(4) Mediating type I hypersensitivity The Fc of IgE could be recognized by the FcR on basophils and mast cells, thus to participate in the development of type I hypersensitivity.

3. Crossing the placenta and mucosa In human, IgG is the only Ig that can cross the placenta and reach the fetus, thus it forms the natural immunity for newborns. The IgA can cross the respiratory and gastrointestinal mucosa, and it is key to our local immunity.

4. Features and functions of different types of Ig Due to differences in the variable domain, the heavy chain is divided into five types: γ, α, μ, δ and ε. Accordingly, they

5种；与此对应的Ig分为5类，即IgG、IgA、IgM、IgD和IgE。各类Ig存在于机体的不同部位，并发挥着不同的效应功能。IgM主要在感染早期起作用；IgA是黏膜表面的主要免疫球蛋白，是机体防御外来抗原的第一道屏障；IgE主要结合在肥大细胞上，是机体防御外来抗原的第二道屏障；而IgG主要存在于血液和组织液中，通过激活补体和调理作用促进吞噬细胞对病原体的吞噬作用。不同部位的Ig各司其职，在同一部位不同种类的Ig协同发挥作用，最终将病原体清除。

define classes of Ig as IgG, IgA, IgM, IgD and IgE. Different Ig distribute in different parts of the body, and also exert different effector functions.IgM exerts its function in the early phase of infection; IgA mainly distributes on mucosal surface, and plays as the body’ s first barrier against foreign antigens; IgE mainly bounds to mast cells, acting as the second barrier defending foreign antigens; as for IgG, it is found in the blood and tissue fluid, and functions to promote phagocytosis of pathogens via complement activation as well as immune conditioning. In different parts of the body, immunoglobulins perform their own duties, and in the same part of the body, different immunoglobulins act synergistically to remove pathogens from the body.

2 单克隆抗体

2 Monoclonal Antibody（McAb）

2.1 单克隆抗体概述

1890年Behring和北里柴三郎等人发现了白喉抗毒素，并建立了血清疗法，开抗体制药之先河。动物来源的抗血清用于肺炎、白喉、麻疹等传染病的早期治疗，显示了一定效果，但毒副作用大，特别是异源蛋白易产生过敏反应。后来由于磺胺类药物和抗生素的出现，抗血清的治疗逐渐被废弃。

直至1975年，Kohler和Milstein等首次利用B淋巴细胞杂交瘤技术制备出单克隆抗体（monoclonal antibody，McAb），开创了抗体技术的新时代。单克隆抗体具有高度特异性、均一性，又有来源稳定、可大量生产等特点，为抗体的制备和应用提供了全新的手段，还促进了

2.1 Introduction of McAb

In 1890, Behring, Kitasato Shibasaburo and others found diphtheria antitoxin, and established serum therapy. It was the beginning of antibody for biopharmaceuticals. Antiserum from animals has been used with limited efficacy in the early therapy of pneumonia, diphtheria, measles and other kinds of contagion, and with significant toxicity, especially allergic reaction caused by the heterogeneous proteins. With the advent of sulpha-drugs and antibiotics, the antiserum-therapy has fallen into disuse.

Till the year 1975, Kohler, Milstein and others have first developed hybridoma technique for producing monoclonal antibody (McAb). This began a new era of technique for antibody. Monoclonal antibody (McAb) is highly specific, homogeneous, and has a steady supply, high-yield. McAb not only provides a new method of preparation and application of antibody, but also promotes preclinical medicine, clinical medicine and other subjects' development. McAb as antibody preparation is mostly used for diagnosis and treatment of diseases in clinical

基础医学和临床医学等众多学科的发展。单克隆抗体作为抗体制剂，在临床实践中主要用于疾病的诊断和治疗两个方面。如利用单克隆抗体检测与某些疾病有关的抗原，辅助临床诊断，或用放射性元素标记单克隆抗体进行肿瘤显像，做免疫定位诊断。单克隆抗体制剂也用于临床治疗，如1986年，美国FDA批准了世界上第一个单抗治疗性药物——抗CD3单抗OKT3进入市场，用于治疗移植物抗宿主病（GVHD）。

practice. Such as, to detect antigens associated with some disease, as the assistance index of diagnosis; or use the McAb labeled with radionuclide in tumor imaging for immune anatomical diagnosis. McAb is also used in clinical practices. In 1986, anti-CD3 monoclonal antibody-OKT3, the first McAb therapeutic drug, was approved by FDA of U.S.A. for marketing as a treatment drug of graft-versus-host disease (GVHD).

但是，随着使用单抗进行治疗的病例数增加，鼠源单抗用于人体的毒副作用也越来越明显。由于大多数单抗均为鼠源性，在人体内反复应用会引起人抗鼠抗体（HAMA）反应，从而降低疗效，甚至可引起过敏反应。人们对于治疗用单抗的热情开始下降。之后，噬菌体展示技术、核糖体展示技术、酵母展示技术、转基因鼠技术等的发展，使人源性基因工程抗体的获得不再是难题。这又激发了单抗药物的研究热情。1994年，美国批准第二个抗体药物上市。之后，抗体药物不断上市。2006年利用能够产生人类抗体的转基因小鼠XenoMouse技术首个完全人源化单克隆抗体药物Panitumumab上市，标志着人源化抗体技术的发展达到了新的水平。

However, with the increase of the amount of treatment with McAb, the mouse derived McAb' s the toxic effect for human body was more and more obvious. Most McAbs were all derived from mouse, so repeated injection of eventually leaded to the production of human anti mouse antibody or induced anaphylaxis. The enthusiasm for treatment with McAb began to remit because of this problem. Later, with the development of Phage Display, Ribosome Display, Yeast Surface Display, Transgenic Mouse Technique, etc., acquirement of human derived antibody became easily. The enthusiasm of researching the McAb drugs was burned again. In 1994, the second antibody drug was approved by U.S.A. for marketing, then, the number of antibody drugs on the market were gradually increasing. In 2006, thanks to the transgenic mouse technique-XenoMouse-which could produce human antibody, the first human derived antibody drug Panitumumab was on the market. This meant that the development of human derived antibody technique had reached a new level.

解决多克隆抗体特异性不高的理想方法是制备单一表位特异性的抗体。如能获得仅针对单一表位的浆细胞克隆，使其在体外扩增并分泌抗体，就有可能获得单一表位特异性的抗体。

Prepare specific antibody with mono-epitope was the ideal way to solve the problem about the low specificity of polyclonal antibody. If we can get a clone of plasma cells with specific mono-epitope, and let it amplify in vitro and secrete antibody, in this way, we may get the specific antibody with mono-epitope. But, in vitro, plasma cells have a short life and are hard to

然而，浆细胞在体外的寿命较短，也难以培养。为克服此缺点，Kohler 和 Milstein 将可产生特异性抗体但短寿的 B 细胞与不应抗原特异性 Ab 但长寿的骨髓瘤细胞融合，建立了可产生单克隆抗体的杂交瘤细胞和单克隆抗体技术（图 6-3）。通过该技术融合形成的杂交细胞系（杂交瘤，hybridoma），既有骨髓瘤细胞大量扩增和永生的特性，又具有免疫 B 细胞合成和分泌特异性抗体的能力。每个杂交瘤细胞由一个 B 细胞融合而成，而每个 B 细胞克隆仅识别一种抗原表位，故经筛选和克隆化的杂交瘤细胞仅能合成及分泌抗单一抗原表位的特异性抗体，是为单克隆抗体。其优点是结构均一、纯度高、特异性强、效价高、血清交叉反应少或无、制备成本低；缺点是其鼠源性对人具有较强的免疫原性，人体反复使用后可诱导产生人抗鼠的免疫应答，从而削弱了其作用，甚至导致机体组织细胞的免疫病理损伤。

culture. To overcome this shortcoming, Kohler and Milstein fused the specific antibody secreting and short-life B cell and nonspecific antibody secreting but long-life myeloma cell, and established a hybridoima cell and monoclonal antibody technique which could secrete monoclonal antibodies (Figure 6-3). The hybrid cell line (hybridoma)which is derived from the new technique has the ability not only to amplify and live forever as the myeloma cell, but also to synthesis and secrete specific antibody as the B cell. Every hybridoima cell is fused by one B cell, and every B cell only recognize one epitope, so after screening and cloning, the obtained hybridoima cell can synthesis and secrete specific antibody with mono-epitope, and this kind of antibody is named monoclonal antibody. The advantages of the monoclonal antibody are homogeneous-structure, high purity, strong specificity, with few or no cross reaction and cheapness; but for human, mouse derived has a very high immunogenicity, repeated injection of eventually leaded to the production of human anti mouse antibody, so this weakens its effect, even causes the immune pathological injure to tissue cells.

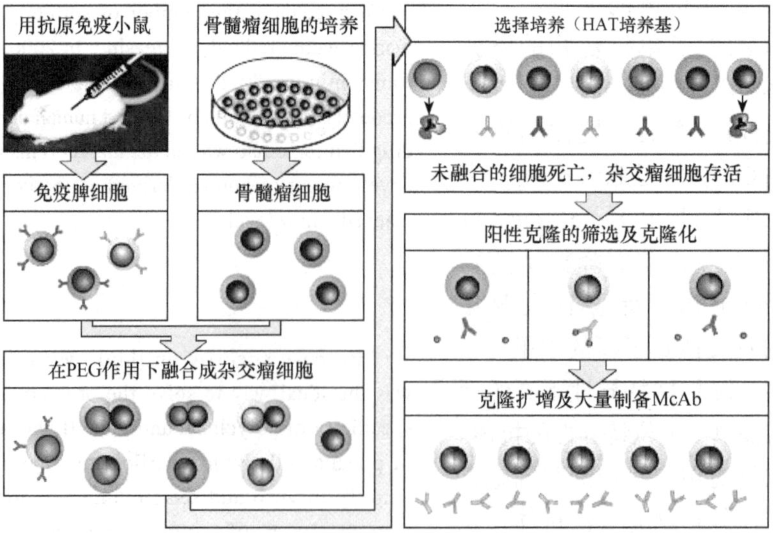

图 6-3　单克隆抗体制备示意

2.2 单克隆抗体的制备

2.2.1 抗原的制备

制备抗体用的抗原在种类上几乎没有限制，病毒、细菌等病原微生物及对免疫动物来讲为异种的外来物质，凡是具有抗原性质的物质均可制备单克隆抗体，抗原可以是全病毒、亚单位组分或者是纯化分子，也可以是蛋白质、多糖、脂类等。制备单克隆抗体的免疫抗原，从纯度上说虽不要求很高，但高纯度的抗原使得到所需单抗的机会增加，同时可以减轻筛选的工作量。因此，免疫抗原是越纯越好，应根据所研究的抗原和实验室的条件来决定。一般来说，抗原的来源有限，或性质不稳定，提纯时易变性，或其免疫原性很强，或所需单抗是用于抗原不同组分的纯化或分析等，免疫用的抗原只需初步提纯甚至不提纯。但抗原中混杂物很多，特别是如果这些混杂物的免疫原性较强时，则必须对抗原进行纯化。在进行筛选时，应尽可能采用均一的所期望的抗原来检测，可提高获得特异性杂交瘤的概率，提高工作效率。

在免疫接种前，检查抗原中是否含有潜在病原体是必要的。这适用于用细胞系、细胞系产生的蛋白质，或者在有血清存在的情况下制备的蛋白质。筛选试验，称为小鼠抗体生产筛查试验（mouse antibody production screening，MAPS），包括用每种特定抗原（2×10^7 个细胞或 10～100μg 蛋白质）接种经检疫的无特定病原体（SPF）动物，观察 4 周。4 周后，收

2.2 Preparation of Monoclonal Antibody

2.2.1 Preparation of Antigen

There are no strict limitations for the type of antigens. Virus, bacteria, pathogens, even foreign substances for immunized animals, any substance that has antigenic properties can be used to prepare mAbs. The antigen can be whole virus, subunits or purified molecules, and it can also be proteins, polysaccharides and lipids. Generally, the purity requirement of immunogenic antigens used for mAb preparation is not very high, but antigens with higher purity could increase the chance of obtaining desired mAbs, and would reduce the workload of mAb screening. Thus, the purity of antigen should not be determined by “the purer the better”, but by the properties of the antigen and the conditions of the laboratory. Generally, if the antigens are with limited sources, unstable, easy to denature during purification, with high immunogenicity, or if the desired mAbs are used to purify or analyze different antigen components, the antigens used for mAb preparation need only primary purification, or even need not to be purified. On the other hand, if the antigens are contaminated by many impurities, especially when these impurities are with high immunogenicity, purification must be performed before mAb preparation. During mAb screening, uniform antigens should be used as possible, because this could increase the chance of obtaining specific hybridomas, thereby improving the work efficiency.

Before immunization, checking whether the antigen contains potential pathogens is very important. This can be applied to cell lines, cell line-derived proteins, or proteins prepared under the existence of serum. The screening test, also named mouse antibody production screening (MAPS), requires to immunize specific pathogen-free (SPF) animals with each specific antigen (2×10^7 cells or 10-100 μg protein), and to observe them for 4 weeks. After 4 weeks, serum from the immunized animals will be collected and compared with the serum before immunization to detect the reactions of known pathogens. Positive results indicate that the antigen has

集免疫后动物的血清，和免疫前的血清比较对一系列已知病原体的反应。阳性结果意味着抗原已被污染，应当抛弃；阴性结果表明抗原是安全的，可用于体内注射。

already been contaminated and should be abandoned, while negative results suggest the antigen is safe, and could be used for *in vivo* injection.

2.2.2 动物的选择

根据所用的骨髓瘤细胞可选用小鼠或大鼠作为免疫动物，因为所用的供杂交瘤技术用的小鼠骨髓瘤细胞系均来源于BALB/c小鼠，所用的大鼠骨髓瘤细胞都来源于LOU/c大鼠，所以一般的杂交瘤生产都是用这两种纯系动物作为免疫动物。最常用是BALB/c小鼠，在选择动物时应考虑到动物品系的免疫应答基因对抗原免疫应答的影响。如果BALB/c小鼠对所用抗原不能产生良好的免疫应答时，应改用其他品系小鼠或大鼠。就小鼠而言，初次免疫时间以8～12周龄为宜，雌性鼠较便于操作。亚美尼亚仓鼠是独特的模型，因为从系统发生学上它们的亲缘关系和小鼠相差甚远，所以能够对小鼠、大鼠或人类来源的蛋白质产生良好的免疫应答，而且能够和小鼠骨髓瘤细胞系发生有效的融合。更重要的是，亚美尼亚仓鼠产生的抗体在小鼠体内是无免疫原性的，这使它们成为一些疾病的理想的体内模型。这种模型已经广泛用于制备高亲和力的针对小鼠细胞因子、受体和转录因子等的抗体。

2.2.2 Animal Selection

Based on the selected myeloma cells, mice or rats are often used as models to be immunized. This is because all mouse myeloma cell lines used for hybridoma production are derived from the BALB/c mouse, and all rat myeloma cell lines are derived from the LOU/c rat. Thus, the above two animals are commonly used for many hybridoma productions. The BALB/c mouse is the most frequently used immunized animal. When selecting animals, it should be concerned that whether the animal' s immune response genes would affect the response of antigens. If the BALB/c mice could not develop desired immune responses, mice or rats of other strains should be considered. As for mice, the optimal time for primary immunization would be 8-12 weeks, and females can be handled more easily. The Armenian hamster is a unique model. According to phylogenetics, the hamsters have a very far genetic relationship with mice, so they can develop good immune responses to proteins that are derived from mice, rats or human. Besides, the hamsters also show effective infusion with the mouse myeloma cell lines, and most importantly, antibodies generated from the Armenian hamsters are with no immunogenicity in mice, which made the hamster an ideal *in vivo* model for some diseases. Nowadays, the Armenian hamster model has been widely used in the preparation of antibodies with high affinity to mice cytokines, receptors, transcription factors, etc.

一种替代细胞融合制备单克隆抗体的方法是使用淋巴细胞条件转化鼠（immorto-mouse）。这些转基因小鼠携带SV40 T抗原的温度敏感突变体，该基因由H-2K^{b}启动子控

Another method that can substitute cell fusion for mAb preparation is to use the immorto-mouse. These transgenic mice carry a temperature-sensitive mutant of the SV40 T antigen, which can be regulated by the H-2K^b promoter. After immunization of these mice, their spleen cells will express thermally unstable SV40 T antigen and further induces B cell transformation only

制。免疫上述转基因动物后，脾细胞在允许的33℃的培养条件下能够表达热不稳定的SV40 T抗原，导致B细胞的转化，从而不需要与永生细胞系融合，而且还避免了由于异核体细胞（杂交瘤细胞）的产生而出现的不稳定现象。从这些不断生长的B细胞中能筛选到分泌特异性抗体的细胞株。

under the allowed temperature of 33℃. This process does not require the fusion with immortalized cell lines, and avoids unstable circumstances caused by heterocaryon cells (hybridomas). Finally, cell lines secreting specific antibodies could be screened from the continuously growing B cells.

在完善的杂交瘤技术的基础上努力开发更加有效的完全人源化抗体，促使了Abge-nix公司XenoMouse的发展。这些转基因小鼠完全缺乏小鼠*JH*和*Ck*基因，致使它们缺乏所有类型的小鼠免疫球蛋白。这些缺乏*JH*和*Ck*基因的小鼠（Ig基因−/−小鼠）也可通过转基因的方式导入人类*IgH*及*Igk*链基因。因此，转基因小鼠XenoMouse具有类似于成年人那样的能够产生多样性初始免疫应答能力的遗传成分，在接受抗原刺激后能够产生强有力的再次应答。常规免疫这些转基因小鼠，将免疫小鼠脾细胞与小鼠源骨髓瘤细胞融合，是制备一系列高亲和力的能够和任何数量蛋白质反应的完全人源化抗体的新策略。这些抗体具有中和活性，很多已被批准用于人类疾病的治疗。

On the sound basis of hybridoma technology, researching of fully humanized antibodies with even higher efficiency has facilitated the development of the XenoMouse from Abge-nix Company. These transgenic mice are knocked out with the *JH* and *Ck* genes, making them lack all types of mice immunoglobulin. However, these mice lacking the *JH* and *Ck* genes (Ig -/- mice) could be introduced with human *IgH* and *IgK* genes via gene modification, so that they, like adult human, can gain the genetic components that produce diverse initial immune responses, and can generate strong secondary responses upon receiving antigen stimulation. By routine immunization of these transgenic mice and by fusing their spleen cells with mice myelomas, a series of fully human antibodies that are with high affinity and can react with any amounts of proteins could be produced. These antibodies are with neutralizing activities, and many have been listed for the treatment of human diseases.

2.2.3 免疫方法及免疫程序

免疫方法是单抗制备过程中的重要环节之一，其目的在于使B淋巴细胞在特异性抗原刺激下分化、增殖，以利于细胞融合形成杂交细胞，并增加获得分泌特异性抗体的杂交瘤的机会。迄今为止尚没有一种适用于各种抗原的免疫方法，但

2.2.3 Immunization Methods and Procedures

Immunization is an important part of mAb preparation. The target of immunization is to stimulate B-lymphocytes with specific antigens, and to induce cell differentiation and proliferation, thereby facilitating the fusion of cells into hybridomas and increasing the chance of obtaining hybridomas that can secrete specific antibodies. By now, there is no one immunization method can suit all antigens, but commonly used immunization methods are listed in below.

常用的有以下几种。

1. *体内免疫法* 适用于免疫原性强、抗原量较多时应用，一般用8～12周龄的雌性鼠。颗粒性抗原（如细菌、细胞抗原）的免疫原性强，可不加佐剂，直接注入腹腔10^7个细胞进行初次免疫，间隔1～3周，再追加免疫1～2次。可溶性抗原则按每只小鼠10～100μg抗原与弗氏完全佐剂等量混合后腹腔内注射，进行初次免疫，间隔2～4周，再用不加佐剂的原抗原追加免疫1～2次。一般在采集脾细胞前3d由静脉注射最后一次抗原，其目的是使对应的B淋巴细胞克隆受到可靠的最大限度的刺激，使其迅速地增殖分裂，因为细胞融合后增殖最好的细胞是迅速分裂的细胞。在免疫后第三天淋巴细胞的增殖率最高，由静脉注射就是为了提高抗原浓度，刺激更多的B细胞。

2. *脾内免疫法* 即在麻醉条件下直接把0.1～0.2ml的抗原注入脾脏进行免疫。脾内免疫法可提高小鼠对抗原的免疫反应性，节省抗原用量，细胞抗原只需10^5个左右，可溶性抗原只需10μg左右。但多数人认为脾内免疫属初次免疫应答，产生IgM类抗体居多，故主张在常规免疫的基础上用脾内免疫法做追加免疫为佳。

3. *体外免疫法* 这在不能采用体内免疫法的情况下，如制备人源性单克隆抗体，或者抗原的免疫原性极弱且能引起免疫抑制时使用。体外免疫法的优点很多：所需抗原量少，一般只需几个微克；免疫期短，仅

1. *In vivo* immunization This method is often applied when the antigen is with high immunogenicity and large amounts. Female mice aged 8-12 weeks are often used as models. Particulate antigens (such as bacteria and cell antigens) are with high immunogenicity, so they can be directly used for intraperitoneal injection at an amount of 10^7 cells with out the adding of adjuvants. The interval of immunization could be 1-3 weeks, and booster immunization could be carried out for 1 or 2 times. As for soluble antigens, they can be injected into the mice abdominal (10-100 μg per mouse) with equal amounts of Freund's complete adjuvant for primary immunization. The interval of immunization could be 2-4 weeks, and booster immunization could be performed 1-2 times without the using of adjuvants. Generally, antigens are intravenously injected for the last time at 3 days before spleen cell collection, and this aims at maximizing the stimulation of B-lymphocyte cloning and proliferating, because after cell fusion, cells with the best proliferation are those under rapid cell division. Lymphocytes normally show the highest proliferation rate at 3 days after immunization, so intravenous injection can increase the antigen concentration and induce the production of more B cells.

2. Splenic immunization This method directly injects 0.1-0.2 ml of antigen into the animal's spleen under anesthesia conditions. Splenic immunization can increase mice immunoreactivity and requires less amount of antigen, as only 10^5 cell antigens or 10 μg of soluble antigens are needed. However, many scholars believe that splenic immunization only triggers primary immune responses and most produced antibodies are IgM, so that splenic immunization is recommended as a method of booster immunization after animals have been routinely immunized.

3. *In vitro* immunization This method is performed when *in vivo* immunization could not be performed, especially when preparing human mAbs, or when antigens are with extremely low immunogenicity, and could activate immunosuppression. *In vitro* immunization has many advantages. It needs fewer antigens, several micrograms in general, has a short

4～5d；干扰因素少，已成功制备出针对多种抗原的单克隆抗体，但融合后产生的杂交瘤细胞株不够稳定。其基本方法是用4～8周龄BALB/c小鼠的脾脏制成单个细胞悬液，再加入适当抗原使其浓度达0.5～5μg/ml，在5%CO_2，37℃下培养4～5d，再分离脾细胞，进行细胞融合。

在设计免疫程序时，应考虑到抗原的性质和纯度、抗原量、免疫途径、免疫次数与间隔时间、佐剂的应用及动物对该抗原的应答能力等。没有一个免疫程序能使用于各种抗原。现用的免疫程序中多数是参照制备常规多克隆抗体的方法。表6-1列举了目前常用的免疫程序。免疫途径常用体内免疫法，包括皮下注射、腹腔或静脉注射，也采用足垫、皮内、滴鼻或点眼。最后一次加强免疫多采用腹腔或静脉注射，目前尤其推崇后者，因为可使抗原对脾细胞作用更迅速而充分。在最后一次加强免疫后第3天取脾融合为好。

immunization period, only 4-5 days, and is with less interfering factors. Currently, many mAbs specific for various antigens have been prepared, but the hybridoma strains formed after cell fusion are unstable. A basic strategy for this is to sample the spleen of BALB/c mice (4-8 weeks) and make it into single cell suspensions; then add corresponding antigens to a concentration of 0.5-5μg/ml, and culture the cell at 5% CO_2 and 37℃ for 4-5 days; afterwards, spleen cells can be isolated for cell fusion.

When designing the immunization procedure, the property, purity and amount of antigen, the pathway, time and interval of immunization, the use of adjuvant and animal's response to the antigen, all these factors need to be considered. No one immunization procedure can be used for all antigens. Nowadays, most immunization procedures are designed in reference to the routine preparation of mAbs. Table 6-1 lists some commonly used immunization procedures, including subcutaneous, intraperitoneal or intravenous injection, and sometimes antigens are injected via animal's footpad, skin, nose or eyes. For the last booster immunization, intraperitoneal or intravenous injection are recommended, particularly the later one, because these pathways can enhance the reaction between antigens and spleen cells, and spleen sampling is recommended at 3 days after the last immunization.

表6-1 不同免疫抗原的免疫程序

免疫原特性	抗原量	接种次数	间隔时间	单抗滴度	单抗亲和性
免疫原性强（如细胞、细菌和病毒等）	10^6～10^7个细胞或1～10μg	2～4	2～4周	高	中等至强
免疫原性中等	10～100μg	2～4	2～4周	中等或高	中等或强
免疫原性弱	A 20～400μg	2～4	每月	中等	强
		随后2～3	2～3月		
	B 10～50μg	2	每月	中等	中等
	其后200～400μg	其后	每天		
	C 10～100μg	2	每月	中等	中等或强
		其后4	10d		
		其后“休息”	1～2月		
		最后加强			

2.2.4 骨髓瘤细胞的选择

在B淋巴细胞杂交瘤技术中，主要使用多发性骨髓瘤细胞作为亲本细胞，小鼠的骨髓瘤细胞系，最初都来源于由矿物油诱导Balb/c小鼠产生的浆细胞瘤（MOPC21），在体外经8-氮杂鸟嘌呤培养（P3K）和选择后，获得嘌呤核苷酸补救合成途径中酶缺陷的克隆（$HGPRT^-$），同时也不分泌免疫球蛋白的重链或轻链。这些骨髓瘤细胞在培养基中可以连续生长。然而，在有选择试剂（如氨基蝶呤、甲氨蝶呤或重氮丝氨酸）存在的情况下，核苷酸的从头合成途径被这些试剂阻断，这些缺乏HGPRT酶的杂交瘤细胞由于不能使用核酸合成补救合成途径而无法生存。因此，这些骨髓瘤细胞就成了能用选择培养剂筛选杂交瘤的非常好的模型。目前最常用的骨髓瘤亲本细胞种类及性状见表6-2。

2.2.4 Selection of Myelomas

In B cell hybridoma technology, multiple myeloma cells are usually used as parent cells. For mice myeloma lines, all of them are derived from the plasmacytoma (MOPC21) generated by mineral oil-induced Balb/c mice, which obtains a defect gene ($HGPRT^-$) for the salvage synthesis of purine nucleotides after selective culture with 8-azaguanine (P3K), and does not secrete the heavy or the light chain of immunoglobulins. Mice myeloma cells can grow continuously on the culturing plate, but under the existence of some selective agents, such as aminopterin, methotrexate or azaserine, the *de novo* pathway of nucleotide synthesis would be blocked, and myelomas lacking the HGPRT enzyme could not survive because the salvage pathway for nucleotide synthesis could not be initiated. Thus, these myelomas have become an ideal model for selective culturing and screening of hybridomas. Commonly used parent myeloma cells and their traits are listed in Table 6-2.

表 6-2 骨髓瘤细胞系

全名	简称	来源	耐受	表达自身IgG分子
P3-X63-Ag8	X63	BALB/c	8-Ag	$Ig\text{-}r_1 \cdot \kappa$
MOPC11-X45-6TG	X45	BALB/c	6-TG	$Ig\text{-}r_2b$
P3-NS1-Ag4.1	NS-1	BALB/c	8-Ag	κ（非分泌型）
P3-X63-Ag8.653	P3.653	BALB/c	8-Ag	—
SP2/0-Ag14	SP2/0	BALB/c	8-Ag	—
Fast-Zero	FO	BALB/c	8-Ag	—
Y3-Ag1，2，3	Y3	Lou	8-Ag	κ链
YB2-3.0-Ag20	YB2	（Lou × AO）F1	8-Ag	—

注：8-Ag指8-杂氮鸟嘌呤；6-TG指6-巯基嘌呤。

无论采用小鼠或大鼠杂交瘤技术，应尽可能选择自身不合成或至少不分泌

Regardless of the use of mice or rat hybridoma techniques, a principle for parent cell selection is to choose

任何免疫球蛋白分子或片段的骨髓瘤细胞作为亲本细胞。无论选用哪种骨髓细胞，其细胞必须处于良好的生长状态，处于对数生长期的瘤细胞最佳。一般认为骨髓瘤细胞浓度低于 1×10^4/ml 时生长缓慢，大于 1×10^6/ml 时细胞分裂逐渐停止，因此，一般细胞传代选择（1～5）$\times10^5$/ml 的浓度，在融合前一天倍比传代一次后用于融合。融合时，活细胞数应至少大于 90%。

myelomas that would not synthesize or at least do not synthesize any immunoglobulin molecules or fragments. No matter what myeloma is used, it should be guaranteed that cells are under good growth state, and normally cells in the logarithmic growth phase have the optimal growth rate. It is found that myeloma would grow slowly under a cell concentration of 1×10^4/ml, and would stop proliferation if the cell concentration is greater than 1×10^6 /ml. Thus, cell passaging should be performed at a concentration of (1-5) $\times 10^5$/ml. Cells should be passaged at 1 day before fusion, and the percentage of living cells should be greater than 90% when used for cell fusion.

2.2.5 细胞融合

1. *饲养细胞的制备* 细胞融合后选择性培养过程中，由于大量骨髓瘤细胞和脾细胞相继死亡，此时单个或少数分散的杂交瘤细胞多半不易存活，通常必须加入其他活细胞使之繁殖，这种被加入的活细胞称为饲养细胞（feeder cells）。饲养细胞促进其他细胞增殖的机制尚不明了，一般认为它们可能释放非种属特异性的生长刺激因子，为杂交瘤细胞提供必要的生长条件；也可能是为了满足新生杂交瘤细胞对细胞密度的依赖性。常用的饲养细胞有胸腺细胞、正常脾细胞和腹腔巨噬细胞，其中以小鼠腹腔巨噬细胞的来源及制备较为方便，且有吞噬消除死亡细胞及其碎片的作用，因此使用最为普遍。一般饲养细胞在融合前一天制备，一只小鼠可获得 5×10^6～8×10^6 个腹腔巨噬细胞，若用小鼠胸腺细胞作为饲养细胞时，细胞浓度为 5×10^6 个 /ml，小鼠脾细胞为 1×10^6 个 /ml，小鼠的成纤维细胞（3T3）1×10^6 个 /ml，均为 100μl/ 孔。

2.2.5 Cell Fusion

1. Preparation of feeder cells After cell fusion, a large number of myeloma or spleen cells would die during selective culturing, and single or dispersed hybridoma cells can hardly survive under such conditions. Normally, other living cells would be added to facilitate the proliferation of hybridomas, and these living cells are called the feeder cells. The mechanisms by which feeder cells promote the proliferation of other cells are still unknown, but it is believed that they can release some non-species-specific growth factors and provide essential conditions for the growth of hybridomas. But others suggest that the adding of feeder cells might provide the cell density that is needed for hybridomas to grow. Commonly used feeder cells include thymus cells, normal spleen cells, and peritoneal macrophages. The mice peritoneal macrophages are the most widely used feeder cells, because they can be gained and prepared easily, and can clear the dead bodies or debris of other cells. Normally, feeder cells are prepared a day before cell fusion. A total of 5×10^6-8×10^6 peritoneal macrophages can be obtained from one mouse. For the concentration of different feeder cells, mice thymus cells are used at 5×10^6/ml, spleen cells at 1×10^6/ml, and mice fibroblast cells at (3T3) 1×10^6 /ml (for medium 100 μl/well).

2. *骨髓瘤细胞悬液的准备* 骨髓瘤细胞要用无血清培养基在 4℃条件下洗 3 次，因为血清中蛋白质会干

2. Preparation of myeloma suspension Myeloma cells need to be washed at 4℃ for 3 times with serum-free medium, because proteins in the serum may interfere

扰膜融合。用无血清的 DMEM 重悬骨髓瘤细胞，对其进行计数，调整浓度为 1×10^7 个 /ml，放在 4℃。骨髓瘤细胞的活细胞比例应大于 95%。

3. *免疫脾细胞悬液的制备* 血清效价有意义（>1：1000）的免疫后动物可以提供抗原反应性 B 细胞用于融合。这些细胞可以从脾脏或淋巴结中获得。动物用吸入 CO_2 处死，在无菌操作下取组织，放在 4℃无血清培养基中，最多可放 1h。用镊子或针头轻轻打开组织脾脏或淋巴结的外包膜获得单细胞悬液，不能用毛玻璃磨碎组织取细胞，因为那样可能损害细胞膜的完整性。在准备骨髓瘤细胞的同时，脾细胞或淋巴结细胞用无血清培养基洗 3 次（4℃，1000r/min 离心 10min），计数并用 20ml 无血清培养基重悬，放入 50ml 聚丙烯离心管，置于冰上。通常，1 只免疫后小鼠可收获 1×10^8 个脾细胞。

4. *细胞融合* 诱导细胞融合的方法有生物学方法（仙台病毒）、物理学方法（电场诱导、激光诱导等）和化学法（聚乙二醇、溶血磷脂等）。其中最常用的是聚乙二醇（PEG），一般来说，PEG 的相对分子质量和浓度越大，其促融率越高。但其黏度和对细胞的毒性也随之增大。目前常用的 PEG 浓度为 40%～50%，相对分子质量以 4 000 为佳。但相对分子质量 400～6000，浓度在 10%～60% 范围之间的 PEG 都能使细胞发生融合。细胞融合的基本方法是，取适量脾细胞（1×10^8）与骨髓瘤细胞（2×10^7～3×10^7）进行混合，在聚乙二醇（polyethylene glycol，PEG）作用下诱导它们融合，时间控制在

with the fusion of membranes. After washing, myelomas are re-suspended by serum-free DMEM for cell counting. The concentration of cells is adjusted to 1×10^7/ml, and the suspension is then stored at 4℃ for future treatment. The percentage of living myelomas should be greater than 95%.

3. Preparation of spleen cell suspension Immunized animals (titer > 1 : 1000) can provide immunoreactive B cells for fusion, and these B cells can be obtained from animal spleen or lymph nodes. Animals are killed by inhaling of CO_2, and tissues are sterilely sampled and placed into serum-free medium at 4℃ (at most for 1 hour). Single cell suspension could be obtained by carefully opening the outer membrane of spleen or lymph nodes with tweezers or needles, but grading with a frosted glass cannot be used because this may damage the integrity of the cell membrane. While myeloma suspension are being prepared, spleen or lymph node cells are washed with serum-free medium for 3 times (4 ℃, then centrifuged for 10 min at 1000 rpm), resuspended in 20 ml of serum-free medium, counted, transferred into a 50-ml EP tube, and placed on ice for future treatment. Normally, about 1×10^8 spleen cells can be harvested from one immunized mouse.

4. Cell fusion Cell fusion induction includes biological (sendai virus), physical (electric field or laser) and chemical (polyethylene glycol or lysophospholipids) methods. Among all, polyethylene glycol (PEG) is the most commonly used agent for cell fusion. The higher relative molecular weight and concentration of PEG are, the greater fusion rate will be, but the viscosity and cytotoxicity of PEG will also increase with its weight and concentration. The commonly used concentration of PEG is 40%-50%, and the optimal molecular weight is around 4000. Yet, all PEG with a molecular weight of 400-6000 and a concentration range of 10%-60% can facilitate the fusion between cells. The basic method for cell fusion is to mix spleen cells (1×10^8) with proper amounts of myelomas (2×10^7-3×10^7), and then use the PEG to induce cell fusion. The time of induction should be controlled within 1 min, and afterwards, cells are diluted with medium. Myelomas used for cell fusion should have high fusion rate, should not secrete antibodies by themselves, and should form hybridomas with strong and stable ability in antibody secretion.

1min 以内，然后用培养液将 PEG 融合液缓慢稀释。用于细胞融合的骨髓瘤细胞应具备融合率高、自身不分泌抗体、所产生的杂交瘤细胞分泌抗体的能力强且长期稳定等特点。

2.2.6 阳性杂交瘤细胞的筛选及克隆

2.2.6 Screening and Cloning of Positive Hybridomas

1. *杂交瘤细胞的选择* 由于融合剂的作用为非特异性的，即细胞管中杂交瘤细胞有骨髓瘤细胞 -B 淋巴细胞、B 淋巴细胞 -B 淋巴细胞、骨髓瘤细胞 - 骨髓瘤细胞融合的细胞。因此，必须选择出既能连续生长又具有分泌抗体的融合细胞，即骨髓瘤细胞 – B 淋巴细胞融合的杂交瘤细胞。

1. Screening of hybridomas Since the fusion agent is non-specific, fused cells contained in the tube include myeloma-B cells, B-B cells, and myeloma-myeloma cells. Thus, cells with the ability to grow continuously and secrete antibodies, i.e. myeloma-B fusion cells, should be screened out.

HAT 培养液含有特殊成分的培养基，在这种培养液中，骨髓瘤细胞是不能生长的，而只有骨髓瘤细胞与 B 淋巴细胞的融合细胞能增殖生长；此外，正常 B 淋巴细胞一般情况下也是不能连续培养。为此，在接种 96 孔板培养时，加入一定量的 HAT 选择培养液，经过 1～2 周的培养后，骨髓瘤细胞与 B 淋巴细胞的杂交瘤细胞即被选出。

The HAT medium contains a special component that can inhibit the growth of both B cells and myelomas, but not the hybridomas fused by myelomas and B cells. Thus, when inoculating cells into the 96-plate well, by adding the HAT selective medium and culturing for 1-2 weeks, the hybridomas fused by B cells and myelomas could be screened out effectively.

2. *特异性抗体的筛选* HAT 培养基选择性地培养出骨髓瘤细胞 -B 淋巴细胞融合细胞，但并不是所有的这类融合细胞都能分泌所期望的抗体，为此需对有杂交瘤生长的细胞进行特异性抗体的筛选。由于细胞产生抗体为分泌型，即在细胞培养上清中随着细胞浓度的增大，其中所含的抗体浓度也增大，为此可通过对细胞培养上清中抗体进行测定，即可达到筛选出具有分泌特异性抗体的杂交瘤细胞的目的。有关筛选检测的方法很

2. Screening of specific antibodies Though myeloma-B cell hybridomas could be screened out by HAT selective culture, it cannot be determined whether all these hybridomas produce desired antibodies. Thus, further screening of specific antibodies should also be carried out. Since hybridomas produce secretive antibodies, the concentration of antibody in the culture would increase with the proliferation of cells, and thus screening of hybridomas secreting specific antibodies could be realized by detecting the antibodies in cell culture supernatant. There are many screening methods provided, and they could be chose according different antigen properties and laboratory conditions. Generally, if the concentration of antibodies reaches a microgram per milliliter level, they can be detected by many conventional

多，需根据不同实验室的条件及抗原的情况选择合适的筛选方法。一般来讲，细胞培养上清中的抗体在每毫升微克级水平，即目前一般的检测方法敏感性均能达到。尽管如此，目前实验室最常用的仍是酶联免疫的方法，即将抗原固相在微孔板上，细胞培养上清培育一定时间后，用酶标的抗鼠二抗，通过酶促底物的颜色变化了解是否有针对该抗原的抗体存在。这种方法，简便易行且可大规模操作，目前大部分实验室均采用这种方法筛选。

methods. But still, ELISA is the most commonly used. For ELISA, the antigen is fixed on a microplate and incubated with the cell culture for a certain period of time. Then, enzyme-labeled anti-mice secondary antibody is added, and the existence of antibodies specific for this antigen could be determined by the color change of enzymatic substrates. ELISA allows for easy handling and large-scale operation, thus most laboratories use this method for antibody screening.

3. *杂交瘤细胞的克隆* 从原始孔中得到的杂交瘤细胞，可能来源于两个或多个杂交瘤细胞，因此它们所分泌的抗体是不同质的。为了得到完全同质的单克隆抗体，必须对杂交瘤细胞进行克隆化。另一方面，杂交瘤细胞培养的初期是不稳定的，有的细胞丢失部分染色体，可能丧失产生抗体的能力。为了除去这部分已不再分泌抗体的细胞，得到分泌抗体稳定的单克隆杂交瘤细胞系（又称亚克隆），也需要克隆化。另外，长期液氮冻存的杂交瘤细胞，复苏后其分泌抗体的功能仍有可能丢失，因此也应作克隆化，以检测抗体分泌情况。通常在得到预定抗原的杂交瘤细胞以后需连续进行2～3次克隆化，有时还需进行多次。所谓克隆化是指使单个细胞无性繁殖而获得该细胞团体的整个培养过程。克隆化的方法很多，如有限稀释法、软琼脂法、单细胞显微操作法、单克隆细胞集团显微操作法和荧光激活细胞分离法（FACS）。有限稀释法是把杂交瘤细胞悬液稀释后，加

3. Cloning of hybridomas Hybridomas obtained from the original wells might be derived from two or more hybridomas, thus the antibodies secreted by these hybridomas are heterogeneous. In order to obtain completely homogeneous mAbs, the obtained hybridomas must be cloned. Besides, hybridomas are very unstable during early culturing, and some cells may lose chromosomes and thereby fail to produce desired antibodies. In order to remove cells that do not secrete antibodies and obtain stable antibody-secreting monoclonal hybridomas (subclones), hybridomas also need to be cloned. Furthermore, due to long-term cryopreservation in liquid nitrogen, hybridomas may lose the ability to secrete antibody after recovery, thus cloning of hybridomas is also needed to detect antibody secretion. Generally, 2-3 times of cloning, sometimes more, are needed after obtaining hybridomas that can secrete desired antibodies. Cloning refers to the culturing from single cells to cell groups via asexual propagation. There are many methods for hybridoma cloning, such as limiting dilution, soft agar culturing, single cell micromanipulation, monoclonal cell group micromanipulation, and fluorescent activated cell sorting (FACS). The limiting dilution method dilutes hybridoma suspensions and inoculates cells into a 96-well plate, making each well contain only 1 cell theoretically. The HT medium is used during the first cloning, and HT-free RPM11640 medium could be used for subsequent clones. Since single cell can hardly survive, sometimes feeder cells are also added to improve hybridoma growth during

入 96 孔细胞培养板中，使每个孔中在理论上只含有一个细胞。第一次克隆化时也要应用 HT 培养液，以后的克隆化可用不含 HT 的 RPM11640 培养液。由于单个细胞难以存活，克隆化时也需加入饲养细胞辅助其生长。软琼脂法是在培养液中加入 0.5% 左右的琼脂糖凝胶，细胞分裂后形成小球样团块，由于培养基是半固体状态，可用毛细吸管将小球吸出，团块经打碎后，移入 96 孔板中继续培养。用这种方法可以吸出大量克隆细胞进行培养，因初代细胞很不易增殖，所以用软琼脂法进行克隆化容易成功。

cloning. The soft agar method adds 0.5% of agar gel into the medium, and after proliferation, cells can form small ball-like clumps that can be aspirated by capillary pipettes. Then, the clumps are shattered and seeded into a 96-well plate for further culturing. This method could aspirate a large number of cloned cells for future culture, thereby facilitating the cloning because it is relatively difficult for primary cells to proliferate.

2.2.7 单克隆抗体的生物特性鉴定

2.2.7 Biological Characterization of Monoclonal Antibodies

1. 杂交瘤细胞的染色体分析 对杂交瘤细胞进行染色体分析可获得其是否是真正的杂交瘤细胞的客观指标之一。杂交瘤细胞的染色体数目应接近两种亲本细胞染色体数目的总数，正常小鼠脾细胞的染色体数目为 40，小鼠骨髓瘤细胞 SP2/0 为 62～68，NS-1 为 54～56；同时骨髓瘤细胞的染色体结构上反应两种亲本细胞的特点，除多数为端着丝点染色体外，还出现少数标志染色体。另一方面，杂交瘤细胞的染色体分析对了解杂交瘤细胞分泌单抗的能力有一定的意义，一般来说，杂交瘤细胞染色体数目较多且较集中，其分泌抗体能力则高，反之，其分泌单抗能力则低。

1. Chromosome analysis of hybridomas Chromosome analysis could provide objective indicators about whether the obtained cells are desired hybridomas. Theoretically, the number of hybridoma chromosomes should be close to the total number added by the chromosomes of two parent cells. As for normal mice spleen cells, the number of chromosomes should be 40. As for mice myeloma SP2/0 and NS-1, the number of chromosomes should be around 62-68 and 54-56 respectively. Also, hybridomas should have characteristic chromosomal structures that are derived from their parent cells. Specifically, most of the chromosomes should be telocentric, and some marker chromosomes should be detectable. Moreover, chromosome analysis of hybridomas is of certain significance for us to understand the ability of mAb secretion. Generally, hybridomas with large number of and more concentrated chromosomes may have higher ability in antibody secretion, and vise versa.

检查杂交瘤细胞染色体的方法最常用秋水仙素法，其原理是应用秋水仙素特异地破坏纺锤丝而获得中期分裂相细胞；再用 0.075mol/L KCl 溶液等低渗处理，使细胞膨胀，体积增

Colchicine is used most commonly in detecting the number of chromosomes in hybridomas. The principle of this method is that colchicine can specifically inhibit the formation of spindle fibers and thereby cells in the metaphase could be obtained. Then, cells are treated with hypotonic solutions such as 0.075 mol/L of KCl, and this would make cells to expand, the volume to increase, and

大，染色体松散；经甲醇 - 冰醋酸溶液固定，即可观察检查。

the chromosomes to loose. Afterwards, cells are fixed by the methanol - acetic acid solution, and the morphology of chromosomes could thereby be observed.

2. 单抗免疫球蛋白的类别和亚类鉴定　抗体的类和亚类对决定提纯的方法有很大的帮助。除采用特殊的免疫方法和检测方法，最经常得到的单抗是 IgM 和 IgG，分泌 IgE 的杂交瘤细胞很少见，而分泌 IgA 的杂交瘤细胞通常只有在用于融合的淋巴细胞来自肠道相关淋巴组织才能得到。鉴定单抗 IgG 类和亚类的方法主要有两种，一种是免疫扩散，另一种是 ELISA。目前国外 Roche 公司的单抗亚型鉴定试剂盒，利用乳胶凝集实验原理，只需 3～5min 便可知检测结果。

2. Identification of mAb types and subtypes The type and subtype of antibody determine the method of antibody purification. If special immunization or detection methods are not used, IgM and IgG are the most commonly obtained antibodies. Hybridomas that can secrete IgE are rarely seen, and hybridomas secreting IgA often derive from gut-associated lymphoid tissues. There are two methods for detecting IgG types and subtypes: one is immunodiffusion, and another ELISA. Currently, the mAb identification kit produced by the Roche Company uses the latex agglutination test and requires only 3-5 min for mAb identification.

3. 分泌抗体稳定性分析　杂交瘤细胞质量另一重要指标即为分泌抗体能力的稳定性，细胞建株后需对细胞株进行分泌抗体稳定性测定，如发现分泌抗体能力下降，提示应对杂交瘤细胞进行再克隆，直至 100% 培养孔细胞的上清抗体分泌阳性后，再进行稳定性测定。常用的方法为抗体连续传代法。即将细胞在体外培养，定期传代、收集细胞上清液，用筛选抗体的方法测定上清中的抗体，必要时可进行半定量比较，即对每次收集的上清进行适当的稀释，以了解不同时间收集的上清中抗体浓度的差异。稳定性测定一般需连续传代 3 个月以上。

3. Stability analysis of the secreted antibodies Another indicator for the quality of hybridomas is the stability of antibody secretion. After hybridoma strains are established, detection of secretion stability should be performed, and once cells are found with decreased antibody secretion, cloning of hybridomas is required until 100% of the cell showed positive antibody secretion. Then, another round of stability analysis should be performed again. Commonly used methods for stability analysis include serial passaging. Specifically, cells are *in vitro* cultured and regularly passaged, and the antibodies in the supernatant are collected and measured. If necessary, semi-quantification of the antibody will be performed. That is, the supernatant collected each time are properly diluted, and the difference in antibody concentration will be compared to reflect hybridoma secretion at different culturing time. Normally, cells are passaged more than 3 months for stability analysis.

4. 外源因子检查　无论所研制的单克隆抗体用于何种目的，杂交瘤细胞的外源污染因子的检查是必不可少的，尤其是对该单克隆抗体的药用开发研究，这也是国家药品监督管理局技术审评中所要求的。

4. Detection of exogenous factors No matter for what purposes the mAbs are prepared, the examination of exogenous factors should not be ignored, especially for those developed as pharmaceutical products, for which detection of exogenous factors is required by the technical review of the State Drug Administration.

2.2.8 单克隆抗体的大量制备

单克隆抗体的生产分为体内法和体外法两种。前者是通过动物体内培养杂交瘤细胞生产单抗。后者是则采用转瓶或生物反应器对杂交瘤细胞进行大规模培养并表达相应的单抗。

1. 体内培养法　将杂交瘤细胞注入BALB/c小鼠或LOU大鼠等组织相容性动物腹腔内，使其在腹腔内生长和表达，产生的抗体浓度可达到体外法的1000倍。其具体过程如下：先向动物腹腔注射0.5ml降植烷或液状石蜡使之致敏，8～10d后再向腹腔内接种1×10^6～1×10^7个杂交瘤细胞，其后5～7d内可见腹部肿胀，此时视动物健康状况，一次性或反复多次抽取腹水或将动物处死后一次抽取。每只小鼠可得到10ml腹水，大鼠可得到50ml以上。腹水和血清均可经离心、沉淀及超滤处理制成粗品单抗，也可冻干后4℃或－20℃贮存。动物体内诱生法操作简便，比较经济，所得单克隆抗体量较多且效价较高，可有效地保存杂交瘤细胞株和分离已经污染杂菌的杂交瘤细胞株，缺点是小鼠腹水中混有来自小鼠的多种杂蛋白，给纯化带来难度。

2. 体外培养法　当培养孔底部长满细胞时，即转移至$25cm^2$耐培养瓶中扩大培养，然后$80cm^2$、$175cm^2$……逐级放大，使细胞密度保持在1×10^5～1×10^6个/ml。如采用转瓶培养，单抗产量为5～10mg/L；如进行悬浮培养，产量可以达到10～100mg/L；而有报道采用微载体方法进行生产，单抗产量可以高达0.1～1g/L。

2.2.8 Mass Production of MAbs

The mass production of mAbs includes *in vivo* and *in vitro* methods. The former uses animals to culture hybridomas and produce mAbs, and the later uses roller bottles or bioreactors to culture hybridomas and express corresponding mAbs.

1. *In vivo* culture In this method, hybridomas are injected into the abdominal cavity of BALB/C mice, LOU rats or other histocompatible animals, and are allowed for growth and antibody expression. The concentration of antibodies produced by this method could be as high as 1000 times of that produced by *in vitro* methods. For detailed handling, animals are sensitized by intraperitoneal injection of 0.5 ml pristane or liquid paraffin. After 8-10 days, 1×10^6-1×10^7 hybridomas will be inoculated intraperitoneally into the animal, and abdominal swelling could be observed after 5-7 days. Based on the health condition of animals, ascites could be collected repeatedly or for once after killing the animal. For mice, 10 ml of ascites could be obtained, and for rats, 50 ml or more could be obtained. The sampled ascites or serum could be made into crude mAb after centrifugation, sedimentation and ultrafiltration, and they can also be preserved at 4℃ or –20℃after lyophilization. *In vivo* culture of hybridomas is an easy and economic way to produce mAbs with large amounts and high titers. Also, it could effectively preserve the hybridoma strains and isolate contaminated hybridomas, but since mice ascites contain many other proteins, the purification of antibodies would be much more difficult.

2. *In vitro* culture When cells reached 100% confluence in the culturing well, they are transferred into a 25-cm^2flask for expanding culture, then into 80 cm^2, 175 cm^2 and even larger flasks. The concentration of cells is maintained at 1×10^5-1×10^6/ml. For roller bottles, mAbs could be yielded at 5-10 mg/L, and for suspension culture, mAbs could be yielded at 10-100 mg/L. Some studies report mAb production with microcarriers, and the concentration of yielded mAbs could be as high as 0.1-1 g/L.

2.3 单克隆抗体的纯化

由于单克隆抗体的Ig的类别和亚类的不同，纯化的方法亦不同，因此在纯化之前必须明确其Ig的类与亚类。另外根据用途不同也应选择不同的纯化方法，如用于体外诊断试剂，IgG类抗体应采用沉淀处理结合亲和层析的方法，IgM类抗体应采用沉淀处理结合凝胶过滤的方法。若制备体内诊断试剂或治疗用药，则应注意去除内毒素、病毒、核酸等微量的污染物，必须经亲和层析和阴离子交换层析处理。动物体内诱生法制备的单克隆抗体的具体纯化方法及一般过程如下。

1. 澄清和沉淀处理　由于小鼠腹水中含有红细胞、细胞碎块、纤维蛋白凝块及脂质等，应首先用1000g离心5min，去除沉淀物；再用20 000g高速离心30min，去除残留的小颗粒物质；然后用0.2μm的微孔滤膜过滤，除掉污染的细菌、支原体和脂质；最后用饱和硫酸铵沉淀抗体，50%饱和硫酸铵能回收90%以上的单克隆抗体。

2. 分离　根据用途不同可选用不同的方法，主要分离方法有凝胶过滤、阴离子交换层析、亲和层析等方法。

（1）凝胶过滤。凝胶过滤用于IgG和IgM类单克隆抗体的分离纯化。常用Sephadex G200作分离介质，可收集到三个峰，最高峰为IgM，另外两个峰为IgG，抗体回收率达50%～80%，能去除污染的微量杂蛋白，抗体纯度可达95%以上。

（2）阴离子交换层析。用于IgG类单克隆抗体的分离纯化。在pH 7.4条件下，IgG1和IgG2能结合在

2.3 mAbs Purification

Different types and subtypes of mAb Ig require different methods of purification. Thus, before purification, the types and subtypes of Ig must be determined. Also, mAbs used for different purposes may require different purification methods. For example, if used as *in vitro* diagnostic agents, IgG antibodies should be purified with sedimentation and affinity chromatography, while IgM antibodies should be purified by sedimentation and gel filtration. If used for *in vivo* diagnostic and therapeutic purposes, affinity chromatography and anion exchange chromatography must be performed in order to remove toxins, viruses, nucleic acids, and other trace contaminants. General processes used for purification of mAbs prepared from *in vivo* culture are as follow.

1. Clarification and sedimentation Since mice ascites contain red blood cells, cell fragments, fibrin clots and lipids, the ascites should be centrifuged at 1000 *g* for 5 min to remove the precipitate and at 20000 *g* for 30 min to remove residual small particles. Then, the supernatant is filtrated by 0.2 μm microporous membrane to remove bacteria, mycoplasma and lipids, and the antibody is finally precipitated by saturated ammonium sulfate. Saturated ammonium sulfate (50%) could recover more than 90% of the mAbs.

2. Separation Based on different purposes, different separation methods could be chosen. Methods used for separation mainly include gel filtration, anion exchange chromatography and affinity chromatography.

(1) Gel filtration　Gel filtration is often used for the separation and purification of IgG and IgM. In most cases, the Sephadex G200 will be used as the separation medium, and 3 peaks could be observed. The highest peak is IgM, and the other two are IgG. By using this method, the recovery of antibody could reach 50%-80%, and trace contaminating proteins could be removed, resulting in the antibody purity higher than 95%.

(2) Anion exchange chromatography　Anion exchange chromatography is often used for the purification of IgG mAbs. The IgG1 and IgG2 could bind to the DEAE column at pH 7.4. Under low ionic

DEAE填料上，其中IgG2a可在较低离子强度条件下洗脱下来，其纯度很高；IgG2b、IgG3在低离子强度下易发生沉淀，可增加离子强度以提高产量，但其纯度相对较低。在pH5.5条件下，把杂交瘤细胞培养的上清液加到琼脂糖柱上时，所有的抗体都能结合到柱上，而55%的杂蛋白可被洗脱掉，再用不同离子强度的洗脱剂洗下。在最适离子强度及pH条件下，以离子交换层析分离单克隆抗体可以提高纯度25～100倍。高容量、高流速、化学稳定的新型离子交换剂适宜单克隆抗体的大规模纯化。

strength conditions, IgG2a could be eluted with high purity, while IgG2b and IgG3 would precipitate under such condition. Though the yield of IgG2b and IgG3 could be improved by increasing the ionic strength, the purity of these antibodies is still very low. However, at pH 5.5, by injecting hybridoma supernatant into the agar gel column, all antibodies could bind to the column, and 55% of the contaminating proteins could be eluted. Then, antibodies could be eluted with different ionic strengths. By selecting the optimal ionic strength and pH conditions, ion exchange chromatography could increase the purity of mAbs by 25-100 times. New ion exchanger with high capacity, flow rate and chemical stability could be used for large-scale purification of mAbs.

（3）亲和层析。多用蛋白A亲和层析，适用于IgG类单克隆抗体的纯化。IgG与蛋白A的结合与pH有密切关系，鼠源性单克隆抗体在pH 8.0时，IgG2b、IgG3几乎全部结合于蛋白A柱上，IgG1稍差。用pH3.5缓冲液可洗脱下IgG2b，pH4.0缓冲液可洗脱下IgG3，pH4.5缓冲液可洗脱下IgG2a，pH6.0缓冲液可洗脱下IgG1。用蛋白A亲和层析收获的抗体纯度很高，但也有极微量的杂蛋白。

(3) Affinity chromatography Protein A affinity chromatography is commonly used, and it suits for purification of IgG mAbs. The binding of IgG with protein A is closely related to the pH value. At pH 8.0, mice IgG2b and IgG3 could almost all bound to the protein A column, but IgG1 showed less strong affinity. Eluents at pH 3.5, pH 4.0, pH 4.5 and pH 6.0 could elute IgG2b, IgG3, IgG2a and IgG1 respectively. Antibodies harvested by protein A affinity chromatography are with very high purity, but trace contaminating proteins could not be completely removed by this method.

当今免疫学、分子生物学能够取得令人瞩目的成就，与单克隆抗体的广泛应用密不可分。抗原的精制和构造分析，细胞发生、分化及细胞功能的阐明，特别是单克隆抗体与抗癌药物、毒素或放射性物质偶联的导向作用，为临床疾病的诊断、预防和肿瘤的治疗开辟了新的前景。但是目前所用的单克隆抗体多为鼠源性，因此，减轻鼠源性抗体诱发的免疫应答反应已成为了新的课题。

Nowadays, remarkable achievements made by immunology and molccular biology could not bc realized independent of the extensive application of mAbs. Purification and structural analysis of antigens, understanding of cell genesis, differentiation and cell function, and the targeting function of mAbs conjugated with anti-cancer drugs, toxins or radioactive substances, all these breakthroughs opened up new prospects for the diagnosis, prevention and treatment of diseases such as cancer. However, most of the mAbs used currently are murine-derived, thus how to reduce murine antibody-induced immune responses has become a new research direction.

3 基因工程抗体

3 Gene Engineered Antibody

尽管单克隆抗体和多克隆抗体受到了广泛的应用，但它们还是有许多难以克服的缺点。其中最主要的原因是它们是异种蛋白，不能直接用于人体内，否则会产生很强的免疫应答，为此在20世纪80年代诞生了应用分子生物学技术制备的基因工程抗体，主要包括人-鼠嵌合抗体（chimeric antibody）、人改型抗体（reshaped humanized antibody）、小分子抗体以及双特异性抗体（bispecific antibody）等。

Although, McAb and polyclonal antibody are widely used nowadays, they still have some formidable shortcomings. The main problem is that they are alloproteins, so they cannot be directly used in human body, or there will be a strong immune response. Considered of it, engineered antibody prepared by molecular biology technology was turned out in 1980s. It mainly includes: chimeric antibody, reshaped humanized antibody, bispecific antibody and so on.

鼠单克隆抗体C区可作为异源蛋白诱发免疫反应，产生人抗小鼠抗体。嵌合抗体是将小鼠单克隆抗体C区用人抗体C区代替拼接而成，使Fc段人源化，但V区保留的鼠源性骨架区序列及其立体结构仍能诱发HAMA。利用互补决定区的技术将鼠单克隆抗体CDR移植到人单克隆抗体的V区框架上，使人单克隆抗体获得鼠单克隆抗体抗原结合特异性，可进一步减少异源性，即人改型抗体。抗体的小分子化是基因工程抗体的又一特色。通过重组DNA技术可以合成任何所需抗体的特定片段，包括Fab片段、Fc片段和含有轻、重链可变区的Fv片段。这些小分子片段分子量小、免疫原性弱、渗透力强，并可有药物导向和中和毒素等用途。同时，又可进一步构建分别针对靶抗原和效应分子或效应细胞的双特异性抗体；在基因水平将目的蛋白基因同Ig部分片段基

The C region of the mouse McAb, as an alloprotein, can induce immune reaction and secrete human anti mouse antibody. Chimeric antibody replaces the C region of the mouse McAb with the C region of the human antibody, to make the Fab section humanized construction, but reserve the mouse framework sequence and its stereoscopic structure so as to induce HAMA. To use the complementary determining region technique, transplant the CDR of mouse McAb to the framework of the V region of human McAb. This method makes human McAb have the specificity of mouse McAb, so as to reduce the heterogencity. The aimed product is chimeric antibody. The engineered antibody is small molecule, and this is another feature. We can get any specific fragment of antibody through DNA recombinant technique, such as, Fab section, Fc section, and the variable region-Fv section which includes light and heavy chain. These small-molecule fragments have low molecular weight, poor immunogenicity, good penetration, drug targeting and ability to counteract toxin. At the same time, we can construct the bispecific antibody against both target antigen and effector molecule or effector cell further; at gene level, connect the gene of aimed protein with gene of Ig section, and the Ig-fusion protein with the above-mentioned structure domain can be expressed in the eukaryotic or prokaryotic cells; the antibody can be used

因相连，并在真核或原核细胞中表达具有上述结构域的Ig-融合蛋白；利用抗体作为载体，携带前药的抗体导向酶等其他抗体衍生形式。

as a carrier, antibody-directed enzyme with prodrugs and other alternative forms.

既然鼠抗体人源化难度很大，那么能不能在没有鼠抗体的情况下直接制备人抗体呢？这样就避免了使用鼠抗体的种种弊端。从20世纪90年代人们开始用基因工程的方法直接制备人抗体，其中一种主要的方法是用噬菌体展示技术从人抗体库中调取抗体。所谓的噬菌体展示技术，就是用噬菌体的基因作为载体，用抗体基因与其融合，在噬菌体表面表达出抗体。目前最常使用的载体来自丝状噬菌体（filament phage）。抗体库的 V_H 和 V_L 基因可以从免疫过的、没有免疫过的人淋巴细胞或人工合成的渠道得到。在构建表达载体时，人们经常在抗体基因和噬菌体载体之间插入一个琥珀终止密码子。这样当琥珀密码子被抑制时，抗体就会以融合蛋白形式在噬菌体表面表达，便于从抗体库中筛选特异性抗体。而当琥珀密码子没有被抑制时，抗体便以分泌形式表达。所筛选出来的抗体如果亲和力较低，还可以通过易错PCR和链替换（chain shuffling）的方式模仿天然抗体在体内的亲和成熟过程，增加其亲和力。噬菌体抗体库技术克服了人体不能随意免疫的缺点，用人的外周血制备抗体库，从而获得人源性基因工程抗体；而且将抗体基因的克隆和表达融为一体，同在一种载体上进行，将抗体基因表达在载体表面，用固相化抗原对表达产物的载体进行淘筛（panning），在数日内就可筛选出阳性克隆，从而能构建出大库容（10^8）文库囊括天然全套抗体基因。人们完全可以

Since it is difficult to humanize the mouse antibody, can we prepare the humanized antibody without mouse antibody? It can avoid the disadvantages of mouse antibody. In 1990s, we began to prepare humanized antibody by the methods of genetic engineering. One of the main methods is to select antibodies from the human antibody library using the phage display technique. The Phage Display is to use the genes of phage as the carrier, fuse the genes of antibody with them, and eventually the antibody can be expressed on the surface of phage. Nowadays, the most commonly used carrier is derived from filament phage. The genes of V_H and V_L in the antibody library can be obtained through human lymphocytes or artificial synthesis. When constructing the expression vectors, we usually insert amber termination codon into the genes of antibody and phage vector. The antibody can be expressed as a fusion protein on the surface of phage when the amber codon is suppressed, so as to select the specific antibody from antibody library. The antibody will be secreted expressed when amber codon is not suppressed. If the selected antibody has a low affinity, we can simulate the course of affinity maturity of the natural antibody in vivo to improve its affinity by the way of Prone PCR and chain shuffling. The Phage Display has solved the problem that is not optionally immune in human body. We can use the human peripheral blood to prepare antibody library so as to obtain the humanized antibody. Furthermore, the cloning and expression of the antibody gene are in the same vector, and express the antibody gene on the surface of vector. We use the immobilized antigen to do panning to vectors of the expressed products, and we can select the positive clones in few days. Because of this, we can construct a large library (10^8) included all genes of antibody, so as to prepare any high-specific antibody by gene engineering.

用基因工程方法研制任何一种具有高度特异性的抗体，使抗体工程的设想成为现实。

4 抗 体 制 药

4 Antibody Therapeutic Drug

由杂交瘤技术生产的各类单克隆抗体，作为诊断剂或检测剂，在过去20年内已被广泛地应用于农业、生物、医药及医疗行业。近年来其作为治疗剂，特别是用于肿瘤、病毒性感染、类风湿性关节炎、心血管疾病以及器官移植中的抗排异反应等疾病的治疗及预防方面，又再度成为生物技术药物领域最引人注目的研究热点。

The monoclonal antibodies (McAbs) which were produced by hybridoma technique as diagnostic or detective reagents were widely used in agriculture, biological and medical industry in the passed 20 years. Otherwise, in recent years, McAbs were also widely used as therapeutic agent in the treatments and prevention for tumors, viral infection, and rheumatoid arthritis, cardiovascular diseases, rejections caused by organ transplant, and other diseases. So, McAbs were again drawn attentions in the area of biotechnology-derived pharmaceuticals.

4.1 临床诊断试剂

4.1 Reagents for Clinical Diagnosis

单克隆抗体技术的应用最成功的领域是医疗诊断。自从单抗问世以来，因其具有特异性强、重复性好、操作简便易行和能够大量生产的特点，所以它已在大部分常规血清学检查中取代了多克隆抗体而广泛地应用于免疫学诊断。最初，单克隆抗体主要作为体外诊断的试剂。在众多的单克隆抗体诊断试剂中可以用到的是怀孕检测的试剂、诊断许多病原微生物的试剂、测量各种药物的血药浓度的试剂、匹配组织相容性抗原的试剂；另外，用放射性同位素标记的单抗也用于体内检测并跟踪肿瘤抗原，为病人体内一些原发性或转移性肿瘤的早期诊断提供可能。例如，乳腺癌细胞的碘-131单克隆抗体，进入血液检测扩散到肿瘤淋巴结的区域。这种单

The monoclonal antibody technique is most successfully used in the medical diagnosis. Since McAbs advent, because of its features, such as, high specificity, good repeatability, and easy operation, high production, they replace polyclonal antibodies as the diagnostic reagents in the most conventional serological identification. Initially,monoclonal antibodies were used primarily as in vitro diagnostic reagents. Among the many monoclonal antibody diagnostic reagents now available are products for detecting pregnancy, diagnosing numerous pathogenic microorganisms, measuring the blood levels of various drugs, matching histocompatibility antigens, and detecting antigens shed by certain tumors Radiolabeled monoclonal antibodies can also be used in vivo for detecting or locating tumor antigens, permitting earlier diagnosis of some primary or metastatic tumors in patients. For example, monoclonal antibody to breast-cancer cells is labeled with iodine-131 and introduced into the blood to detect the spread of a tumor to regional lymph nodes. This monoclonal imaging technique can reveal breast-cancer metastases that would be undetected by other, less sensitive scanning techniques. Now, there

克隆成像技术可以揭示其他不太敏感的扫描技术难以发现的乳腺癌转移。现在市面上已有商品化的单抗试剂盒出售，可对过敏源、癌症、性病、怀孕、排卵等进行早期诊断。此外，单抗能准确识别活细胞正在产生哪些蛋白质，哪些蛋白质正处于激活状态等，有助于科学家深入了解人体发育及工作机制。

are commercialized McAbs kits on the market and these McAbs can be used in the early diagnosis of allergens, cancers, and venereal diseases, pregnancy, ovulation, and other diseases. Furthermore, McAbs can accurately identify what proteins the living cells are producing and which proteins are active, and so on. They help researchers to deeply understand development of human and working mechanisms.

4.2 单抗治疗药物

单抗作为诊断剂或检测剂，过去20年内在医学和生物学领域得到了广泛应用，近年来其作为治疗剂又再度成为生物技术药物领域最引人注目的研究热点。

4.2 Monoclonal Antibody Therapeutic Drug

As a diagnostic or detection agent McAb have been widely used in fields of biology and medicine in the past 2 decades. Recently, the therapeutic property of McAbs has again made them the most striking research hotspot in the field of biopharaceutics.

4.2.1 单抗治疗药物的发展历程和市场规模

1986年第一个治疗性鼠源性单克隆抗体OKT3上市，然而治疗效果并不乐观，抗体治疗在1988年到1993年间陷入低谷。为了解决免疫原性问题，诞生了各种人源化技术。2002年，第一个全人源化抗体Humira上市，一直到2006年FDA批准了21种单克隆抗体药物上市（表6-3），其发展又重新进入轨道。单克隆抗体是近年来复合增长率最大的一类抗体药物，1999年全球单克隆抗体药物的市场销售额只有12亿美元，2002年已经达到40亿美元，2004年全球销售额大约在103亿美元，其中8种销售额最大的单克隆抗体药物销售额达到101亿美元，其年均增长率超过150%。随着已上市品种的不断增长以及新品种的上市，单克隆抗体药物市场将会迅速攀升。

4.2.1 Development History and Market Scale of McAbs

In 1986, the first mouse derived monoclonal antibody OKT3 was approved for marketing, but it did not achieve a good therapeutic effect. The antibody therapy has fallen into the low ebb during 1988 to 1993. Then, various humanized-construction techniques emerged as required solving the problem of immunogenicity. In 2002, the first complete humanized antibody Humira came to on the market. Till 2006, 21 kinds of McAb drugs were approved by FDA for marketing(Table 6-3), its development again went into orbit. McAbs' compound growth rate was highest among the antibody drugs. In 1999, market sales all over the world of McAb drugs was only 1.2billion dollars, but in 2002, market sales already reached 4 billion dollars, and in 2004, market sales all over the world was about 10.3 billion dollars, especially the market sales of 8 biggest-sales McAb drugs was 10.1 billion dollars, and the average annual growth rate was more than 150%. As the increasing of varieties and new variety on the market, McAb drugs market will expand rapidly.

表 6-3 部分美国 FDA 批准的单克隆抗体药物
（Therapeutic monoclonal antibodies approved by the US FDA）

学名（Generic name）	商品名（Trade name）	生产厂家（Sponsor company）	抗体类型（Type）	批准日期（Approval date）
Muromonab-CD3	Orthoclone	Ortho Biotech	鼠源（Murine）	1986
Murine Mab（B72.3）	OncoScint	CYTOGEN	鼠源（Murine）	1992
Abciximab	ReoPro	Centocor	嵌合（Chimeric）	1994
MurineMAb（PMSA）	ProstaScint	CYTOGEN	鼠源（Murine）	1996
Murine Fab′fragment（Anti-CEA）	CEA-Scan	Immunomedics	鼠源（Murine）	1996
T99 nofetumo mab merpentan Murine MAb	Verluma	DuPont Merck	鼠源（Murine）	1996
Rituximab	Rituxan	Genentech	嵌合（Chimeric）	1997
Daclizumab	Zenapax	Hoffman-La Roche	人源化（Humanized）	1997
Basiliximab	Simulect	Novartis	嵌合（Chimeric）	1998
Palivizumab	Synagis	MedImmune	人源化（Humanized）	1998
Infliximab	Remicade	Centocor	嵌合（Chimeric）	1998

4.2.2 单抗治疗癌症的发展趋势

单克隆抗体现已被确立为一种关键治疗方式，用于一系列疾病的治疗。因单抗制剂能选择性地针对肿瘤细胞，迄今为止单抗治疗主要集中在癌症。

1. *已批准的抗癌单抗* 从 1980 年到 2005 年，世界各地的商业公司针对种种癌症迹象，在临床试验中对共计 206 个独特的治疗性单克隆抗体进行了研究。这些候选抗体包括得到授权的以及自己产生的单克隆抗体，有可能代表了在政府、学术和商业机构产生的最有前景的抗癌单克隆抗体。到目前为止，这些抗癌单克隆抗体中有 12 个已经至少在一个国家被允许进行市场销售。抗癌单克隆抗体被用于临床试验 15 年的时间，直到其中的一个抗体依决洛单抗（edrecolomab）在 1995 年时才在德国被允许市场销售。这个首次批准是一

4.2.2 Development Trends for Monoclonal Antibody Cancer Therapeutics

Monoclonal antibodies are now established as a key therapeutic modality for a range of diseases. Owing to the ability of these agents to selectively target tumour cells, cancer has been a major focus of development programmes for monoclonal antibodies so far.

1. Approved anticancer mAbs From 1980 to 2005, a total of 206 unique therapeutic mAbs were studied in clinical trials by commercial companies worldwide for a variety of cancer indications. The candidates included in-licensed as well as self-originated mAbs, and probably represented the most promising anticancer mAbs produced in government, academic and commercial settings. To date, 12 of these anticancer mAbs have been approved for marketing in at least one country. Anticancer mAbs had been in clinical study for 15 years before one—edrecolomab— was approved in Germany in 1995. This first approval was a harbinger because, with the exception of 1999, at least one anticancer mAb has been approved each year since 1997. The successes have come with setbacks, however; edrecolomab was withdrawn from the market when a postapproval study demonstrated that the treatment was inferior to the standard of care.

个先驱，因为除1999年外，自1997年开始每年至少有一个抗癌单克隆抗体被批准销售。但是成功总是伴随着挫折，后来得到证明的一项研究揭示出依决洛单抗的疗法劣于护理标准，于是依决洛单抗被从市场召回。

得到批准的产品是两个主要研究领域多年工作的结果，这对单克隆抗体能成功成为定向的抗癌疗法是非常重要的：四种主要单克隆抗体（小鼠、嵌合体、人性化、人）的设计和生产工艺的发展；研究的行为模式（例如，通过辐射、细胞毒素、激活免疫系统组件或受体/配体封锁的肿瘤细胞毒性）。临床试验最初集中在小鼠单克隆抗体（在20世纪80年代最容易得到的一种），以及用放射性元素或细胞毒素杀死肿瘤细胞的免疫结合。但是，只有两种小鼠免疫结合（都是放射性同位素标记）目前为止允许销售。当前得到批准的抗癌单克隆抗体，人性化产品是四种里面数量最多的，运用未经修改的或“裸”单克隆抗体的激活免疫系统组件或受体/配体封锁是最常用的行为模式。

The approved products are the result of years of work in two major research areas which are critical to the success of mAbs as targeted cancer therapeutics: development of engineering and production methods for the four main categories of mAbs (murine, chimeric, humanized and human), and study of modes of action (for example, tumour cell toxicity via radiation, cytotoxins, activation of immune system components or receptor/ligand blockade). Clinical studies initially focused on murine mAbs-the category most readily available in the 1980s-and on immunoconjugates that used radioactive elements or cytotoxins to kill tumour cells. However, only two murine immunoconjugates (both radiolabelled) have been approved to date. Of the currently approved anticancer mAbs, humanized products are the most numerous of the four categories and activation of immune system components or receptor/ligand blockade using unmodified or ‘naked’ mAbs are the most common modes of action.

2. *抗癌单抗生产方法的发展* 由公司进行的临床研究的抗癌单抗的总数一直呈现稳定的增长。事实上，由商业机构提供的新的抗肿瘤的单克隆抗体进入临床研究的比率自从八十年代以来增长了3倍。从20世纪80年代每年4.3个到90年代的每年8.3个再到2000～2005年的13.3个。增加对单克隆抗体的投入的其中一个原因是检测技术的进步。最初的生产单克隆抗体的方法见于1975年的报道，

2. Evolution of production methods The total number of anticancer mAbs in clinical study by companies has risen steadily over time. In fact, the rate at which new oncology mAbs entered clinical study sponsored by commercial organizations has more than tripled since the 1980s, from 4.3 per year in the 1980s to 8.3 per year in the 1990s and 13.3 per year for 2000–2005. One reason for the increased investment in mAbs was the evolution of the discovery technology. The initial method for producing mAbs, first described in 1975, involved use of mouse-derived hybridomas. However, in humans, murine mAbs were commonly (although not always) ineffective as cancer therapeutics. Studies revealed

包括了鼠源性杂交瘤的应用。然而，对于人类而言，鼠源性的单抗通常（并非总是）对治疗癌症疗效欠佳。研究显示，鼠源性的各种单抗半衰期较短且病人易产生针对性的抗体，限制了它们的应用。另外需说明的是，只有特定的几种鼠源同种型单抗显示出有效结合并激活人类免疫系统，激活细胞毒作用的能力。在实践中，鼠源性单抗在作为放射性元素的指示物或细胞毒性作用指示物时受到了很大的限制。相对于传统的鼠源性模式来讲，由基因工程的巨大发展所引发的设计单抗的途径可以提供更强更有效的单克隆抗体。适用于基因嵌合的技术连同包含了人源和鼠源性蛋白质序列的人类化了的单抗在20世纪80年代中后期发展起来了。另外，噬菌体展示（将抗体或肽表位展示于噬菌体表面）来源的和转基因小鼠源性的人单抗在20世纪90年代得到了应用。值得一提的是人单抗还可以经人类杂交瘤得到，但是这种生产方法不具备商业价值，原因是细胞株不足以生产足够数量的单克隆抗体。

that murine candidates often had short circulating half-lives7 and patients frequently developed antibodies to the mouse-derived proteins, which limited their utility8, 9. In addition, only certain murine mAb isotypes have been shown to effectively bind to and activate elements of the human immune system, thereby triggering cytotoxic effector functions10. In practice, murine mAbs were mostly limited to acting as targeting agents for radioactive elements or cytotoxins that might kill targeted tumour cells.Advances in genetic engineering over the years have provided numerous ways to design mAbs that are more robust and efficacious compared with the original murine versions11. Techniques suitable for generating chimeric and humanized mAbs that contain protein sequence from both human and murine sources were developed in the mid-to late 1980s. In addition, human mAbs derived from phage display and transgenic mice became available in the 1990s. It should be noted that human mAbs were also made from human hybridomas, but this production method was not commercially viable because the cell lines did not reliably produce sufficient quantities of the desired mAb.

技术的进步对用于临床研究的单抗的类别趋势有明显的影响，这种影响从20世纪80年代持续到现在（图6-4）。在20世纪80年代，86%的临床研究用的为鼠源性单抗，从90年代起就没有一种单抗在研究中占绝对的优势，因为可用于试验的单抗的种类有很多，人化的单抗进入临床试验从80年代到90年代间增幅明显（从0.1到2.8每年）。因为人化单抗的发现技术滞后于人化单抗技术，在20世纪90年代人化单抗进入

Technological advances have had a clear impact on the trends in mAb category in clinical study in the 1980s, 1990s and so far this decade (Figure 6-4). Murine candidates constituted 86% of the total in clinical study in the 1980s. No single category has dominated since the 1990s because of the availability of a variety of mAb categories. The rate at which humanized mAbs entered clinical study increased dramatically between the 1980s and the 1990s (from 0.1 to 2.8 mAbs per year). Because human mAb discovery technology lagged behind that for humanized mAbs, the rate that human mAbs entered clinical study was still slow in the 1990s (0.9 mAbs per year). However, since 2000, humanized and human mAbs have been entering clinical study at approximately the same rate (4.3 versus 4.5 mAbs per year, respectively).

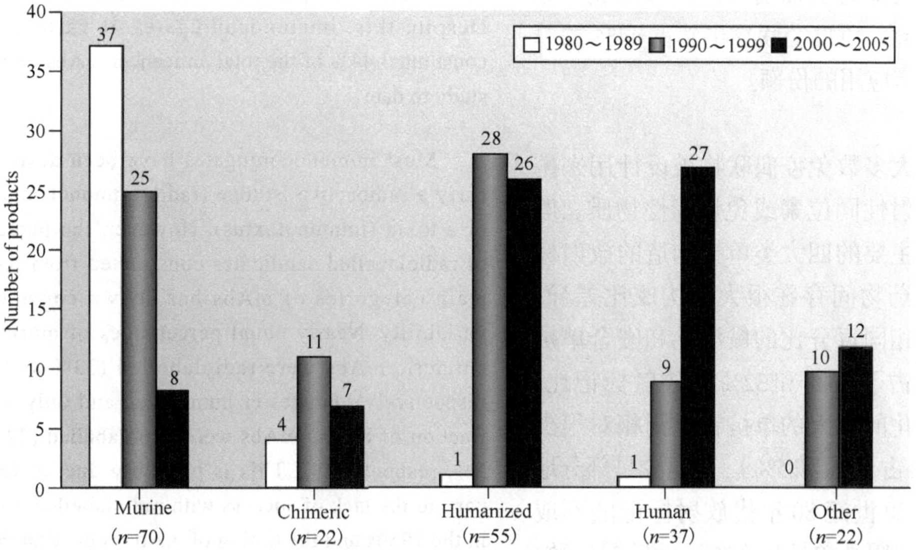

图 6-4 1980—1989 年、1990—1999 年和 2000—2005 年进入临床试验的治疗肿瘤的单抗

临床试验比率还是很低（0.9 单抗每年）。然而，2000 年后，人化和人类单抗进入临床试验的比率大约保持在同一水平（每年为 4.3～4.5 单抗）。

3. *抗癌单抗的作用方式* 单抗具有诸多潜在功能。一个有效的抗肿瘤单抗，必须能与适量的抗原结合，从而介导疾病相关反应。当肿瘤细胞表面抗原被靶向时，单抗药物介导的反应最终导致靶细胞的破坏。此方式可以是直接性（如偶联放射性同位素或毒素，或抗体诱导细胞凋亡）的，也可以是间接性的（如活化免疫补体或阻断关键受体）。对于可溶性抗原来说，单抗药物应能阻断靶抗原。单抗的复杂性取决于药效不同的方式，如免疫偶联物必须携带放射性元素、细胞毒素或细胞因子，而在其他情况下，未经修饰的单抗也会有治疗效果。

3.Modes of action mAbs are potentially capable of multiple functions. Efficacious anticancer mAbs must bind to an appropriate antigen in quantities sufficient to mediate a disease-relevant response. When tumour cell-surface antigens are targeted, the response should ultimately lead to destruction of the targeted cell. The mode of destruction can be direct or indirect . In the case of soluble antigens, the mAb should sequester the target. The complexity of the mAb varies with the mode of action—immunoconjugates must be constructed to carry radioactive elements, cytotoxins or cytokines, whereas unmodified mAbs will suffice in other cases.

偶联的单抗可以增加化疗和放疗的特异性，但是它们也有不足之处；它们更加难以制造并且与裸露的部分

Conjugated mAbs can increase the specificity of chemo-or radiation therapy and improve the efficacy of immunotherapy, but have some drawbacks; they are more difficult to manufacture and may have greater

相比可能有更加安全的内部构成。尽管如此，免疫交联物仍然占44%的单抗临床应用的份额。

safety issues compared with their naked counterparts. Despite this, immunoconjugates of various kinds constituted 44% of the total anticancer mAbs in clinical study to date .

大多数免疫偶联物被设计用来携带放射性同位素或免疫毒性物质。但是由主要的四大类单抗制造的放射标记的药物间存在很大的构成比差异。大约相同百分比的鼠源性和嵌合单抗各自占了33%和32%的放射标记比，而人化的和人的单抗在此则相对很少（各自占22%和5%）。这至少是部分地因为20世纪80年代放射标记的不成功和时限性单抗生产方法的发展失败。直到临床研究揭示了放射性标记在抗癌治疗中的局限性后，人化的和人的抗癌单抗直到20世纪90年代和2000年以后才大批量进入临床（如单抗无法携带有效的放射性物质治疗剂量，尤其对实质肿瘤更是如此，经常需要添加化疗药物，而且对正常组织也有损害）。临床研究中常用的放射性同位素是Y-90（半衰期2.7d）和I-131（半衰期8.0d）。替伊莫单抗（免疫调节药）、I-131托西莫单抗（抗肿瘤药）和一种嵌合剂I-131 ch-TNT共计3类放射性标记药物中，只有14%目前仍用于临床。

Most immunoconjugates have been designed to carry a radioactive isotope (radio-immunoconjugates) or a toxin (immunotoxins). However, the percentage of radiolabelled candidates constructed from the four main categories of mAbs has shown considerable variability. Nearly equal percentages of murine and chimeric mAbs were radiolabelled (33% and 32%, respectively), but fewer humanized and only a small fraction of human mAbs were radiolabelled (22% and 5%, respectively). This is probably due, at least in part, to the lack of success with radiolabelled products in the 1980s and the timing of mAb production method development. Humanized and human anticancer mAbs did not enter the clinic in large numbers until the 1990s and 2000, well after clinical studies revealed the limitations of radio-immunoconjugates as cancer therapeutics (for example, mAbs did not deliver effective radiation doses, especially to solid tumour sites20, 21, complex chemistry was often required for conjugation, and there were potentially toxic effects on normal tissues). The two radioisotopes most commonly conjugated to mAbs studied in the clinic were Y-90 (2.7-day half-life) and I-131 (8.0-day half-life). Three radiolabelled mAbs — two murine, Y-90 ibritumomab tiuxetan (Zevalin; Biogen Idec) and I-131 tositumomab (Bexxar; Corixa/GSK), and one chimeric, I-131 ch-TNT (Shanghai Medipharm Biotech) — have been approved for non-Hodgkin's lymphoma or lung cancer, but only 14% of anticancer mAbs currently in the clinic are radio-immunoconjugates.

组成广泛的偶联物的蛋白质和小分子细胞毒物质现已进入了临床研究。用于此目的的蛋白质包括广泛的改版了的假单胞菌外毒素、葡萄球菌肠毒素、新抑癌蛋白和植物源性的蓖麻毒素重链和白树毒素。与治疗型免疫毒结合的小分子物质包括长春碱、氨甲蝶呤、阿霉素、刺胞霉素和美坦生源性的物质。药物一般都通过化学键与单抗分子连接，

Immunotoxins composed of mAbs that are conjugated to a wide variety of either protein or small-molecule cytotoxins have also been studied in the clinic18. The protein cytotoxins used in this way include various modified versions of Pseudomonas exotoxin, Staphylococcus enterotoxin, neocarzinostatin, and the plant-derived molecules ricin (and ricin A chain) and gelonin. Small molecules incorporated in therapeutic immunotoxins included vinblastine, methotrexate, doxorubicin, calicheamicin and maytansine derivatives (auristatin immunoconjugates entered the clinic in 2006,

但是蛋白型的免疫毒性物质可以通过化学连接和基因工程的方法连接。这两种方法都面临着挑战，因为蛋白性的免疫毒性物质可以导致机体产生免疫，而药物毒性物质则缺乏携带足够剂量的药物到达肿瘤位置的潜力。所以这两种情况下产品的效果都有所降低。伴随化学性结合的免疫毒性物质出现了进一步的效果和安全性的局限性，这有可能是因为连接抗体和毒性物质的键出现了问题。伴随细胞毒性出现的另外一个局限性是目标的选择和抗体上，它们要求细胞的内吞作用以实现对肿瘤细胞的细胞毒性。对于44种自1980～2005年应用于临床研究的免疫毒素来说，27种（61%）是蛋白-单抗结合的，17种（39%）是药物-单抗结合的，两者各有6种目前还应用于临床（图6-5）。

so are not included in our analysis). Drugs are attached to mAbs through chemical linkers, whereas protein immunotoxins can be produced through either chemical conjugation or genetic engineering22. Development of both types of immunotoxins has posed challenges because the protein toxins can be immunogenic and the drug toxins may lack potency at the doses delivered to the tumour site17. In both cases the effectiveness of the product is reduced. Further potential limitations in efficacy and safety arise with chemically conjugated immunotoxins, which might be subject to decay of the linker used to join the antibody to the toxin. Another limitation in the choice of targets and antibodies arises with immunotoxins, which require internalization in order to have a cytotoxic effect on tumour cells. Of 44 immunotoxins that entered clinical study from 1980 to 2005, 27 (61%) were protein–mAb conjugates and 17 (39%) were drug–mAb conjugates, although an equal number of drug and protein immunoconjugates (6 candidates each) are now in the clinic (Figure 6-5).

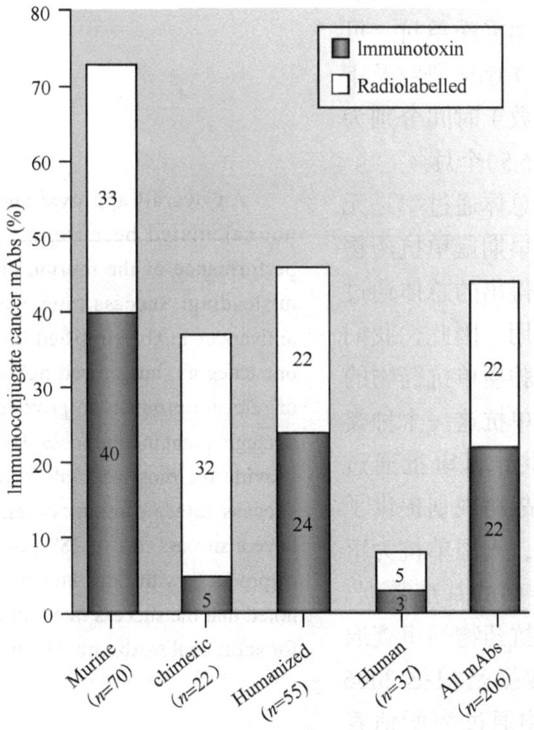

图 6-5 1980—2005 年临床试验的免疫结合物治疗肿瘤的方式

4. *抗癌单抗发展规划* 到2005年为止，在美国已有9个抗癌单克隆抗体被批准。对所有9个产品进行统计，平均（中位数）的临床开发和审批阶段所需时间分别为90.8个月（82.5）和11.4（6.6）个月。但是，不同单抗临床开发和审批阶段所需时间跨度比较大，临床阶段50.5～140.3个月，审批阶段4.7～33.4个月。奇怪的是，大多数美国批准的产品可能会被认为是“非典型”的。特别注意的是两种放射性免疫偶联物因为它们的放射性有着特殊的发展考虑，这种药物免疫结合物（吉姆单抗和奥佐米星）因为临床疾病稀少、显著限制了病人的数量而没有临床Ⅲ期试验。两种未修饰单抗（阿仑珠单抗和西妥昔单抗）至少部分由于各种业务问题而进行了较长的临床试验。余下的四种“典型”的抗癌单抗（利妥昔单抗、曲妥珠单抗、贝伐单抗及panitumumab）临床研究及审批阶段的平均（中位数）时间分别为71.3（76.1）个月、6.2（5.5）个月。

4. Planning for the future To 2005, nine anticancer mAbs have been approved in the United States. With all nine products included in the calculations, the mean (median) clinical development and approval phases were 90.8 (82.5) months and 11.4 (6.6) months, respectively. However, the ranges were remarkably broad— 50.5 to 140.3 months for the clinical phase and 4.7 to 33.4 months for the approval phase. Oddly enough, most US-approved products might be considered atypical. Specifically, the two radio-immunoconjugates had extraordinary development considerations due to their radioactivity, the drug immunoconjugate (gemtuzumab ozogamicin) had no Phase III trials because rarity of the disease would have significantly limited patient accrual, and two unmodified mAbs (alemtuzumab and cetuximab) had lengthy clinical phases that were due, at least in part, to various business issues. The remaining four ' typical' cancer mAbs (rituximab, trastuzumab, bevacizumab and panitumumab) had mean (median) clinical development and approval phases of 71.3 (76.1) months and 6.2 (5.5) months, respectively.

单抗药物审批的总体通过率是无法计算的，因为鉴于早期鼠单抗药物疗效不佳，通过其所得出的总体通过率也将会产生误导作用。因此，我们计算了不同种类的抗肿瘤单抗药物的通过率，但只有人源单抗这一个种类有足够数量的药物供计算审批通过率，此结果对之后的战略规划提供了依据。在默认情况下，人源单抗为审批通过率提供了最准确的基准数据。目前，人源抗肿瘤单抗药物的审查通过率为18%，这一数据是根据2006年中40%的单抗药物通过率所估算而来的。需要注意的是，药物审查通

An overall approval success rate for mAbs was not calculated because, given the historically poor performance of the murine mAbs, the result would be misleading. Success rates were instead calculated for anticancer mAbs stratified by category. However, only one category, humanized mAbs, had a sufficient number of relevant products to provide a success rate useful for strategic planning purposes. By default, humanized mAbs provide the most accurate benchmark data on approval success rates. Humanized anticancer mAbs currently have a success rate of 18%, based on results for the 40% of products with fates known by mid-2006. It should be noted that the success rate will fluctuate over time as fates for additional mAbs are determined.

过率可能随时间波动，因为新通过的单抗药物会影响之后药物的审查。

1997年首个抗肿瘤抗体药物Rituximab进入临床，并在治疗非霍金奇淋巴瘤上获得成功。此后，Cetuximab、Trastuzumab等治疗性抗体陆续进入临床，截至2014年12月，用于癌症治疗的抗体药物总计25种。除此之外，尚有160多种抗肿瘤抗体药物处于临床研究阶段。药物靶点涵盖血液分化抗原、细胞生长因子、肿瘤坏死因子配基、血管内皮生长因子等。

In 1997, Rituximab became the first anti-tumor antibody drug in clinical, and has obtained satisfactory outcome in the treatment of non-Hodgkin's lymphoma. After that, other therapeutic antibodies such as Cetuximab and Trastuzumab entered the clinical successively, and by the time of December 2014, a total of 25 antibody drugs have been used for cancer treatment. Apart from these, more than 160 anti-tumor antibody drugs are under clinical studies. The targets of these drugs include cluster of differentiation antigens, cytokines, tumor necrosis factor ligands, vascular endothelial growth factors etc.

4.2.3 单抗在医药领域的应用实例

1. 抑制免疫系统 ①**MuromonabCD3**（OKT3）和两个人源化的抗CD3单克隆抗体，结合在T细胞表面的CD3分子上，用于预防器官的急性排斥反应，如肾脏移植等。人源化的抗体有可能成功地抑制I型糖尿病自动免疫过程中破坏β细胞的过程。②**因福利美**（Remicade®）和**阿达木单抗**（Humira®），与肿瘤坏死因子-α（TNF-α）结合，在抵抗一些炎症性疾病如类风湿关节炎（钝化Th1细胞活性）方面可能取得成功。副作用是可以使潜伏期的结核病转化为活动性结核病，可以诱导自身抗体的形成（通过促进Th2细胞的发育）。③**奥玛左单抗**（Xolair®），与免疫球蛋白E结合，阻止其与肥大细胞结合，对抵抗过敏性哮喘可能取得成功。④**赛尼派**（Zenapax®），与白细胞介素-2受体暴露在激活的T细胞表面部分结合，由于预防肾脏移植的急性排斥反应，有希望用于抵抗T细胞淋巴瘤。

4.2.3 Some Monoclonal Antibodies That Have Been Introduced into Human Medicine

1. suppress the immune system ① Muromonab-CD3 (OKT3) and two humanized anti-CD3 monoclonals. Bind to the CD3 molecule on the surface of T cells. Used to prevent acute rejection of organ, e.g., kidney, transplants. The humanized versions show promise in inhibiting the autoimmune destruction of beta cells in Type I diabetes mellitus. ② **Infliximab** (Remicade®) and **adalimumab** (Humira®). Bind to tumor necrosis factor-alpha (TNF-α). Show promise against some inflammatory diseases such as rheumatoid arthritis (by blunting the activity of Th1 cells). Side-effects: can convert a latent case of tuberculosis into active disease; can induce the formation of autoantibodies (by promoting the development of Th2 cells). ③ **Omalizumab** (Xolair®). Binds to IgE thus preventing IgE from binding to mast cells.Shows promise against allergic asthma. ④ **Daclizumab** (Zenapax®). Binds to part of the IL-2 receptor exposed at the surface of activated T cells.Used to prevent acute rejection of transplanted kidneys. Has also showed promise against T-cell lymphoma.

2. 杀死或抑制恶性肿瘤细胞 ①**利妥昔单抗**（商品名Rituxan®）与大多数B细胞中存在的CD20分子结合，用于治疗B细胞淋巴瘤。②**伊脖单抗**®用于抵抗B细胞（和淋巴瘤）上的CD20分子与放射性同位素铟-111（111In）或放射性同位素钇90（90Y）的结合。这两种同位素都被用药于淋巴瘤的患者，111In首先被给予，然后再给予^{90}Y（每种情况下都要补充Rituxan）。③**托西莫单抗**（Bexxar®）抵抗CD20与放射性同位素碘-131（131I）的结合，也用于淋巴瘤的治疗。虽然**Bexxar**®和**Zevalin**®都能杀死正常的B细胞，但是它们并不损害B细胞的前体细胞，因为它们并不表达CD20。因此，在时间上，前体细胞可以和成熟的B细胞一起调控机体。2005年2月3日，新英格兰医学杂志报道指出，患有B细胞淋巴瘤的患者应用131I－托西莫单抗进行单一治疗时，59%的人在5年内不会发病。④**西妥昔单抗**（Erbitux®）阻断HER1，一种发现于一些肿瘤细胞中（一些乳腺癌、淋巴瘤）的表皮生长因子（EGF）受体。⑤**Mylotarg**®结合CD33的偶联单克隆抗体，CD33是在急性骨髓性白血病（AML）表达的细胞表面分子，但是在正常的干细胞中并没有发现。⑥**阿仑单抗**（MabCampath®）与CD52分子结合，这是一个在白细胞中发现的分子，能够导致慢性淋巴细胞白血病的完全缓解。⑦**Lym-1**（Oncolym®）结合到HLA-DR上，HLA-DR编码组织相容性抗原，它在高水准的淋巴细胞中能够表达。⑧**易普力姆码**（Yervoy®）能够增强身体自身对肿瘤的免疫反应。

2. To kill or inhibit malignant cells ① **Rituximab** (trade name=Rituxan®). Binds to the CD20 molecule found on most B-cells and is used to treat B-cell lymphomas. ② **Zevalin**®. This is a monoclonal antibody against the CD20 molecule on B cells (and lymphomas) conjugated to either the radioactive isotope indium-111 (111In) or the radioactive isotope yttrium-90 (90Y). Both are given to the lymphoma patient, the 111In version first followed by the 90Y version (in each cases supplemented with Rituxan). ③ **Tositumomab** (Bexxar®). This is a conjugate of a monoclonal antibody against CD20 and the radioactive isotope iodine-131 (131I). It, too, is designed as a treatment for lymphoma. Although both Bexxar® and Zevalin® kill normal B cells, they don't harm the B-cell precursors because these do not express CD20. So, in time, the precursors can repopulate the body with healthy B cells. On 3 February 2005, the New England Journal of Medicine reported that 59% of patients with a B-cell lymphoma were disease-free 5 years after a single treatment with 131I-tositumomab. ④ **Cetuximab** (Erbitux®). Blocks HER1, a receptor for epidermal growth factor (EGF) that is found on some tumor cells (some breast cancers, lymphomas). ⑤ **Mylotarg**®. A conjugate of a monoclonal antibody that binds CD33, a cell-surface molecule expressed by the cancerous cells in acute myelogenous leukemia (AML) but not found on the normal stem cells. ⑥ **Alemtuzumab**(MabCampath®). Binds to CD52, a molecule found on white blood cells. Has produced complete remission of chronic lymphocytic leukemia. ⑦ **Lym-1** (Oncolym®). Binds to the HLA-DR-encoded histocompatibility antigen that can be expressed at high levels on lymphoma cells. ⑧ **Ipilimumab** (Yervoy®) that acts to enhance the body's own immune response to tumors.

3. 抑制血管生成 ①**维他心**结合

3. Angiogenesis Inhibitors ① **Vitaxin.** Binds to

到血管整合素（α-v/β-3）上，血管整合素（α-v/β-3）在肿瘤的血管中发现，但是并没有在正常组织供应的血管中发现。在Ⅱ期临床试验中，维他心可以使一些实体肿瘤收缩而没有副作用。②**贝伐单抗**（Avastin®）与血管内皮生长因子（VEGF）结合，阻止它与其受体结合。2004年2月已经被美国食品药品管理局批准用于治疗结直肠癌。

a vascular integrin (alpha-v/beta-3) found on the blood vessels of tumors but not on the blood vessels supplying normal tissues. In Phase II clinical trials, Vitaxin has shown some promise in shrinking solid tumors without harmful side effects. ② **Bevacizumab** (Avastin®). Binds to vascular endothelial growth factor (VEGF) preventing it from binding to its receptor.Approved by the US FDA in February 2004 for the treatment of colorectal cancers.

4. 其他 **阿昔单抗**（ReoPro®）通过结合在它们表面的受体来抑制血小板的聚集，这些受体通常由纤维蛋白原连接在表面上。对于预防施行了冠状动脉血管成形术的患者冠状动脉的反复堵塞有很大的帮助。

4. Other **Abciximab** (ReoPro®). It inhibits the clumping of platelets by binding the receptors on their surface that normally are linked by fibrinogen. It is helpful in preventing reclogging of the coronary arteries in patients undergone angioplasty.

4.2.4 单抗存在的问题及解决的方法

4.2.4 Problems With Monoclonal Therapy

鼠抗体被人类免疫系统识别为外来蛋白，并且人类患者会对其发生免疫应答反应，产生HAMA（“人抗鼠抗体”）。这些不仅会引起治疗性的抗体在人体内被快速的排除，而且会形成免疫复合物，引起对肾脏的损伤。但是，遗传工程的应用，使鼠-人杂交抗体在减少HAMA的问题方面成为可能。①嵌合抗体。该抗体结合了鼠抗体抗原结合部分（可变区）和人类抗体的效应区域（恒定区），如因福利美、利妥昔单抗、阿昔单抗。②人源化抗体。该抗体仅结合了小鼠（或大鼠）抗体的氨基酸的抗原结合位点（高变区）和人类抗体分子的其他部分，因此取代了它自己的高变区，如赛尼哌、维他心、吉妥单抗、赫赛汀以及奥玛珠单抗。

Mouse antibodies are “seen” by the human immune system as foreign, and the human patient launches an immune response against them, producing HAMA (“human anti-mouse antibodies”). These not only cause the therapeutic antibodies to be quickly eliminated from the host, but also form immune complexes that cause damage to the kidneys.However, using genetic engineering it is possible to make mouse-human hybrid antibodies to reduce the problem of HAMA. ① **Chimeric antibodies**. It comtains the antigen-binding parts (variable regions) of the mouse antibody with the effector parts (constant regions) of a human antibody. Infliximab, rituximab, and abciximab are examples. ② **Humanized antibodies**. This type of antibody comtains only the amino acids responsible for making the antigen binding site (the hypervariable regions) of a mouse (or rat) antibody with the rest of a human antibody molecule thus replacing the hypervariable regions. Daclizumab, Vitaxin, Mylotarg®, Herceptin, and Xolair® are examples.

（编者 吴胜昔，蔡家利）

(Compiled by Shengxi Wu & Jiali Cai)

参 考 文 献
References

郭亚军 .2015. 基于单克隆抗体的肿瘤免疫疗法研究进展 . 生物工程学报，31 (6) : 857-870

霍华德 G.C.，凯瑟 M.R. 著，张权庚，张玉祥等主译. 2010. 抗体制备与使用实验指南. 北京：科学出版社

何维. 2005. 医学免疫学. 北京：人民卫生出版社

沈倍奋. 2002. 抗体药物研究进展. 第二军医大学学报，23（10）：1047-1049

夏焕章，熊宗贵. 2006. 生物技术制药. 北京：高等教育出版社

熊宗贵. 1999. 生物技术制药. 北京：高等教育出版社

杨贵贞. 2002. 医学免疫学. 北京：高等教育出版社

姚文兵. 2010. 生物技术制药概论. 北京：中国医药科技出版社

甄永苏，邵荣光. 2010. 抗体工程药物. 北京：化学工业出版社

Goldsby *et al*. Immunology fifth Edition

Janice M. Reichert, Viia E. Valge-Archer. 2007.Development trends for monoclonal antibody cancer therapeutics. Nat Rev Drug Discov，6（5）：349-356

Mohit Trikha，Li Yan, Marian T Nakada. 2002.Monoclonal antibodies as therapeutics in oncology. Current Opinion in Biotechnology (13): 609-614

Mohit Trikha，Li Yan, Marian T Nakada. 2002.Monoclonal antibodies as therapeutics in oncology. Current Opinion in Biotechnology (13)：609-614

Milenic，D. E，Brady，E. D, Brechbiel，M. W. 2004.Antibody-targeted radiation cancer therapy. Nature Rev. Drug Discov (3): 488-499

Sang Jick Kim，Youngwoo Park, Hyo Jeong Hong. 2005.Antibody Engineering for the Development of Therapeutic Antibodies. Mol. Cells，20(1)：17-29

Sharkey，R. M，Goldenberg，D. M. 2005.Perspectives on cancer therapy with radiolabeled monoclonal antibodies. J. Nucl. Med (46): 115S-127S

第七章　生物技术疫苗

Chapter 7　Biotech Vaccine

疫苗是一种免疫生物制品，诱导机体产生免疫保护力，预防特异的引起疾病的传染病病原的感染，阻断或减少病原在易感群体间传播，从而保护宿主不感染此传染病。同时疫苗也调节或者增强宿主针对某一抗原的特异免疫力，从而防治疾病或增强康复。生物技术疫苗与传统疫苗不同，是利用现代生物技术制备的疫苗，包括基因工程活载体疫苗、基因工程亚单位疫苗、病毒样颗粒疫苗、转基因疫苗以及核酸疫苗等。

A vaccine is an immunogenic biologicals that induces host's protective immunity to prevent specific infectious pathogen from causing disease and interrupt or reduces the pathogen' s transmission within the susceptible population by inducing hurd immunity. A vaccine also modifies or augments a host' s in vivo antigen-specific immunity to prevent the progress of a disease or enhance the recovery propess. Different with conventional vaccine,the biotech vaccines are made by genetic engineering technique,including genetic engineering live carrier vaccine,genetic engineering subunit vaccine,transgenetic vaccine, DNA vaccine, virus like particle vaccine etc .

1　传统疫苗学

1　Conventional Vaccinology

两个多世纪以来，疫苗的发展都始终遵循着相同的基本原则。1796 年爱德华·詹纳（Edward Jenner）给詹姆斯·菲普斯（James Phipps）接种牛痘病毒诱导产生免疫保护，来抵御与人紧密相关的天花病毒，大约一个世纪之后，巴斯德（Pasteur）研制了狂犬病减毒活疫苗，疫苗发展的基本原则逐步建立起来。这些方法作为疫苗发展的指导方针一直延续到二十世纪，预防了许多种致死的传染病。事实上，所有现存疫苗的发展都利用了下面列出的至少一种方法：死（灭活）疫苗，减毒活疫苗或亚单位疫苗（包括与蛋白结合的荚膜多糖、类毒素、无细胞提取物、重组蛋白以及独立的荚膜多

Vaccine development followed the same basic principles for more than two centuries. When Edward jenner inoculated James Phipps with a bovine poxvirus to induce protection against the closely related human pathogen smallpox virus in 1796 and then, almost a century later, Pasteur developed alive attenuated vaccine against rabies, the basic principles for vaccine development were established . These approaches served as guidelines for the development of vaccines throughout the twentieth century, allowing the protection against many once lethal infectious diseases. In face, all existing vaccines were developed using at least one of the following approaches: killed (inactivated), live attenuated or subunit (including the protein-conjugated capsular polysaccharides, toxoids, cell-free extracts, recombinant proteins and stand-alone capsular polysaccharides) vaccines. Because of those basic principles, several infectious diseases can be prevented by vaccines. Table 7-1 shows the vaccines licensed for immunization and distribution in USA, all of them obtained basically by

糖）。由于这些基本原则，一些传染性疾病可以通过疫苗来预防。表 7-1 列出了已在美国获得报批的用作免疫和分发的疫苗，这些疫苗基本上都是通过传统方法获得。传统方法在疫苗史上获得了极大地成就，如天花病毒的根除以及其他疾病真正意义上的消失，包括白喉、破伤风、百日咳、小儿麻痹症、麻疹、腮腺炎、风疹和侵入性流感嗜血杆菌 b，大大提高了人们的生活质量和期望。传统方法为疫苗学提供了基础，占据了重要的地位。但是传统方法的研究耗时长，可能要耗费几年甚至几十年的精力。多年来，灭活和减毒活疫苗始终是第一选择。但是一些微生物难以体外培养，并且减毒活疫苗甚至可能造成有害的或不希望的免疫应答这一事实，都表明这些方法在一定程度上是不切实际的。在许多情况下，即使某种特定抗原的纯化产物也不能作为保护性候选疫苗。然而传统方法的使用往往只能鉴别出表达最丰富但也最多变，却并不适宜作为疫苗的候选物。传统疫苗学尽管成功地抵御了多种病原体的侵袭，但仍有许多疫病不能通过接种疫苗的手段来控制。考虑到新的疾病会通过突变和基因交换的方式演化出现，并且种间相互作用使人始终暴露于新的环境中，一个可及时应对这些威胁的更快更可靠的方法亟需建立。

conventional approaches. Conventional approaches led to great achievements such as the eradication of smallpox and the virtual disappearance of diseases like diphtheria, tetanus, poliomyelitis, pertussis, measles, mumps, rubella and invasive *Haemophilus influenzae* b, increasing the life quality and expectancy. It was also important to provide the basis of vaccinology but showed to be time-co nsuming, leading to years or even decades of research. Inactivation and attenuation were the first choice for many years, but the difficulty in cultivating some microorganisms in vitro and the fact that even attenuation may result in detrimental or unwanted immune responses showed that these approaches are impractical in some instances. Even the purification of specific antigens failed in many cases in providing a protective vaccine candidate,however, the methods usually used led to the identification of the most abundant but also most variable and less suitable vaccine candidates. Although successful for many pathogens, conventional vaccinology still left many diseases uncontrolled by vaccination. Considering that even new diseases are sure to emerge through evolution by mutation and gene exchange, interspecies transfer or human exposure to novel environments, a faster and more reliable approach must be available to promptly respond to those threats. A new promising group of vaccine candidares. Nevertheless, reverse vaccinology seemed to be a powerful tool that could help researchers to overcome the obstacles of conventional vaccinology and lead to the discovery and development of novel vaccines for the most concerning emerging diseases.

自爱德华·詹纳（18 世纪 90 年代，天花疫苗）和路易斯·巴斯德（19 世纪 80 年代，狂犬疫苗）早期的疫苗试验以来，疫苗已经成为很重要的能预防那些造成每年几百万死亡的疾病的方法

Since the earlier vaccine trials by Edward Jenner (1790s, smallpox vaccine) and Louis Pasteur (1880s, rabies vaccine), vaccine has become an important method to prevent diseases that cause multi-millions death every year (Walsh 2003). It is recognized as the top of 10 greatest public health achievements of the last century and is the number one contributor to

表 7-1　在美国已获批的预防性及上市疫苗

疫苗或目标	类型	疫苗或目标	类型
炭疽热	亚单位疫苗	肺炎球菌	亚单位疫苗
白喉	亚单位疫苗	小儿麻痹症	灭活疫苗
b 型流感嗜血杆菌	亚单位疫苗	狂犬病	灭活疫苗
甲型肝炎	灭活疫苗	轮状病毒	活疫苗
乙型肝炎	亚单位疫苗	风疹	活疫苗
人乳头瘤状病毒	亚单位疫苗	天花	活疫苗
流感病毒	灭活疫苗，活疫苗	破伤风	亚单位疫苗
日本脑炎	灭活疫苗	肺结核	活疫苗
麻疹	活疫苗	伤寒	活疫苗，亚单位疫苗
脑膜炎球菌	亚单位疫苗	水痘	灭活疫苗
流行性腮腺炎	活疫苗	黄热病	活疫苗
百日咳	亚单位疫苗	带状疱疹	活疫苗
鼠疫	灭活疫苗		

（Walsh2003）。它在20世纪最伟大的十个公共健康成就中名列第一（CDC1999；Bonanni1999），而且在降低全球发病率和死亡率方面的贡献是冠踞首位。在美国20世纪许多致死疾病，尤其那些儿童传染病，已被彻底清除或发病率降低百分之九十以上。全球疫苗市场现有几十个疫苗产品，还有更多的新产品包括抗各种传染病和癌症的疫苗正在研发中。市场上的疫苗预防众多疾病，包括白喉、破伤风、百日咳、脊髓灰质炎、轮状病毒、麻疹、腮腺炎、风疹、水痘、B型肝炎、b型流感嗜血杆菌（Hib）、肺炎、人乳头状瘤病毒、禽流感、天花、炭疽杆菌、霍乱和伤寒。在美国被食品药物检验局（FDA）批准的疫苗产品陈列在FDA的网站上（www.fda.gov/BiologiesBloodVaccines/ApprovedProducts/ UCM093833）。

decreased global morbidity and mortality (http://www.cdc.gov/vaccines/recs/default.htm; Bonanni 1999). Many deadly diseases, especially those infect children, have eliminated or declined more than 90% in the past century in the United States due to extensive vaccine program. There are several dozen vaccines on the global market, and many more in pipeline targeting various infectious and cancer diseases. Marketed vaccines include those against diphtheria, tetanus, pertussis, poliovirus, rotavirus, measles, mumps, rubella, varicella, hepatitis B, Haemophilus influenzae type b (Hib), pneumococcal disease, Human Papillomavirus (HPV), seasonal and pandemic influenza, smallpox, anthrax, cholera, typhoid, and many other diseases. The list of vaccines that are approved by US FDA is presented in FDA website (www.fda.gov/BiologiesBloodVaccines/ApprovedProducts/UCM093833).

市场上的传统疫苗主要有三种：减毒活疫苗、灭活疫苗、纯化的组分疫苗或亚单位疫苗（Walsh 2003；

There are traditionally three types of vaccines, live attenuated, killed/inactivated organism and purified organism components/subunit vaccines (Walsh 2003; Ulmer *et al.* 2006a). They are made from whole or

Ulmer 2006a），它们都是从细菌或病毒制备而来的。减毒细菌或病毒丧失了致病能力但仍能生长代谢，减毒是将致病菌或病毒长时间置于特殊生长条件下获得的，这个过程能筛选那些适应了特殊生长条件却丧失了在正常条件下生长能力的突变体。人类免疫缺陷型病毒（HIV）疫苗的开发使用了与之类似的一种病毒，类人猿免疫缺陷型病毒（SIV）作为缺失变异体方法研究的模型（Berkhout 2011）。其他方法的抗 HIV 病毒的疫苗也正在研究中。由于减毒活疫苗能在正常宿主体内短暂生长，它具有很强的免疫原性并能产生记忆 T 细胞，因此一次接种也就足够了。然而，减毒疫苗却有可能转化为致病体的缺点，它也有被其他病毒污染的可能。

component of bacteria or viral particles. Live attenuated bacteria or viruses lost their ability of causing disease (pathogenicity) but retain their capability of growth. Attenuation is usually achieved by growing a pathogenic organism for prolonged time under special culture conditions. This procedure selects mutants that adapted to the special growth condition but do not grow in normal culture condition. The development of attenuated human immunodeficiency virus (HIV) vaccine used a closely related virus, Simian immunodeficiency virus (SIV) as model to study the deletion strategy (Berkhout 2011). Other strategies of making HIV vaccine have been tested. Due to their capability of transient growth in host, attenuated vaccine is strong in immunogenicity and produce memory T cells, hence only one immunization is needed. However attenuated virus reverting to virulence was observed which caused acquired disease over time in vaccinated animals. Contamination of other virus that may present in attenuated vaccine is another concern.

灭活疫苗是被加热或化学方法处理过的丧失了复制能力的细菌或病毒。Salk 脊髓灰质炎和百日咳疫苗就是由甲醛灭活而获得的（Walsh 2003）。灭活过程不能太剧烈，否则蛋白会变性（例如表面抗原）。抗原决定基结构需要被保存以用来引发免疫反应。灭活疫苗主要诱导抗体免疫，与减毒疫苗相比它的细胞免疫诱导效果要差一些。

Killed and inactivated vaccines are those bacteria or viruses that have been heated or chemically treated so that they have lost their capability of replication. Examples are Salk polio and pertussis vaccines that are produced by formaldehyde inactivation (Walsh 2003). The inactivation should not be too harsh to denature the protein (such as surface antigens of organism). The epitope structure need be maintained to elicit immune response. Killed vaccines mainly induce antibodies and are less effective in inducing cell mediated immune response than attenuated vaccines.

纯化的组分疫苗包括毒素、多聚糖和抗原蛋白（亚单位疫苗）。细菌毒素之所以作为有效疫苗是因为它能诱导被感染宿主产生抗体。经纯化和甲醛灭活过的白喉和百日咳内毒素疫苗称作类毒素（Walsh 2003）。从荚膜细菌提纯的多聚糖能引起抗体免疫反应，但是 T- 细胞免疫反应却很差。疫苗对婴儿和小于两岁的幼儿的作用很有限，其原因是因为他（她）们的免疫系统

Purified organism components include toxin, polysaccharide and antigen proteins (also called subunit vaccine). Toxins from many bacteria can induce antibodies from infected host and therefore is an effective vaccine. Diphtheria and tetanus exotoxins are purified and inactivated by formaldehyde to form a vaccine so called toxoid (Walsh 2003). Polysaccharide purified from capsular bacteria can elicit antibody immune (humoral) response, but poor in T-cell mediated immune response. Limitation in infant and children less than 2-year old was observed due to immature immune systems in this age group. To overcome this, often a carrier protein is

不成熟。为了克服这些缺陷，一般会用一个载体蛋白结合到多聚糖疫苗上来促进T-细胞反应。举例来说，b型流感嗜血杆菌疫苗就是纯化的多聚糖结合几个不同的蛋白载体制成的。抗原疫苗（亚单位疫苗）是从不同的表达系统，包括细菌、酵母、昆虫细胞和哺乳动物细胞表达系统中产生的重组蛋白。亚单位疫苗的优点在于良好的安全性、可大量生产和基于简单小分子特性而易于监测分析。但是它们的免疫原性很差，一般只诱导抗体免疫，而不诱导T细胞免疫。第一个抗原疫苗是默克公司生产的从酵母菌中表达的表面抗原（HBsAg）作为B型肝炎疫苗（Recombivax HB，FDA，1981）。随后GlaxoSmithKline生产了类似的B型肝炎疫苗（Engerix-B，FDA，1989）。

conjugated to polysaccharide vaccine to increase the T-cell response. Hib as an example is purified polysaccharides chemically conjugated with several different protein carriers. Antigen (subunit) vaccines are recombinant proteins that expressed from chosen host, such as bacteria, yeast, insect or mammalian cell culture. Advantages of subunit vaccines include better safety, large quantity of production, and better characterization of the product due to its simple and small molecule nature. However, they are very poor on immunogenicity and induce mostly humoral antibody but not T cell mediated response. The first such vaccine is hepatitis B vaccine (Recombivax HB) that contains hepatitis B virus antigen (HBsAg) expressed from recombinant yeast cells by Merck and followed by GlaxoSmithKline.

2 反向疫苗学

2 Reverse Vaccinology

1995年，首个细菌基因组的测序引领疫苗的发展进入了一个崭新的时代，开启了疫苗发展指南的新篇章。忽然之间，由微生物编码的所有蛋白都触手可得。并且在传统疫苗发明两个多世纪之后，首次不使用常规疫苗的原则来甄别候选疫苗成为可能。这种新的方法，称为反向疫苗学，使得人们可以充分接触到微生物可以编码的所有蛋白。并且通过计算机分析，以相反的方式——从基因组而不是微生物开始，来甄别潜在的表面裸露蛋白成为可能。关于微生物难

The sequencing of the-first bacterium genome in 1995 led the vaccine development to enter a new era and to open a new chapter in the vaccine development guidebook. Suddenly, all proteins encoded by a microorganism were available and for the first time, aftermore than two centuries, it was possible to identify vaccine candidates without usingthe conventional vaccinology principles. This new approach, named reverse vaccinology, gives full access to all the proteins that a microor-ganism can encode and, by computer analysis, it is possible to identify potential surface-exposed proteins in a reverse manner, starting from the genome rather than the microorganism. Problems related to noncultivable microorganisms and also to antigens that are not expressed in in vitro conditions, conferring the most important obstacles for vaccine development,

以培养以及在体外条件下抗原不能表达的问题，给疫苗发展带来的严重阻碍，都将可以通过反向疫苗的方法避免。反向疫苗学方程式：获得 *n* 个基因组序列，加上先进的生物学信息技术以及高通量的免疫保护筛选系统，即可等于疫苗的发展。这种方法的可行性有赖于免疫保护的高通量筛选系统的有效性，以及与免疫保护的良好的相关性（图 7-1）。

could be avoided by the reverse vaccinology approach. The reverse vaccinology equation: the availability of *n* genome sequences, associated to advanced bioinformatic tools and reliable high-throughput systems for protective immune screening have proved to be a balanced equation for vaccine development. The feasibility of this approach relies on the availability of a high-throughput system for protective immunity screening and also on good correlates of protection (Figure 7-1).

图 7-1　反向疫苗学方程式

反向疫苗最大的限制在于其不能识别非蛋白抗原，如多糖——许多成功的疫苗的成分，以及糖脂——一个新的有希望的候选疫苗群。然而，反向疫苗就像是一个十分有力的工具，可以帮助研究者们克服传统疫苗的阻碍，并引导了新兴疾病的新颖疫苗的发现与发展。

The greatest limitation of the reverse vaccinology approach is the inability in identifying nonproteic antigens such as polysaccharides—components of many successful vaccines, and glycolipids—a new promising group of vaccine candidares.Nevertheless, reverse vaccinology seemed to be a powerful tool that could help researchers to overcome the obstacles of conventional vaccinology and lead to the discovery and development ofnovel vaccines for the most concerning emerging diseases.

2.1　经典的反向疫苗学方法

多年来，科学家集中大量的精力发展抗血清 B 型脑膜炎奈瑟菌（脑膜炎球菌）的疫苗。在感染脑膜炎球菌的案例中，B 型脑膜炎奈瑟菌导致约 50% 的败血症和脑膜炎症状。尽管治疗手段的不断发展，它仍有 5%～15% 的死亡率。脑膜炎奈瑟氏球菌是一种革兰氏阴性细菌，根据其荚膜多糖的

2.1 The Classical Reverse Vaccinology Approach

For many years great efforts were concentrated in the development ofa vaccine against serogroup B *N. meningitidis* (meningococcus), a bacterium that causes about 50% of meningococcal cases of sepsis and meningitis and has a mortality rate of 5%-15% despite the advances in therapeutics. *N. meningitidis* is a Gram-negative capsulated bacterium that can be classified in 13 serogroups on the basis ofthe chemical composition of the capsule polysaccharide. Among them, only 5 serogroups

化学组成可分为13种血清型。其中，与流行性脑膜炎和败血症相关的仅5种血清型：A、B、C、Y和W135。对于血清型A、C、Y和W135，利用其荚膜多糖作为疫苗成分已表明是可以为成人和婴幼儿提供免疫保护的。特别是当多糖成分连接到载体蛋白上时，可以刺激T细胞依赖的免疫应答从而获得长效免疫保护。不幸的是，这种方法对B血清型来说是不可行的。因为B血清型的荚膜糖蛋白是一个2，8-N-乙酰神经氨酸聚合物，在哺乳动物的组织中也存在，在疫苗中使用可能会导致自身免疫反应。

have been associated with meningococcal meningitis and sepsis: A, B, C, Y and W135. For sero-groups A, C, Y and W135 the use of capsular polysaccharides as vaccine components has shown to confer protection in adults and infants, especially when the polysaccharide is conjugated with a carrier protein. Eliciting a T-cell dependent immune response and conferring long-term protection. Unfortunately, this approach was not feasible for serogroup B, since its capsular polysaccharide is a polymer of a (2-8)-linked N-acetylneuraminic acid that is also present in mammal tissues and its use in a vaccine could lead to autoimmune responses.

当许多其他的常规方法未能研制出针对B型脑膜炎球菌的有效疫苗时，反向疫苗作为符合逻辑和有希望的可替代研制疫苗的形式出现。在脑膜炎奈瑟氏球菌的基因组序列仍处在匹配的阶段时，可以通过计算机分析的手段，预测蛋白质的表面裸露程度或同源性，以此来了解与毒力和发病机理相关因素，从而得到570个潜在的疫苗候选物。已有350个蛋白在大肠杆菌中成功克隆并表达，将其纯化后用于免疫小鼠。之后，免疫血清通过ELISA、FACS以及免疫印迹方法分析，从而来确定它们的表面位点。也可通过补体介导的抑菌试验来检测它们的免疫原性和保护效力，该试验是与人类的免疫保护相关的。其中81种蛋白在上述至少一种试验中显示强阳性结果，28种蛋白在所有试验中都显示阳性。为考察利用这些候选蛋白作为疫苗来针对异源性毒株的可能性，检测了针对一组由世界范围内多年来分离出的31株脑膜炎奈瑟

When many other conventional approaches failed to produce an effective vaccine against meningococcus B, the reverse vaccinology approach appeared as a logical and promising alternative to deliver a vaccine. While the *N. meningitidis* genome sequence was still been assembled, computer analysis allowed the prediction of proteins that could be surface-exposed or homologous to known factors associated with virulence and pathogenesis, leading to the selection of 570 potential vaccine candidates. Successful cloning and expression was obtained for 350 proteins in *Escherichia coli,* which were then purified and used to immunize mice. lmmune sera were then tested in ELISA, FACS and Western Blotting analysis in order to confirm their surface-exposed localization and also in complement-mediated bactericidal assays to test their immunogenicity and protective activity, since this assay correlates with protection in humans. Out of 81 proteins found to be strongly positive in at least one of the mentioned assays, 28 showed to be positive in all ofthem. To confirm the possibility ofusing these candidates in a vaccine able to protect against heterologous strains, their presence and conservation were tested in a panel of 31 strains of *N.meningitidis* isolated worldwide and over many years. Out of the 28 proteins tested, five showed to be strikingly conserved in the panel, a result not quite expected since they are surface-exposed proteins. In less then two years, reverse vaccinology achieved what the conventional

菌，看其是否具有保守性。在被测的28种蛋白中，5种蛋白是高度保守的。鉴于它们都属于表面裸露的蛋白，结果显得并不理想。在不到两年的时间里，反向疫苗就获得了传统疫苗方法追求了几十年的成果——B型脑膜炎奈瑟菌的表面裸露蛋白，可以在远缘毒株和血清型中诱导交叉保护作用，适合制成针对脑膜炎球菌的通用疫苗。从基因组信息中提取有利之处并将其应用到新型疫苗的发展中，将使疫苗研发变得简单可行。

vaccinology approaches pursuit by decades: surface-exposed proteins in *N. meninyitidis* B, able to induce protection and cross-reactivity among distantly related strains and serotypes, suitable to be used in a universal vaccine against this microorganism. Simple by extracting the benefits from genome information and applying them in the development of novel vaccines.

2.2 比较基因组分析——反向疫苗学的第二阶段

针对B型脑膜炎奈瑟菌获得的成功经验促进了反向疫苗学应用于其他病原体，如肺炎链球菌、牙龈卟啉、肺炎衣原体、炭疽杆菌、无乳链球菌、化脓性链球菌、肠外致病性大肠杆菌（Extraintestinal pathogenic Escherichia coli，ExPEC）等等，已逐步成为疫苗开发的常规经典方法（表7-2）。但是，当对无乳链球菌（B组链球菌）的强毒株的全基因组进行测序时，意想不到的事情发生。在美国以及欧洲，无乳链球菌是引发新生儿细菌性败血症、肺炎以及脑膜

2.2 Comparative Genome Analysis: The Second Phase of Reverse Vaccinology

The success obtained with the N. meningitidis B experience prompted the application of the reverse vaccinology to other pathogens, such as Streptococcus pneumoniae, Porphyromonasgingi-va/is, Chlamydia pneumoniae, Bacillus anthrads, Streptococcus agalactiae, Streptococcus pyogenes, ExPEC and many others, becoming a routinely classical approach for vaccine development (Table 7-2). But something unexpected happened when the complete genome of a virulent isolate of S. *agalactiae* (group B streptococcus) was sequenced. S. *agalactiae* is one of the leading causes of bacterial sepsis, pneumonia and meningitis in neonates in the USA and Europe and also an emerging cause of infection in the elderly. Conjugate vaccines based on the five major capsular polysaccharides were currently under development, but they were not able to cover all available serorypes, Since conventional approaches

表7-2 利用反向疫苗方法开发的疫苗举例

病原体	疾病	疫苗研究现状	参考文献
B型脑膜炎奈瑟菌	脑膜炎和败血症	二期临床	7
无乳链球菌	败血症、肺炎和脑膜炎	临床前	9，10
肺炎链球菌	肺炎	临床前	11
炭疽杆菌	炭疽病	临床前	12
衣原体肺炎	咽炎、支气管炎和肺炎	临床前	13
牙龈卟啉单胞菌	牙周炎	临床前	14

炎的主因之一，也是中老年人感染的新兴源头。基于与五大荚膜多糖结合疫苗目前正在研发中，但它们不能涵盖所有的血清型。这是由于传统方法不能为大部分受感染群体提供一个通用的有效疫苗。同时，由于2002年才得到两个无乳链球菌毒株的全部基因组序列，利用经典的反向疫苗学的做法听起来才是合乎逻辑的。但是，经验告诉我们无乳链球菌疫苗的研制将是更具挑战性。为了验证无乳链球菌基因组的多样性，并为未来研制出针对这种病原体的通用疫苗提供信息，需利用比较基因组杂交技术——利用测序菌株作为参考。人们发现，测序菌株编码的大约18%的基因在其他19种无乳链球菌中并不存在。比较基因组杂交技术存在的问题是，它只能提供已在菌株之间共享的基因组序列信息。因此，在其他菌株不存在于已知基因组测序的特定基因是无法检测的。这是获得通用疫苗的阻碍，使得经典的反向疫苗学不断演化发展。仅测序一株菌株的基因组，是不足以提供开发通用疫苗的所需信息，特别是在观察到基因组的变异程度高时。为了得到所要求的信息，需要确定无乳链球菌的另六组额外的基因组序列。结果表明，有1806个基因为所有无乳链球菌菌株所共享，作为“核心基因组”，约占各个菌株所编码基因数量的80%。核心基因组主要编码促成主要代谢途径的功能因子，即所谓的管家基因，通常可限定某种物种的特性。互补基因不存在于所有菌株基因组中，相当于“非必需基因组”，可能与菌株的适应性有关，

failed in providing a universal and efficient vaccine for the most affected group and since the complete genome sequence of two S. *agalactiae* strains was available in 2002, the classical reverse vaccinology approach sounded logical. But the S. *agalactiae* experience was going to be more challenging. In order to verify the diversity of S.*agalactiae* genome and provide information for a future universal vaccine against this microorganism, comparative genomic hybridization was applied, using the sequenced strain as reference. It was found that approximately 18% of the genes encoded in the sequenced strain was absent in at least one of the other nineteen S. *agalactiae* tested. The problem is that comparative genome hybridization is able to provide information only for the genome sequence that is shared among the strains. Therefore, specific genes in the other strains that are absent in the sequenced genome could not be detected. It could be a problem for a universal vaccine achievement, leading the classical reverse vaccinology to evolve. Sequencing the genome of only one strain could not be enough anymore to provide the information needed for a universal vaccine development, especially when a high variability is observed To provide the information requested, another additional six genome sequences of S. *agalactiae* were determined. This new information showed that 1806 genes are shared by all strains of S. *agalactiae,* representing the “core genome” that corresponds to approximately 80% of the average number of genes encoded in each strain. The core genome mainly encodes factors for functions that contribute to the major metabolic pathways, the so called housekeeping genes that usually define the identity of a species. The complementary set of genes, absent in at least one strain, correspond to the “dispensable genome”, probably related to the adaptation of strains in specific environmental conditions by conferring selective advantages. Mathematical extrapolation predicted that, no matter how many strains have been sequenced, each new genome should provide genes that have never been found before.Sequencing additional genomes allowed the estimationTof s. *agalactiae* pan-genome size (the set of genes that will be observed at least once if an infinite number of different strains would to be sequenced), corresponding to 2713 genes, of which 907 belong to the dispensable genome and also allowed the prediction that the pan-genome is going to grow about 33 new genes every time a new strain is sequenced. This profile is completely different of those observed for other microorganisms, such as *Bacillus antbracis.* Eight genome

在特定的环境条件下赋予菌株以选择性优势。通过数学外推法预测，不管有多少株已经被测序，每个新基因组应当提供以前从未找到的基因。对另外基因组测序结果，可用来判断无乳链球菌的泛基因组大小（如果无限量的不同菌株有待测序，至少观察改组基因一次）。无乳链球菌的泛基因组对应于2713个基因，其中907个基因属于非必需基因。并且可以预测到，每次新序列测序完成，该泛基因组将增加33个新基因。这一描述完全不同于从其他微生物中观察到的现象，如炭疽杆菌。已确定了该微生物的八条基因组序列，但在第四条基因组序列之后，经证实加入到泛基因组的新基因数量已迅速收敛到零。

sequences were determined for this microorganism, but after the fourth it was verified that the number of new genes added to the pan-genome rapidly converged to zero.

比较基因组分析也可提供必要的信息，来满足研制针对无乳链球菌的通用疫苗的需求。计算算法预测到589种表面相关蛋白，其中396种由核心基因组编码，另193种不存在于每个基因组。检测每种蛋白的保护效力，发现有四种抗原在动物模型上可以引起免疫保护应答。这里的动物模型不仅有被动免疫并攻毒的新生小鼠模型，还有通过主动母体免疫并攻毒的出生48h的子代小鼠模型。这四种保护性抗原都不能纳入通用疫苗行列，因为其中三种不存在于被测菌株中，而第四种则属于核心基因组，在一些菌株中存在氨基酸残基表面可及性的缺陷。将四种最佳候选物组合给药，可针对一组12种无乳链球菌给予59%～100%的保护效力，包括主要的血清型菌种，以及来自不常见的

Comparative genome analysis also provided the information necessary to face the quest of providing a universal vaccine against S. *agalactiae*. Computational algorithms predicted 589 surface-associated proteins, of which 396 belong to the core genome and 193 were absent in at least one strain. Each of this protein was tested for protection and four antigens were able to elicit protective immune responses in the animal model, not only by passive immunization and challenge in newborn mice but also active maternal immunization and challenge of offspring within the first 48 hours of life. None of these protective antigens could be classified as universal, because three of them were absent in a fraction of the tested strains and the fourth, belonging to the core genome, had a deficient surface accessibility in some strains.17 The cocktail combining the four best candidates conferred 59%-100% protection against a panel of 12S. *agalactiae* isolates, including the major serotypes, as well as two strains from a less common serotype. Without the determination of additional genome sequences, a universal vaccine against S. *agalactiae* would have been compromised. The comparative genome analysis provided new concepts in delivering universal vaccines by the reverse vaccinology approach, even for microorganisms in which a high variabilitycan be observed, opening the

血清型的两种菌种。没有额外的基因组序列的测定，无乳链球菌的通用疫苗将大打折扣。比较基因组分析通过反向疫苗学方法为通用疫苗的研发提供了新的概念，针对可以观察到高变异性的微生物，疫苗的发展开启迈向泛基因组的反向疫苗学的时代。

pan-genomic reverse vaccinology era.

尽管经典的反向疫苗方法十分有效地筛选出了针对B型脑膜炎奈瑟菌的候选疫苗，但深入分析表明，NadA——最佳候选物之一，仅有50%来自于脑膜炎患者分离的菌株和25%健康人群携带的菌株有该物质存在。这都暗示了微生物的泛基因组与无乳链球菌的泛基因组有相似的特征。脑膜炎球菌的新基因序列的可用性可以提供更多的候选物，来提高已获得的疫苗的保护范围及免疫效力。

Although the classical reverse vaccinology approach was efficient in providing protective candidates against serogroup B *N. meningitidis,* subsequent analysis showed that NadA, one of those best candidates,is present in only 50% of strains from patients with meningococcal disease and only 25% of strains carried by healthy people, indicating that also the pan-genome of this microorganism could have the same features of S. *agalactiae* pan-genome. It is possible that the availability of novel genome sequences of *N.meningitidis* could provide even more candidates that could increase the coverage and efficiency already achieved.

2.3 消减基因组分析——反向疫苗学的第三阶段？

2.3 Subtractive Genome Analysis: Third Phase of Reverse Vaccinology?

1995年首个基因组测序完成之后，人们把大量的精力集中于其他致病菌的基因组序列的测定中来，不断引导以反向疫苗学方法为基础的新型疫苗的发展，如B型脑膜炎奈瑟菌和无乳链球菌。病原体毒株的比较基因组分析（第二代反向疫苗学）致力于比较一物种的病原体毒株，寻找构成通用疫苗时可引起最大范围的免疫保护的识别抗原。但是关于非致病性菌株呢？为了获得针对那些致病性毒株的疫苗而对非致病性菌株基因组进行测序，这听起来好像很奇怪。但是非病原性菌株的基因组可以提供必要

After the sequencing of the first genome in 1995, great efforts have been concentrated in the determination of the genome sequences of other pathogenic bacteria that could lead to the development of novel vaccines based on the reverse vaccinology approach such as the *N.meningitidis B and S.* agalactiae experiences. Comparative genome analysis of pathogenic strains (the second gene ration of reverse vaccinology) have concentrated their efforts in comparing pathogenic strains of a species looking for the identification of antigens that could lead to a max imumcoverage in a universal vaccine. But what about nonpathogenic strains? It could sound strange to sequence a nonpathogenic strain genome in order to obtain a vaccine against those pathogenic ones, but a nonpathogenic genome could provide the information necessary for the identification of antigens that really could make the difference

的识别抗原，可以真正提供与致病性菌株有差异的信息——主导了最严苛的宿主 - 病原体的相互作用。在消减比较基因组分析中，一个相同或甚至相关物种的致病性和非致病性菌株的保守基因序列是可以舍弃的，从而减少了候选物的数量，因此缩短了获得疫苗的时间。由于引发疾病的能力往往与某些基因的完整性有关，演算时必须考虑到某些基因的失活如移码突变。当然，通常有一整套的因素导致菌株具有致病性。因此，同样在这种情况下，只有一个非致病菌株基因组的测序结果并不能充分了解致病性机理。在某些情况下，寻找某些不存在的候选物可以得到通过比较致病性毒株不能得到的答案。

inpathogenesis, responsible for the most strictly host-pathogen interactions. In a subtractive comparative genomeanalysis, genes conserved between pathogenic and nonpathogenic strains of a same or even related species could be discarded, reducing the number of candidates and, consequently, reducing the time for the delivery of a vaccine. Since the ability of causing diseases is frequently related to the integrity of some genes, algorithms must take into account some gene inactivation process such as frame shifting. Of course a whole set of factors are usually responsible for pathogenicity, therefore,also in this case the sequencing of only one nonpathogenic genome could not be enough for a complete understanding of pathogenicity. In some cases, looking for candidates where they do not exist could provide the answers that the comparison between pathogenic strains could not give.

由于术语“共生”通常是与非致病株甚至是益生菌相关联的，共生菌株的概念也明确了这一方法的大获成功。共生菌通常是那些寄生于个体而不引起个体疾病的菌株。在共生菌群中，我们发现“致病性共生”——当机体脆弱时共生菌有引发疾病的能力。不混淆那些“致病性共生”（如肺炎链球菌和 b 型流感嗜血杆菌）和“非致病性共生”（如某些种类的益生菌），这一点是非常重要的。

The concept of commensal strains must be also clarified for the success of this approach, since the term “commensal” is usually associated with nonpathogenic strains or even probiotics. Commensal bacteria are those that usually colonizean individual without causing disease. Within the commensal flora we can find the “pathogenic commensals ”that have the ability to cause disease when the organism is vulnerable. It is important not to confuse those“pathogenic commensals”(like S. pneumoniae and H. injluenzae B) with the“nonpathogenic”ones (like some species of Lactobacillus).

菌群的结构和组成是在微生物和宿主水平上筛选的结果，促使这个复杂的系统相互之间合作稳定，使得菌群组成天然屏障并形成多面保护（病原体位移、养分竞争、受体竞争等），多种结构（构筑屏障、诱导产生 IgA、免疫系统发育等）以及多重代谢效应（饮食中致癌物质的代谢、

The structure and composition of the flora is a result of selection at both microbial and host level, leading to a mutual cooperation and stability of this complex system, allowing the flora to form a natural barrier that has numerous protective (pathogen displacement, nutrient competition,receptor competition,etc), structural (barrier fortification, induction of IgA, immune system development, etc) and metabolic effects (metabolization dietary carcinogens, synthesis of vitamins, ion absorption, etc.). Hence the delivering of vaccines against

维生素的合成、离子吸附等）。因此研制出针对属于共生菌群的微生物的疫苗，并不是一项简单的任务，即便是针对“致病性共生”。已证实，针对肺炎链球菌的七价多糖疫苗可以作为无疫苗血清型的替代物，但仍有可能引起感染。另一方面，针对b型流感嗜血杆菌的与多糖结合的疫苗则导致了F型的发生率。由于竞争水平的下降可促进其他“致病性共生”因子的生长，针对共生菌研制出疫苗的效果总是出乎意料的。考虑到毒力、适应度和分布因素可能是重叠的，以及对生态环境条件的依赖性，理解“非致病性”共生可以促成疫苗的开发，给共生菌群以最小的冲击。特别是那些两种表型都具备的微生物，如大肠杆菌。

microorganisms that belong to the commensal flora is not an easy task, even for the pathogenic commensals. It has been shown that polysaccharide heptavalent vaccine against S.pneumoniae can lead to the replacement with nonvaccine serotypes, but still able to cause infection and that the polysaccharide-conjugated vaccine against H. Injluenzae B led to the increase of serotype F incidence. The consequences of a vaccine against commensal flora components are usually unexpected, since reducing the level of competition could promote the emerging of other “pathogenic commensal”Considering that virulence , fitness and colonization factors could be overlapping and dependent of the niche and environmental conditions, understanding the“nonpathogenic”commensals could prevent the development of vaccines with a minimum impact to the commensal flora, especially for microorganisms that possessboth phenotypes, such as *E. coli.*

2.4　结论

迄今为止，已测定了1327种细菌的基因组，其中520种的基因组序列已经完成。表7-3显示的是与人类疾病相关的细菌数量，这些都是由基因测序计划完成的。对于这些疾病，大多数都没有有效的疫苗可以使用。在那种情况下，以基因组为基础的比较分析不仅能更好地理解菌株的多变性和泛基因组评估，还能促成以单一或联合抗原组成的通用疫苗的研制。这一领域是传统疫苗方法没能达到的，而却被B型脑膜炎奈瑟球菌和无乳链球菌的成功经验（经典的泛基因组反向疫苗学）所征服。即使已有上市的疫苗，在有需要时，仍可通过这些方法纳入新型的抗原，从而来提高和增加

2.4 Conclusion

To date, 1327 bacterial genomes have been sequenced, of which 520 genome sequences have already been completed. Table7-3 shows the number of genome sequencing projects of bacteria that could be associated to human diseases. For most of them, no efficient vaccine is available. In those cases, a comparative genome-based analysis would permit not only the better understanding of strain variability and pan-genome estimation, but also the delivering of a single or combined-antigen universal vaccines in a field where conventional approaches have failed as proved by N. meningitidis B and S. agalactiae experiences (classical and pan-genome reverse vaccinology). Even for vaccines already available, these approaches could permit the improvement and increment of protection by inclusion of novel antigens, when required. Moreover, the knowledge of antigen gene variability could drive strategies of rational engineering to improve the coverage with proven efficacy.Considering the fact that nonpathogenic strains are present for many genera, sequencing of nonpathogenic genomes could be also important in highlighting the antigens that really could make the difference in

表 7-3　与人类疾病相关的传染性细菌种类的基因组测序计划

微生物名称	基因组计划
阿氏疏螺旋体 B.afzelii、艰难梭状芽孢杆菌、具核梭杆菌、幽门螺旋杆菌、牛型结核菌、生殖道支原体、立克次体、痢疾志贺氏菌、弗累克斯讷氏杆菌、副溶血性弧菌	3
鲍氏不动杆菌、博氏疏螺旋体、泰国伯克菌、鼠伤寒沙门氏菌、腐败希瓦菌、南极希瓦菌	4
洋葱伯克菌、肺炎衣原体、嗜肺性军团病杆菌、微小脲原体	5
猪链球菌	6
绿脓杆菌	7
产气荚膜梭状芽孢杆菌、结核分枝杆菌、无乳链球菌、解脲脲原体	8
肉毒杆菌	9
鼻疽伯克霍尔德氏菌、贝纳特氏立克次体	10
炭疽杆菌、空肠弯曲杆菌	11
化脓性链球菌	13
肺炎链球菌	14
兔热病杆菌、霍乱弧菌	15
蜡样芽孢杆菌、金黄色酿脓葡萄球菌、鼠疫耶尔辛杆菌	16
流感嗜血杆菌	19
类鼻疽杆菌	21
单核细胞增多性李斯特菌、沙门氏菌	24
大肠杆菌	25

免疫保护。此外，认识抗原基因的多变性，对合理地构建比已证实的免疫效力更广的疫苗起到决策性作用。考虑到许多物种多为非致病性菌株这一事实，为了突出致病性抗原发病机制是与非致病菌株存在差异的，非致病性基因组的测序也是十分重要的。致病性与非致病性基因组的比较基因组学分析，可直接提示与发病机理有关的基因，这些基因主导了病原体与宿主最直接的相互作用。因此，好的疫苗也可以从中发现（图 7-2）。

pathogenesis. Comparative genome analysis between pathogenic and nonpathogenic genomes would prompt straight to the genes responsible for pathogenesis and, in most cases, responsible for the closest interactions between pathogen and host and, therefore, good vaccine candidates (Figure 7-2).

总之，许多疾病仍不能通过接种疫苗的方法来控制，而且还有更多的

In conclusion, many diseases can not be still controlled by vaccination and more diseases are expected

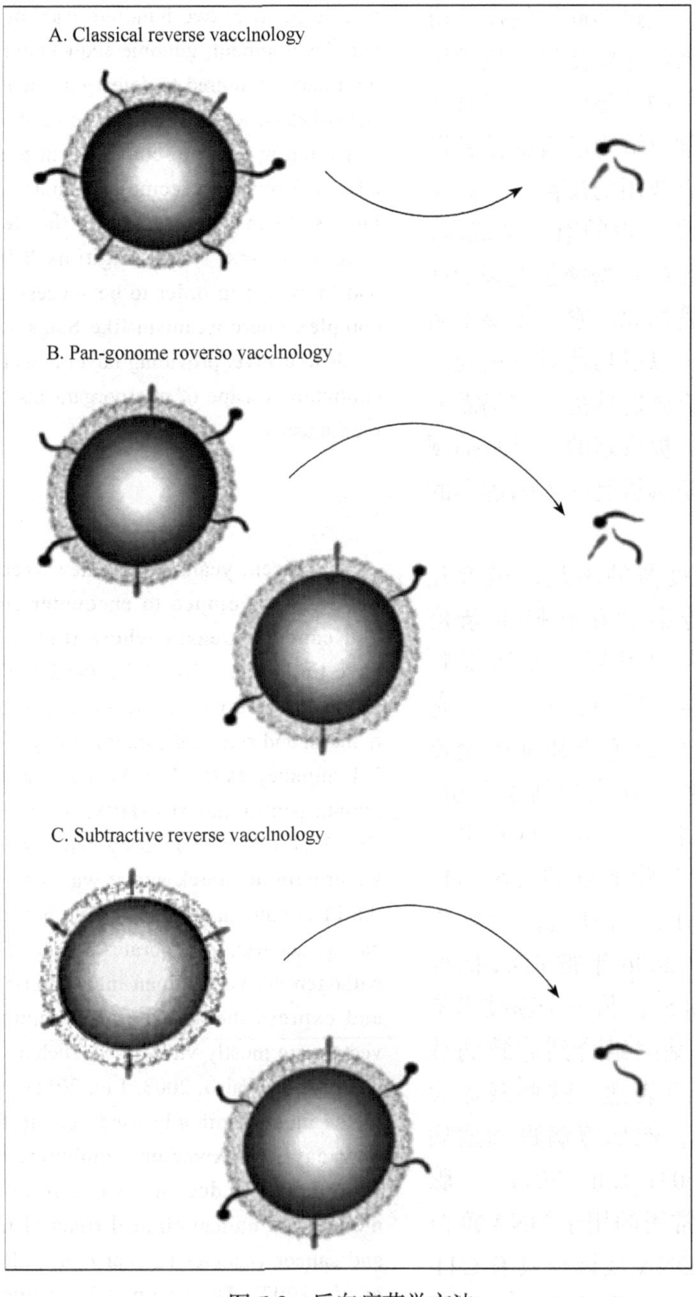

图 7-2 反向疫苗学方法

A. 经典的反向疫苗学方法在于挖掘一基因组序列，从而推定其表面裸露的抗原，并鉴定其是否适宜作为候选疫苗；
B. 泛基因组的反向疫苗学方法比较了不同菌株的不同基因组序列，扩大药用范围，避免因抗原性变异导致的免疫逃逸；
C. 消减反向疫苗学方法在于将致病性与非致病性基因组序列作比较，从而筛选出直接导致疾病的抗原疫苗将会出现。经过200年的疫苗研究与开发，基因组测序提供了战胜传染性微生物所需的关键信息。尽管如此，我们需要学习解密已有的信息以便将其转化成有效用的疫苗。经典反向疫苗学方法所取得的第一个成就，已成功研发出针对B型脑膜炎奈瑟球菌的疫苗。尽管如此，它必须要不断演化发展，才可以成功地对抗更为复杂的微生物如无乳链球菌。它仍需要不断演化发展，提供新的工具来抑制那些可从宿主防御机制中不断逃逸的微生物。

to emerge.After two hundred years of vaccine research and development, genome sequencing provided crucial information required to defeat infectious microorganisms. Nevertheless, we need to learn to decipher the information generated in order to convert it in protective vaccines. Classical reverse vaccinology approach made the first moves, been successful for the development of a vaccine against N. meningitidis B.In spite of that,it had to evolve in order to be successful against a more complex microorganism like S.agalactiae. And it still needs to evolve, providing novel tools that will avoid the continuing escape of microorganisms from host defense mechanisms.

鉴于传统疫苗的不足，最近几年针对慢性传染病和肿瘤的新疫苗技术正在大力发展。类病毒颗粒（VLP）技术就是其中之一。它是能同时诱导体液免疫和细胞免疫（CTLs）的有效而安全的新型疫苗。第一个类病毒颗粒疫苗是自我组装的抗4种人乳头状瘤病毒（6、11、16和18）的L1蛋白疫苗，由默克公司生产并于2006年被FDA批准（Buckland，2005）。另一个新技术基因疫苗把致病基因组合到非致病载体上并在接种后表达。这些载体大多数源于病毒，例如腺病毒和痘病毒（Walsh，2003；Liu，2011）。腺病毒是现今最常用的用于DNA疫苗的载体之一，DNA载体与具有CTL诱导作用的亚体疫苗结合后在动物和人体的抗感染与抗癌症的临床试验中显示了大好前景（Ulmer *et al.*，2006b；Rottinghaus *et al.*，2003）。Dendreon的Provenge是FDA于2010年四月批准的抗前列腺癌症的自身免疫疗法疫苗，属于抗癌症疫

In recent years several new vaccine technologies are being developed to encounter chronic infections and cancer diseases where traditional approaches have limitation. Virus-Like Particle (VLP) is one that producing safe vaccine while efficiently inducing both humoral and cell-mediated immunity including Cytotoxic T Lymphocytes (CTLs). Vaccine against four types of human papillomavirus (HPV, 6, 11, 16 and 18) is the first VLP that produced by self-assembly of the HPV L1 protein at Merck and it was, approved in 2006 by the FDA (Buckland, 2005). Another technology, Gene-based vaccines, incorporates a gene/cDNA coding for a pathogen-derived antigen into a non-pathogenic vector, and express the antigen after immunization. These vectors are mostly viral-based, such as adenoviruses and poxviruses (Walsh, 2003; Liu, 2011). Adenovirus is one of the most commonly used vectors for DNA vaccine nowadays. DNA vectors combining subunit vaccines with the CTL induction have shown promise in animal models and human clinical trials of infectious disease and cancer vaccines (Ulmer *et al.*, 2006b; Rottinghaus *et al.*, 2003). Dendreon' s Provenge, an autologous cellular immunotherapy vaccine targeting prostate cancer approved by FDA in April 2010, serves as a pioneer of cancer vaccine. It consists of autologous antigen-presenting cells and a fusion protein composed of prostatic acid phosphatase and granulocyte macrophage colony-stimulating factor (GM-CSF). Instead of attacking cancer cell, it “trains” human body' s own immune cells to do the job (Schlom *et al.*, 2007).

苗的先驱，它含有自我抗原表达细胞和前列腺酸性磷酸酶与粒状巨噬细胞刺激因子（GM-CSF）的融合蛋白，它训练人体的自身免疫细胞来抗癌，而不是用外源注射的疫苗靶向癌细胞（Schlom *et al.*，2007）。

免疫佐剂常用于增强重组抗原的免疫性从而降低对剂量的要求。常用的佐剂有免疫刺激和免疫载体两种（Reed *et al.*，2008）。第一种包括能够刺激免疫反应的细胞因子和细菌外毒素，第二种包括微量元素（例如铝盐、AS04）、乳胶（MF59）和脂质体，其作用在于控释抗原和增强特定免疫反应。铝盐类的免疫佐剂常常被使用，在许多上市产品中显示了其巨大的功效，然而铝盐在引发 CD4 和 CD8 T 细胞反应上还远远不够。Novartis 用的水包油乳胶（MF59）和 GSK 的 AS03 显示了比铝盐在降低剂量上更大的潜力，从而被批准用于人体。更多的能够提供强有力且引起长效体液和 T 细胞免疫疫苗的免疫佐剂正在研究中。

Adjuvant is used for vaccines to enhance the immune response to the recombinant antigens which result in reduced dose requirement. There are two types of adjuvant, immunostimulants and vehicles (Reed *et al.*, 2008), and the first class includes cytokines and bacterial exotoxins which can stimulate immune response, while the second class includes minerals (such as Aluminum slat：Alum, AS04), emulsions (MF59) and liposomes, which controls release of antigen or depot delivery system to increase the specific immune response. Aluminum-based adjuvant is very commonly used and has been shown great effectiveness for many commercial products. However, Alum is not sufficient to trigger immunity based on CD4 and CD8 T-cell response. Oil-in-water emulsions used by Novartis (MF59) and GlaxoSmithKline (AS03) have shown a greater dose-reduction potential than aluminum salts and have been approved for human use. More new adjuvants are being developed to deliver strong and long lasting humoral and cellular immune response vaccines.

3 病毒样颗粒疫苗

3 Virus Like Particle Vaccine

病毒性疾病的疫苗传统上通过弱化病毒株或灭活传染性病毒的方法制得。基于表达外源病毒蛋白研制的亚单位疫苗已经可以有效地抵御一些病原体，但由于蛋白质的不正确折叠与修饰使得疫苗的免疫原性差。本节内容专注于一类特殊的病毒亚单位疫苗，这类疫苗模仿了

Vaccines against viral diseases had traditionally relied on attenuated virus strains or inactivation of the pathogenic viruses. Subunit vaccines based on viral proteins expressed in heterologous systems had been effective for some pathogens, but often suffered from poor immunogenicity due to incorrect protein folding or modification. In this chapter we focus on a specific class of viral subunit vaccine that mimics the overall structure of virus particles and thus preserves the native antigenic conformation of the immunogenic proteins. These virus-like particles (VLPs) have been produced for

病毒颗粒的整体结构，因此保留了病毒先天的免疫原性。这类病毒样颗粒已经广泛地应用于类别和结构都不同的病毒中，并且呈现出比以往更安全、免疫原性更强的独特优势。随着乳头瘤 VLP 疫苗开始进入市场，可以认为已经到达必须考虑一个可行的疫苗策略的时候了。

a wide range of taxonomically and structurally distinct viruses, and have unique advantages in terms of safety and immunogenicity over previous approaches. With new VLP vaccines for papillomavirus beginning to reach the market place we believe that this technology has now "come-of-age"and must be considered a viable vaccine strategy.

目前，仍然有很多传染性的病毒威胁着公众健康（表 7-4）。凡有效的疫苗接种通常可以在防止病毒传播方面提供最符合成本效益的长期保护。因此，疫苗通过使用一个零星的"诱饵"，模仿致命的病原体来刺激机体的自然防御。越类似天然病毒的疫苗，在随后的免疫应答中起到的效果就越好。一般情况下，抗病毒感染取决于病毒粒子呈现在细胞表面或病毒感染细胞时抗原的免疫应答的发展。因此发展有效疫苗的第一步就是鉴定保护性抗原。

There are many infectious viruses that remain major threats to public health (Table 7-4). Where an effective vaccine exists, vaccination is usually the most cost-effective long-term protection against disease and spread for most viruses. The principle of vaccination is to generate sufficient immunity to protect from infectious disease. Thus the vaccine stimulates the body' s natural defenses against disease through use of a benign "decoy"that mimics the virulent pathogen. The more similar a vaccine is to the natural pathogen, the better the immune response to the subsequent exposure of the pathogen. In general, resistance to virus infection depends on the development of an immune response to antigens present on the surface of virions or virus-infected cells. Therefore identification of protective antigens is the first step in the development of effective viral vaccines.

表 7-4 主要威胁人体健康的病毒

病毒	疾病	病毒	疾病
HIV（艾滋病病毒）	AIDS（艾滋病）	Epstein Barr Virus(疱疹病毒）	淋巴瘤，鼻咽癌
RSV（呼吸道合胞病毒）	呼吸感染	HPV（人乳头状瘤病毒）	宫颈癌
Hepatitis B（乙肝病毒）	肝癌	Measles（麻疹）	肺炎
Hepatitis C（丙肝病毒）	肝硬化 / 肝癌	Influenza（流感病毒）	肺炎

目前，很多成功的病毒疫苗已被开发并正在使用。这些疫苗主要是灭活疫苗和弱毒疫苗。弱毒疫苗，如麻疹、腮腺炎、风疹、脊髓灰质炎、天花、水痘和黄热病的疫苗都是通过弱化野生病毒的形式研制的。这些疫苗接种后在宿主体内有限的复制，产生与自然感染相似的免疫应答且在单次

Currently many successful viral vaccines have been developed and are placed in use. These vaccines are predominantly based on live attenuated or inactivated viruses. The live attenuated vaccines such as measles, mumps, rubella, oral polio, smallpox, varicella and yellow fever are a weakened form of the "wild" viruses. These attenuated virus vaccines rely on limited replication of the virus in the host following vaccination. Immune responses induced are similar to those from natural infections and often these vaccines are effective after a single dose. However,

剂量免疫后就会产生免疫效果。然而，这种疫苗有可能在某些患者身上产生严重的反应，这往往是返毒的结果。相反，灭活的疫苗不能复制，且它们的遗传物质或整体结构遭到破坏，因此，这类疫苗比活疫苗安全，但免疫效果差，一般在抗体滴度下降前需要3～5个剂量。

such vaccines may cause severe reactions in some patients, often the result of the limited replication of the attenuated virus following vaccination. In contrast to attenuated live virus vaccines, inactivated (or killed) vaccines can not replicate, as their genetic material or overall structure are purposely and fully destroyed. These vaccines are safer than live vaccines but generally not as effective, requiring 3-5 doses as antibody titer falls overtime.

它们缺乏活疫苗自我复制的能力，正因为这样，它们是安全的，避免了与病毒复制相关的危险。这些疫苗是由全细胞（如流感、小儿麻痹症、狂犬病、甲肝疫苗）或部分细胞或者亚单位疫苗构成的，比如乙肝疫苗。亚单位疫苗基于运载有限数量的病毒蛋白，这些蛋白通常可以提供保护性免疫的衣壳蛋白或包膜蛋白。它比灭活疫苗安全，因为它能独立培养复制病毒。事实上，若疫苗在安全方面存在任何的灭活不完全或批次间的差异，这种疫苗都该被淘汰。然而，亚单位疫苗历来有一个严重的缺点：仅表达纯化一种蛋白作为免疫原，它的效果一般没有掺入病毒的疫苗免疫效果好。原因可能是，蛋白的构象存在错误的折叠，因此，需要加大免疫剂量才能实现相同级别的免疫保护。

They lack the self-boosting qualities of live attenuated vaccines; they are safer in the sense that the inherent dangers associated with virus replication are avoided. These vaccines are made as whole cell vaccines (such as Influenza, polio, rabies, and hepatitis A) or as fractional or subunit vaccines such as hepatitis B. Subunit vaccines are based on the delivery of only a limited number of viral proteins, often the major protein in the capsid or envelope that is sufficient to confer protective immunity. These vaccines are an order of the magnitude safer than inactivated vaccines because the subunit vaccines can be prepared independent to the culture of replicating virus. Indeed, any remaining possibility of incomplete inactivation or batch to batch variation in the safety of the vaccine is eliminated. However, subunit vaccines have traditionally suffered from one important drawback; often single proteins when expressed and purified in the absence of other viral components are less immunogenic than those that are incorporated into infectious virus. This is probably due to a fraction of this protein presented in a misfolded conformation relative to the native protein. Thus, higher doses and amounts of antigen are required to achieve the same level of protection.

在提高亚单位疫苗的免疫原性方面，如今有一个很大的进步，就是可以利用酵母、昆虫或哺乳细胞表达系统来生产病毒样颗粒（VLP）。VLP是一种不含病毒基因组，模仿了病毒的天然构象，免疫效果好的亚单位疫苗。事实上，VLPs缺乏病毒的全部基因组，但却有同灭活疫苗一样的结构蛋白的表面构象，避免了病毒复制以及灭活疫苗的任何风险。

A major advance in subunit immunogen production has been assembly of proteins as virus-like particles (VLPs) using protein expression technology in yeast, insect or mammalian cells. VLPs are a highly effective type of subunit vaccines that mimic the overall structure of virus particles without any requirement that they contain infectious genetic material. Indeed, many VLPs lack the DNA or RNA genome of the virus altogether, but have the authentic conformation of viral capsid proteins seen with attenuated virus vaccines, without any of the risks associated with virus replication or inactivation.

VLP制剂是体外表达的衣壳蛋白能自发地组装成结构上类似真实病毒的粒子。在实际应用中，VLPs模仿了病毒的空间结构就意味着它应该能引发强效的体液免疫甚至低剂量的亚单位疫苗足以引发相似的保护反应。它们除了能刺激B细胞产生体液免疫反应外，还被证明能够非常有效地刺激CD4细胞的增殖以及CTL反应。VLP疫苗在该领域的特色可能归功于它的有效性。从它本身，我们也可以更清楚地看到精确的初免-加强免疫战略在疫苗接种预防疾病方面起着重要的作用。因此，作为各种疾病装备军的VLPs可以采用初免-加强免疫战略。

VLP preparations are all based on the observation that expression of the capsid proteins of many viruses leads to the spontaneous assembly of particles that are structurally similar to authentic virus.':' In practical terms, the fact that VLPs mimic the structure of virus particles usually means that VLPs should elicit strong humoral response and that lower doses of antigen relative to subunit vaccines are sufficient to elicit similar protective response. In addition to their ability to stimulate B cell mediated immune responses, VLPs have also been demonstrated to be highly effective at stimulating CD4 proliferative and= (CTL) responses.This feature of VLP vaccines is likely to be a major contribution to their effectiveness in the field. It is also becoming increasingly clear that precise prime-boost strategy can be important to how effective vaccination is as a strategy to control disease. Therefore, the addition of VLP to the 'arsenal' of vaccine strategies for any disease extends the type of prime-boost regime that can be employed.

到目前为止，已经生产了很多种感染人类或其他动物病毒的VLPs（表7-5）。它们的一个显著特点是极其多样化的单个病毒结构，比如它们有的是单个病毒蛋白的衣壳蛋白组装成的，有的是多个衣壳蛋白组装成的，甚至有的衣壳蛋白不包含脂质膜。显然，并不是所有生成的VLPs都能按预期成为合适的疫苗，一些已经形成了一种促进病毒装配过程、形态发生的基本原理和体系结构。然而，很重要的一点是，主要的病毒粒子并不影响VLP的成功生产，虽然很多的表达系统被应用于VLPs的表达，但本节主要集中讲解昆虫细胞表达系统用于开发候选疫苗VLPs的情况。之所以选择昆虫细胞表达系统，是因为在所有的表达系统中，昆虫细胞/杆状病毒表达系统在VLP技术研发疫苗中是最有前途的一个系统（图7-3）。

To date, VLPs have been produced against many different viruses that infect humans and other animals (Table 7-5). One of the most striking features of this group is that it is extremely diverse in terms of the structure of the individual viruses. It includes viruses that have a single capsid protein, multiplecapsid proteins and those with and without lipid envelopes. Clearly not all of the VLPs that are generated to date are appropriate vaccine targets, some VLPs have been generated to facilitate in fundamental understanding of virus assembly process, morphogenesis or architecture of viruses. However, an important point remains that the structure of the target virion is not limiting to the success of VLP production. Although various expression systems have been employed for VLP production, this chapter will mainly focus on insect cell culture produced VLPs that are being developed as candidate vaccines. The rationale behind this is that among all expression systems, insect cells, together with baculovirus expressing system, appear to be one of the most promising for VLP technology for development of viral vaccines (Figure 7-3).

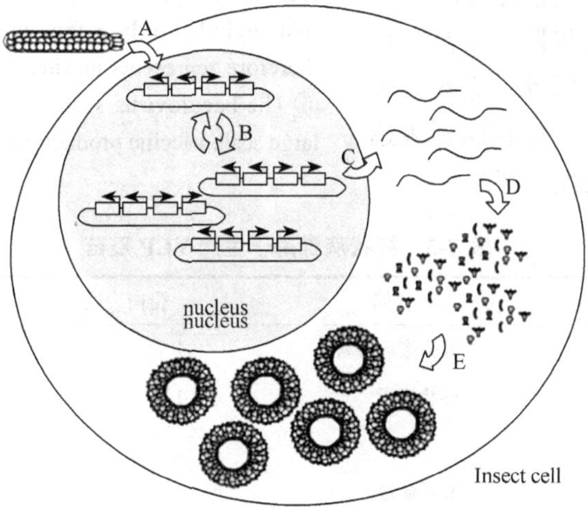

图 7-3　应用杆状病毒表达系统 VLPs 在细胞内形成的关键步骤

A. 杆状病毒作为载体，在昆虫细胞核内将编码重组蛋白 DNA 有效释放；B. 在细胞核内病毒 DNA 裸露并复制；C. 通过病毒强后启动子驱动重组蛋白表达；D. 应用病毒 mRNA 合成重组蛋白；E. 在细胞质内蛋白质相互作用整合成 VLPs

3.1　昆虫细胞 / 杆状病毒表达系统是 VLP 生产的首选系统

多种表达系统能够表达重组蛋白和颗粒。但对于生产预防性的乳酸标准疫苗，特别是针对人类的病毒感染，用哪种表达系统就需要慎重考虑。为了使 VLP 成为一种可实现的候选疫苗，它需要一个安全的表达系统，且易于大规模生产。

如表 7-5 所示，已经证明应用昆虫细胞 / 杆状病毒生产的 VLP，具有很高的免疫原性，是一种潜在的疫苗候选者。用昆虫细胞表达蛋白对于 VLP 的构建有很多的优点。①在高密度细胞培养的条件下，真核细胞能够对表达的大量蛋白进行正确的折叠修饰；②杆状病毒表达系统能够同时表达一个重组病毒的多个外源蛋白，在每一个感染的昆虫细胞中促进衣壳蛋白的自我组装；③培养昆虫细胞不需要哺乳动物细胞所需的添加剂，就减

3.1 Insect Cells and Baculovirus Expression System as Preferred System for VLP Production

As stated above, a variety of protein expression systems are available to express recombinant proteins and particles. However certain criteria for generation of VLPs as prophylactic vaccines, particularly for human viral infection, must be considered. In order for a VLP to be a realistic vaccine candidate, it needs to be produced in a safe expression system that is easy to scale up to large-scale production.

Table 7-5 shows baculovirus expressed/insect cell produced VLPs that have been demonstrated to be highly immunogenic and potential vaccine candidates. This insectcell-based protein production system has many advantages for VLP production. ① Extremely large amounts of correctly folded recombinant proteins can be produced in high-density cell-culture conditions in eukaryotic cells. ② Baculovirus expression systems have been developed for expression of multiple foreign proteins simultaneously from a single recombinant virus facilitating capsid assembly in each infected cell. ③ As the insect cells that are used for vaccine production can be cultured without the need for manunalian cell derived supplements. The risks of coculture of opportunistic pathogens are minimized. ④ The baculovirus used for

少了共培养条件下产生致病菌的风险；④杆状病毒宿主界限非常高，只感染少数的鳞翅目，因此，它对接种的个体不存在威胁；⑤杆状病毒表达系统适合疫苗的大规模生产。

recombinant protein expression has a narrow host range that includes only a few species of Lepidoptera and therefore represents no threat to vaccinated individuals. ⑤ The baculovirus system is amenable to scale-up for large scale vaccine production.

表 7-5 杆状病毒用于生产 VLP 疫苗

VLP	科属	蛋白表达	疫苗测试
乳头瘤病毒	乳头瘤病毒科	1	人体（有效）
诺沃克及诺沃克类病	杯状病毒	1	鼠，猫，人体
猫的杯状病毒			（第一阶段）
戊型肝炎病毒	肝炎病毒	1	食蟹猴，鼠
猪细小病毒	细小病毒	1	貂，猪，犬
貂肠炎细小病毒			
犬细小病毒			
腺相关细小病毒			
B19			
鸡传染性贫血病病毒	环状病毒	1，2	鸡
猪圆环病毒			
SV40	多瘤病毒科	1	鼠，兔（体外）
JC 病毒			
鼠多瘤病毒			
脊髓灰质炎病毒	小核糖核酸病毒	1（聚合体）	—
蓝舌病病毒	呼肠孤病毒	4（蓝舌病）	绵羊（蓝舌病）
轮状病毒		2～3	鼠，猪
丙型肝炎病毒	黄病毒科	3	鼠，非狒狒
HIV，SIV，FIV	逆转录病毒	2	鼠，豚鼠
绵羊髓鞘脱落病毒			
猫白血病病毒			
牛白血病病毒			
劳斯肉瘤病毒			
新城疫病毒	副黏病毒科	1	鸡
SARS 冠状病毒	冠状病毒	3	鼠（体外）
汉坦病毒	布尼亚病毒科	3	鼠
甲型流感病毒	正黏病毒科	2～4	鼠
鸡传染性法氏囊病毒	双 RNA 病毒科	1	鸡

3.2 简单的无包膜病毒 VLP 的生产

对于一些无膜的衣壳蛋白，由 1 个或 2 个主要蛋白构成，用异源系统表达是相对简单的，比如乳头瘤病毒、细小病毒、多瘤病毒和戊型肝炎病毒都是由主要的衣壳蛋白形成的病毒样颗粒。所有的这些病毒都无包膜，且病毒粒子的主要结构成分是单一的病毒编码蛋白。乳头瘤病毒样颗粒疫苗研究的 VLP 最彻底，且在该领域处于高级阶段。它的 VLP 是由过表达衣壳蛋白 L1 形成的，有高效的免疫原性，能引发细胞免疫和体液免疫。人类乳头瘤病毒（HPV）是宫颈癌发生的主要原因。在全国范围内，大约 70% 的宫颈癌与 HPV 的两种血清型有关：HPV-16 和 HPV-18。用昆虫细胞生产 VLP 已经成功地进入临床试验 I 阶段和 II 阶段，且被证明高度有效。此外，葛兰素史克公司研发的宫颈癌疫苗（Cervarix："161"）在世界范围内有 30 000 多名女性接受三期临床试验。三期临床的随机、双盲试验在丹麦、爱沙尼亚、芬兰、希腊、荷兰和俄罗斯联邦多个试验中心进行。所有的接种者收到的 HPV VLP 如下：158 名 10～14 岁的健康女孩和 458 名 15～25 岁的健康女性接受了 VLP 候选疫苗的接种，接种后按照 0/1/6 月的时间表对抗体滴度进行检测。7 个月后，两组 100% 都出现了 HPV-16 和 HPV-18 的阳性血清，但两种类型的平均抗体滴度是 10～14 岁女孩的两倍。该疫苗适用于所有的患者且并未出现与疫苗

3.2 VLPs Produced for Structurally Simple Non-Enveloped Viruses

For a number of nonenveloped viruses viral capsids are formed by only one or two major proteins and thus are relatively easy to manipulate for generation of VLPs by heterologous expression systems. Examples of these are the VLPs formed by the expression of the major capsid protein of Papillomaviruses,Parvoviruses,Calcivi ruses,Circovirses,Polyomaviruses and Hepatitis E virus. All of these viruses are nonenveloped and have a single, virally encoded protein that forms the major structural component of the virion. Papilloma virus VLPs are among the most completely studied of this collection of VLPs and are at the most advanced stage with respect to production of a useful vaccine. VLP of Papillomaviruses are formed from the over expression of the major capsid protein Ll. These particles are highly immunogenic and are able to stimulate both humoral and cell mediated immune responses. HumanPapillomavirus (HPV) is the leading cause of cervical cancer. Globally, approximately 70% of all cervical cancer cases are associated with two serotypes of HPY, HPV-16 and HPV-18. VLPs produced in insect cells have been used successfully for Phase I and II human clinical trials in large numbers and were shown to be highly efficacious. Moreover, GlaxoSmithKline's cervical cancer vaccine candidate (Cervarix) targeting HPV 16/18 is currently undergoing PhaseIII clinical trials involving more than 30,000 women worldwide. In this Phase III randomized, double-blinded trial conducted in multiple centres in Denmark, Estonia, Finland, Greece, the Netherlands and the Russian Federation. All vaccinees received the HPV VLPs as follows: 158 10-14 years old healthy girls and 458 15-25 years old young women received the candidate VLP vaccine at a 0, 1, and 6 month schedule and anti-HPV antibody titers were assessed. At month seven 100% seropositivity was achieved in both groups for HPV 16 and 18 although average antibody titers for both HPV types were at least two -fold higher in 10-14 year-old girls. The vaccine was tolerated by all patients and no vaccine related serious adverse effects were detected. Further, the follow-up study clearly demonstrated the sustained efficacy of HPV-16/18 VLPs up to 4.5 years. In conclusion, the bivalent HPV vaccine is highly immunogenic and safe and induces a high degree of protection against HPV-16 and HPV-18 infection and

相关的副作用；此外，后续的研究显示，它对 HPV-16/18 的免疫保护平均可以达到 4.5 年。总之，HPV 的二价苗具有很高的免疫原性和安全性，在 HPV-16/18 的感染以及宫颈病变中可以提供高度保护。

associated cervical lesions.

这些研究不仅证明了 HPV VLP 的有效性，还突出了昆虫细胞可以作为一种选择来制作疫苗，预防人类的疾病。还应该提及的是默克公司利用酵母细胞生产的一种四价疫苗，在 2006 年 6 月已经通过 FDA 的许可，应用于 9～26 岁的女性中。

These studies are not only an important demonstration of the effectiveness of HPVVLP vaccine, and that multi-serotype VLPs are effective, but also highlight the fact that insect cell produced VLPs are a realistic alternative as human vaccines against viral disease. It should also be mentioned at this point that a tetravalent (HPV-6/11/16/18) VLP vaccine, Guardasil (Merk), produced in yeast cells was approved by FDA in June 2006 for use in women aged 9-26.

虽然目前还没有其他经历过像 HPV 那样大规模的临床试验的报道，但各种由细小病毒感染导致疾病的 VLP 疫苗的研制已进入后期阶段。利用昆虫细胞表达犬细小病毒（CPV）和猪细小病毒（PPV）的主要结构蛋白 VP2 实现了 VLPs 的装配。CPV VLPs 和 PPV VLPs 的免疫试验的结果令人鼓舞。在犬的功效实验中，只需接受一点点如 10μg 或 25μg 的 CPV VLP 就能完全应对病毒感染及恶性病毒攻击。此外单独的 3μg HPV VLP 混合 50μg ISCOM 佐剂，皮下注射能够保护貂应对抗原相似的病毒，例如貂肠炎病毒（MEV）。同样据最近报道，0.7μg 的 PPV（猪株）VLPs 的单次免疫接种能使目标动物产生应对感染性 PPV 的抗性。事实上，在母猪身上的微克剂量的 VLPs 不仅会产生高度免疫原性，也能有效地防止胎盘病毒传播且显著降低生殖失败的次数。此外，在符合欧洲药典的条件下，可行的安全大规模生产的猪细小病毒 VLP 疫苗的方案已经建立。

VLP vaccines for various diseases caused by parvovirus infections are also at an advanced stage although as yet none have undergone such large scale trials as those reported for HPV. Synthesis of major structural proteins VP2 of canine parvovirus (CPV) and porcine parvovirus (PPV) led to assembly of VLPs in insect cells. Vaccination trials of CPV VLPs in dogs and PPV VLPS in pigs were highly encouraging. In one efficacy assay dogs that received as little as or 10 μg or 25 μg of CPV VLP were completely protected from virus infection when challenged with virulent virus. Furthermore a single subcutaneous dose 3μg same CPV VLP with 50μg ISCOM adjuvant was able to protect mink against challenge with the anti-genically similar virus, mink enteritis virus (MEV). Similarly it has been reported recently that a single immunization with 0.7μg of PPV (porcine strain) VLPs yielded complete protection in targeted animals against infectious PPV strains. Indeed microgram doses of VLPs in gilts were not only highly immunogenic, but were also very efficient in preventing trans-plancental virus transmission and significantly reduced the number of reproductive failures. In addition, the feasibility of safe large-scale production of the porcine parvovirus VLP vaccine has been established complying with the European Pharmacopoeia requirements.

Calicivirus studies have relied heavily on the

由于杯状病毒无法在细胞中培养，因此杯状病毒的研究很大程度上主要依赖于在异源系统中表达蛋白。因此，VLP 对于诺沃克类病毒的抗原诊断，监测疾病的暴发极为有用，同时在小鼠体内免疫试验还证明 VLP 能够刺激 IgG、IgA 的分泌及其体液免疫应答。人体内初步的一期临床试验检测了昆虫细胞表达的诺沃克类病毒 VLP 疫苗的安全性和免疫原性，结果证实，VLP 能够安全有效地刺激机体产生 IgG、IgA 反应。

production of proteins in heterologous systems mainly due to the fact that it is not yet possible to grow the virus in cell culture. Thus, VLP to Norwalk-like viruses have been extremely useful as sources of diagnostic antigen to monitor disease outbreaks. Norwalk virus VLP have also been shown to be effective at stimulating IgG, IgA and humoral responsesin mice. Preliminary Phase I trials in humans to test the safety and inun400nogenicity of insect cell expressed Norwalk virus VLPs has confirmed that they are both safe and effectivce to stimulate IgG and IgA response.

戊型肝炎病毒的 VLP 是由病毒衣壳蛋白截短的形式组装成的。鼠内免疫研究显示，口服给药后，VLP 能诱导小鼠的全身免疫和黏膜免疫。此外，给予食蟹猴口服肝炎病毒 VLP，可以诱导 IgM、IgA 和 IgG 反应，足以预防感染与抵抗病毒攻击。因此，将 VLP 作为一种疫苗预防戊型肝炎很有潜力。

VLPs for Hepatitis have been assembled using a truncated from of the virus capsid protein. In immunilization studies in mice these VLPs were able to induce systemic and mucosal immune responses fellowing oral administration. Furthermore,oral administration of the Hepatitis E VLPs to cynomologous monkeys induced IgM,IgA and IgG responses and was sufficient to protect against infection and disease on challenge with virus.30Thus there is potential for the application of these VLPs as a vaccine for hepatitis E.

兔免疫接种含佐剂的人 JC 病毒 VLP 疫苗，产生的高免血清能有效地中和传染性的病毒，然而，在没有佐剂的情况下就不存在应答。这种应答的模式是不寻常的，一般情况下，没有佐剂，VLP 也能产生强效的反应。一种小鼠多瘤病毒 VLP 在没有佐剂辅助的情况下注射 610ng 就可产生强效的免疫应答。有趣的是，这些颗粒还特别稳定，在常温储藏 9 周后，没有改变病毒的颗粒形态以及降低免疫原性。

Vaccination of rabbits with YLPs for human JC virus in the presence of adjuvant allowed production of a hyperimmunle serum that effectively neutralized infectious virus preparations.31 However, in the absence of adjuvant there was no response.This pattern of response is unusual for VLPs in general which often stimulate strong immune responses even in the absence of adjuvant.Indeed, VLPs of murine polyoma virus were able to stimulate a strong immune response in the absen of adjuvant when administered as a 610 ng dose. Intriguingly,these particles appear to be particularly stable with no alteration of particle morphology or reduction in immunogenicity even after 9 weeks storage at room temperature.

3.3 结构复杂的多个蛋白层构成的 VLP

3.3 VLPs of Structurally,Complex Viral Capsids with Multiple Protein Layers

在表达含有多个交互的衣壳蛋白的病毒颗粒时，往往比那些由一个或两个主要衣壳蛋白构成的病毒粒子

Viral particles that contain multiple interacting capsid proteins present more of a technical challenge than those that are formed by one or two major capsid proteins. Particularly,it is far more difficult if the assembling

要求更高的技术挑战。特别是，如果衣壳的组装蛋白是由多个独立的mRNA编码，然而在小核糖核酸病毒存在的情况下无法加工单一的多聚蛋白时，就更为困难了。这是由于一个VLP能够有效组装，其相互作用的衣壳蛋白必须靠近彼此表达，换句话说——必须在同一细胞中表达。对脊髓灰质炎病毒和传染性法氏囊病病毒来说，利用杆状病毒表达系统，通过多蛋白的加工装配成为VLP已经实现。针对呼肠孤病毒群，更为复杂的多层多蛋白VLP的组装也已有效地生产出来。这些病毒具有由不同的衣壳蛋白组成的同心层的衣壳。2～4个衣壳蛋白在昆虫细胞中的协同表达，有赖于病毒和粒子组成——使得VLP产物不含分段dsRNA病毒基因组——然而在另一方面表明其与真实的病毒粒子有区别。呼肠孤病毒科的第一个合成VLP的病毒是蓝舌病毒（BTV），一种昆虫传播给动物的病毒。经测试，利用杆状病毒表达系统表达不同的VLP和重组单一抗原亚基的免疫原，该病毒群仍然最具多样性。此外，在该系统、同一昆虫细胞中，病毒衣壳蛋白有效的协同表达这一要求使得杆状病毒基因表达载体不断发展。本部分将集中于这个系统的一些细节——突出研究VLP疫苗的有效性以及一些已用于具有复杂结构的VLP的生产的先进技术。

proteins of capsids are encoded by multiple discrete mRNAs,but not processed a single polyprotein as in the case of picornaviruses. This is due to the fact for effective assembly of a VLP, the interacting capsid proteins must be expressed in the vicinity to each other,in other words in the same cell. Assembly of VLPs by processing of polyproteins have been achieved both for poliovirus33 and for Infectious Bursal disease virus34 using the baculovirus expression system. More complex assembly of multilayered,multiprotein VLPs have also been efficiently produced for the members of the Reovirida. These viruses have capsids made up of concentric layers of different capsid proteins. Co-expression in insect cells of2-4 of these capsid proteins depending on the virus and the particle made, has allowed the production of VLPs which are empty of the segmented dsRNA viral genome,but are otherwise indistinguishable from authentic viral particles. The first member of the Reoviridae for which VLPs were described is Bluetongue virus(BTV),an insect transmitted animal virus.This remains the system in this family for which the largest variety of different VLPs and recombinant single antigen subunit immunogens made by baculavirus expression systems has been tested. In addition,the requirementfor efficient co-expression of viral capsid protein in the same insect cell in this system has resulted in the development of baculovirus multigeneexpression vectors.We will focus on this system in some detail as it highlights both the effectivenes of VLP vaccines and some of the technological advances that have been made for the production of VLP with complex architecture.

蓝舌病主要作用于牛和羊，在欧洲是新兴的一种疾病。该病的病原体是蓝舌病毒（BTV），二十面体对称，主要由7种蛋白构成（VP1～VP7）。有三种结构蛋白（VP1、VP4、VP6）

Bluetongue disease affects mainly sheep and cattle and is classified as an emerging disease in Europe. The disease is caused by bluetongue virus,BTY. which has a multi-layered icosahedral structure rormed by nonequimolar amounts of seven viral proteins (VP1-VP7). Three of these structural proteins (VP1,VP4, VP6) are

在 VLP 形成的过程中可有可无，因为它们只在病毒转录中起到机械酶的作用。剩下的四种蛋白（VP2、VP3、VP5、VP7）组成两种衣壳蛋白，内在衣壳是外在衣壳的支架，负责细胞的进入和中和病毒的作用。

利用杆状病毒表达系统可以同时表达 BTV 的 4 种蛋白质，共感染几个杆状病毒，每个杆状病毒表达一个蛋白。这种方法的优势是，在每个细胞中都实现了感染，因而 VLP 的装配是更有效的，因为表达的控制是在细胞水平，而不是作为混合感染的情况下的培养水平。BTV 的 VLP（图 7-4）与病毒颗粒在结构上没有区别，但没有分割，在感染性的病毒中通常存在双链的 RNA。

通过纯化得到的 BTV 病毒样颗粒给予了针对同源血清型 BTV 的抗体很高的抗体水平，在随后的临床试验中 1 岁的美利奴羊接种了 BTV 病毒样颗粒 10 个不同量（10～200μg）的血清型。所有接种疫苗的动物的后续反应表明：当接种疫苗 4 个月后，所有的动物都能产生中和抗体，且完全脱离了疾病。相比之下，未接种疫苗的对照动物呈现出典型的 BT 临床症状。与此同时，甚至在 VLP 剂量低至 10μg 时，免疫产生的抗体依然足够为接种动物提供针对该类型疾病的保护。来自两个不同的血清型的 VLP 联合接种相同动物进行了更进一步的疗效测试。在接种 VLP 疫苗的动物中，两种血清型疫苗为接种疫苗的动物提供了完整的保护，对非血清型疫苗为接种疫苗的动物提供了部分保护进行了挑战。在这些试验中，接种疫苗后产生的保护

dispensable for the formation of VLPs as they play only an enzymatic role in the virus transcription machinery. The remaining four structural proteins (VP2,VP3, VP5 and VP7) areorganized in two capsids.The inner capsid acts as a scaffold for the assembly of outer capsid that is responsible for cell entry and hence contains the major candidate late for virus neutralisation.

Expression of all four major structural proteins of BTV was achieved by conscrucing a baculovirus that simultaneously expressed all four proteins.The advantage of this approach over co-infection with several baculoviruses each expressing a single protein is that equivalent conditions are achieved in all infected cells. Thus assembly of VLP is more efficient as expression is controlled at the level of the cell,rather than the level of the culture as in the case with mixed infections. BTV VLPs(Figure 7-4) are structurally indistinguishable from virus particles but lack the segmented,double-stranded(ds) RNA virus genome normally present in infectious virus.

Antibodies raised to purified BTV VLPs gave high levels of neutralizing antibodies against the homologous BTV serotype. In subsequent clinical trials 1 year-old Merino sheep were vaccinated with various amounts (10-200μg) of VLPs for BTVserotype. 10 All vaccinated animals developed demonstrable neutralizing antibodies and when challenged with virulent virus after four months of vaccination were completely protected from disease.In contrast,unvaccinated control animals developed typical BT clinical symptoms. Even at doses as low as 10μg VLP was suffcient to protect animals from any signs of disease. Further efficacy tests were performed where VLPs from two different serotypes were combined to vaccinate the same animal. In these animals VLPs vaccination provided complete protection against the two vaccine serotypes and also partial protection from challenge with related nonvaccine serotypes.The protective efficacy of vaccination in these trials extended over along(14 months) period.This observation raises the possible that a broad spectrum vaccine against all 24 BTV serotypes is a possibility by combining VLPs from a relatively small number of serotypes.

作用，被延长到了14个月长的周期。这个观察提出一种通过和相对小的数目的血清型的VLPS结合而产生一种广谱疫苗24 BTV血清型的可能性。

A

B

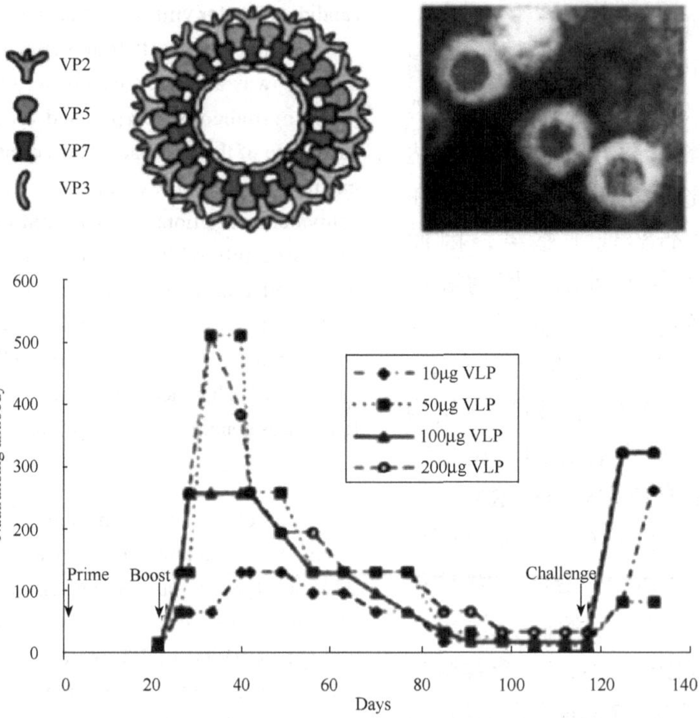

C

Inoculum	CRI	viraemia (days post challenge)
10μg VLP	0.0	none
50μg VLP	0.0	none
100μg VLP	0.0	none
200μg VLP	0.0	none
Saline	0.0	Days 4-14

图 7-4　蓝舌病 VLP 的动物实验

A. 左图为蓝舌病毒（BTV）结构图，其表面有多个 VLPs 结构，右图为 BTV VLPs 负染色电镜照片；B. 每只用10～200μg 的 VLPs 疫苗接种的美利奴绵羊均产生中和抗体，用强毒 BTV 攻击免疫绵羊，显示中和抗体效价达 117d；C. 用各种剂量的 VLP 接种的绵羊和对照组显示的临床指标和病毒血症结果：VLP 接种的动物无临床症状和病毒血症

基于解离的抗原或组装的重组抗原的亚单位疫苗的不同，该BTV系统还展示了VLPs疫苗的免疫效率。此

The BTV system also demonstrates the efficiency of VLP vaccines relative to immunization with subunit vaccines based on dissociated antigens or unassembled

外，还介绍了组装VLPs的BTV外衣壳，VP2和VP5的两个组成部分，并在疫苗接种研究中测试。而100μg VP2，主要血清型确定抗原，只能对病毒攻击的部位提供短时间的保护（75d），50μg的VP2联合25μg的VP5也具有保护性。与此相反，10μg VLPs（只含有1～2μg VP2），可以在更长的持续时间提供更好的保护水平。这些研究表明组成VLPs的抗原比远离的蛋白具有更有效的免疫原性。

recombinant antigens.In addition the assembled VLPs the two components of the BTV outer capsid,VP2 and VP5,were also prepared and tested in vaccination studies. While 100μg VP2,the major serotype determining antigen,was only partially protective for a short duration (75 days) against virulent virus challenge,50 μg of VP2 combined with 25μg VP5 was protective.In contrast,10μg VLPs (containing only 1-2μg VP2) afforded a better level of protection for a much longer duration.These studies demonstrate that assembly of antigens into VLPs results in a more effective immunogen than delivery of separately isolated proteins.

除了BTV，VLP也已产生呼肠孤病毒的一种——轮状病毒。有趣的是，由两个单独的轮状病毒衣壳的两个内部结构蛋白形成的VLPs已在动物模型被证明是有效的免疫原。事实上，在小鼠甚至直肠内的黏膜局部免疫诱导应答足以抵抗轮状病毒感染。这些免疫试验的数据是令人鼓舞的，可能的轮状病毒VLP可以为轮状病毒的活病毒疫苗提供一个可行的替代。

In addition to BTV,VLP have also been produced for rotavirus,another member of the Reoviridae. Intriguingly,VLPs formed from the two inner structural proteins alone of the rotavirus capsid have been shown to be effective immunogens in animal models.Indeed in mice even intrarectal immunisation which induces a local mucosal response is sufficient for protection from rotavirus infection. The data from these immunogenicity experiments are encouraging and it is possible rotavirus VLP may provide a viable alternative to the live virus vaccine for rotavirus.

3.4 病毒样颗粒与脂质膜

许多致病性病毒如流感病毒、艾滋病毒和丙型肝炎病毒都是由一个宿主细胞产生的双层脂质，并有病毒糖蛋白嵌入的膜包裹着。这些糖蛋白是中和抗体的靶点，而且是病毒必需的基本组成。由于脂质包膜的固有性质，对那些产生多个衣壳的病毒来说，在昆虫细胞中装配VLPs是一个不同类型的挑战。尽管如此，还是有一些包膜病毒在昆虫细胞中有效形成VLPs的报道。例如，有几个逆转录病毒，如丙型肝炎病毒、SARS病毒与A型流感病毒的VLPs已经证明具有含糖蛋白的脂质膜的正确装配。对于逆转录病毒，人们已经能够生产出

3.4 VLPs from Viruses with Lipid Envelopes

Many pathogenic viruses such as Influenza, HIV and Hepatitis C are surrounded by an envelope that consists of a lipid bilayer derived from the host cell,inserted with virus glycoprotein spikes.These proteins are the targets of neutralizing antibodies and are essential components of vaccine. Due to the inherent properties of lipid envelope,assembly of VLPs in insect cells for these viruses is a different type of technical challenge to those produced for viruses with multiple capsids. Nevertheless,efficient formation of VLPs of a number of enveloped viruses insect cells has been reported. For example,VLPs of Hepatitis C virus. Several retroviruses,SARS Coronavirus and influenza A have demonstrated correct assembly of the lipid envelope with the glycoproteins inserted. Indeed, for retroviruses, it has been possible to produce hybrid VLPs that contain the gag capsid protein from one virus (SIV) and the envelope protein from another

由 SIV 病毒衣壳中的 gag 蛋白与 HIV 中的膜蛋白混合组装的 VLPs。虽然没有任何逆转录病毒产生的 VLPs 进入临床疫苗试验中使用的阶段，但在动物模型最初的试验中是有前途的。

(HIV) in insect cells.Although none of the retrovirus derived VLPs are yet at the stage that they are being used in clinical vaccine trials,initial experiment in animal models are promising.

SARS 病毒的 VLPs 作为疫苗接种的基础在 2002～2003 非典的时候被快速地生产。然而 SARS 冠状病毒的流行病学控制措施，已经严重限制了 SARS 病毒样颗粒作为疫苗的发展。尽管如此，小鼠在对抗昆虫细胞产生的 SARS 病毒的 VLPs 时，产生的抗血清能够中和伪类型的逆转录病毒的 SARS S 蛋白（图 7-5）。

VLPs for SARS Coronavirus as a basis for vaccination were produced rapidly following the SARS outbreak in 2002—2003. However the control of SARS Coronavirus by epidemiological measures,continued lack of re-emergence of the virus,and difficulties working directly with the virus have severely limited the development of SARS VLPs as vaccine. Despite this,anti-serum raised in mice against insect cell derived SARS VLPs were able to neutralize a retrovirus pseudo- typed with the SARS S protein(Figure 7-5).

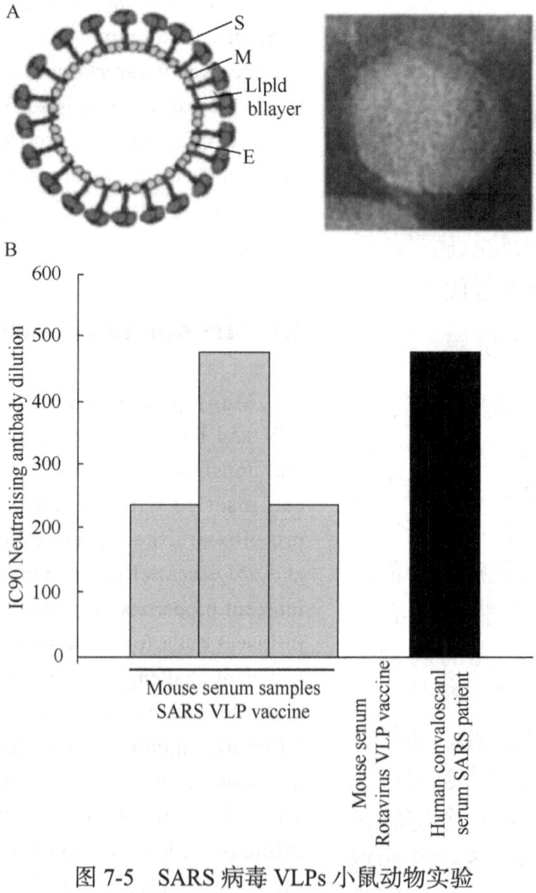

图 7-5 SARS 病毒 VLPs 小鼠动物实验

A. 左图为 SARS 冠状病毒的结构图，右图为表达 SARS 病毒 E、M 和 S 蛋白的 VLPs 电镜照片，VLPs 在小鼠中产生抗血清能保护小鼠感染 SARS S 蛋白，S 蛋白和慢病毒整合；B. 免疫小鼠产生 SARS S 慢病毒 IC90 中和抗体，用 SARS VLP 免疫血清、轮状病毒 VLP 血清以及 SARS 康复病人血清作为对照丙型肝炎VLPs已经对小鼠和狒狒进行测试，结果证明对刺激细胞和体液免疫反应有效。在一项实验中，6～8周龄的雌性BALB/c小鼠肌内免疫3次，每3周用20μg的HCV VLP。接种小鼠最终免疫3d用vv HCV.S进行攻毒，5d后收获感染小鼠的卵巢，并进行牛痘病毒滴度测定，7只小鼠中的5只在卵巢上没有检测到牛痘病毒。剩余的2只与对照小鼠比较有较低的牛痘滴度。此外，本研究能够证明所述VLPs的功效主要基于$CD4^+$和$CD8^+$T细胞应答的刺激。进一步狒狒试验研究已经表明，病毒样颗粒的耐受性良好，并能广泛、持久地有针对性地刺激免疫反应。

The Hepatitis C VLPs have been tested in mice and baboons and shown to be effective at stimulating both cellular and humoral immune responses.In one experiment, 6-8 week old female BALBlc mice were immunized intramuscularly three times. At three week intervals with 20μg insect cell derived HCV VLP,produced by co-expressing HCV coreEI-E2. Because of the lack of a suitable animal model for HCV infections a recombinant vaccinia virus expressing HCV structural proteins (vvHCV.S) was used as a model system. Vaccinated mice were challenged three days after the final immunization with vvHCV.S and then five days later the ovaries of infected mice were harvested and the vaccinia virus titre determined.Five out of seven vaccinated animals had no detectable vaccinia virus in the ovaries at this point.The remaining two mimals had five logs lower vaccinia titres compared to control mice.In addition, this study was able to demonstrate that the VLPs efficacy was based largely on its stimulation of $CD4^+$and $CD8^+$ T-cell responses.A further study in baboons has demonstrated that the VLPs are well tolerated and can stimulate broad and long-lasting HCV targeted immune responses.

迄今为止，流感是用最复杂的结构包膜病毒粒子所产生的VLP。甲型H9N2和H3N2已产生了其他群体。这些研究表明单独的主要结构蛋白M1的表达在昆虫细胞即可构建病毒样颗粒。共表达流感病毒的M1、M2、NA、HA蛋白也能形成VLP，其中M1-HA和M1-HA-NA型的VLP能够保护小鼠免于同型流感病毒的感染。利用单个的重组杆状病毒表达系统共表达H3N2病毒的HA、NA、M1、M2蛋白，成功组装成VLP。到目前为止，这些流感病毒的VLP并没有在人的体内进行检测，然而，HA和NA不需要组织培养就能直接整合成VLP的潜力，在控制快速改变的流感病毒上有着不可估量的优势。

To date,the most structurally complicated enveloped virus particle that has been used to generate VLP is influenza. VLPs forInfluenza A H9N2 and H3N2 have been produced by other groups. These studies have shown that expression of the major structural protein M1 alone is sufficient result in the budding of virus-like vesicles from insect cells.s2Also,co-expression of M1 with M2,HA and NA leads to the assembly of influenza VLP and M I-HA and M 1- HA-NA VLPs confer protection from lethal challenge with the same type influenza A in mice. VLP production was also successfully achieved by co-expressing HA,NA,M1 and M2 from influenza virus A/Udorn/72 (H3N2) using a single recombinant baculovirus.11 To date none of these influenza VLP have been tested in humans. However the potential that HA and NA could be incorporated directly into these VLP from circulating influenza strains without passage in tissue culture has particular advantage for the control of rapidly changing influenza A virus.

4 疫苗生产未来展望

VLPs除了用作直接的免疫原，它们可以刺激细胞和体液免疫的特性，使得它们成为携带抗原表位、DNA、小分子的候选者。抗原表位暴露在VLP表面的设计，为VLP的免疫应答提供便利。许许多多的VLP已经发展成为药计疫苗并且在它们自己的范围内早已成为其他颗粒的运载系统。将这些有衣壳包裹的病毒原颗粒作为载体分子或者其他疾病的表位都是不受限制的，并且具有一定的经济学效果。在疫苗表位传送过程中，病毒微粒结构被接纳进抗原呈递细胞中，并能够很好的长期存在于CTL抗体反应中，这是许多VLPs成为极好的载体分子的原因。当然大多数VLP疫苗可以激发细胞与体液免疫反应。这些研究证明细菌的颗粒表位系统是有效的刺激物，所以VLP在其他疾病里作为表位决定系统具有巨大的潜能。这个方法唯一的缺点是衣壳蛋白组装时经常包含不相关的序列，这些序列会拴在VLP中。有一种方法可以克服这个约束条件，就是衣壳蛋白某种程度上需要从里面和外面扩展N或者C目标蛋白。当然，这仅仅适合于一个或两个暴露于同一表面或者外面的目标蛋白。到目前为止，已知的VLP中还没有能完全利用这种潜在的方法；但是已经成功运用了一种蛋白依赖的类似颗粒结构，是在B细胞和T细胞的刺激下产生的与潜在蛋白相互作用的颗粒。

尽管有很多潜在的证据都证明

4 Future and Alternative Directions

In addition to the use of VLPs as direct immunogens,the efficiency with which they stimulate cellular and humoral responses has made them prime candidates as carrier molecules for the delivery of epitopes, DNA and small molecules targeting other diseases.This has been facilitated by the excellent structural information that is often available for virus particles allowing rational design of vaccines where epitopes are exposed on the surface of the VLP.Many of the VLPs that have been developed as vaccines in their own right have also been tested as delivery systems for other molecules. The use of VLPs as carrier molecules for epitopes for other diseases is not limited to those VLPs that are fanned from the capsids of economically significant viruses.The reason that many VLPs make excellent carrier molecules for the delivery of epitopesin vaccines is most likely because the particulate VLP structure is readily taken up into antigen presenting cells and thus is able to prime long lasting CTL responses in addition to antibody responses.Certainly accumulated evidence on VLP vaccines suggests that they arc efficienc at stimulate both cellular and hwnoral immune responses. These studies demonstrate that like bacterial epitope display systems VLPs are efficient stimulators of MHC class I and class II responses. Thus VLPs have great potential as epitope display systems for other diseases. The only major drawback for this approach is that the requirement of the capsid protein to assemble often constrains the size of the foreign sequence that can be tethered to the VLP. One approach that may be of use to overcome this constraint would be to link foreign protein sequences to capsid proteins in such a way that they extend the N or C termini of the protein and extend either inside or outside to partied. Of course, this is only suitable where one or both termini of the protein are exposed on the inside or outside face of the capsid. So far,there are no VLP that we are aware of that have fully exploited the potential of this approach but it has been successfully employed for other protein-based particulate structures that are similar to VLPs in their stimulation of B-cell and T-cell responses and requirement for complex protein-protein interactions for particle assembly.

Despite the accwnulated evidence of the potential

VLP 可以作为一种强有力的免疫原，但用 VLP 作为一般疫苗接种的策略还存在一定的阻力。这在某种程度上是由于一些在早期 VLP 的发展中产生的令人失望的结果，例如根据 TY 制成的艾滋病病毒的 VLPs 就没有效果。这个例子给早期 VLPs 疫苗的设计者们敲响了警钟。总体上，VLP 可以有效地刺激细胞和体液免疫应答，但是作为任何一种疫苗，它们都依赖于长期的实践才会变得更有效果。VLP 的设计工作是对免疫力低下的个体提供有效地免疫保护，与其他免疫方法需要克服类似的挑战。事实上，积累的经验表明 VLP 比其他亚单位疫苗更有效，因为它们相对于其他疫苗更具有真正构象以及更安全。VLP 产品不局限于某一类型的病毒或病毒家族，也不像病毒疫苗那样复杂。

of VLPs as potent inununogens for many viral systems that we have discussed. There remains some resistance to the VLP approach as a general vaccination strategy for diseases caused by viruses.In part this is due to some high profile disappointing result for VLP vaccines in the earthly stages of development, for example an ineffective early vaccine for HIV based on Ty VLPs.This is example raises a point of caution for VLP vaccine designers.In general. VLPs stimulate efficient cellular and humoral inunune responses, but as with any vaccine,they rely on the long term host response to be effective.VLPs designed to work in immunocompromised individuals need to overcome the same challenges to efficient immune response as any other vaccine approach.The notion that VLPs are ineffective vaccines is clearly a myth that is exploded by the inuninenc release of two new VLP-based HPV vaccines.Indeed,the accwnulated data from the field suggests that VLPs are more effective than many other types of subunit vaccines because they are more conformationally authentic, and are safer than many live virus preparations because they are usually free of viral genetic material. VLP production does not appear to be limited to any one type of virus or virus family, nor is it significantly limited by the complexity of the virus particle.

利用昆虫细胞作为蛋白表达系统，提供了在相同细胞中组装成构象真实的 VLPs 的机会。该系统优于其他系统的地方在于，能够实现工业大规模表达多个蛋白，且昆虫细胞具备储存大量外源基因的能力。因此，在表达系统中呈现出通常加工感染病毒的细胞机制，且能够自动装配成真正的 VLPs.

The use of insect cells as a protein expression system offers exciting opportunities for the synthesis of conformationally authentic VLPs that are formed from the intracellular assembly of multiple proteins expressed in the same cell.The advantage of this system over others used for protein expression is its capacity for industrial scale synthesis of large and multiple proteins and the fact that insect cells are the natural replication reservoir for many pathogenic viruses.Thus the basic cellular machinery that normally processes the infection form of the virus is present within the expression system and available to produce authentic VLPs.

（编者　蔡家利）

(Compiled by Jiali Cai)

参 考 文 献
References

Andre F E. 2003. Vaccinology: past achievements, present roadblocks and future promises. Vaccine (21): 593-595

Fraser C M, Rappuoli R. 2005. Application of microbial genomic science to advanced therapeutics. Annu Rev Med (56): 459-474

FDA. 2007. Vaccines licensed for immunization and distribution in the USA. http: //www.fda. gov /cber/ vaccine/licvacc.htm

Purcell A W, Mc Cluskey, Rossjohn J. 2007. More than one reason to rethink the use of peptides in vaccine design. Nat Rev Drug Discov (6): 404-414

Plotkin S A. 2005. Vaccines: past, present and future. Nat Med , 11（4 Suppl): S5-S11

Pizza M, Scarlato V, Masignani V, et al. 2000. Identification of vaccine candidates against serogroup B meningococcus by whole-genome sequencing. Science (287): 1816-1820

Rappuoli R. 2000. Reverse Vaccinology. Curr Opin Microbiol (3): 445-450

第八章 核 酸 药 物

Chapter 8 Nucleic Acid Drugs

目前，基因治疗中使用的核酸药物种类较多，可大致区分为长片段的核酸分子和短的核苷酸片段即寡核苷酸。前者主要是重组质粒 DNA 分子和病毒载体，后者则种类相对繁杂，包括反义寡核苷酸、核酶、脱氧核酶、siRNA 等。下面分别予以介绍。

Currently, various nucleic acid drugs have been used in gene therapy. The commonly used nucleic acid drugs may be classified into two groups, i.e. large nucleic acid molecules and short nucleotide fragments (oligonucleotides). The former mainly include recombinant plasmid DNA and viral vectors whereas the latter represents a variety of oligonucleotides including antisense oligonucleotides, ribozyme, DNAzyme, and siRNA etc.

1 重组质粒 DNA 和病毒载体

1 Recombinant Plasmid DNA and Viral Vectors

1.1 重组质粒 DNA

1.1.1 重组质粒 DNA 的特征及其设计

重组质粒 DNA 是包含一个编码特定蛋白质的治疗性基因的高分子量双链 DNA 分子。在分子水平上，质粒 DNA 分子可以被视为药物前体，一旦被细胞摄取后即可利用胞内的转录和翻译机制而合成具有治疗作用的蛋白质。基因治疗即利用了质粒 DNA 的这种特性，把治疗性基因导入细胞内进而产生蛋白质而发挥治疗作用。质粒 DNA 需要在进入细胞质后再进入细胞核才能最终发挥作用。

除了治疗性基因外，重组质粒 DNA 分子还包括一些调节信号如启动子和增强子序列；此外，还包含有剪接和多聚 A 位点以便于转录后 mRNA 的正确加工。一些载体还含有内含子以提高 pre-mRNA 的加工和核转运。

1.1 Recombinant Plasmid DNA

1.1.1 The Characteristics and Design of Recombinant Plasmid DNA

Recombinant plasmids are high molecular weight, double-stranded DNA constructs containing therapeutic genes, which encode specific proteins. On a molecular level, plasmid DNA molecules can be considered pro-drugs that upon cellular internalization employ the DNA transcription and translation apparatus in the cell to biosynthesize the therapeutic entity, the protein. Gene therapy involves the use of plasmid DNA to introduce therapeutic genes into cells that inherently lack the ability to produce the protein that the therapeutic gene is programmed to generate. The mechanism of action of plasmid DNA requires that the plasmid molecules gain access into the nucleus after entering the cytoplasm.

In addition to the therapeutic gene of interest, recombinant plasmid DNA molecules typically contain several regulatory signals such as promoter and enhancer sequences that play an important role in regulating gene expression. In addition, splicing and polyadenylation sites are present in the transgene construct that help in the correct processing of the mRNA generated after transcription. Some vectors also have introns that may increase pre-mRNA processing and nuclear transport.

启动子序列在启动基因转录时起着重要作用。启动子序列提供了RNA聚合酶的识别位点以启动转录过程。可以通过对质粒改造、添加更强的组织特异性启动子而获得更高的效率。常用的启动子序列来源于病毒如巨细胞病毒（CMV）和劳氏肉瘤病毒启动子，或人自身如肌动蛋白α启动子。然而，一旦发生免疫刺激时，启动子则会丧失活性。人源启动子一般比病毒性启动子对免疫应答激活具有更好的弹性。例如，在乙肝表面抗原基因疫苗小鼠试验中，与常用的CMV启动子相比，肌肉肌酸激酶启动子没有免疫刺激效应。

Promoter sequences play a vital role in initiating gene transcription. Promoter sequences offer recognition sites for the RNA polymerase to initiate the transcription process. Higher efficiency can be obtained by engineering the plasmid with strong tissue-specific promoters. Commonly used promoter sequences are derived from viral origins such as cytomegalovirus (CMV) and roux sarcoma virus, or are obtained from human origins such as alpha actin promoter. However, sometimes promoters can lose their activity upon immune stimulation. Human promoters have been shown to have greater resiliency against immune response activation than viral promoters. For example, it was demonstrated that human muscle creatine kinase promoter has no immunostimulatory effect in mice, as opposed to the commonly used CMV promoter sequence, during the expression of a gene vaccine encoding the hepatitis B surface antigen.

质粒DNA中的增强子区域可使目的基因的表达增强多达几百倍。增强子可以是组织特异性的，可位于启动子的上游或下游。目的基因的转录效率可被合适的增强子明显提高。例如，来源于免疫球蛋白基因的增强子可用于提高造血细胞的转染率以及提高病毒性载体在治疗血液恶性肿瘤时的特异性。

Enhancers are regions in the plasmid DNA that enhance the production of the gene of interest by as much as several hundred times. Enhancers can be tissue specific and can be present on the plasmid locus either upstream or downstream from the promoter region. Transcription efficiency can be substantially improved by the choice of suitable enhancers. For example, promoters and enhancers derived from immunoglobulin genes have been used to increase gene transduction in hematopoietic cells and to improve specificity of viral vectors useful in the treatment of hematological malignancies.

1.1.2 重组质粒DNA的制备

1.1.2 The Preparation of Recombinant Plasmid DNA

质粒DNA通常在研究级别从各种细菌细胞中提取和纯化。然而，制备工业级别的能够符合药物产品苛刻纯度标准的质粒DNA仍然比较困难，相关资料也鲜有发表。生产用于基因治疗试验的质粒DNA的普通方法参见图8-1。在制备DNA之前，研究者需要构建一个携带治疗基因的合适载体并转入生产微生物（如大肠杆菌）。后续的常规大规模质粒生产

Plasmid DNA is routinely extracted and purified from various microbial cells at the research level. However, industrial-scale manufacture to the exacting standards of purity demanded of pharmaceutical products is a pursuit still in its infancy, and little has been published on the subject. The overall generalized approach used to produce plasmid DNA for the purposes of gene therapy trials is presented in Figure 8-1. Prior to its manufacture, researchers would have constructed an appropriate vector housing the therapeutic gene and introduced it into a producer microorganism, such as *E. coli* Routine large-scale plasmid manufacture then entails culture of a batch of producer microorganisms by fermentation, followed by plasmid extraction and purification. In this regard, the overall approach used

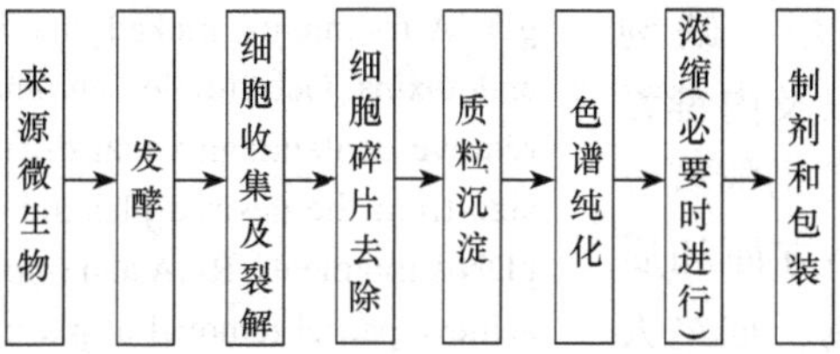

图 8-1 质粒 DNA 的大规模生产流程示意图

尚需对大批量生产微生物进行培养发酵，然后提取纯化。此处使用的总体方法和重组治疗蛋白的大规模生产方法是类似的。

resembles the approaches taken in the large-scale manufacture of recombinant therapeutic proteins.

工业级别的微生物发酵（即上游处理）已经在第二章进行了描述。发酵促进微生物细胞的繁殖和大量质粒的生物合成。发酵之后，通过离心或微量过滤法收集微生物细胞。再将细胞重悬于小体积的缓冲液中，然后破碎细胞以释放质粒。微生物细胞的裂解最常用的方法就是添加含有 NaOH 和 SDS（十二烷基磺酸钠）的裂解液。高 pH 和去污剂联合作用，使微生物细胞壁和细胞膜破裂，以及细胞内容物的最终释放。除了所需的质粒 DNA 之外，这种粗提混合物还包含各种杂质，这些杂质在后续的下游处理步骤中必须去除。需要注意的一些杂质包括：细胞壁碎片、一些不完整的细胞、蛋白质、基因组 DNA 和 RNA 小分子量的代谢物、内毒素。

Industrial-scale microbial fermentation (upstream processing) has also been described in Chapter 2. Fermentation promotes microbial cell replication and thus the biosynthesis of large quantities of plasmid. Subsequent to fermentation, the microbial cells are harvested (collected) by either centrifugation or microfiltration. Following resuspension in a low volume of buffer, the cells must be disrupted in order to release the plasmids therein. This appears to be most commonly achieved by the addition of a lysis reagent, consisting of NaOH and SDS (sodium dodecyl sulphate). The combination of high pH and detergent action disrupts the microbial cell wall and membranes, with consequent release of the intracellular contents. In addition to the desired plasmid DNA, this crude mixture will also contain various impurities that must be removed by subsequent downstream processing steps. Notable impurities include: Cell wall debris and some intact cells、Proteins、Genomic DNA、RNA、Low molecular mass metabolites and Endotoxin.

细胞裂解完成之后，下一步就可添加高盐中和溶液如醋酸钾。该步骤可促使基因组 DNA 和 SDS- 蛋白质复合物形成聚集物，这些聚集物可通过离心或过滤予以去除。然后再通过添加合适的溶剂（通常是异丙醇或乙醇）使质粒 DNA 从溶液中沉淀出来。重悬质粒 DNA 沉淀后，质粒制品可进行色谱纯化。主要的污染物

After lysis is complete, the next step can entail the addition of a high-salt neutralization solution, such as potassium acetate. This promotes formation of aggregates of genomic DNA (gDNA) and SDS–protein complexes, which can subsequently be removed by centrifugation or filtration. The plasmids can then be themselves precipitated from the resultant solution by the addition of appropriate solvent (usually either isoproponal or ethanol). Upon resuspension, the plasmid preparation can then be subjected to chromatographic purification. The major contaminants likely still present include RNA,

如RNA、基因组DNA片段、缺刻或其他形式的质粒DNA以及内毒素可能仍然存在。凝胶过滤色谱能有效地去除那些在形状、大小方面和质粒DNA明显不同的污染物，如绝大部分基因组DNA片段、RNA和绝大部分内毒素；同时也能部分地将一些开环的质粒从主要质粒制备物（即超螺旋质粒）中去除。离子交换色谱则能去除很多蛋白质污染物以及RNA。但是，基因组DNA和内毒素通常和质粒DNA一起被纯化。其他一些基于反相和吸附系统的色谱方法已经至少在实验室水平上被开发出来。

gDNA fragments, nicked or other plasmid variants and endotoxins. Gel filtration chromatography can effectively remove contaminants that differ substantially in shape/size from the desired plasmid. These can include most gDNA fragments, RNA and (most) endotoxins. It can also achieve partial removal of plasmid variants, such as open circular plasmids, from the main (supercoiled) plasmid preparation. Ion exchange chromatography can remove many protein contaminants, as well as RNA. However, gDNA and endotoxins generally co-purify with the plasmid DNA. Additional chromatographic approaches based upon reverse-phase and affnity systems have been developed at laboratory scale at least.

使用捕获色谱进行质粒纯化（即涉及质粒与色谱珠子的结合）时，其显著特点是低的质粒结合能力。商业化的捕获色谱介质的孔径尺寸通常较小，不足以使质粒DNA进入，仅限于结合在珠子表面。因此，与使用相同的介质去纯化治疗蛋白（比质粒DNA小）相比，质粒DNA的结合能力则要低至1/100或更低。

A significant feature of plasmid purification employing capture chromatography (i.e. involves plasmids binding to the chromatographic beads) is the low plasmid-binding capacities observed. The pore size of commercially-available capture chromatographic media is insuffciently large to allow entry of plasmids, restricting binding to the bead surface. Binding capacities can, therefore, be 100-fold or lower than those observed when the same medium is used to purify (much smaller) therapeutic proteins .

纯化的质粒可以通过各种方法进行分析。是否污染了核酸和/或蛋白质可以通过电泳方法进行评价。通常也要进行内毒素和无菌测试。纯化的质粒DNA还必须进一步进行配制以生产出最终的非病毒性药物传递系统。与此相关的药物配制研究仍是一个需要进一步研究的领域。迄今报道的绝大部分工作都和基于lipoplex的基因传递系统的配制和稳定技术有关。这些水性混悬液式的非病毒性系统会在几分钟至几小时内发生快速聚集。为此，在较早的临床试验中，最终的传递系统实际上通常都是在患者

Purified plasmids may then be analyzed using various analytical techniques. Freedom from contaminating nucleic acid and/or proteins can be assessed electrophoretically. Endotoxin and sterility tests would also be routinely undertaken. The purified plasmid DNA must next be formulated to yield the final non-viral delivery system. Formulation studies relating to such systems remain an area requiring further investigation. Most work reported to date relates to formulating and stabilizing lipoplex-based gene delivery systems. Aqueous suspensions of these (and other) non-viral based systems tend to quickly aggregate (in a matter of minutes to hours). In order to circumvent this problem, the final delivery systems were often actually formulated at the patient' s bedside in earlier clinical trials.

床旁进行配制。

目前关于鉴定合适的稳定辅料和制剂形式的研究仍在进行。因为冷冻剂型不易发生聚集，所以可以使用单纯的冷冻。但是，冷冻过程尤其是缓慢冷冻本身会诱发聚集。使用快速冷冻方法（如浸入液氮）可以使聚集的发生最小化，但这种方法在工业级别可能不太可行。添加冷冻保护剂可能对此有所帮助，初步研究表明各种糖类物质（如葡萄糖、蔗糖和海藻糖）具有一些作用。另外的研究则和最终的冻干产品的生产有关。其他一些问题，如工业级别冷冻干燥器的相对较低的缓慢冷冻过程，使快速冷冻仍难付诸实践应用。

Research aimed at identifying appropriate stabilizing excipients and formulation formats is ongoing. Simple freezing is an option, particularly as frozen formulations would be immune to agitation-induced aggregation. However, the process of freezing, particularly slow freezing, in itself induces aggregation. This can be minimized by flash freezing (e.g. by immersion in liquid nitrogen), although this approach may not prove practicable at an industrial scale. The addition of cryoprotectants may help minimize this problem and initial studies indicate that various sugars (e.g. glucose, sucrose and trehalose) show some potential in this regard. Another avenue under investigation relates to the generation of a final freeze-dried product. Again issues, such as the (relatively) slow freezing process characteristic of industrial-scale freeze-driers, complicate attaining this goal in practice.

1.2 重组病毒载体

1.2 Recombinant Viral Vectors

1.2.1 重组病毒载体的设计

关于重组病毒载体的设计，其基本策略和方法与重组质粒 DNA 非常类似，尤其是表达盒的设计。例如，重组病毒载体中的表达盒也包括一个外源治疗性基因以及一些调控序列如启动子和增强子元件等。

1.2.1 The Characteristics and Design of Recombinant Viral Vectors

As for the design of recombinant viral vectors, it is very similar to that of recombinant plasmid DNA, especially regarding the expression cassette. For example, in recombinant viral vectors, the expression cassette also contains an exogenous therapeutic gene as well as some regulatory sequences such as promoter, enhancer elements.

1.2.2 重组病毒载体的制备

目前用于基因治疗的病毒载体大规模生产的文献报道仍然寥寥可数。因为主要是一些从事基因治疗产品研发的公司在进行这些大规模生产方法的研制，因此，方法的具体细节仍然是商业秘密。但其总体方法和治疗蛋白的生产大体一致，涉及胞内合成、回收、浓缩、纯化和制剂步骤。逆转录病毒载体的大致生产流程如图 8-2 所示。其他病毒性载体的生产也与此类似。

1.2.2 Manufacture of Recombinant Viral Vectors

Literature reports describing the large-scale manufacture of viral vectors for gene therapy application are few and far between. Development of manufacturing protocols has largely been undertaken by companies engaged in gene therapy product development and, consequently, protocol details remain confidential. The overall approach likely taken is not too dissimilar to that of therapeutic protein manufacture. It involves synthesis in cells, recovery, concentration, purification and formulation steps. A likely generalized manufacturing scenario for retroviral-based vectors is outlined in Figure 8-2. The manufacture of alternative viral vectors likely follows a similar approach.

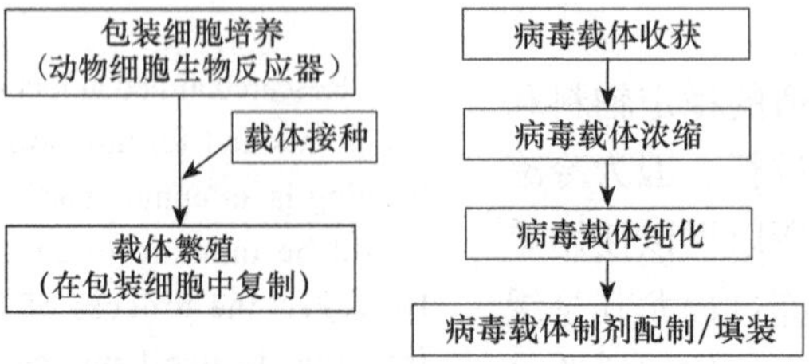

图 8-2　用于临床基因治疗的逆转录病毒载体药物的大规模生产

该过程首先要在合适的动物细胞生物反应器中培养包装细胞。动物细胞培养的具体原理和方法参见第二章。迄今为止，已经使用的生物反应器的尺寸为 100L 或稍小，该尺寸足以满足临床试验需要。然后，将复制缺陷型病毒接种至包装细胞，使病毒载体繁殖。进而采用微量过滤的方法收获病毒，该法将不完整的包装细胞或细胞碎片从包含病毒载体的产物中分离。病毒载体的浓缩采用超滤法。后续的载体纯化策略包括与纯化蛋白类似的各种色谱法。离子交换色谱和各种各样的亲和色谱使用最为广泛。目前对色谱介质和病毒表面分子的相互作用的具体细节仍不清楚，仍在研究之中。纯化之后，进行终产物分析和制剂制备，这些方法的细节目前也几乎没有公开发表的资料。

The process is initiated by the culture of packing cells in suitable animal cell bioreactors. The principles and practice of animal cell culture have been overviewed in Chapter 2. To date, bioreactor size of 100 L or less have been used, which are suffcient to satisfy clinical trial demand. The packing cells are then seeded with the replication-deficient virus, allowing vector propagation. Viral harvest may then be undertaken by methods of microfiltration, which separates intact packing cells/cell debris from the vector-containing product stream. Viral vector concentration can then be undertaken by ultrafiltration. Subsequent vector purification strategies employed include chromatographic approaches similar to those used to purify proteins. Ion exchange and various forms of affnity chromatography have received most attention. A detailed understanding of the interactions between chromatographic media and viral surface molecules is currently lacking and hence is the subject of ongoing research. After purification, final product analysis and formulation is undertaken. Again, few details of these steps have been openly published.

2　寡核苷酸

2　Oligonucleotides

2.1　常见的寡核苷酸药物

根据其分子作用机制不同，寡核苷酸药物可以分为反义寡核苷酸、核酶和脱氧核酶以及小干扰 RNA 等。

2.1　Common Oligonucleotide Drugs

Based on the molecular mechanism of action, oligonucleotide drugs may be classified into several groups including antisense oligonucleotides, ribozymes and DNAzymes, and small interfering RNAs etc.

2.1.1　反义寡核苷酸

1. 分子机制　反义寡核苷酸是短的单链 DNA 片段，一旦被摄取入

2.1.1　Antisense Oligonucleotides

1. Mechanism of Action　Antisense oligonucleotides are short single-stranded segments of DNA that upon cellular internalization can selectively inhibit the

细胞后即可选择性地抑制单个蛋白质的表达。绝大多数情况下，反义寡核苷酸通过和mRNA或pre-mRNA结合形成二聚体而抑制翻译或抑制pre-mRNA的加工，最终抑制蛋白质的生物合成。此外，寡核苷酸还可以进入细胞核，和双链的基因组DNA结合形成三链聚合体而抑制蛋白质表达的翻译和转录过程（这种寡核苷酸有时被特称为反基因寡核苷酸）。

expression of a single protein. In most cases, antisense oligonucleotides interact and form a duplex with the mRNA or the pre-mRNA and inhibit their translation or processing, consequently inhibiting protein biosynthesis. Additionally, oligonucleotides might enter the cell nucleus, form a triplex with the double-stranded genomic DNA, and inhibit the translation as well as the transcription processes of the protein (this kind of oligonucleotides are specifically called as antigene oligonucleotides).

反义寡核苷酸一旦与其靶mRNA相互结合作用后即可导致靶蛋白的翻译阻滞。反义靶识别是通过寡核苷酸和靶mRNA之间Watson-Crick碱基配对而实现的。目前在分子水平上已经提出多种寡核苷酸的作用机制。“占位”机制认为，寡核苷酸和靶mRNA中关键序列的结合可抑制靶mRNA与蛋白质合成相关的分子或细胞器相互作用。例如，寡核苷酸和起始密码子或剪接序列的结合能通过类似机制导致翻译阻滞。此外，寡核苷酸还能通过激活RNase H引起靶mRNA的降解，进而抑制蛋白质合成。因为反义技术的靶分子位于细胞质内，因此反义寡核苷酸通常无需进入细胞核。

Antisense oligonucleotides achieve translational arrest of their target proteins upon interaction with their mRNA targets. Antisense target recognition is facilitated by Watson-Crick base pairing between the oligonucleotide and the target mRNA. On the molecular level, numerous mechanisms have been proposed to explain the basis of oligonucleotide action. Occupancy-based mechanisms suggest that binding of oligonucleotides to key sequences in the target may inhibit the target's ability to interact with the cellular machinery required for protein synthesis. For example, oligonucleotide interaction with the initiation codon can cause translational arrest by this route. Alternatively, oligonucleotides can function by activation of RNase H, an enzyme that degrades the targeted mRNA, thereby inhibiting protein biosynthesis. Since the targets for antisense applications are in the cytoplasm, oligonucleotides usually do not need to enter the cell nucleus.

2. *反义寡核苷酸的设计* 设计对于寡核苷酸的临床有效性至关重要，需要考虑的因素主要包括长度、骨架修饰、杂交能力以及二级构象。

2. Design of Antisense Oligonucleotides Design is critical for the clinical efficacy of oligonucleotides and should include consideration of length, chemistry, conformation, and ability to hybridize with the target mRNA.

（1）长度。建议具有有效反义活性的寡核苷酸的最佳长度通常为12～28个碱基。一般认为，太短的序列可能特异性差，而超过25个碱基的长序列又会不容易被细胞摄取。随着分子长度增加，碱基对间的氢键和

(1)Length The suggested optimal length for oligonucleotides with efficient antisense activity ranges from 12 to 28 bases. There is a consensus that very short sequences are likely to be nonspecific and sequences longer than 25 bases would experience difficulty in cellular uptake. As the length of the molecule increases, hydrogen bonding between the base pairs and stacking

堆积作用力跟着增加，从而导致寡核苷酸与靶序列总体结合力增加。然而，对于较长的寡核苷酸，因为其分子尺寸大且易于发生自身杂交（self-hybridize），因此难于进入细胞，进而影响其吸收。即便在细胞培养实验中，建议的最佳寡核苷酸长度也有很大的变化区间。

interactions increases, leading to an increased overall affinity for the target. However, longer oligonucleotides are inherently difficult to import into cells because of their size and tend to self-hybridize, thereby affecting their uptake. Even in cell culture experiments, the optimum length suggested for oligonucleotides varies considerably.

（2）骨架修饰。寡核苷酸骨架含有磷酸二酯键，易于被核酸酶降解，故而限制了其应用空间。鉴于此，各种各样的化学修饰被用于提高寡核苷酸的稳定性。最常用的化学修饰是向寡核苷酸骨架中引入硫代磷酸和甲基磷酸连接。硫代磷酸类似物对核酸酶具有很强抗性，甲基磷酸骨架则具有较强疏水性和易于跨膜扩散。与没有修饰的磷酸二酯键寡核苷酸相比，硫代磷酸骨架寡核苷酸的生物学半衰期显著提高。目前，第二代混合骨架寡核苷酸如GEM231正在癌症治疗临床试验中使用。

(2) Backbone Modifications　Oligonucleotides having the endogenous phosphodiester backbone are susceptible to degradation by nucleases and hence have limited use for antisense applications. Various chemical modifications to the backbone have been used to improve oligonucleotide stability. The most common modifications include the introduction of phosphorothioate and methyl phosphonate linkages in the backbone. Phosphorothioate analogs are chosen for their stability against nucleases and the methylphosphonate backbone for its relative hydrophobicity and ease of diffusion across membranes. Phosphorothioate backbone oligonucleotides can have significantly increased biological half-life compared to their corresponding unmodified phosphodiester oligonucleotides. Currently, second-generation mixed-backbone oligonucleotides such as GEM231 are being used in clinical trials for cancer.

（3）杂交能力。反义寡核苷酸主要对其与靶mRNA的杂交能力进行设计。mRNA具有复杂的二级结构和三级空间结构，这就限制mRNA中某些局部片段与寡核苷酸的杂交结合。因此，尽管理论上来讲可以对mRNA的任意区段设计相应的反义寡核苷酸，但并非所有寡核苷酸能有效地抑制蛋白质合成。目前正在使用的一些预测mRNA分子中易杂交位点的策略有：预测mRNA结构的算法、基因步移（gene-walking）、RNase H作图（RNase H mapping）和联合筛选。

(3) Hybridization Capacity　Antisense oligonucleotides are designed primarily for their ability to hybridize with the mRNA of interest. The mRNA has a complex secondary and tertiary spatial structure that restricts the accessibility of certain segments of the molecule to hybridize with the oligonucleotide. Consequently, although it is theoretically possible to design antisense constructs to virtually any part of the mRNA, not all oligonucleotides are efficient in inhibiting protein synthesis. Complex algorithms predicting mRNA structure, “gene-walking”, RNase H mapping, and combinatorial screening are some of the strategies that are being explored to predict hybridization-accessible sites on mRNA molecules.

（4）二级构象。寡核苷酸中一些特异序列的存在使其具有一种偏好的

(4) Secondary Conformation　The presence of some specific sequences in the oligonucleotide can

构象。例如，已经证实：因为连续的鸟嘌呤可在细胞内形成超级结构，所以寡核苷酸 3'- 末端连续鸟嘌呤的出现可使其不易降解。这种超级结构可抵抗核酸酶活性，能有效地被细胞摄取。寡核苷酸的二级结构可受到细胞中常见的二价阳离子如 Ca^{2+} 和 Mg^{2+} 的影响，进而影响其活性或细胞的摄取。

3. 反义寡核苷酸的临床应用

对于治疗性目的，反义寡核苷酸能被用于选择性地阻断与疾病发生相关的蛋白质的表达。随着动物模型中蛋白质表达反义抑制的成功实施，1998 年首个反义药物福米韦生钠（fomivirsen sodium）被批准用于 AIDS 患者巨细胞病毒性视网膜炎的治疗。福米韦生钠剂型为玻璃体内注射液（pH 8.7 的碳酸氢钠缓冲液中）。反义寡核苷酸如 MG98 和 ISIS 5132 设计用来分别抑制 DNA 甲基化酶和 c-raf 激酶的生物合成，已进入肿瘤治疗临床试验阶段。靶向 ICAM-1 的反义寡核苷酸药物 ISIS 2302，用于溃疡性结肠炎的治疗研究。另外两个反义寡核苷酸药物 Affinitak 和 Alicaforsen，也处于三期临床试验阶段，分别用于非小细胞肺癌和克罗恩病的治疗。Aventis 和 Genta 公司研发的反义寡核苷酸药物 Genasense，处于二期临床试验阶段，联合化疗药物用于各种癌症如恶性黑色素瘤、慢性淋巴细胞白血病和多发性骨髓瘤的治疗。

2.1.2 核酶和脱氧核酶

核酶是具有催化作用的 RNA 分子，能够对 mRNA 分子进行序列特异性的剪接，在自然界存在于多种有

allow it to have a preferred conformation. For example, it was demonstrated that the presence of continuous guanines at the 3'-end of an oligonucleotide made it resistant to degradation because of the ability of these guanines to assemble into hyperstructures in vivo. These hyperstructures were resistant to nuclease activity and were efficiently taken up by cells. The secondary structure of oligonucleotides can be affected by divalent cations such as Ca^{2+} and Mg^{2+} that typically occur in the cellular environment, thereby possibly affecting their activity or cellular uptake.

3. Clinical Applications of Antisense Oligonucleotides For therapeutic purposes, antisense oligonucleotides can be used to selectively block the expression of proteins that are implicated in diseases. With successful antisense inhibition of proteins in animal models, the first antisense drug, fomivirsen sodium (Vitravene, Isis Pharmaceuticals, Carlsbad, CA) was approved for the treatment of cytomegalovirus retinitis in AIDS patients in 1998. Fomivirsen sodium is formulated as an intravitreal aqueous injection in sodium bicarbonate buffer at pH 8.7. Antisense oligonucleotides such as MG98 and ISIS 5132, designed to inhibit the biosynthesis of DNA methyltransferase and c-raf kinase, respectively, are in human clinical trials for cancer. ISIS 2302, targeting ICAM-1, is being investigated for the treatment of ulcerative colitis. Two other drug candidates, Affinitak and Alicaforsen (Isis Pharmaceuticals), are in phase 3 clinical trials for non-small cell lung cancer and Crohn's disease, respectively. Genasense (oblimersen sodium), developed by Aventis (Bridgewater, NJ) and Genta (Berkeley Heights, NJ), is being investigated in advanced phase 2 trials in combination with other chemotherapy regimens for a range of cancers including malignant melanoma, chronic lymphocytic leukemia, and multiple myeloma.

2.1.2 Ribozymes and DNAzymes

Ribozymes are catalytic RNA molecules that are capable of sequence-specific cleaving of mRNA molecules, exist in nature in multiple organisms. They can selectively bind to target mRNAs and form a duplex,

机体中。核酶能选择性地与靶 mRNA 结合形成双链复合物，其中 mRNA 的构象高度扭曲变形而易于被水解，由此即可实现对特定基因的靶向抑制。目前在临床治疗中研究和使用最为广泛的核酶主要有两种，即垂头型核酶和发夹型核酶。核酶分子中 RNA 骨架易于被广泛存在的 RNA 酶作用而降解，因此核酶分子在体内非常不稳定，在临床应用时需要进行化学修饰。迄今为止，仅有一个核酶进入临床试验阶段。该核酶靶向作用于 VEGF 受体，用于难治性实体瘤的血管生成抑制治疗。

which includes a highly distorted conformation that is easily hydrolyzed. The hydrolysis of the mRNA can be used for targeted suppression of specific genes. Two types of ribozymes, the hammerhead and hairpin ribozymes have been extensively studied for therapeutic applications. The presence of the RNA backbone in ribozymes makes them easy targets for degradation from the ubiquitous RNases. Thus these molecules are biologically unstable in vivo and require chemical modifications for clinical application. So far only one ribozyme has reached the stage of clinical trials. It is directed against a VEGF receptor to reduce angiogenesis in refractory solid tumours.

脱氧核酶与核酶类似，也具有催化作用，但主要为体外筛选获得。同样地，脱氧核酶与核酶一样能够序列特异性地剪切 RNA 分子，因此同样可以作为抑制基因表达的合适工具。值得注意地是，和核酶不同，脱氧核酶的骨架为 DNA，因此稳定性更高。DNA 序列还更易于修饰、合成，从而产生了作用更强的第二代类似物。因此，在疾病治疗应用方面，脱氧核酶作为基因表达抑制剂具有更大应用潜力。

A close relative of ribozymes are the also autocatalytically operating DNAzymes, which have been obtained by in vitro selection. Consequently, as DNAzymes can cleave RNA also sequence specifically just as ribozymes they should equally be appropriate tools for silencing gene expression. Noticeably, the RNA backbone chemistry is replaced by the DNA motifs that confer improved biological stability. DNA sequences are also easy to modify synthetically, thereby generating even stronger second-generation analogs. Thus, DNAzymes have tremendous potential as gene suppression agents for a variety of therapeutic applications.

2.1.3 核酸适配体

核酸适配体是一类能够与蛋白质直接相互作用的小分子单链或双链核酸片段。核酸适配体的这种相互作用可以用来干扰那些参与疾病发生的蛋白质或者是参与转录或翻译过程蛋白质的分子功能，进而抑制该靶蛋白的生成。基于其特异性、非免疫原性和作为药物制剂的稳定性，与抗体相比，人们更倾向于使用核酸适配体来抑制蛋白质。核酸适配体通常是使用

2.1.3 Aptamers

Aptamers are small single-stranded or double-stranded nucleic acid segments that can directly interact with proteins. Aptamer-assisted interaction can be used to interfere with the molecular functions of disease-implicated proteins or those that participate in the transcription or translation processes, thus inhibiting the processing of target therapeutic proteins. Aptamers are preferred over antibodies in protein inhibition owing to their specificity, nonimmunogenicity, and stability of pharmaceutical formulation. Aptamers are usually obtained by selection procedure through the systematic evolution of ligands by exponential enrichment (SELEX)

联合核酸文库、采用SELEX技术筛选获得。这不仅可以作用于蛋白质，也可以是肽甚至是非肽分子。目前，核酸适配体的研究还主要是前导物的鉴定和靶向验证。已经证明具有良好治疗应用前景的核酸适配体，如作用于HIV-1整合酶的DNA适配体和作用于HIV-1转录酶的RNA适配体。由Eyetech公司和Pfizer公司共同研发的抗VEGF适配体Macugen已经进入三期临床试验，用于抑制与年龄相关的黄斑退行性改变的黄斑血管的病理性过度增生。

with combinatorial nucleic acid libraries. They are selected not only directed against proteins, but also peptides and non-peptide molecules. Currently, most aptamer research is directed toward lead identification of inhibitors of protein function and target validation. Aptamers that have demonstrated promise in intervention of pathogenic protein biosynthesis include DNA aptamers against the HIV-1 integrase enzyme and RNA aptamers against HIV-1 transcriptase. The anti-VEGF aptamer Macugen (pegaptanib sodium) co-developed by Eyetech (New York, NY) and Pfizer (Groton, CT) is in advanced phase 3 clinical trials for age-related macular degeneration to inhibit excessive pathological blood vessel growth in the macula.

2.1.4 小干扰RNA

RNA干扰（RNA interference, RNAi）主要指的是真核生物中短链RNA通过转录后基因沉默机制对基因表达调节的作用。这些短链RNA包括短链干扰RNA（short interfering RNA, siRNAs）和微RNA（micro RNAs, miRNAs）。虽然miRNA可能也具有很大的治疗应用潜力，但迄今为止以siRNA开展的临床治疗应用最为广泛。

siRNA主要用以通过RNAi机制下调致病基因的表达。siRNA是一种短的21~23核苷酸长度的双链RNA片段，其序列和编码蛋白质的mRNA序列的某一区域互补。一旦被摄入细胞后，siRNA分子参与形成RNA诱发的沉默复合物（RNA-induced silencing complexes, RISCs），后者和靶mRNA结合而引发mRNA降解机制（如核酸酶活性），进而导致特定靶基因的沉默。siRNA分子既可被化学合成后传递至靶细胞，也可以在细胞内生成。细胞内siRNA的产生主

2.1.4 Small Interfering RNAs

RNA interference (RNAi) is the central role which short RNAs play in eukaryotic biology by the regulation of gene expression through post-transcriptional gene silencing. The short RNAs consist of short interfering RNAs (siRNAs) and micro RNAs (miRNAs). So far, only siRNA are most explored for clinical therapeutic applications although miRNAs might also have great therapeutic potentials.

siRNAs are mainly used for downregulation of disease-causing genes through RNA interference. siRNAs are short double-stranded RNA segments with typically 21to 23nucleotide bases that are complementary to the mRNA sequence of the protein. Upon administration, siRNA molecules are incorporated into RNA-induced silencing complexes (RISCs), which bind to the mRNA of interest and stimulate mRNA degradation mechanisms, such as nuclease activity, that lead to silencing of the particular gene. siRNA molecules can either be chemically synthesized and delivered to the cells or generated in vivo. In vivo production of siRNAs is achieved by using long double-stranded RNA molecules, which are processed by the Dicer enzyme to generate siRNAs, or by transfection with plasmids and viral vectors that biosynthesize siRNAs by transcription.

要通过使用长的双链 RNA 分子实现，后者被 Dicer 酶处理而产生 siRNA，或者也可通过转染质粒或病毒载体而实现，这些载体可通过转录而在细胞内合成 siRNA。

siRNA 的治疗性应用仍处于早期开发阶段。目前，siRNA 正在被研究用来治疗包括癌症和传染病在内的多种疾病。与质粒 DNA 和其他寡核苷酸分子相比，siRNA 具有多种优势。siRNA 在多个方面均表现出众，如对靶 mRNA 的高度特异性、无免疫原性、高度耐受核糖核酸酶。siRNA 也不会整合入基因组，因此比质粒 DNA 等分子更为安全。而且，siRNA 也无须转移通过核膜而发挥活性，因此也不需要复杂的药物传递系统。此外，因其分子量小，故可以在一个单一的药物传递系统中同时传递靶向多个致病基因的多种 siRNA，以治疗诸如癌症等多基因功能失常导致的复杂性疾病。

The use of siRNA for therapeutic applicatioin is still in its infancy. Currently, siRNAs are being investigated to several disearses including cancer and infectious diseases. These potent molecules offer several advantages over plasmid DNA molecules and other oligonucleotides. Compared with other gene ablation technologies such as antisense oligonucleotides, siRNAs are remarkably superior thanks to their high degree of specificity to mRNAs, their nonimmunogenic nature, and their high resistance to ribonucleases. Since siRNAs do not integrate into the genome, they offer greater safety than plasmid molecules. Furthermore, siRNAs do not have to transfer through the nuclear membrane for their activity and therefore require less sophisticated delivery systems, promising faster development and higher efficiencies compared with plasmid DNA. In addition, because of their small size, it might be possible to deliver a cocktail of siRNAs targeting multiple disease-causing genes in a single delivery system to control complex diseases such as cancer where several genes are malfunctioning.

2.2 寡核苷酸的生产

和前面介绍的生物药物（重组蛋白和用于基因治疗的质粒）不同，寡核苷酸通常是以直接化学合成的方式生产。寡核苷酸是一种广泛使用的分子生物学试剂，因此其有机合成方法得以不断的发展、优化和商业化。其基本合成策略和多肽合成的 Merrifield 法非常类似。需要的核苷酸（根据需要进行修饰或不修饰）首先和一个保护性化学基团反应。然后，每一个被保护的核苷酸依次被偶联到核苷酸链延伸的末端，核苷酸链被连接至一固相。偶联完成后，最初的保护性基团被去除。整条链合成完

2.2 Manufacture of Oligonucleotides

In contrast to the biopharmaceuticals thus far discussed (recombinant proteins and gene therapy plasmids), oligonucleotides are usually manufactured by direct chemical synthesis. Organic synthetic pathways have been developed, optimized and commercialized for some time, as oligonucleotides are widely used reagents in molecular biology. The basic synthetic strategy is very similar in concept to the means by which peptides are synthesized via the Merrifield method. The nucleotides required (themselves either modified or unmodified, as desired) are first reacted with a protecting chemical group. Each protected nucleotide is then coupled in turn to the growing end of the nucleotide chain, itself attached to a solid phase. After coupling, the original protecting group is removed and, when chain synthesis is complete, the bond anchoring the chemical to the solid phase is hydrolysed, releasing the free oligo. This may then be

成后，锚定至固相的化学键被水解，从而释放游离的寡核苷酸。寡核苷酸进而可通过 HPLC 纯化。最常用的合成方法是磷酰胺酸法，该法使用二甲氧三苯甲基（DMTr）保护性基团和四唑作为偶联剂。目前已经实现商业化自动合成，能快速和廉价地合成超过 100 个核苷酸长度的寡核苷酸。

purified by HPLC. The most common synthetic method used is known as the phosphoramidite method, which uses a dimethoxytrityl (DMTr) protecting group and tetrazole as the coupling agent. Automated synthesizers are commercially available which can quickly and inexpensively synthesize oligos of over 100 nucleotides.

3　核酸药物传递系统

3　Delivery System for Nucleic Acid Drugs

核酸药物的传递系统已经从实验室逐渐发展成熟至可以进行临床试验和应用。药物传递系统在实现核酸疗法的治疗潜力上起着至关重要的作用。作为一种理想的核酸药物传递系统，其特性包括：高转染效率和高度的靶细胞特异性、低的毒性和免疫原性、生物可降解性以及药物制剂稳定性。此外，这种理想的核酸药物传递系统还应该是简便易于程式化，并能被修改以用于专门的核酸释放、传递和表达。当前常用的核酸药物传递系统分为两类：病毒载体介导和非病毒载体介导的药物传递系统。

Cellular delivery of nucleic acid drugs assisted by delivery systems has matured from a laboratory science into a methodology suitable for use in clinical trials. Delivery systems will play a critical role in the realization of the therapeutic potential of nucleic acid therapeutics. As an ideal delivery system for nucleic acid drugs, its properties include high transfection efficiency with a high degree of target cell specificity, low occurrence of toxicity and immunogenicity, biodegradability, and stability of the pharmaceutical formulation. In addition, the ideal delivery system would be simple to formulate and would lend itself to easy modification for customized nucleic acid release, delivery, and expression. Currently available delivery systems for nucleic acid drugs can be classified into two types: viral DNA delivery systems and non-viral delivery systems.

3.1　病毒载体介导的传递系统

目前，在世界范围内，超过 70% 的人类基因治疗临床试验采用病毒作为核酸药物传递系统。病毒载体介导的传递系统在各种疾病如肌肉萎缩、AIDS 和肿瘤等的治疗中均取得了重大进展。目前唯一批准的基因治疗方案也是采用重组腺病毒载体来传递药物。当前核酸疗法中常用的病毒包括逆转录病毒、腺病毒、慢病毒和腺相

3.1　Viral Delivery Systems

Viruses are currently used in more than 70% of human clinical gene therapy trials worldwide. Gene therapy using viral systems has made considerable progress for the treatment of a wide range of diseases, such as muscular dystrophy, AIDS, and cancer. The only approved gene therapy treatment (Gendicine) delivers the therapeutic gene using a recombinant adenoviral vector. The currently used viruses in nucleic acid therapeutics include retroviruses, adenoviruses, lentiviruses, adeno-associated viruses, and the herpes simplex virus. Among those, retroviruses and adenoviruses are the most

关病毒以及单纯疱疹病毒，其中以逆转录病毒和腺病毒最为常用。

commonly used.

3.1.1 逆转录病毒传递系统

逆转录病毒为包膜病毒，其基因组为单链RNA，大小5～8kb。一旦感染进入宿主细胞后，病毒RNA被反转录并最终生成双链DNA。该双链DNA最终将整合入宿主细胞基因组。基本的逆转录病毒基因组包含至少三个结构基因，即*gag*、*pol*和*env*，分别编码核心病毒蛋白、逆转录酶和病毒包膜蛋白。病毒基因组的两个末端为长末端重复序列（long terminal repeats，LTRs），含有强的启动子和增强子以及促进病毒基因组整合入宿主细胞基因组的必需序列。紧接着5'-LTR的是包装序列（ψ），为病毒RNA包装所必需。

构建逆转录病毒载体时需要将参与正常病毒复制的内源性病毒基因用外源性的目的基因进行替换（图8-3）。这些基因的去除将导致构建的病毒载体不能进行自行复制。为了产生包含载体核酸的成熟的病毒颗粒，载体核酸就必须导入一个包装细胞，包装细胞是经过改造的包含*gag*、*pol*和*env*结构基因的重组细胞。这样，包装细胞就能产生包含目的基因的成熟但复制缺陷的病毒颗粒（见本章1.2.2重组病毒载体的制备）。这种病毒颗粒在作为基因转移系统发挥作用时，正是所谓的“每次单一打击”的模式。

3.1.1 Retroviral Delivery System

Retroviruses are enveloped viruses. Their genome consists of single-stranded RNA (ssRNA) of approximately 5-8 kb. Upon entry into sensitive cells, the viral RNA is reverse-transcribed and eventually yields double-stranded DNA. This subsequently integrates into the host cell genome. The basic retroviral genome contains a minimum of three structural genes; *gag* (codes for core viral protein), *pol* (codes for reverse transcriptase) and *env* (codes for the viral envelope proteins). At either end of the viral genome are the long terminal repeats (LTRs), which harbour powerful promoter and enhancer regions and sequences required to promote integration into the host DNA. Also present, immediately adjacent to the 5'-LTR, is the packing sequence (ψ). This is required to promote viral RNA packaging.

The construction of retroviruses to function as gene vectors entails replacing the endogenous viral genes, required for normal viral replication, with the exogenous gene of interest(Figure 8-3). Removal of the viral structural genes means that the resulting vector cannot itself replicate. In order to generate mature virion particles harbouring the vector nucleic acid, this genetic material must be introduced into a “packing cell”. These are recombinant cells that have previously been engineered to contain the *gag*, *pol* and *env* structural genes. In this way, packing cells are capable of producing mature but replication-deficient viral particles, harbouring the gene to be transferred (section 8 1.2.2 Manufacture of viral vectors). These viral particles function as so-called “ one-time, single-hit” gene transfer systems.

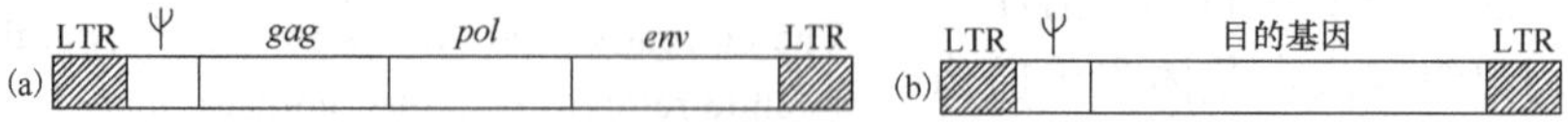

图8-3 逆转录病毒载体及基因组示意图

（a）基本逆转录病毒的前病毒基因组结构；（b）被改造的携带目的基因的逆转录病毒载体

逆转录病毒作为基因治疗载体具有多方面优点。总结如下：①逆转录病毒的相关研究很多，其生物化学和分子生物学机制清楚。②绝大多数逆转录病毒可将其前病毒 DNA 整合入宿主细胞基因组。③基因转移效率对于绝大多数敏感细胞类型来讲都非常高，通常高达 100%。④整合的 DNA 可持续长期和相对高水平的基因表达。⑤前病毒 DNA 随机整合入宿主细胞染色体。⑥逆转录病毒宿主广泛，可感染各种分裂细胞。⑦前病毒 DNA 可随着细胞分裂传递给子代细胞。⑧可生产好的高滴度的复制缺陷的逆转录病毒颗粒。⑨目前已经在各种动物上进行了安全研究。

Retroviruses have multiple advantages as vectors in gene therapy. These may be summarized as follows: ① Retroviruses as a group have been studied in detail and their biochemistry and molecular biology is well understood. ② Most retroviruses can integrate their proviral DNA only into actively replicating cells. ③ The effciency of gene transfer to most sensitive cell types is very high, often approaching 100%. ④ Integrated DNA can be subject to long-term, relatively high-level expression. ⑤ Proviral DNA integrates randomly into the host chromosomes. ⑥ Retroviruses are promiscuous in that they infect a variety of dividing cell types. ⑦ Complete copies of the proviral DNA are passed on to daughter cells if the original recipient cell divides. ⑧ Good, high-level titre stocks of replication-incompetent retroviral particles can be produced. ⑨ Safety studies using retroviral vectors have already been carried out on various animal species.

但是，逆转录病毒的一些缺点也限制了其应用。①很多情况下，它只感染分裂细胞，这明显地限制了其应用范围。对感染的分裂细胞类型没有选择性也限制了其应用。逆转录病毒也并非感染所有分裂细胞，这取决于靶细胞表面是否有合适的病毒受体。因为目前对绝大多数逆转录病毒受体仍不清楚，因此在基因治疗时很难预测逆转录病毒可能感染的全部细胞种类。外源基因在靶细胞之外的细胞表达时，会导致并发症的出现。②逆转录病毒载体的另外一个缺陷就是被转移的基因会随机地整合入受体细胞的染色体。当被整合入细胞内功能至关重要的基因中时，会不可逆转地损害细胞功能。例如：重要代谢酶的破坏会导致细胞死亡；而抑癌基因的破坏又可能引起细胞转化；此外，若前病毒 DNA 整合到静息的细胞原癌基因附近时，则可能导致该原癌基因的激

However, some disadvantages of ritroviruses curtail the its application as gene therapy vectors. ① In most instances, their ability to infect only dividing cells clearly restricts their use. Their lack of selectivity in terms of the dividing cell types they infect is also a disadvantage. They will not infect all dividing cell types—the entry of any specific retrovirus being dependent upon the existence of an appropriate viral receptor on the surface of a target cell. As the identity of most retroviral receptors is still unknown, it remains difficult to predict the entire range of cell types any retrovirus is likely to infect during a gene therapy protocol. Integration and expression of the exogenous gene in cells other than target cells could result in physiological complications. ② An additional drawback with regard to retroviral-based vectors is the propensity of the transferred gene to integrate randomly into the chromosomes of the recipient cells. Integration of the transferred DNA in the middle of a gene whose product plays a critical role in the cell could irrevocably damage cellular function, e.g. disruption of a central metabolic enzyme could cause cell death, while disruption of a tumour suppresser gene could give rise to cellular transformation; in addition, integration of the proviral nucleic acid to sites adjacent to quiescent cellular proto-

活。③另外一个限制逆转录病毒应用的缺陷是其病毒颗粒的相对不稳定性。因此，虽然逆转录病毒相对易于繁殖，但通常会在后续的纯化和浓缩受到损伤，而这些纯化和浓缩步骤又恰恰对于临床应用是必需的。

oncogenes could result in their activation. ③ Another impediment to routine use of retroviral vectors is the relatively labile nature of these particles. Thus, while retroviruses are relatively easy to propagate, they are often damaged by subsequent purification and concentration—steps essential for their clinical use.

3.1.2 腺病毒传递系统

腺病毒是相对较大的无包膜双链DNA病毒。其基因组比逆转录病毒更大也更复杂（约35kb）。多数情况下，仅有一小部分基因组区域在构建腺病毒载体时被去除。一旦感染细胞后，腺病毒DNA会定位于细胞核内，但不整合入宿主细胞的基因组DNA。

作为基因治疗中一种有潜力的载体，腺病毒具有多种优缺点（表8-1）。其主要优点是能感染非分裂细胞和外源基因的表达量高。然而，腺病毒DNA无法整合入宿主细胞基因组，导致基因表达的持续性有限。将携带各种标志基因（即其基因表达产物容易检测）的腺病毒载体导入动物后，发现标志基因的表达可以在包括心脏、肝脏、肌肉、骨髓、中枢神经系统和内皮细胞等各种组织细胞中检测到，表达持续时间2～3周至几个月。

3.1.2 Adenoviral Delivery System

Adenoviruses are relatively large, non-enveloped structures, housing double-stranded DNA as their genetic material. Their genome is much larger (approximately 35 kb), and more complex than those of retroviruses. In most instances, only a small fraction of this genome is removed when constructing an adenovirus-based vector. Upon cellular infection, adenoviral DNA becomes localized in the nucleus, but does not integrate into the host cell genomic DNA.

As potential vectors for gene therapy, adenoviruses display a number of both advantages and disadvantages(Table8-1). Their major advantage relates to their ability to effciently infect non-dividing cells and the usually observed expression of large quantities of the desired gene products. However, the failure of the adenoviral-based DNA to integrate into the host cell genome generally means that its survival, and hence the duration of gene expression, is limited. Adenovirus-based vectors carrying various marker genes (i.e. a gene whose expression product is easily detected) have been administered to animals. Marker gene expression has been subsequently noted in various tissues, including heart, liver, muscle, bone marrow, CNS and endothelial cells. Duration of marker gene expression ranged from 2-3 weeks to several months.

表 8-1 腺病毒药物传递系统的优缺点

优点	缺点
可传递基因至非分裂细胞	对人具有高免疫原性
易于进行大量培养繁殖	转染基因的表达水平持续时间短
基因表达水平高	感染后可导致细胞裂解
属于相对稳定的病毒	对细胞类型具有选择性

虽然短期高水平的表达对于某些基因治疗应用非常适合，但对于某些疾病如遗传性疾病的治疗就用处不

While short-term, high-level gene expression may be appropriate for some gene therapy applications, it would be of less use for the treatment of, for example, genetic diseases, where long-term gene expression

大，因为此时需要长期持续的基因表达。理论上来讲，这可以通过腺病毒载体的重复给药来解决。但是，腺病毒会激发很强的免疫反应，这又恰恰限制了重复给药的效力。

would be required. This could be achieved in theory by repeat administration of the adenoviral vector. However, adenoviruses prompt a strong immune response, which limits the efficacy of repeat administration.

3.2 非病毒载体介导的传递系统

虽然病毒性载体介导的药物传递系统是当前的主流，但目前也有20%～25%的临床试验使用的是非病毒性载体介导的药物传递系统。非病毒性载体介导的核酸药物传递系统能避开病毒性载体的一些问题，因此在某些情况下也是一个很好的选择。非病毒性传递系统的最显著的优点就是无免疫应答、易于配制。常用的非病毒性药物传递系统有两类，即物理性和化学性方法。

3.2 Non-viral Delivery Systems

Although viral-mediated gene delivery systems currently predominate, some 20%-25% of current clinical trials use non-viral-based methods of gene delivery. Non-viral delivery systems can circumvent some of the problems associated with viral vectors and are emerging as favorable alternatives to viral vectors. Among the greatest advantages of non-viral delivery systems are lack of immune response and ease of formulation and assembly. Commonly used non-viral delivery systems for nucleic acid drugs can be classified into two major types: physical and chemical methods.

3.2.1 物理方法

物理方法主要是机械或电学方法，包括显微注射法、电穿孔法、颗粒轰击法、超声波法等。其中，对于显微注射法来说，因为每次仅操作一个细胞，故而其效率很高，但其精确性是以浪费时间为代价取得的。DNA-金微粒的冲击转移可通过颗粒轰击设备如基因枪的方法实现，但因其需要靶组织的直接暴露，因此该法主要局限于真皮、肌肉或黏膜组织。电穿孔法使用高压电流实现核酸药物的传递，该法可导致细胞的大量死亡，故不适用于临床应用。虽然使用物理性方法也取得了很好的转染效率，但此类方法的缺点是很难在临床中实现标准化，费力、不实用且具有损伤性。

3.2.1 Physical Methods

Physical methods are mainly mechanical and electrical; including microinjection, electroporation, particle bombardment, and ultrasonic methods. Microinjection is highly efficient since one cell at a time is targeted for nucleic acid transfer; however, this precision is achieved at the expense of time. Ballistic transfer of DNA-gold microparticles can be achieved using particle bombardment equipment such as the gene gun. However, since direct exposure of target tissues is required, particle bombardment is restricted to local expression in the dermis, muscle, or mucosal tissue, unless surgical exposure of the target is permitted. Electroporation uses high-voltage electrical current to facilitate nucleic acid transfer. This technique results in high cell mortality and therefore is not suitable for clinical use. Though significant transfection efficiencies have been achieved using physical techniques, they are extremely difficult to standardize in a clinical setting and are considered laborious, impractical, and invasive.

3.2.2 化学方法

化学方法包括DNA-磷酸钙共沉淀法、脂质体法、受体介导的基因转移等，其中以脂质体核酸药物传递系统的应用最为广泛。

脂质体目前已经成为一种最为通用的核酸药物传递工具。它是一些由磷脂双层包裹而成水性小室组成的小囊泡。如果围绕着一个核室以同心圆状形成多重双层脂质，这种复合物就被称为多层脂质体（multilamellar vesicles，MLV）。可通过聚碳酸酯膜拉伸多层脂质体而产生特定尺寸（100～500 nm）的小单层脂质体（small unilamellar vesicles，SUV）。也可以通过对多层脂质体或大的小单层脂质体进行超声而产生更小尺寸的小单层脂质体（50～90 nm）。亲水和疏水性的药物均可包裹在脂质体中，例如脂质体和药物/脂质复合物已经被用于抗癌药物阿霉素和柔红霉素的传递。

脂质体在用于核酸药物传递时，既可将核酸药物包裹在水性中心内，或者也可将核酸药物和脂质体复合成磷脂层。与病毒载体介导的核酸药物传递系统相比，脂质体具有更为明显的优点。例如：脂质体没有蛋白质成分，故而一般无免疫原性。因为脂质体双层中的磷脂成分种类很多，因此，脂质体传递系统也很容易进行改造以产生符合各种需要的尺寸、表面电荷、组成和形状。一些脂质体的阳离子带电表面还有助于核酸药物的复合物形成和传递。脂质体还保护核酸药物不受核酸酶的降解和提高核酸药物的稳定性。脂质体也可用于一些专门的基因药物传递（如长半衰期、持

3.2.2 Chemical Methods

Chemical methods DNA-calcium phosphate precipitation method, liposome method, and receptor-mediated gene transfer method etc. Among those methods, liposomal delivery system is the most commonly used.

Liposomes have emerged as one of the most versatile tools for the delivery of nucleic acid drugs. Liposomes are vesicles that consist of an aqueous compartment enclosed in a phospholipid bilayer. If multiple bilayers of lipids are formed around the primary core in a concentric fashion, the complex assemblies that are generated are known as multilamellar vesicles (MLV). Small unilamellar vesicles (SUV) of specific size (100-500 nm) can be produced by extrudeing MLV through polycarbonate membranes. SUV (50-90 nm) can also be produced by sonication of MLV or larger SUV. Both hydrophilic and hydrophobic drugs can be entrapped in the liposomes. Liposomes and drug/lipid complexes have been used for the delivery of the anticancer drugs doxorubicin and daunorubicin.

Liposomes can be used as nucleic acid drug delivery systems either by entrapping the nucleic acid drugs inside the aqueous core or complexing them to the phospholipid lamellae. Liposomes offer significant advantages over viral delivery options for the delivery of nucleic acid drugs. For example, liposomes are generally nonimmunogenic owing to lack of proteinaceous components. Since the phospholipid composition in the liposome bilayers can be varied, liposomal delivery systems can be easily engineered to yield a desired size, surface charge, composition, and morphology. The cationic surface charge of some liposomes can help in the complexation and delivery of nucleic acid drugs. Liposomes can offer substantial protection to the nucleic acid drugs from nucleases and improve their biological stability. Liposomes can also be used for specialized gene delivery options (such as long circulation half-life, sustained, and targeted delivery). A variety of cationic, anionic, synthetically modified lipids, and combinations thereof have been used to deliver a wide range of nucleic acid drugs.

久和靶向传递)。迄今为止,各种各样的阳离子、阴离子、人工合成的修饰性脂质以及其不同组合物均已经被用于各种各样的核酸药物的传递。

4 结 束 语

4 Concluding Remarks

作为一种新的疾病治疗模式,基因治疗以及核酸药物已经受到了越来越多的关注。尽管绝大多数的核酸药物仍处于临床试验的早期阶段,但其在各种疾病的治疗中均显示出了非常令人鼓舞的结果,这些疾病包括癌症、AIDS、帕金森病和阿尔茨海默病等神经性疾病,以及心血管疾病等(表 8-2)。

As a novel approach to human disease treatment, gene therapy as well as nucleic acid drugs had received more and more attentions. Although most of the DNA-based drugs are in early stages of clinical trials, this class of compounds has emerged in recent years to yield extremely promising candidates for drug therapy for a wide range of diseases, including cancer, AIDS, neurological disorders such as Parkinson' s disease and Alzheimer' s disease, and cardiovascular disorders(Table8-2).

表 8-2 处于临床阶段的核酸药物(Ⅲ期及以后)*

药物名	公司	类别	状态	分子机制	应用
今又生	深圳赛百诺基因技术公司	重组腺病毒	中国食品及药物管理局批准	编码抑癌基因 *p53* 的腺病毒	头颈部鳞状细胞癌
Advexin	Introgen Therapeutics 公司	重组腺病毒	Ⅲ期,单独或与顺铂和 5- 氟尿嘧啶联用	编码抑癌基因 *p53* 的腺病毒	难治性头颈部鳞状细胞癌
Vitravene(福米西林钠)	Isis Pharmaceuticals 公司	反义寡核苷酸	美国 FDA 批准	人巨细胞病毒立早基因区 2(IE2)的抑制剂	AIDS 病人的巨细胞病毒视网膜炎
Affinitak	Isis Pharmaceuticals 公司	反义寡核苷酸	Ⅲ期,与卡铂和紫杉醇联用	抑制 PKC α 表达	Ⅲ b 或Ⅳ期非小细胞肺癌
Alicaforsen	Isis Pharmaceuticals 公司	反义寡核苷酸	Ⅲ期	抑制 ICAM-1 表达	克罗恩病
Macugen(哌加他尼钠)	Eyetech 和 Pfizer 公司	核酸适配体	Ⅲ期单独或与光学疗法联用	抑制 VEGF 表达	老年性黄斑变性
Genasense(奥利默森纳)	Aventis 和 Genta 公司	反义寡核苷酸	Ⅲ期后期,与地塞米松联用	抑制 Bcl-2 的表达	恶性黑色素瘤
Genasense(奥利默森纳)	Aventis 和 Genta 公司	反义寡核苷酸	Ⅲ期,与氟达拉滨和环磷酰胺联用	抑制 Bcl-2 的表达	慢性淋巴细胞性白血病
Genasense(奥利默森纳)	Aventis 和 Genta 公司	反义寡核苷酸	Ⅲ期,与达卡巴嗪联用	抑制 Bcl-2 的表达	多发性骨髓瘤

* 上述临床试验均在美国 FDA 备案。

与当前使用的小分子药物相比，核酸类药物的一个显著优点就是对分子靶点和途径良好的选择性识别。这类药物能预防性地或在早期阶段缓解疾病状况，故而可防止疾病进展和并发症发生。例如，对功能失常的基因进行校正时，经典的基因治疗通过导入并表达相应的正常基因拷贝，产生该基因的单一的蛋白质产物。类似地，在使用核酸药物进行基因去除（gene ablation）时，仅仅被选择的基因被关闭，故能保证疾病治疗的特异性。因此，与特异性相对较差的常规药物相比，核酸类药物的副作用至少从理论上说可被最小化。

尽管目前在临床试验中取得了很多较好的结果，但核酸类药物在人体上的使用仍有局限，成功的例子仍然很少。该类药物研发的迟缓部分可归因于体内应用时细胞对核酸药物的摄取率低。此外，核酸类药物的生物学稳定性也很差，且半衰期短，导致其药代动力学难于评估和预测。而核酸分子进入细胞后还要经受胞内降解，且进入核内时也受到很大的限制。因为很多核酸类药物仍处于试验阶段，其对人体的效应还未被全面研究，故其长期应用的可行性仍有待评估确认。目前对于核酸药物的细胞摄取、分布及代谢仍知之甚少。然而，人们坚信，随着来自科学家的不断努力，基因治疗和核酸药物将最终走向临床并革新现代医学。

（编者　卜友泉，宋方洲）

One of the significant advantages of DNA-based drugs over currently available low molecular weight pharmaceuticals is their selective recognition of molecular targets and pathways, which imparts tremendous specificity of action. These drugs can be used to mitigate disease states either prophylactically or at a very early stage, thereby preventing disease progression and its complications. For example, gene therapy typically involves correction of a malfunctioning gene by the introduction and expression of its correct copy, thus resulting in a single protein product. Similarly, for DNA-based therapeutics intended for gene ablation, only selected genes are switched off, thus ensuring specificity in controlling the disease status. Therefore, at least theoretically, the potential for negative side effects of DNA-based therapeutics may be minimized compared with conventional pharmaceuticals that typically have less specificity.

Despite many favorable characteristics and signs of possible clinical victories, the introduction of nucleic acid-based drugs for human use can be best described as limited, with rare successes. The inertia in the development of these drugs can be attributed, in part, to their poor cellular uptake profile in vivo. In addition, poor biological stability and a short half-life result in unpredictable pharmacokinetics. Furthermore, nucleic acid molecules that do manage to enter the cell are subsequently subjected to intracellular degradation along with stringently restricted nuclear access. Since many DNA-based compounds are new pharmaceutical drug candidates and the effects of their human exposure have yet to be completely investigated, the feasibility of their long-term use remains to be determined. In addition, very little is known about their cellular uptake, distribution, and metabolism. However, it is strongly believed that, with the continuous efforts from scientists, gene therapy and nucleic acid drug will eventually be applied to clinics and revolutionize the current medical science.

(Compiled by Youquan Bu & Fangzhou Song)

缩略语　Abbreviations

ACIP	Advisory Committee on Immunization Practices（免疫实施指导委员会）
ANVISA	Agência Nacional de Vigilância Sanitária（National Health Surveillance Agency Brazil）（巴西国家卫生监督局）
API	Active Pharmaceutical Ingredient（活性制药组分）
AcMNPV	Autographa californica Multicapsid Nucleopolyhedro Virus（加州型多壳核多角体病毒）
BCG	Bacillus Calmette-Guerin（卡介苗）
BEVS	Baculovirus Expression Vector System（杆状病毒表达载体系统）
BHK-21	Baby Hamster Kidney Cells（仓鼠肾细胞系 -21）
BLA	Biologic License Application（生物制品认证申请）
CBER	Center for Biological Evaluation and Research（生物制品评价研究中心）
CDACF	Chemically Defined，Animal Component Free（无特定化学成分动物）
CDC	Center for Disease Control and Prevention（疾病预防控制中心）
CGMP	Current Good Manufacturing Practice（现代药物质量管理规范）
CHO	Chinese Hamster Ovary（中国仓鼠卵细胞）
CMC	Chemistry，Manufacturing and Control（化学药物生产与控制）
CMO	Contract Manufacturing Organization（合约生产组织）
CMV	*Cytomegalovirus*（巨细胞病毒）
CPP	Critical Process Parameters（关键生产参数）
CQA	Critical Quality Attributes（关键质量因素）
CTL	Cytotoxic T Lymphocytes（细胞毒性 T 淋巴细胞）
DHFR	Dihydrofolate reductase（二氢叶酸还原酶）
DNA	Deoxyribonucleic Acid（去氧核糖核酸）
DO	Dissolved Oxygen（溶解氧）
DSC	Differential Scanning Calorimetry（示差扫描量热法）
E. coli	*Escherichia coli*（埃希氏大肠杆菌）
EM	Electronic Microscopy（电子显微镜）
EMA/EMEA	European Medicines Agency（欧洲医药委员会）
EPO	Erythropoetin，Erythropoietin or Erithropoyetin（成红细胞生成素）
EU	European Union（欧洲联盟）
FBS	Fetal Bovine Serum（胎牛血清）
FDA	Food and Drug Administration（食品药品管理局）
FDAMA	The United States Food and Drug Administration Modernization Act（美国食品药品管理现代行动组）
FFDCA	Federal Food，Drug and Cosmetic Act（联邦食品药品和美容产品法案）
GLP	Good Laboratory Practice（实验室管理规范）

GM-CSF	Granulocyte Macrophage Colony-stimulating Factor（颗粒巨噬细胞集落刺激因子）
GMP	Good Manufacturing Practice（药品生产质量管理规范）
GOI	Gene Of Interest（目的基因）
GS	Glutamine Synthetase（谷氨酰胺合成酶）
HA	Hemagglutinin（血凝素）
HAE	Hereditary Angioedema（遗传性血管水肿）
HBsAg	Hepatitis B Virus Antigen（乙型肝炎病毒表面抗原）
HEK 293	Human Embryonic Kidney 293（人胚肾细胞系 293）
Hib	Heamophilus Influenza type b（副嗜血杆菌流感 b 型）
HIC	Hydrophobic Interaction Chromatography（疏水交换层析）
HIV	Human Immunodeficiency Virus（人免疫缺陷病毒）
HPLC	High Pressure Liquid Chromatography（高压液相色谱）
HPV	Human Papillomavirus（人乳头瘤病毒）
HVAC	Heating，Ventilation，and Air-Conditioning（冷暖空调）
ICH	International Conference on Harmonization（国际合作研讨会）
IEX	Ion exchange（离子交换层析）
IL-15	Interleukin 15（白细胞介素 15）
IND	Investigational New Drug（在研新药）
IP	International Pharmacopeia（国际药典）
IPTG	Isopropyl β-D-1-thiogalactopyranoside（异丙基 β-D-1 硫代半乳糖苷）
kb	kilobases（千碱基）
MAb	Monoclonal Antibody（单克隆抗体）
MCB	Master Cell Bank（主细胞库）
MCC	Medicines Control Council（药物控制委员会）
MSX	Methionine Sulphoximine（蛋氨酸）
MTX	Methotrexate（甲氨蝶呤）
M. bovis	Mycobacterium bovis（牛支原体）
NIR	Near-infrared（近红外线）
NMR	Nuclear Magnetic Resonance（核磁共振）
NRA	National Regulatory Authorities（国际管理权威机构）